NONLINEAR AND OPTIMAL CONTROL SYSTEMS

NONLINEAR AND OPTIMAL CONTROL SYSTEMS

THOMAS L. VINCENT
Department of Aerospace and Mechanical Engineering
University of Arizona

WALTER J. GRANTHAM
School of Mechanical and Materials Engineering
Washington State University

A Wiley-Interscience Publication
JOHN WILEY & SONS, INC.
New York • Chichester • Weinheim • Brisbane • Singapore • Toronto

This text is printed on acid-free paper.

Library of Congress Cataloging in Publication Data:

Vincent, Thomas L.

Nonlinear and optimal control systems / Thomas L. Vincent, Walter J. Grantham.

p. cm.

Includes index.

ISBN 0-471-04235-8 (cloth : alk. paper)

1. Feedback control systems. 2. Nonlinear systems. I. Grantham, Walter J. (Walter Jervis), 1944– . II. Title.

TJ216.V56 1997 96-37129

CONTENTS

PREFACE

This text provides an introduction to the analysis techniques used in the design of nonlinear and optimal feedback control systems. The fundamental topics of stability, controllability, and optimality are developed and presented in a unique unified fashion that establishes strong connections between all three topics.

There are two important approaches to the design of feedback controllers for nonlinear systems: Lyapunov stability methods and optimal control theory. On the surface, these two approaches might not appear to have much in common. Lyapunov stability does not address optimality, while optimal control theory generally provides only open-loop control, which need not provide stability. In between these two approaches there is a large, and until now, relatively unexplored area of function minimizing feedback controllers. These controllers address optimality, but on an instantaneous basis, yielding closed-loop feedback controllers without having to solve an optimal control two-point boundary value problem. Instantaneous function minimization can be expected to miss global aspects of optimality that are provided in an optimal control solution. However, when function minimization is combined with Lyapunov stability methods, the resulting controllers do contain global information. These function minimizing Lyapunov controllers have proven to be highly effective and also robust with respect to uncertainties in system models, noisy inputs, and so on.

With the recent growth in interest in nonlinear systems, many engineering departments are beginning to add courses on nonlinear control systems. At the same time, optimal control theory has matured to the point where optimal controls courses can be modified to include other topics in nonlinear control systems design.

This text provides an integrated approach to both nonlinear and optimal control systems. The emphasis is on the fundamental topics of stability, controllability, and optimality, and on the corresponding geometry associated with these topics. In Chapter 1 we discuss background material on nonlinear dynamical systems and various phenomena that can occur only in nonlinear systems, such as limit cycles, chatter, and chaotic motion. In addition, we present numerical algorithms for solving systems of nonlinear differential equations, so that students can quickly begin simulating nonlinear systems, using these routines or commercial software, such as MATLAB. Chapter 2 presents an introduction to nonlinear control systems. In Chapter 3 we present basic results in nonlinear parameter optimization and parametric two-player games, results that are used in later chapters. Chapter 4 covers Lyapunov stability theory, and Chapter 5 applies these results to the design of Lyapunov optimizing feedback controllers. Chapter 6 is devoted to controllability concepts associated with nonlinear control systems. In particular, the controllability minimum principle is developed, which allows for the determination of the boundary of the set of points that are controllable to a specified target. This leads directly to a development of Pontryagin's minimum principle in Chapter 7. This principle is used to solve optimal control problems and for the design of optimal controllers in Chapter 8. The same geometric ideas introduced in Chapters 6 and 7 also lead to necessary conditions for two-player differential games, discussed in Chapter 9.

This text is designed for one-semester introductory senior- or graduate-level nonlinear and optimal control systems courses and does not presume any background on the student's part beyond differential equations. In particular, we do not assume that the student has taken a linear control systems course. Although our focus is on nonlinear systems, this does not preclude us from referring to linear systems or using a number of linear systems (with bounded controls, which make these systems nonlinear) as examples. Any linear control systems background that is required to understand any allusion to linear systems is presented directly in the text. Throughout the text we focus on a variety of example systems. Viewing such systems in terms of stability, controllability, and optimality provides the student with an opportunity to compare the same systems in different design contexts. In these examples we consider not only physical dynamical systems, but also biological systems, economic systems, and so on. Thus this text is suitable for students from a wide range of disciplines.

There are a variety of approaches to structuring separate one-semester nonlinear or optimal control systems courses using this text. Chapter 1 provides mathematical background on the nature of solutions to nonlinear dynamical systems and on some classical analysis techniques for such systems. Depending on the objectives of the course, this material could be covered either briefly or in detail. Chapters 1–6 provide the basis for a nonlinear controls course. We have found that it is also possible to cover at least some of the material in Chapter 7 as well, creating a nonlinear and optimal controls course with an emphasis on nonlinear controls and an introduction to optimal control systems. Chapters 2, 3, and 6–9 form the basis for an optimal controls course, with the emphasis on optimization, calculus of variations, optimal control, and two-player differential games.

However, by design, this text is structured to allow for combined nonlinear and optimal controls courses. For such courses there are at least two approaches. For an emphasis on nonlinear controls one could skip the material on differential games, covering (perhaps only the first half of) Chapter 1 briefly, but covering Chapters 2–7 in detail. For an emphasis on optimal controls one could skip the second half of Chapter 4, dealing with the details of constructing Lyapunov functions and estimating stability regions. The remaining chapters can then be covered in detail, with Chapter 1 (or the first half only) being covered briefly.

Each chapter contains several examples and a variety of exercises to aid the student. In addition, several sections contain numerical algorithms. In particular, the algorithms that we present for nonlinear simulation and nonlinear parameter optimization form the basis for the current state-of-the-art algorithms in these areas. We have found these and other algorithms (such as the routines in MATLAB) to be useful, and we hope that the reader will also benefit from them. Finally, it is hoped that this text will stimulate the reader to further study in nonlinear and optimal control system design and to the application of these methods to practical problems of interest.

THOMAS L. VINCENT
WALTER J. GRANTHAM

Tucson, Arizona
Pullman, Washington
June 1997

1

NONLINEAR DYNAMICAL SYSTEMS

1.1 FUNDAMENTALS

System Models

In this text we consider nonlinear dynamical control systems of the form

$$\dot{\mathbf{x}} = \mathbf{f}(\mathbf{x}, \mathbf{u}), \tag{1.1-1}$$

where $\mathbf{x} \in R^{n_x}$ is the n_x-dimensional **state** vector, $\mathbf{u} \in \mathcal{U} \subseteq R^{n_u}$ is the n_u-dimensional **control** vector input (to be chosen), and $(\dot{\ })$ denotes differentiation with respect to time t. The set $\mathcal{U}$ is a **control constraint set**, reflecting the fact that control inputs typically have some constraints imposed upon them, such as bounded thrust on a rocket or bounded turning rate for a vehicle. The vector function $\mathbf{f}(\cdot)$ and its partial derivatives with respect to $\mathbf{x}$ and $\mathbf{u}$ are assumed to be continuously differentiable functions of $\mathbf{x}$ and $\mathbf{u}$, although the control input function $\mathbf{u}(\cdot)$ may be discontinuous, subject to certain restrictions that we discuss shortly. In general we are concerned with developing a **feedback control** law $\mathbf{u}(\mathbf{x})$ to make the system (1.1-1) behave in some specified fashion. We assume, throughout, that the state information needed for the control law $\mathbf{u}(\mathbf{x})$ can be measured precisely.

The dot notation used above to indicate the time derivative is not to be confused with the notation $(\cdot)$, which is used in place of a function argument, as in $\mathbf{f}(\cdot)$, when we wish to refer to the functional relationship itself rather than its value.

Target Sets There may be several design objectives involved in the choice of a control algorithm, but common among them is a specified target set $\mathcal{X}$ in state space. For example, an objective might be to achieve a certain target and maintain it, as in a rendezvous with an orbiting satellite, or simply to hit a target, as in an aircraft-missile intercept problem. In all cases the underlying basic controllability objective is one of finding a feedback control function $\mathbf{u}(\mathbf{x})$ that will transfer the state to a specified **target** set $\mathcal{X} \subset R^{n_x}$, usually defined by a system of n_g equalities

$$\mathcal{X} = \{\mathbf{x} | \mathbf{g}(\mathbf{x}) = 0\}, \tag{1.1-2}$$

with $\mathbf{g}(\mathbf{x}) = [g_1(\mathbf{x}) \cdots g_{n_g}(\mathbf{x})]^{\mathrm{T}}$ assumed to be continuous and continuously differentiable. Furthemore we assume that the gradient vectors $\partial g_i(\mathbf{x})/\partial \mathbf{x}, i = 1, \ldots n_g$, are linearly independent on $\mathcal{X}$.

Control Constraints In many applications the elements of the control vector $\mathbf{u}$ may have various constraints imposed upon them, which may limit our ability to transfer the state to the target set. For example, the magnitude of the thrust vector, for a variable thrust rocket, would vary from zero to some upper limit corresponding to the maximum thrust available. Also, the direction of the thrust vector relative to the rocket would generally be limited by the gimbals on which the exhaust nozzles are mounted.

We consider the case in which the control vector is required to satisfy a given set of constraints

$$\mathbf{u} \in \mathcal{U}, \tag{1.1-3}$$

where $\mathcal{U} \subseteq R^{n_u}$ is a specified **control constraint set**. If no control constraints are imposed, then $\mathcal{U} = R^{n_u}$. In the general case we allow for the possibility that the control constraints may depend on the current state, with the control constraint set being defined by a system of n_h inequalities

$$\mathcal{U} = \{\mathbf{u} | \mathbf{h}(\mathbf{x}, \mathbf{u}) \geq 0\}. \tag{1.1-4}$$

Most often, however, we restrict the discussion to state-independent control constraints of the form

$$\mathcal{U} = \{\mathbf{u} | \mathbf{h}(\mathbf{u}) \geq 0\}, \tag{1.1-5}$$

with $\mathbf{h}(\mathbf{u}) = [h_1(\mathbf{u}) \cdots h_{n_h}(\mathbf{u})]^{\mathrm{T}}$ assumed to be continuous and continuously differentiable. We also require that the gradient vectors $\partial h_i(\mathbf{u})/\partial \mathbf{u}$ satisfy a **regularity** assumption, which is satisfied if they are linearly independent. This requirement is discussed in detail in Section 3.2, where we develop first-order necessary conditions governing solutions to nonlinear constrained parameter optimization problems.

The state vector $\mathbf{x}$ might also have constraints imposed upon it, such as a minimum and maximum allowed speed for an aircraft. The case where a portion of a trajectory runs along a state constraint boundary is considered in Section 7.7. However, for the most part, we do not consider state constraints explicitly and, instead, assume that trajectories lie inside the region defined by the state constraints.

Example 1.1-1: An Introductory Optimal Control Problem Consider a drag race in which the objective is to go in minimum time from a standing start to a full stop at the other end of the drag strip. Consider just one car, of mass m, and let $y(t)$ denote the car's position along the drag strip. Under an applied force $F(t)$, which we get to pick, the equation of motion may be written as

$$m\ddot{y} = F(t).$$

The control input $F(t)$ can be either positive (acceleration) or negative (braking) and is bounded by the constraints

$$F_{\min} \leq F \leq F_{\max}.$$

We can convert this problem to state-space form by letting $x_1 = y$ and $x_2 = \dot{y}$. With $u = F/m$ the equations of motion become

$$\dot{x}_1 = x_2$$
$$\dot{x}_2 = u.$$

The optimal control problem is to find $u(t)$ to drive the system in minimum time from the initial state $\mathbf{x}(0) = (0, 0)$, corresponding to $y(0) = \dot{y}(0) = 0$, to a target final state $\mathbf{x}(t_f) = (L, 0)$, corresponding to a full stop $y(t_f) = L$, $\dot{y}(t_f) = 0$ at the far end of the drag strip. The performance measure to be minimized is the travel time

$$t_f = \int_0^{t_f} dt$$

and the control constraints are of the form

$$u_{\min} \leq u(t) \leq u_{\max}.$$

Solution techniques for optimal control problems are presented in Chapter 7. For this example, however, it is intuitively clear that the optimal control is maximum acceleration followed by maximum braking, that is,

$$u(t) = \begin{cases} u_{\max} & \text{if } t < t_s \\ u_{\min} & \text{if } t > t_s \end{cases}$$

with the switching time t_s, that is, the onset of braking, being chosen to bring the vehicle to rest precisely at the end of the drag strip. Note that the switching time will depend on L. A complete solution to this problem involves finding the optimal control as a function of the current state $u(\mathbf{x})$, rather than as a function of time $u(t)$. ■

Control System Concepts

For a given specified control law $\mathbf{u}(\mathbf{x})$ and a given initial point $\mathbf{x}$ in state space, we may think of the solution $\mathbf{x}(t)$ of (1.1-1) as a function that transfers the initial state at time zero to some other state at time t. Our study of nonlinear control systems focuses on the stability, controllability, and optimization properties of this solution. In particular, we are concerned with the relationship between a specified control law and the region of the state space that can be transferred to a designated target set under that control law. In this context:

Stability Is concerned with determining those initial states that, under a specified feedback control law, will be transferred to and maintained at the specified target set.

Controllability Is concerned with determining those initial states where a feedback control exists that will transfer the state to a specified target.

Optimality Has to do with finding the "best" feedback controller, based on some specified performance measure, which will transfer any controllable state to a specified target.

Note that stability implies controllability; however, controllability need not imply stability since controllability does not require that the system be maintained at the target. The remainder of this chapter is devoted to a discussion of the nature of solutions to nonlinear dynamical systems with the control input function $\mathbf{u}(\cdot)$ already specified so that the right-hand side of (1.1-1) is simply $\mathbf{f}(\mathbf{x})$ or $\mathbf{f}(\mathbf{x}, t)$. Chapter 2 removes the assumption that $\mathbf{u}(\cdot)$ has been specified and specifically deals with some of the aspects of nonlinear control systems. Chapter 3 introduces nonlinear optimization techniques needed for the control design methods introduced in the chapters that follow. Stability is discussed in Chapters 4 and 5. Controllability is considered in Chapters 6 and optimality in Chapters 7 and 8. Finally, in Chapter 9, the fundamental concepts are extended to the two-player differential game where two separate players control the dynamical system, as in an air-to-air combat engagement.

1.2 PROPERTIES OF SOLUTIONS

Admissible Controls

A solution $\mathbf{x}(t)$ to (1.1-1) cannot be specified until initial conditions are given for $\mathbf{x}(t)$ and a control law for $\mathbf{u}(\cdot)$ is specified. A control law in which the control input depends explicitly on the state of the system is called a **closed-loop control** or a **feedback control,** $\mathbf{u}(\mathbf{x})$, since the output of the system (the state $\mathbf{x}$) "feeds back" to affect the input, thus creating a "closed-loop," from the output back to the input. A control law of the form $\mathbf{u}(t)$ is called an **open-loop control,** since a control input $\mathbf{u}(t)$ does not depend on the output of the system but, instead, is a prescribed function of time. The difference between open-loop and closed-loop control is made evident with a simple example. Consider the heating of a home using, as the controller, either a thermostat (closed loop) or a timer (open loop). In the former case the thermostat senses the room temperature, the current "state," and turns on the furnace when it is too cold and turns if off when it is too hot. The timer, on the other hand, must be set by experience. The furnace will come on and shut off at the prescribed times independent of the actual room temperature.

Open-loop controls are quite common in applications, such as open-loop robots programmed to expect a certain part to be in a certain place on a conveyor belt at a specified time. However, in order to be effective, open-loop controllers require either that the designer has a perfect ability to predict the resulting motion of the dynamical system, that is, the model (1.1-1) for the system is perfect, or that the resulting motion be suitably insensitive to unmodeled dynamic effects. In this text we generally assume that the model (1.1-1) is in fact a perfect model for the dynamical system, that is, *the model is the system.* Thus while we are primarily concerned with designing closed-loop feedback controllers $\mathbf{u}(\mathbf{x})$, we also consider open-loop controls $\mathbf{u}(t)$. The effects of imperfect models or systems with uncertain inputs can be treated as two-player "differential games," which are discussed in Chapter 9.

For a given closed-loop control law $\mathbf{u}(\mathbf{x})$ or open-loop control $\mathbf{u}(t)$ to be of any use, it must not only yield a solution $\mathbf{x}(t)$ to (1.1-1) for any initial state of interest, but it must also satisfy the specified control constraints $\mathbf{u} \in \mathcal{U}$ for all times of interest. A control that satisfies these two requirements is called an **admissible control**. A more formal definition of an admissible control is given in Section 2.1.

Since our focus in this chapter is on the nature of solutions to (1.1-1), as opposed to the logic behind a particular control law, we consider the control law as being given a priori, either as an open-loop control $\mathbf{u}(t)$ or as a closed-loop control $\mathbf{u}(\mathbf{x})$. Since we allow the control input $\mathbf{u}(\cdot)$ to be discontinuous, we consider each of the two cases separately.

Solutions for Nonautonomous Systems

When the input function $\mathbf{u}(\cdot)$ in (1.1-1) is an explicit function of time [e.g., open-loop control $\mathbf{u}(t)$], the resulting dynamical system is said to be **nonautonomous.** Let $\mathbf{u}(t)$ be a specified input function, with $\mathbf{u}(t) \in \mathcal{U}$ for all $t \in (-\infty, \infty)$ and with $\mathbf{u}(t)$ at worst **piecewise continuous**. That is, $\mathbf{u}(t)$ may be discontinuous at some isolated, discrete times. This assumption excludes functions that are discontinuous on a nonzero-length time interval, such as random functions that take on a different value at each instant of time, or functions that continuously "chatter" between two different values. For many (but not all) control applications, piecewise continuity is usually satisfied. It is a relatively unrestrictive assumption, and it allows us to consider a very wide class of practical control systems, including as a special case the more common situation where $\mathbf{u}(t)$ is continuous.

For a specified $\mathbf{u}(t)$, (1.1-1) become ordinary differential equations

$$\dot{\mathbf{x}} = \mathbf{f}[\mathbf{x}, \mathbf{u}(t)]. \tag{1.2-1}$$

For an initial time t_0 and an initial condition $\mathbf{x}(t_0) = \mathbf{x}_0$, a function $\mathbf{x}(t)$ is a **solution** to (1.2-1), on some time interval $a < t < b$ containing t_0, if

$$\mathbf{x}(t) = \mathbf{x}_0 + \int_{t_0}^{t} \mathbf{f}[\mathbf{x}(\tau), \mathbf{u}(\tau)]d\tau, \tag{1.2-2}$$

for all $t \in (a, b)$, that is, if $\mathbf{x}(t)$ is an **absolutely continuous** function satisfying

$$\mathbf{x}(t_0) = \mathbf{x}_0 \tag{1.2-3}$$

and

$$\frac{d\mathbf{x}(t)}{dt} = \mathbf{f}[\mathbf{x}(t), \mathbf{u}(t)] \tag{1.2-4}$$

for all $t \in (a, b)$ where $\mathbf{u}(t)$ is piecewise continuous [i.e., continuous for "almost all" $t \in (a, b)$.]

A solution plotted in state space is called a **trajectory**. In most cases, the time interval (a, b) over which a solution is defined can be extended to $(-\infty, \infty)$, but as we will see there are cases where this is not possible.

To display the fact that a solution to (1.2-1) depends on both the initial conditions and the initial time, we designate the general solution to (1.2-1) generated by a piecewise continuous control $\mathbf{u}(t)$ by $\boldsymbol{\xi}(t, t_0, \mathbf{x}_0)$ where $\boldsymbol{\xi}(t_0, t_0, \mathbf{x}_0) \triangleq \mathbf{x}_0$ for all t_0.

Since we have assumed that $\mathbf{f}(\mathbf{x}, \mathbf{u})$ is continuous and has continuous first-order partial derivatives in both $\mathbf{x}$ and $\mathbf{u}$, and since $\mathbf{u}(t)$ is at worst piecewise continuous, the **Caratheodory existence theorem** and related results from the theory of ordinary differential equations [Coddington and Levinson, 1955, pp. 1–61] assure us that a unique solution $\mathbf{x}(t) = \boldsymbol{\xi}(t, t_0, \mathbf{x}_0)$ to (1.2-1) exists on some time interval $a < t < b$ containing t_0. Moreover, the solution is continuous in t_0 and $\mathbf{x}_0$, differentiable in $\mathbf{x}_0$, and the time interval (a, b) can be extended at both ends so long as $\|\mathbf{f}[\mathbf{x}(t), \mathbf{u}(t)]\|$ remains bounded, where $\|\cdot\|$ denotes the (Euclidean) norm of a vector. Unless otherwise specified, we generally assume that the solution $\mathbf{x}(t)$ is defined throughout the time interval $t \in (-\infty, \infty)$.

One important characteristic class of solutions are the **equilibrium** or constant-state solutions $\mathbf{x}(t) \triangleq \hat{\mathbf{x}} =$ constant, which correspond to points where $\mathbf{f}[\mathbf{x}, \mathbf{u}(t)] = 0$ for all t. Usually in such situations the control input is also constant $\mathbf{u}(t) \triangleq \hat{\mathbf{u}} =$ constant; however this need not be the case, if the effect of the control input vanishes at the equilibrium state. Equilibrium solutions are characteristics of the system in the sense that they represent **fixed points** for the solution function $\boldsymbol{\xi}(t, t_0, \cdot)$, since at such points $\boldsymbol{\xi}(t, t_0, \hat{\mathbf{x}}) = \hat{\mathbf{x}}$ for all t and all t_0.

Autonomous Systems

When the system (1.1-1) has no control input, or when the control input is a differentiable closed-loop control $\mathbf{u}(\mathbf{x}) \in \mathcal{U}$, we have an **autonomous** system, of the form

$$\dot{\mathbf{x}} = \mathbf{f}(\mathbf{x}), \tag{1.2-5}$$

where $\mathbf{f}(\mathbf{x}) \triangleq \mathbf{f}[\mathbf{x}, \mathbf{u}(\mathbf{x})]$ is continuous and has continuous first-order partial derivatives with respect to $\mathbf{x}$.

Since $\mathbf{f}(\cdot)$ is not an explicit function of time t, the solution to (1.2-5) is of the form

$$\mathbf{x}(t) = \boldsymbol{\xi}(t - t_0, \mathbf{x}_0),$$

where the initial time t_0 is arbitrary. For such systems we take $t_0 = 0$, yielding

$$\mathbf{x}(t) = \boldsymbol{\xi}(t, \mathbf{x}_0). \tag{1.2-6}$$

From our previous results, applied to the special case of autonomous systems, we know that a solution to (1.2-5), satisfying the initial condition $\mathbf{x}(0) = \boldsymbol{\xi}(0, \mathbf{x}_0) \triangleq \mathbf{x}_0$, exists on some time interval $a < t < b$ containing 0, and that the time interval can be extended at both ends as long as $\|\mathbf{f}[\mathbf{x}(t)]\|$ remains bounded. Furthermore, the solution is unique and differentiable in both t and $\mathbf{x}_0$.

One of the most important geometric properties of autonomous systems of the form (1.2-5) is that there is only one solution $\mathbf{x}(t) = \boldsymbol{\xi}(t, \mathbf{x}_0)$ passing through any given initial state $\mathbf{x}(0) = \mathbf{x}_0 \triangleq \boldsymbol{\xi}(0, \mathbf{x}_0)$. Hence trajectories in state space can never intersect each other. Compare this with the case of nonautonomous systems, where the solution $\mathbf{x}(t) = \boldsymbol{\xi}(t, t_0, \mathbf{x}_0)$ depends on the starting time t_0, so that the solution from $\mathbf{x}_0$ at $t_0 = 0$ will generally not be the same as the solution starting from $\mathbf{x}_0$ at some other initial time $t_0 \neq 0$. For nonautonomous systems of the form (1.2-1), trajectories in state space may intersect each other.

Of course a nonautonomous system can always be converted to an autonomous system by treating time as a new state variable, say the last state variable $x_{n_x} = t$ with $\dot{x}_{n_x} = 1$. In the new state space the resulting trajectories will not intersect. The disadvantage of this approach is that any equilibrium solutions for the nonautonomous system (1.2-1), those points $\mathbf{x}$ such that $\mathbf{f}[\mathbf{x}, \mathbf{u}(t)] = 0$ for all t, will no longer exist in the new autonomous system, since x_{n_x} is monotonic and $\dot{x}_{n_x}$ is never zero. Since we are mainly concerned with closed-loop control laws $\mathbf{u}(\mathbf{x})$, we generally consider only autonomous systems.

Continuity with Respect to Initial State

For comparison later with the definition of stability, and with the notion of "chaos," we state the formal definition of continuity with respect to initial conditions. For an admissible control $\mathbf{u}(\cdot)$ the solution $\boldsymbol{\xi}(t, t_0, \mathbf{x}_0)$ is **continuous** in $\mathbf{x}_0$ at (t, t_0) if and only if any $\varepsilon > 0$ there exists a corresponding $\delta(\varepsilon) > 0$ such that the condition

$$\|\boldsymbol{\xi}(t, t_0, \mathbf{x}_0 + \Delta\mathbf{x}_0) - \boldsymbol{\xi}(t, t_0, \mathbf{x}_0)\| < \varepsilon \tag{1.2-7}$$

holds for all $\Delta\mathbf{x}_0$ such that

$$\|\Delta\mathbf{x}_0\| < \delta(\varepsilon). \tag{1.2-8}$$

From this definition, as illustrated in Figure 1.2-1, we see that continuity with respect to an initial state implies that we can make neighboring solutions be arbitrarily close together at time t by making the change in the initial state at time t_0 sufficiently small.

Example 1.2-1: Continuity of Solutions Consider the linear system given by

$$\begin{aligned} \dot{x}_1 &= x_1 \\ \dot{x}_2 &= -x_2. \end{aligned} \tag{1.2-9}$$

Figure 1.2-2 illustrates various state-space trajectories of the system. The gen-

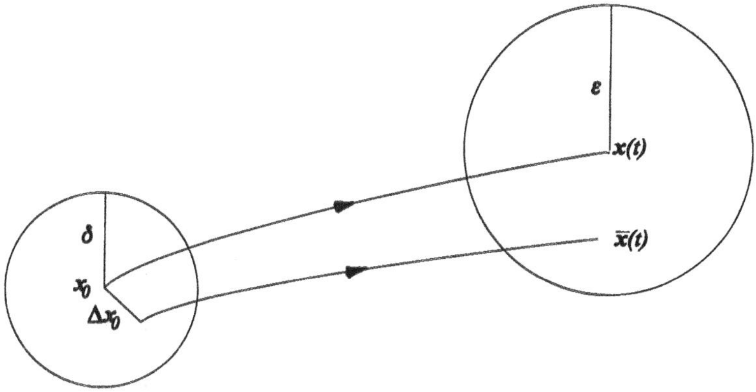

Figure 1.2-1 Continuity of solutions with respect to initial condition.

eral solution for this system is given by

$$
\begin{aligned}
x_1(t) &= x_1(0)e^{t} \\
x_2(t) &= x_2(0)e^{-t}.
\end{aligned}
\tag{1.2-10}
$$

Consider the solution $\mathbf{x}(t) = [0 \quad e^{-t}]^T$ starting from $\mathbf{x}(0) = [0 \quad 1]^T$, and

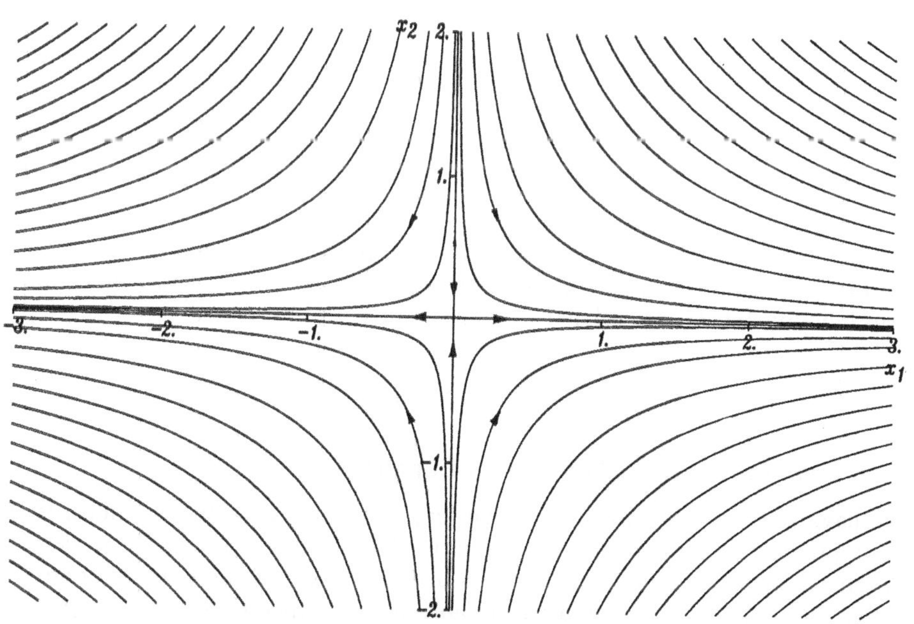

Figure 1.2-2 Trajectories for Example 1.2-1.

a neighboring solution $\bar{\mathbf{x}}(t) = [\delta e^t \quad e^{-t}]^T$ starting from $\bar{\mathbf{x}}(0) = [\delta \quad 1]^T$, with $\delta > 0$ but small. The solution $\mathbf{x}(t)$ approaches the origin asymptotically as t approaches infinity, while the neighboring solution $\bar{\mathbf{x}}(t)$ approaches the positive x_1-axis at infinity as t approaches infinity, regardless of how small we take $\delta > 0$. The solution $\mathbf{x}(t)$ is unstable, since neighboring solutions are not close to $\mathbf{x}(t)$ for all $t > 0$. Nevertheless, the solution $\mathbf{x}(t)$ is continuous in its initial conditions, since for any given time $t > 0$ (no matter how large) and any given tolerance $\epsilon > 0$, we can pick a value $\delta = \delta(\epsilon) > 0$ such that $\|\bar{\mathbf{x}}(t) - \mathbf{x}(t)\| = |\delta e^t| < \epsilon$. ■

Retro-Time Systems

As we have indicated previously, the solution $\mathbf{x}(t) = \boldsymbol{\xi}(t, \mathbf{x}_0)$ to a system of differential equations

$$\dot{\mathbf{x}} = \mathbf{f}(\mathbf{x}), \tag{1.2-11}$$

with initial condition $\mathbf{x}(0) = \boldsymbol{\xi}(0, \mathbf{x}_0) \triangleq \mathbf{x}_0$, can generally be extended both forward and backward in time. Frequently, it will be convenient to examine solutions involving **backward integration** along a trajectory. For example, if we know the state $\mathbf{x}_f = \mathbf{x}(t_f)$ at some time $t_f > 0$ and wish to determine the state at some previous time, we can accomplish this by introducing the **time-to-go**

$$\tau \triangleq t_f - t \tag{1.2-12}$$

and then integrating the **retro-time system**

$$\frac{d\mathbf{x}}{d\tau} = -\mathbf{f}(\mathbf{x}) \tag{1.2-13}$$

to determine the retro-time solution $\mathbf{x}(\tau) = \boldsymbol{\chi}(\tau, \mathbf{x}_f)$ with "initial condition" $\mathbf{x}(0) = \mathbf{x}_f$. The retro-time solution function $\boldsymbol{\chi}(\tau, \mathbf{x}_f)$ and the forward-time solution function $\boldsymbol{\xi}(t, \mathbf{x}_0)$ are related by

$$\boldsymbol{\chi}(\tau, \mathbf{x}_f) = \boldsymbol{\xi}(-\tau, \mathbf{x}_f). \tag{1.2-14}$$

Discontinuous Ordinary Differential Equations

We have seen that the use of an open-loop piecewise continuous control $\mathbf{u}(t)$ poses no real problems as far as solutions to (1.1-1) are concerned. The principle difference between continuous controls $\mathbf{u}(t)$ and piecewise continuous controls is that the solution $\mathbf{x}(t)$ is not differentiable at times when $\mathbf{u}(t)$ is discontinuous. At such times the trajectory may have a **corner** corresponding to a discontinuous change in the "velocity" vector $\dot{\mathbf{x}}(t)$. Across a switching time t we simply change $\mathbf{u}(t-)$ to $\mathbf{u}(t+)$ and then continue integrating (1.1-1). Note that if we were examining a family of trajectories $\mathbf{x}(t)$, say for different initial conditions, then a discontinuity in an open-loop control $\mathbf{u}(t)$ would cause all trajectories to switch at the same time.

Now, suppose we use a closed-loop control $\mathbf{u}(\mathbf{x})$ in which $\mathbf{u}(\cdot)$ is discontinuous. Then (1.1-1) yields an autonomous system of the form

$$\dot{\mathbf{x}} = \mathbf{f}(\mathbf{x}), \tag{1.2-15}$$

where $\mathbf{f}(\mathbf{x}) \triangleq \mathbf{f}[\mathbf{x}, \mathbf{u}(\mathbf{x})]$ in general is neither differentiable nor continuous, since $\mathbf{u}(\mathbf{x})$ is not continuous. This type of discontinuity, in which $\mathbf{u}(\mathbf{x})$ switches as $\mathbf{x}$ crosses some surface in state space, can be fundamentally very different from the case of an open-loop discontinuity in $\mathbf{u}(t)$, where discontinuities occur at certain times rather than at certain states.

Discontinuous systems will often occur in nonlinear control applications, particularly in the implementation of optimal controllers such as minimum-time controllers. The difficulty, from the point of view of the theory of differential equations, and the reason that introductory texts seldom mention discontinuous autonomous differential equations, is that the discontinuities can destroy the uniqueness, continuity, or differentiability of solutions, and can even destroy the notion of the existence of a solution. Fortunately, these problems are often not as bad as they might seem and in most cases discontinuous controllers can be dealt with effectively. The following example illustrates some of the potential difficulties.

Example 1.2-2: Discontinuous Systems Consider a two-dimensional control system of the form

$$\begin{aligned}\dot{x}_1 &= f_1(x_1, x_2, u)\\ \dot{x}_2 &= f_2(x_1, x_2, u)\end{aligned}$$

where the scalar u is determined by

$$u = \begin{cases} +1 & \text{if } \sigma(\mathbf{x}) > 0\\ \in \{-1, 1\} & \text{if } \sigma(\mathbf{x}) = 0\\ -1 & \text{if } \sigma(\mathbf{x}) < 0.\end{cases}$$

The control u is assigned one of two possible values, depending on the sign of some **switching function** $\sigma(\mathbf{x})$. Even if the velocity vector $\mathbf{f} = [f_1 \quad f_2]^T$ is continuous in x_1 and x_2 for a constant u, it will generally be discontinuous across the **switching surface** $\sigma(\mathbf{x}) = 0$. For this example we take $\sigma(\mathbf{x}) = x_2$, so that the switching surface is the x_1-axis. We consider three separate cases for f_1 and f_2.

Case 1: Switching Surface Let $f_1 = u$, $f_2 = -1$, and $\sigma = x_2$. The trajectories for this case are illustrated in Figure 1.2-3*a*. Since $\dot{x}_2 \equiv -1$, the trajectories always penetrate the switching surface $x_2 = 0$ from above. On either side of the switching surface the differential equation $\dot{\mathbf{x}} = \mathbf{f}(\mathbf{x})$ is continuous. A complete solution $\mathbf{x}(t)$ can be constructed by using the state $\mathbf{x}(t_s)$ at a switching time t_s as the initial condition for the time interval $t \geq t_s$. The resulting unique solution through any initial state (including initial states that lie on $x_2 = 0$) exists for all time and is a differentiable function of the initial state.

This case, where $\mathbf{u}(t) \triangleq \mathbf{u}[\mathbf{x}(t)]$ and hence $\mathbf{f}(t) \triangleq \mathbf{f}(\mathbf{x}(t), \mathbf{u}[\mathbf{x}(t)])$ is piecewise continuous along a trajectory, is a common switching behavior in many control systems. As in Example 1.1-1, consider the motion of the read-write head in a computer disk drive. To move from one disk track to another in minimum time, the head must first accelerate at its maximum possible value and then switch to decelerating at the maximum possible deceleration, with the switch occurring so that the head comes to rest at the desired disk track.

Case 2: Dispersal Surface Let $f_1 = -1$, $f_2 = u$, and $\sigma = x_2$. The trajectories for this case are illustrated in Figure 1.2-3*b* and the switching surface is termed

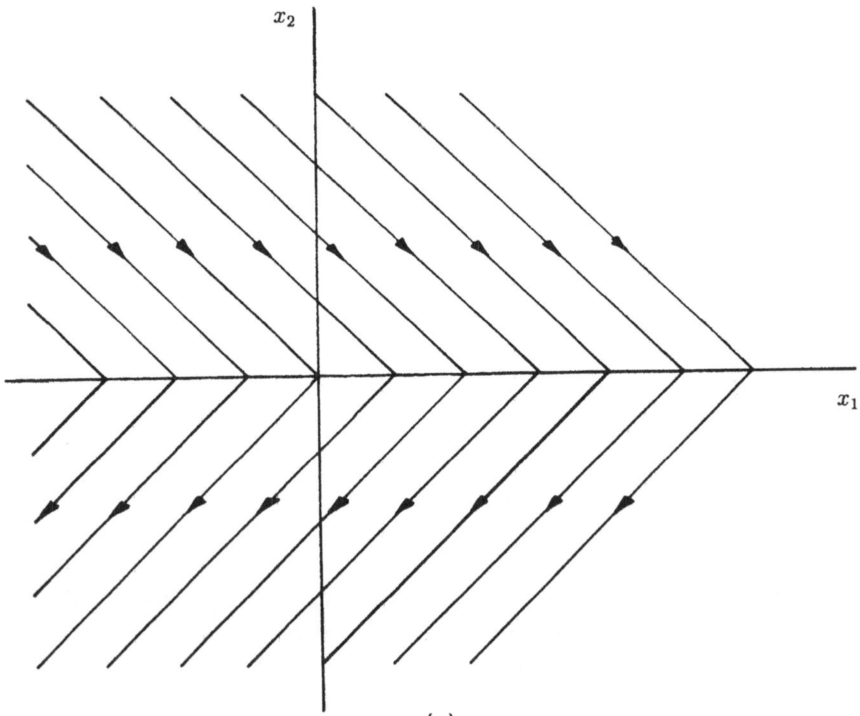

(*a*)

Figure 1.2-3 (*a*) Switching surface. (*b*) Dispersal surface. (*c*) Chatter.

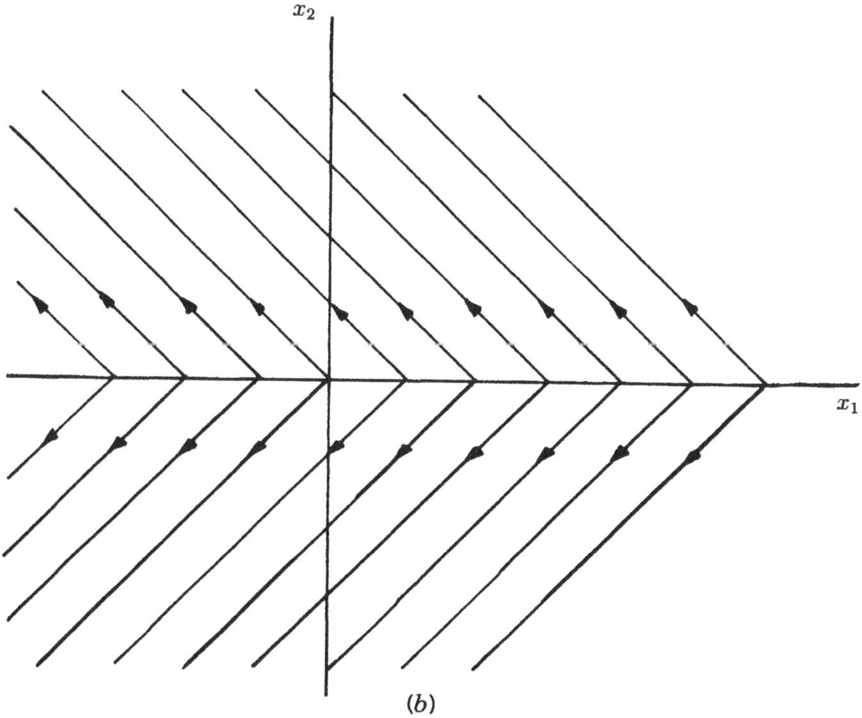

(*b*)

Figure 1.2-3 (*Continued*)

a **dispersal surface**. Since $u = +1$ or -1 on the switching surface $x_2 = 0$, we see that the trajectories immediately leave the dispersal surface and enter a region in which $\dot{\mathbf{x}} = \mathbf{f}(\mathbf{x})$ remains continuous. Thus for all trajectories except those starting on the dispersal surface, the corresponding solution $\mathbf{x}(t)$ forward in time ($t \geq 0$) exists, is unique, and is differentiable in $\mathbf{x}(0)$. However, no solution can be continued backward in time prior to the time at which it leaves the dispersal surface. Note that there are two trajectories through any given initial state on the dispersal surface. This problem could be alleviated by assigning a specific control, say $u = +1$, to points on the dispersal surface, but we would still have the problem that solutions that start on the dispersal surface are not continuous in the initial state.

In control applications, dispersal surfaces are relatively rare. One case where they occur is in optimal control problems at points where two different trajectories under optimal control yield the same optimal cost. In practice, one can more or less ignore the dispersal surface, except when integrating backward in time prior to the time at which the trajectory leaves the dispersal surface.

Case 3: Chatter Let $f_1 = +1$, $f_2 = u$, and $\sigma = -x_2$. The trajectories for this case are illustrated in Figure 1.2-3*c* and the switching surface is termed a **uni-**

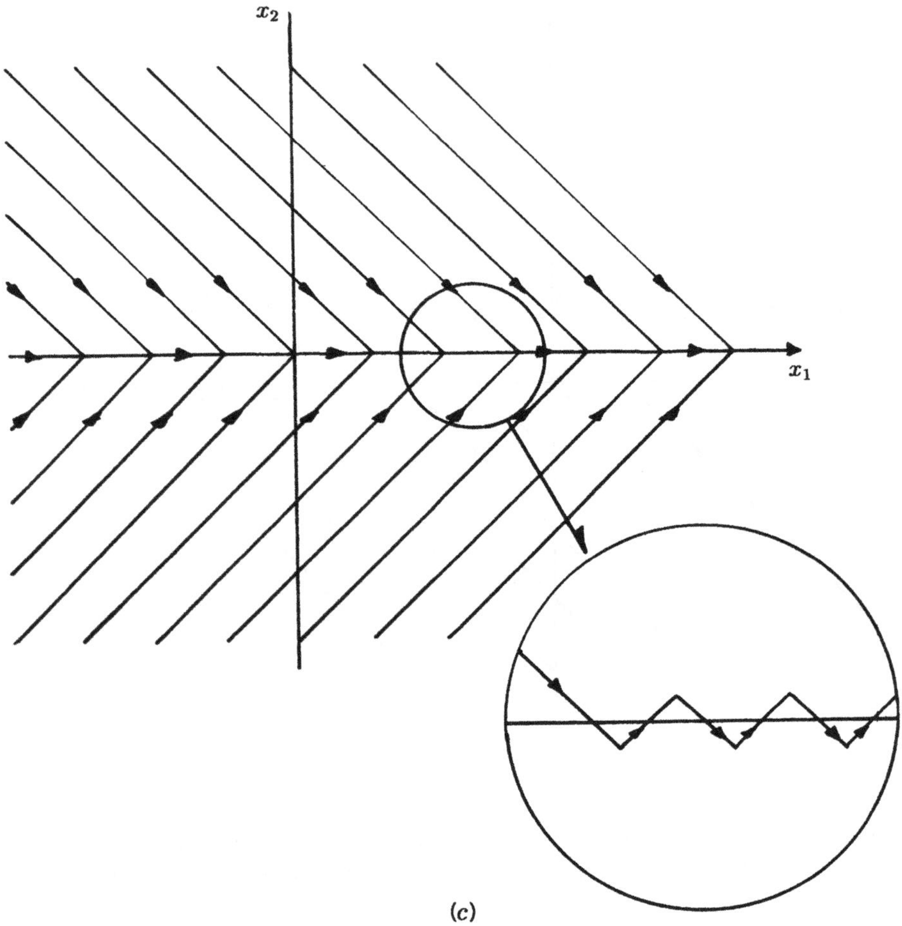

(c)

Figure 1.2-3 (*Continued*)

versal surface or a **chatter surface**. As indicated, the trajectories approach the switching surface and then move along it.

There are two fundamental difficulties associated with this type of dynamic behavior. First of all, while the solutions forward in time are unique, continuous, and differentiable in the initial state, the solutions backward in time, for trajectories whose retro-time "initial" states lie on the switching surface, are not even unique.

But an even more fundamental problem exists, and it concerns the very meaning of a solution to discontinuous differential equations that exhibit **chatter**. Consider a trajectory that approaches the switching surface and suppose that there is a small time delay in the system, so that the trajectory overshoots the switching surface by a small amount, as illustrated in the expanded view shown in Figure 1.2-3*c*. As soon as the trajectory crosses the swtiching surface the velocity vector switches to a new value, which brings the state back toward the switching surface. The trajectory overshoots again by a small amount and again the velocity vector switches and brings the state back to the switching surface, and this process continually repeats itself.

Mathematically, in the absence of any actual time delay, the amount of overshoot is zero and the result is a trajectory that "chatters" or "slides" along the switching surface. For our example the chattering portion of the trajectory is given by

$$\mathbf{x}(t) = [x_1(t) \quad 0]^{\mathrm{T}},$$

where, from inspection of f_1 and f_2, the "velocity" vector along the switching surface must be

$$\dot{\mathbf{x}} = [1 \quad 0]^{\mathrm{T}},$$

which is not equal to $\mathbf{f}(\mathbf{x}, u)$ for either $u = +1$ or $u = -1$! ■

All physical control systems must operate under bounded control. For this reason even a "linear" system with control bounds may chatter when the gain is set too high. For example, consider the linear system

$$\dot{\mathbf{x}} = \mathbf{A}\mathbf{x} + \mathbf{B}u$$

where $\mathbf{A}$ and $\mathbf{B}$ are constant matrices and u is a scalar control. Consider a stabilizing state feedback controller of the form

$$u = -\hat{k}\mathbf{k}^{\mathrm{T}}\mathbf{x}$$

where $\mathbf{k}$ is a feedback gain vector, yielding a control that is a linear combination of the components of $\mathbf{x}$, and $\hat{k}$ is an overall gain often associated with a power amplifier used to drive the system. Applying bounds on the control yields a nonlinear system. For a high value of the gain $\hat{k}$ the system may tend to chatter as the control will often be saturated. In fact as $\hat{k} \to \infty$, $\mathbf{k}^{\mathrm{T}}\mathbf{x}$ becomes a switching function for the system and (under certain conditions, discussed in the next section) the system may chatter about the switching surface $\mathbf{k}^{\mathrm{T}}\mathbf{x} = 0$. In any real system, even if a chattering control is called for, there will be some time delay between switches, so that the mathematical chattering behavior is only approximated by bang-bang switching at some frequency. In some applications chatter may be at a high enough frequency to be heard as a buzzing. In most cases, chatter is an undesirable behavior that can be avoided in linear systems with control bounds by turning down the gain.

A simple physical example of chatter would be an on–off heating system for a room, with the heater being on whenever the temperature is below a desired fixed temperature, and off when the temperature is above the desired temperature. During either mode the room temperature will be an exponential function of time. But, because the system chatters between off and on, the room temperature remains constant at the desired value. This might be all right for an electric heater, but would be very undesirable for a furnace. Chatter in a temperature controller can be avoided by using a "dead zone" controller. The switching temperature is replaced by two separate switching temperatures, so that the heater comes on only after the temperature falls a specified amount below the desired temperature, and then turns off after the temperature becomes higher than the desired temperature.

Filippov Velocity

As we have seen from Case 3 of Example 1.2-2, in a discontinuous dynamical system chatter may produce velocity vectors that cannot occur otherwise. To handle this phenomenon mathematically we need to extend the notion of a solution for a system of discontinuous differential equations. We begin by defining the **forward-time derivative** (or simply the "velocity") vector as

$$\dot{\mathbf{x}}(t) \triangleq \lim_{\substack{\Delta t > 0 \\ \Delta t \to 0}} \frac{\mathbf{x}(t + \Delta t) - \mathbf{x}(t)}{\Delta t} \tag{1.2-16}$$

and letting $\dot{\mathbf{x}}(t-)$ and $\dot{\mathbf{x}}(t+)$ denote the (forward-time) velocity vector just before and after time t, respectively.

A function $\mathbf{x}(t)$ is a **solution** to (1.2-15), in the sense of Filippov [1964], if $\dot{\mathbf{x}}(t)$ is a convex combination of $\dot{\mathbf{x}}(t-)$ and $\dot{\mathbf{x}}(t+)$, that is, if there exists an $\alpha \in [0, 1]$ such that

$$\dot{\mathbf{x}}(t) = \alpha\dot{\mathbf{x}}(t-) + (1 - \alpha)\dot{\mathbf{x}}(t+) = \dot{\mathbf{x}}(t+) + \alpha[\dot{\mathbf{x}}(t-) - \dot{\mathbf{x}}(t+)] \tag{1.2-17}$$

where $\dot{\mathbf{x}}(t-) = \mathbf{f}[\mathbf{x}(t-)]$ and $\dot{\mathbf{x}}(t+) = \mathbf{f}[\mathbf{x}(t+)]$.

To clarify the meaning of (1.2-17), consider (1.2-15) and suppose that $\mathbf{f}(\mathbf{x})$ is discontinuous across a **switching surface** given by

$$\sigma(\mathbf{x}) = 0, \tag{1.2-18}$$

where $\sigma(\cdot)$ is C^1 (continuous and continuously differentiable) and has a nonzero **gradient** (row vector)

$$\frac{\partial \sigma(\mathbf{x})}{\partial \mathbf{x}} \triangleq \left[\frac{\partial \sigma}{\partial x_1} \quad \frac{\partial \sigma}{\partial x_2} \quad \cdots \quad \frac{\partial \sigma}{\partial x_{n_x}} \right], \tag{1.2-19}$$

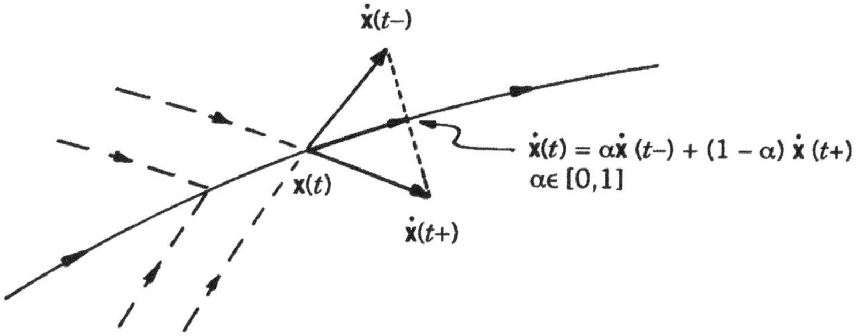

Figure 1.2-4 Solutions in the sense of Filippov.

so that the switching surface has a unique tangent plane at each point, with $\partial\sigma(\mathbf{x})/\partial\mathbf{x}$ being perpendicular to the switching surface and pointing into the region where $\sigma(\mathbf{x}) > 0$.

Off of the switching surface $\mathbf{f}(\mathbf{x})$ is locally continuous, and from (1.2-17) $\dot{\mathbf{x}}(t) = \dot{\mathbf{x}}(t+) = \dot{\mathbf{x}}(t-)$ is the usual time derivative. However, for the case where $\mathbf{x}(\cdot)$ chatters along the switching surface, as illustrated in Figure 1.2-4, (1.2-17) indicates that the actual or **Filippov velocity** vector $\dot{\mathbf{x}}(t)$ is the convex combination of the velocity vectors $\dot{\mathbf{x}}(t-)$ and $\dot{\mathbf{x}}(t+)$ adjacent to the switching surface, with α in (1.2-17) chosen so that $\dot{\mathbf{x}}(t)$ is tangent to the switching surface.

Assuming that neither $\dot{\mathbf{x}}(t-)$ nor $\dot{\mathbf{x}}(t+)$ is tangent to the switching surface at a point where $\mathbf{x}(t)$ meets the surface, chatter requires that the two dot products $(\partial\sigma/\partial\mathbf{x})\ \dot{\mathbf{x}}(t-)$ and $(\partial\sigma/\partial x)\ \dot{\mathbf{x}}(t+)$ have opposite signs.

It should be noted that since chattering systems may produce velocity vectors $\dot{\mathbf{x}}(t)$ other than $\mathbf{f}[\mathbf{x}(t)]$, it is possible for the Filippov velocity to be zero at a point where the system velocity would not be zero under a nonchattering control. That is, the system (1.2-15) may exhibit a **chattering equilibrium** point.

Example 1.2-3: Chattering Equilibrium Consider the discontinuous system

$$\dot{x}_1 = 0$$
$$\dot{x}_2 = -u$$

with

$$u(\mathbf{x}) = \begin{cases} 1 & \text{if } x_2 \geq 0 \\ -1 & \text{if } x_2 < 0. \end{cases}$$

As illustrated in Figure 1.2-5, every trajectory $\mathbf{x}(t)$ that starts off of the switching surface $x_2 = 0$ is a vertical straight line that reaches the switching surface in some finite time. Suppose that $\mathbf{x}(t)$ starts at $x_2 > 0$ and reaches the x_1-axis at

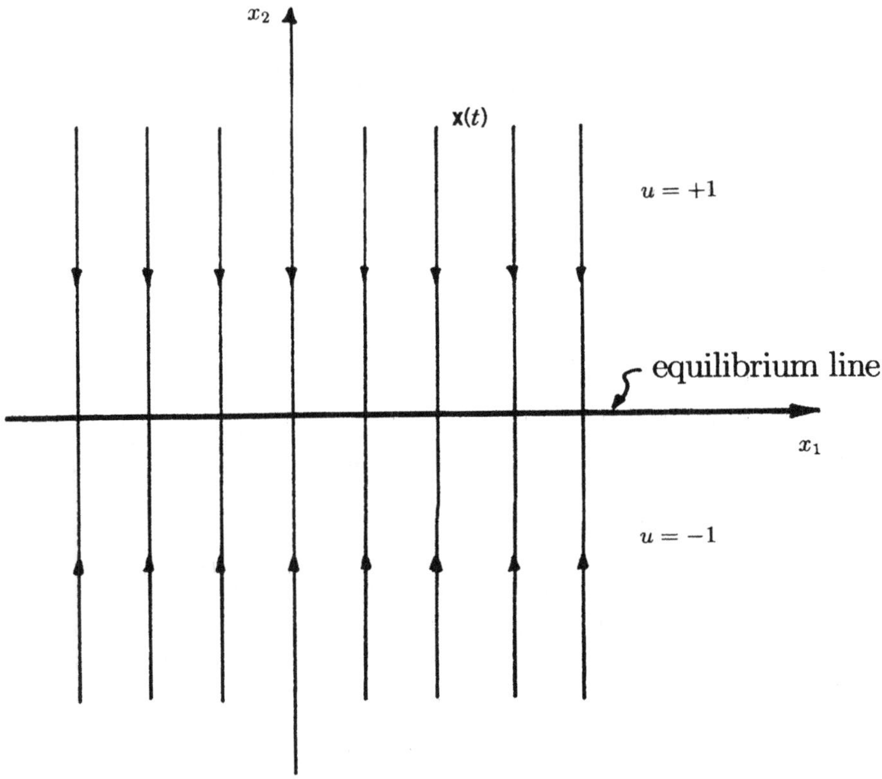

Figure 1.2-5 A line of chattering equilibriums for Example 1.2-3.

a switching time t_s. Then $\dot{\mathbf{x}}(t_s-) = [0 \quad -1]^T$ and, if the trajectory were to overshoot the x_1-axis by an arbitrary small amount, the velocity vector would switch to $\dot{\mathbf{x}}(t_s+) = [0 \quad 1]^T$, which would send the state vertically back to the switching surface. Thus $\mathbf{x}(t)$ chatters and stays on the switching surface, with a Filippov velocity given by

$$\dot{\mathbf{x}}(t_s) = \alpha \begin{bmatrix} 0 \\ -1 \end{bmatrix} + (1 - \alpha) \begin{bmatrix} 0 \\ 1 \end{bmatrix} = \begin{bmatrix} 0 \\ 1 - 2\alpha \end{bmatrix}.$$

For $\dot{\mathbf{x}}(t_s)$ to be tangent to the x_1-axis switching surface, we have $\dot{\mathbf{x}}_2 = 0$, which implies $\alpha = 1/2$. Thus the control u chatters equally between ± 1, with an effective value of zero. This yields $\dot{\mathbf{x}}(t_s) = 0$ and every point on the x_1-axis becomes a chattering equilibrium point, even though $\mathbf{f}[\mathbf{x}, u(\mathbf{x})]$ is never zero!

■

1.3 NUMERICAL SOLUTION ALGORITHMS

In Section 1.2 we have discussed some general mathematical properties of solutions to nonlinear differential equations. However, for most nonlinear systems we are unable to find an algebraic expression for the solution. Instead we must often rely on some numerical integration technique to compute and plot trajectories for the system. In this section we present two such algorithms, leading to a current state-of-the-art algorithm. In addition we also present some other useful results related to plotting trajectories.

Runge–Kutta Integration Algorithms

Consider a system of differential equations of the form

$$\dot{\mathbf{x}} = \mathbf{f}(\mathbf{x}, t), \tag{1.3-1}$$

where $\mathbf{f}(\mathbf{x}, t)$ and the solution $\mathbf{x}(t)$ are n_x-dimensional vectors that are assumed to be continuous and to have continuous partial derivatives of any order required by the ensuring analysis. In particular, for the moment we ignore the possibility of discontinuous systems and assume that the trajectory $\mathbf{x}(t)$ is smooth and has no "corners" anywhere.

The basic idea behind the Runge–Kutta algorithms is to expand both $\mathbf{x}(t)$ and $\mathbf{f}[\mathbf{x}(t), t]$ in Taylor series and then to use the results to construct a numerical integration algorithm that takes into account the curvature of the trajectory $\mathbf{x}(t)$. We illustrate the development process for the case of a second-order Runge–Kutta algorithm.

Second-Order Runge–Kutta Knowing t and $\mathbf{x}(t)$, we may approximate $\mathbf{x}(t + \Delta t)$ by its second-order Taylor series expansion:

$$\mathbf{x}(t + \Delta t) \approx \mathbf{x}(t) + \Delta t \frac{d\mathbf{x}(t)}{dt} + \frac{\Delta t^2}{2} \frac{d^2\mathbf{x}(t)}{dt^2}.$$

Using (1.3-1) to substitute for $\dot{\mathbf{x}}$ yields

$$\mathbf{x}(t + \Delta t) \approx \mathbf{x}(t) + \Delta t \mathbf{f} + \frac{\Delta t^2}{2} \left[\frac{\partial \mathbf{f}}{\partial t} + \frac{\partial \mathbf{f}}{\partial \mathbf{x}} \mathbf{f} \right], \tag{1.3-2}$$

where the last term in brackets is $d\mathbf{f}/dt$ and

$$\frac{\partial \mathbf{f}}{\partial \mathbf{x}} \triangleq \begin{bmatrix} \dfrac{\partial f_1}{\partial x_1} & \cdots & \dfrac{\partial f_1}{\partial x_{n_x}} \\ \vdots & \ddots & \vdots \\ \dfrac{\partial f_{n_x}}{\partial x_1} & \cdots & \dfrac{\partial f_{n_x}}{\partial x_{n_x}} \end{bmatrix}.$$

The expression in (1.3-2) is then compared with $\mathbf{x}(t + \Delta t)$ evaluated by a weighted average of two derivatives

$$\mathbf{x}(t + \Delta t) = \mathbf{x}(t) + \Delta t[a\mathbf{f}(\mathbf{x}, t) + b\mathbf{f}(\mathbf{x} + q\,\Delta t\,\mathbf{f}(\mathbf{x}, t), t + p\,\Delta t)]$$

where a, b, p, and q are constants to be chosen to make this relation accurate. Expanding the second term in brackets yields the following alternate expression:

$$\mathbf{x}(t + \Delta t) = \mathbf{x}(t) + \Delta t(a + b)\mathbf{f}(\mathbf{x}, t) + \Delta t^2 \left[bp\,\frac{\partial \mathbf{f}}{\partial t} + bq\,\frac{\partial \mathbf{f}}{\partial \mathbf{x}}\,\mathbf{f} \right]. \qquad (1.3\text{-}3)$$

Equating like terms in (1.3-2) and (1.3-3), we see that to achieve second-order accuracy we must choose $a + b = 1$ and $bp = bq = 1/2$. One solution for the parameters is given by

$$\begin{aligned} a &= b = \tfrac{1}{2} \\ p &= q = 1, \end{aligned} \qquad (1.3\text{-}4)$$

yielding the **second-order Runge–Kutta algorithm**

$$\mathbf{x}(t + \Delta t) = \mathbf{x}(t) + \frac{\Delta t}{2}\,[\mathbf{f}(\mathbf{x}, t) + \mathbf{f}(\overline{\mathbf{x}}, t + \Delta t)],$$

where

$$\overline{\mathbf{x}} = \mathbf{x}(t) + \Delta t\,\mathbf{f}(\mathbf{x}, t).$$

In summary, the general second-order Runge–Kutta algorithm is of the form

$$\mathbf{x}(t + \Delta t) = \mathbf{x}(t) + \frac{\Delta t}{2}\,(\mathbf{k}_1 + \mathbf{k}_2), \qquad (1.3\text{-}5)$$

where $\mathbf{x}(t)$ is the current state at time t, Δt is a fixed time step for numerical integration, $\mathbf{x}(t + \Delta t)$ is the state at the next time point $t + \Delta t$, and $\mathbf{k}_1$ and $\mathbf{k}_2$ are vectors, of the same dimension as $\mathbf{x}$, given by

$$\mathbf{k}_1 = \mathbf{f}(\mathbf{x}, t) \tag{1.3-6}$$
$$\mathbf{k}_2 = \mathbf{f}(\mathbf{x} + \mathbf{k}_1\,\Delta t, t + \Delta t). \tag{1.3-7}$$

The error between $\mathbf{x}(t + \Delta t)$ and the actual solution is of order Δt^3.

A geometric interpretation for this second-order Runge–Kutta algorithm is illustrated in Figure 1.3-1. Starting from the current state and time and using the current velocity vector, given by $\mathbf{k}_1$, a simple Euler integration step is taken to the trial point

$$\bar{\mathbf{x}} = \mathbf{x} + \mathbf{k}_1 \Delta t$$
$$\bar{t} = t + \Delta t.$$

At the trial point $(\bar{\mathbf{x}}, \bar{t})$ the corresponding velocity vector $\mathbf{k}_2 = \mathbf{f}(\bar{\mathbf{x}}, \bar{t})$ is computed. The difference between the two neighboring velocity vectors $\mathbf{k}_1$ and $\mathbf{k}_2$ is a measure of how much the trajectory $\mathbf{x}(t)$ turns in the vicinity of the point $(\mathbf{x}, t)$. Having determined these two velocity vectors, the actual integration step is taken as an Euler step using the average of the two velocity vectors. For this reason this particular second-order Runge–Kutta algorithm is called the **modified Euler method**.

Fourth-Order Runge–Kutta The same type of procedure used to develop the second-order algorithm can also be used to develop high-order algorithms, by retaining higher-order terms in the Taylor series expansions. The most common of these algorithms is the following **fourth-order Runge–Kutta** algorithm:

$$\mathbf{x}(t + \Delta t) = \mathbf{x}(t) + \frac{\Delta t}{6}\,[\mathbf{k}_1 + 2\mathbf{k}_2 + 2\mathbf{k}_3 + \mathbf{k}_4], \tag{1.3-8}$$

where, for $\mathbf{x} = \mathbf{x}(t)$,

$$\mathbf{k}_1 = \mathbf{f}(\mathbf{x}, t) \tag{1.3-9}$$

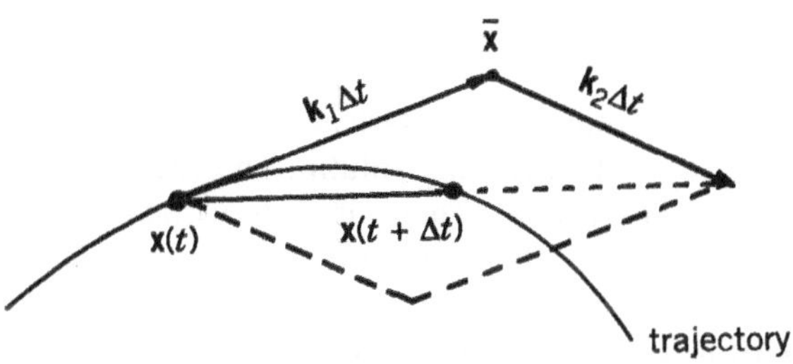

Figure 1.3-1 Second-order Runge–Kutta geometry.

$$\mathbf{k}_2 = \mathbf{f}\left(\mathbf{x} + \mathbf{k}_1 \frac{\Delta t}{2}, t + \frac{\Delta t}{2}\right) \tag{1.3-10}$$

$$\mathbf{k}_3 = \mathbf{f}\left(\mathbf{x} + \mathbf{k}_2 \frac{\Delta t}{2}, t + \frac{\Delta t}{2}\right) \tag{1.3-11}$$

$$\mathbf{k}_4 = \mathbf{f}(\mathbf{x} + \mathbf{k}_3\, \Delta t, t + \Delta t). \tag{1.3-12}$$

The step size Δt can be either positive or negative, corresponding to forward or backward integration, respectively. The error between $\mathbf{x}(t + \Delta t)$ and the actual solution is of order Δt^5.

We need to mention a few words of caution when using numerical algorithms such as these for control systems such as

$$\dot{\mathbf{x}} = \mathbf{f}(\mathbf{x}, \mathbf{u}). \tag{1.3-13}$$

When the control law $\mathbf{u}(\cdot)$ has discontinuities, it is recommended that the control $\mathbf{u}$ be calculated at the current state and time and then held constant over the time interval Δt while propagating to the new state and time. The reason for this is that the algorithms have been developed assuming that the velocity vector does not change discontinuously. Such a change should not occur while we are propagating from one state to the next.

To handle a switching time that occurs between t and $t + \Delta t$, we can either take the time step Δt small enough that we can ignore any error in the switching time, or, once a switch has been detected, we can return to the previous time and use some root-finding algorithm to determine the precise switching time. Then we can change the step size Δt to propagate to this time, switch the control, and then restore the step size to its original value to continue integration using the new control value.

Many available numerical integration software packages use a variable step size Δt, computed to ensure that the integration error is kept smaller than a specified tolerance. When these algorithms detect a "corner," as a large change in the velocity vector, they automatically reduce the step size to achieve specified error tolerances.

Gear's Integration Algorithm

Predictor-corrector algorithms are numerical integration procedures that save certain information about the evolving solution in order to enhance the performance of the integration scheme. There are many such algorithms available, but the one that we present is also well-suited for a particular class of systems, known as stiff systems, that we discuss below.

Gear's algorithm [Gear, 1967] is a fixed step size algorithm for systems of the form

$$\dot{\mathbf{x}} = \mathbf{f}(\mathbf{x}). \tag{1.3-14}$$

The algorithm is a predictor-corrector procedure that is initialized by using the fourth-order Runge–Kutta procedure for the first two time steps. For subsequent steps, the state is updated by first predicting the state to the new time and then correcting the state estimate. The algorithm and the initialization procedure are as follows:

Initialization Starting from $\mathbf{x}_0 = \mathbf{x}(0)$ with step size Δt, compute

$$\begin{aligned} \mathbf{x}_{C2} &= \mathbf{x}_0 \\ (h\dot{\mathbf{x}})_2 &= \mathbf{f}(\mathbf{x}_0)\Delta t. \end{aligned}$$

Take the first Runge–Kutta step, yielding $\mathbf{x}_1 = \mathbf{x}(\Delta t)$, and compute

$$\begin{aligned} \mathbf{x}_{C1} &= \mathbf{x}_1 \\ (h\dot{\mathbf{x}})_1 &= \mathbf{f}(\mathbf{x}_1)\Delta t. \end{aligned}$$

Take the second Runge–Kutta step, yielding $\mathbf{x}_2 = \mathbf{x}(2\Delta t)$, and compute

$$\begin{aligned} \mathbf{x} &= \mathbf{x}_2 \\ (h\dot{\mathbf{x}}) &= \mathbf{f}(\mathbf{x}_2)\Delta t. \end{aligned}$$

For each subsequent time step use the following predictor-corrector sequence:

Predictor

$$\begin{aligned} \mathbf{x}_P &= -18\mathbf{x} + 9\mathbf{x}_{C1} + 10\mathbf{x}_{C2} + 9(h\dot{\mathbf{x}}) + 18(h\dot{\mathbf{x}})_1 + 3(h\dot{\mathbf{x}})_2 \\ (h\dot{\mathbf{x}})_P &= -57\mathbf{x} + 24\mathbf{x}_{C1} + 33\mathbf{x}_{C2} + 24(h\dot{\mathbf{x}}) + 57(h\dot{\mathbf{x}})_1 + 10(h\dot{\mathbf{x}})_2. \end{aligned}$$

Corrector

$$\begin{aligned} \mathbf{d} &= (h\dot{\mathbf{x}})_P - \mathbf{f}(\mathbf{x}_P)\Delta t \\ \mathbf{x}_{C2} &= \mathbf{x}_{C1} - \frac{11}{1440}\mathbf{d} \end{aligned}$$

$$
\begin{aligned}
(h\dot{\mathbf{x}})_2 &= (h\dot{\mathbf{x}})_1 \\
\mathbf{x}_{C1} &= \mathbf{x} + \frac{3}{160}\mathbf{d} \\
(h\dot{\mathbf{x}})_1 &= (h\dot{\mathbf{x}}) \\
\mathbf{x} &= \mathbf{x}_P - \frac{95}{288}\mathbf{d} \\
(h\dot{\mathbf{x}}) &= \mathbf{f}(\mathbf{x}_P)\Delta t.
\end{aligned}
$$

Gear's algorithm is a general purpose sixth-order procedure, with error in $\mathbf{x}(t+\Delta t)$ being proportional to Δt^7. As a result one can use a step size approximately five to ten times larger than the fourth-order Runge–Kutta procedure and still have about the same error magnitude. However, because of the increased number of computations, the overall integration procedure may be not be significantly faster that the Runge–Kutta procedure.

Stiff Systems Gear's algorithm is particularly well-suited for **stiff systems** of the form

$$\dot{\mathbf{x}} = \mathbf{f}(\mathbf{x}, \mathbf{z}) \tag{1.3-15}$$

$$\epsilon\dot{\mathbf{z}} = \mathbf{g}(\mathbf{x}, \mathbf{z}), \tag{1.3-16}$$

where ϵ is a small parameter. These systems exhibit two time scales, with $\mathbf{x}(t)$, governed by (1.3-15), being the state of the **slow subsystem**, and $\mathbf{z}(t)$, governed by (1.3-16), being the state of the **fast subsystem**. This terminology follows from the fact that if (1.3-16) is divided by the small parameter ϵ, then $\mathbf{z}(t)$ changes much more rapidly than $\mathbf{x}(t)$. In the limit as $\epsilon \rightarrow 0$, (1.3-16) becomes an algebraic equation and $\mathbf{z}(t)$ changes instantaneously with changes in $\mathbf{x}(t)$. An example of a stiff system is given by the problem of controlling the position of a rotating flexible steel shaft pinned at one end, such as a flexible robotic manipulator. The slow subsystem would correspond to the motion of the center of mass as the object is rotated, while the fast subsystem would correspond to the vibrational motion of the flexible shaft or perhaps the position of the free end of the vibrating shaft.

To numerically integrate (1.3-15) and (1.3-16) as a single system of the form (1.3-14), with $\epsilon \neq 0$, we must choose the step size small enough to accommodate the rapidly changing fast subsystem. But then the slow subsystem may exhibit very little change over several time steps. Gear's algorithm can handle this problem directly, by treating the slow and fast subsystems separately. The procedure is to "predict" both the slow and fast subsystems at each time step, but only "correct" the slow subsystem after every N time steps. This can achieve a significant computational savings, since the prediction step does not require evaluating either $\mathbf{f}(\cdot)$ or $\mathbf{g}(\cdot)$ and the correction step, which does require function evaluations, is only performed for $\mathbf{f}(\cdot)$ once every N time steps.

Livermore Differential Equation Solver (LSODAR)

LSODAR is a comprehensive code for solving ordinary differential equations in double precision with an automatic switching of techniques for stiff and nonstiff problems and with root-finding routines that are particularly useful for solving control problems with switching functions [Hindmarsh, 1983, pp. 55–64; Petzold, 1983, pp. 136–148]. The user does not have to determine whether the problem is stiff or not, as the solver will automatically choose the appropriate method. It uses an Adams variable-order predictor-corrector algorithm for nonstiff systems and a backwards difference formula for stiff systems. It locates the roots of any of a set of functions $\boldsymbol{\sigma}(\mathbf{x}) = \mathbf{0}$ and finds only those roots for which one or more of the functions $\boldsymbol{\sigma}(\mathbf{x})$ changes sign in the interval of integration. It then returns the solution at the root, if that occurs sooner than the specified stop condition, and otherwise it returns the solution according the specified stop condition.

Root-Finding Algorithms

When numerically integrating discontinuous systems, there is a need to compute a switching time accurately, using some root-finding technique. The LSODAR package does this using the Illinois algorithm, which is related to the false position method below. Some such routine should be included in any code intended for use with control systems involving switching-type controllers. For reference, we present two simple algorithms for finding roots of nonlinear equations.

Regula Falsi For the root(s) of a scalar-valued function of a single variable

$$g(x) = 0,$$

we can use the **regula falsi** (false position) algorithm, shown in Figure 1.3-2.
Suppose that two points x_1 and x_2 have been found such that

$$g(x_1)g(x_2) < 0.$$

Then a root of $g(x)$ must lie between x_1 and x_2, provided $g(\cdot)$ is continuous. We approximate $g(x)$ by a linear function $\overline{g}(x)$, using the two-point form of a line with $\overline{g}(x_1) = g(x_1)$ and $\overline{g}(x_2) = g(x_2)$, and take the point where $\overline{g}(x) = 0$ as a new estimate for the root. This point is given by

$$x = \frac{x_1 g_2 - x_2 g_1}{g_2 - g_1}, \tag{1.3-17}$$

where $g_1 = g(x_1)$ and $g_2 = g(x_2)$. To shrink the interval of uncertainty (x_1, x_2), we compute $g = g(x)$ and compare its sign to that of g_1. If $g_1 g > 0$, we set $x_1 = x$ and $g_1 = g$. Otherwise we set $x_2 = x$ and $g_2 = g$. A new x is then computed from

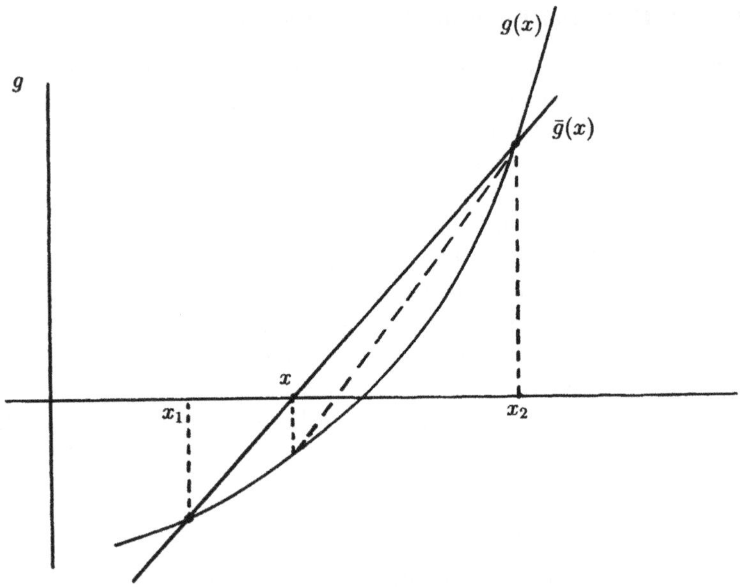

Figure 1.3-2 Regula falsi geometry.

(1.3-17) and the process is repeated until either $|g(x)|$ is sufficiently small or $|x_2 - x_1|$ is sufficiently small. The procedure always converges, provided $g(\cdot)$ is differentiable, and the errors at steps k and $k+1$ are related by $\epsilon(k+1) \sim \epsilon(k)^{1.618}$.

Newton's Method For finding roots of systems of nonlinear equations with more than one unknown, we can use **Newton's method**. Let

$$\mathbf{g}(\mathbf{x}) = \mathbf{0}$$

be a system of n_x equations in n_x unknowns. If $\mathbf{x}_k$ is the current estimate of a root, we can expand $\mathbf{g}(\mathbf{x}_k + \Delta\mathbf{x})$ in a Taylor series, retaining only the linear terms, and then set the result equal to zero. Thus $\Delta\mathbf{x}$ is determined by using Gaussian elimination to solve the system of linear equations

$$\frac{\partial \mathbf{g}(\mathbf{x}_k)}{\partial \mathbf{x}} \Delta\mathbf{x} = -\mathbf{g}(\mathbf{x}_k),$$

and the new estimate for the root is given by

$$\mathbf{x}_{k+1} = \mathbf{x}_k + \Delta\mathbf{x}.$$

The process is repeated until either $\|\mathbf{g}(\mathbf{x})\|$ is sufficiently small or $\|\Delta\mathbf{x}\|$ is sufficiently small. The procedure may not converge unless the initial guess is close to an actual root. But when the procedure converges it is relatively fast, with the errors at steps k and $k+1$ being related by $\epsilon(k+1) \sim \epsilon(k)^2$.

Plotting Function Contours

One use of differential equations is to plot lines of constant value for some specified function. For example, suppose we have a scalar-valued function of two variables, $V(x, y)$, and we wish to plot the contour

$$V(x, y) = c. \tag{1.3-18}$$

Since V is a constant it follows that

$$\frac{dV}{dt} = \frac{\partial V}{\partial x}\dot{x} + \frac{\partial V}{\partial y}\dot{y} = 0.$$

One way to satisfy $dV/dt = 0$ is to solve the differential equations

$$\dot{x} = \frac{\partial V(x, y)}{\partial y} \tag{1.3-19}$$

$$\dot{y} = -\frac{\partial V(x, y)}{\partial x}, \tag{1.3-20}$$

subject to initial conditions that satisfy $V[x(0), y(0)] = c$. For functions of more than two variables, we can generate two-dimensional cross sections simply by holding all of the variables constant except for two, x and y, which we force to satisfy the differential equations (1.3-19) and (1.3-20), respectively.

Plotting Three-Dimensional Trajectories

Since we must usually rely on numerical integration techniques to obtain plots of the trajectories of a nonlinear system, it would be nice if we could not only plot in two dimensions but also in three dimensions.

Any method for plotting three-dimensional objects on a two-dimensional surface must incorporate an algorithm for projecting from three dimensions down to two dimensions. The **parallel projection** algorithm, illustrated in Figure 1.3-3, can be used for this purpose.

Let (x, y, z) denote the coordinates of a point in a three-dimensional coordinate system and let (X, Y) denote plotter coordinates, in a plane perpendicular to a viewing direction along a Z-axis, with the origin of the X-Y system being at $(\bar{x}, \bar{y}, \bar{z})$, shown as the (x, y, z) origin in Figure 1.3-3.

We define a viewing direction by specifying an azimuth angle φ counterclockwise from the positive x-axis in the x-y plane, and an elevation angle γ

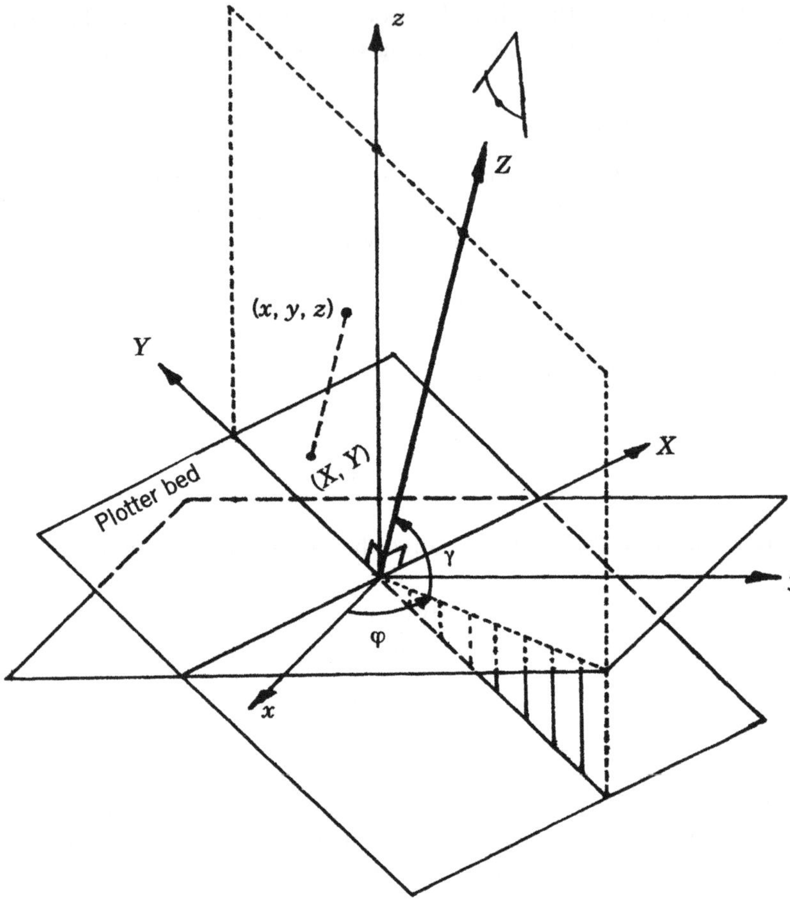

Figure 1.3-3 Parallel projection.

upward from the x-y plane. The two-dimensional plot of an object is generated by projecting each point (x, y, z) on the object onto the X-Y plane along lines parallel to the viewing direction. The transformation is given by

$$\begin{bmatrix} X \\ Y \end{bmatrix} = \begin{bmatrix} -S_x \sin \varphi & S_y \cos \varphi & 0 \\ -S_y \sin \gamma \cos \varphi & -S_y \sin \gamma \sin \varphi & S_z \cos \gamma \end{bmatrix} \begin{bmatrix} x - \bar{x} \\ y - \bar{y} \\ z - \bar{z} \end{bmatrix},$$

where S_x, S_y, and S_z are the distances per unit x, y, and z values, respectively.

1.4 NONLINEAR SYSTEMS PHENOMENA

In this section we focus on differences between linear and nonlinear dynamic systems. In particular, we consider some of the basic phenomena that can occur only in nonlinear systems of the form

$$\dot{\mathbf{x}} = \mathbf{f}(\mathbf{x}), \tag{1.4-1}$$

compared with linear systems of the form

$$\dot{\mathbf{x}} - \mathbf{A}\mathbf{x}. \tag{1.4-2}$$

No Superposition Principle

The superposition principle for linear systems does not apply to nonlinear systems. To verify this, let $\mathbf{x}_1(t)$ and $\mathbf{x}_2(t)$ be any two functions satisfying (1.4-1), without regard to initial conditions. Then, in general, a linear combination of solutions

$$\mathbf{x}(t) = a\mathbf{x}_1(t) + b\mathbf{x}_2(t) \tag{1.4-3}$$

only satisfies (1.4-1) if

$$\mathbf{f}(a\mathbf{x}_1 + b\mathbf{x}_2) = a\mathbf{f}(\mathbf{x}_1) + b\mathbf{f}(\mathbf{x}_2),$$

that is, only if $\mathbf{f}(\cdot)$ is a linear function of the form

$$\mathbf{f}(\mathbf{x}) = \mathbf{A}\mathbf{x}.$$

The inclusion of any nonlinearity destroys the superposition principle. This includes bounding the control input on an otherwise linear system. In linear control systems theory such constraint bounds are usually not considered, but in any real system such bounds will always exist. This means that even "linear" systems have the potential for nonlinear behavior. For example, consider the linear single-input control system

$$\dot{\mathbf{x}} = \mathbf{A}\mathbf{x} + \mathbf{b}u, \tag{1.4-4}$$

with a linear state feedback control law

$$u(\mathbf{x}) = -\mathbf{k}^{\mathrm{T}}\mathbf{x}. \tag{1.4-5}$$

So far, the control system given by (1.4-4) and (1.4-5) is linear and the superposition principle will apply. However suppose, for example, that the scalar control in (1.4-5) is required to satisfy the constraints

$$u_{\min} \leq u \leq u_{\max} \tag{1.4-6}$$

where $u_{\max} > u_{\min}$ are specified constants. Clearly, some modification must be made to the control law given by (1.4-5), since for some sufficiently large $\mathbf{x}$ the control constraints will be violated.

Saturating State Feedback Control A typical approach to handle the bounded control requirement is to let the control (1.4-5) **saturate** when it reaches the control constraint boundary. Under this approach a nonlinear **saturating state feedback controller** is defined as

$$\text{sat}[u(\mathbf{x})] \triangleq \begin{cases} u_{\max} & \text{if } u(\mathbf{x}) > u_{\max} \\ u(\mathbf{x}) & \text{if } u_{\min} \leq u(\mathbf{x}) \leq u_{\max} \\ u_{\min} & \text{if } u(\mathbf{x}) < u_{\min}, \end{cases} \tag{1.4-7}$$

as illustrated in Figure 1.4-1. The superposition principle is no longer valid, since the general superposition solution $\mathbf{x}(t)$ specified by (1.4-3) will cause $u[\mathbf{x}(t)]$ to saturate for sufficiently large parameters a and b.

Despite their nonlinear nature, saturating feedback controllers that are otherwise linear are often very effective controllers. But the reader should be fore-

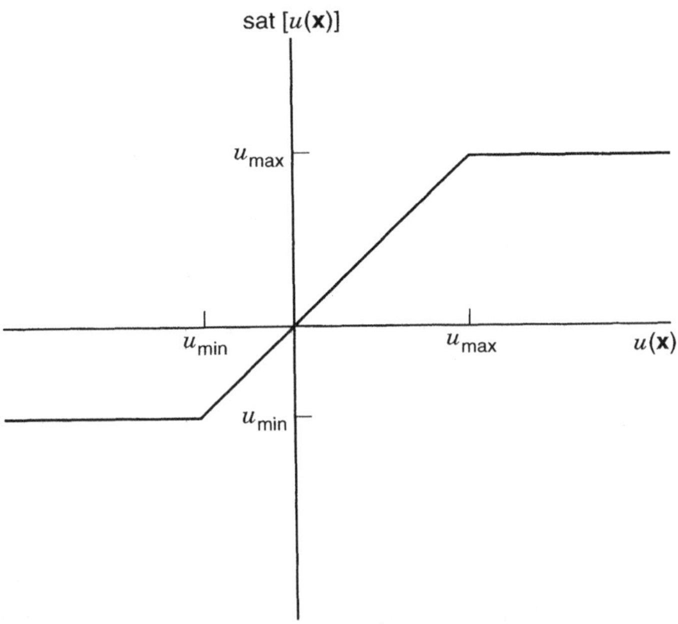

Figure 1.4-1 Saturating feedback control.

warned that most linear design techniques implicitly assume that the superposition principle holds, and one should not be surprised when using saturating feedback control that the system does not yield expected results (see Example 2.2-5).

Multiple Separated Equilibriums

The possible equilibrium points for the linear system (1.4-2) are given by the solutions to

$$\mathbf{A}\mathbf{x} = \mathbf{0}.$$

If $\mathbf{A}^{-1}$ exists (i.e., if rank[$\mathbf{A}$] $= n_x$), then $\mathbf{x} = \mathbf{0}$ is the only equilibrium solution. Otherwise, the set of equilibrium points is a subspace (a line, a plane, etc., through $\mathbf{x} = \mathbf{0}$) of dimension $n_x -$ rank[$\mathbf{A}$]. Thus we see that, for a linear system, the only possible isolated equilibrium point is the origin.

The nonlinear system (1.4-1), on the other hand, may have more than one distinct separated equilibrium solution. This is simply because the equilibrium condition

$$\mathbf{f}(\mathbf{x}) = \mathbf{0}$$

may have more than one isolated solution. Indeed, multiple equilibriums are often associated with the rich dynamic behavior found in various nonlinear systems.

Example 1.4-1: Nonlinear Pendulum To understand the possible effect of multiple isolated equilibrium points, consider the equations of motion for a simple two-dimensional pendulum. The pendulum may be thought of as a point mass on the end of a massless rigid rod free to pivot, without friction, about the origin in the vertical plane. Let x_1 be the angle from the downward vertical position to the current position of the pendulum, let x_2 be the angular velcoity, and let the ratio of the gravitational constant to the length of the rod equal one. The equations of motion for the system in this case are given by

$$\begin{aligned} \dot{x}_1 &= x_2 \\ \dot{x}_2 &= -\sin(x_1) \end{aligned}$$

This nonlinear system has an infinite number of equilibrium positions, given by $(x_1, x_2) = (\pm k\pi, 0)$, $k = 0, 1, \ldots$. related to the number of revolutions made by the pendulum. The inverted equilibriums (k odd) are unstable and the downward equilibriums (0, k even) are stable. If the initial angular velocity $x_2(0)$ is sufficiently high, the system will undergo a constantly spinning motion, instead of oscillating about one of the lower, stable equilibrium positions.

If we linearize the system, by assuming that x_1 is small and replacing $\sin(x_1)$ by x_1, the resulting linear system

$$\dot{x}_1 = x_2$$
$$\dot{x}_2 = -x_1$$

only has one equilibrium position, at the origin, corresponding to the stable $k = 0$ equilibrium for the nonlinear system. As far as this linear system is concerned, all of the other nonlinear equilibriums have vanished. This results in a fundamentally different type of solution. No matter how large we make the initial angular velocity $x_2(0)$, the linear system simply oscillates about the origin, with no hint of the rotational spinning motion that can occur in the nonlinear system. ■

Finite Escape Time

As we indicated in Section 1.2, a solution $\mathbf{x}(t)$ for a nonlinear system may not exist for all $t \in (-\infty, \infty)$. For example, all solutions to the system

$$\dot{x} = 1 + x^2 \tag{1.4-8}$$

exhibit a **finite escape time**. That is, $|x(t)| \to \infty$ as t approaches some finite time T. In particular, for $x(0) = 0$, rearranging (1.4-8) and integrating

$$\int_0^{x(t)} \frac{dx}{1 + x^2} = \int_0^t d\tau$$

yields $\tan^{-1}[x(t)] = t$, or

$$x(t) = \tan(t),$$

and $|x(t)| \to \infty$ as $t \to \pm\pi/2$.

Constant-coefficient linear systems of the form (1.4-2) can never exhibit a finite escape time, since the general solution is of the form

$$\mathbf{x}(t) = \gamma_1(t)e^{\mu_1 t}\boldsymbol{\xi}_1 + \cdots + \gamma_{n_x}(t)e^{\mu_{n_x} t}\boldsymbol{\xi}_{n_x}, \tag{1.4-9}$$

where μ_i and $\boldsymbol{\xi}_i$ are the eigenvalues and eigenvectors [see (1.5-12)] of $\mathbf{A}$, respectively, and the coefficients $\gamma_i(t)$ are at worst polynomials in t.

Limit Cycles

Linear and nonlinear systems may exhibit **periodic** trajectories, defined by $\mathbf{x}(t + T) = \mathbf{x}(t)$, where $T > 0$ (to exclude equilibrium points) is the **period** of the trajectory. Certain nonlinear systems exhibit an even more interesting geometric phenomenon known as a **limit cycle**, which is defined as a periodic trajectory having no periodic neighbors. Trajectories starting near a limit cycle may approach the limit cycle, or diverge from it, or both, but they cannot be periodic themselves. Geared servomotor systems frequently exhibit limit cycle behavior. Because of slack and backlash in the gears, the object being controlled cannot be positioned exactly at the desired location. Instead, it oscillates slightly about the target, in a small limit cycle.

Limit cycles can occur in nonlinear systems of any dimension ≥ 2. However, the limit cycle itself is a one-dimensional object; that is, it is defined by the trajectory that describes it. For example, in a three-dimensional system it is possible to have a limit cycle, but a three-dimensional plot of the limit cycle would look like a string that closes on itself. The limit cycle cannot define a higher-dimensional object, say the surface of a sphere, since to do so would require the crossing of the "string" on itself in order for the string to close on itself. State-space trajectories for systems defined by (1.4-1) that are continuous in $\mathbf{x}$ and that have first-order partial derivatives with respect to $\mathbf{x}$ cannot cross.

We illustrate the limit cycle phenomenon with three examples, two of which involve classical dynamic systems. Example 1.4-2 yields a particularly simple limit cycle, which can be found analytically. Example 1.4-3 is a system in which a limit cycle is not only desirable (in contrast to the case of geared systems), but also forms the basis for the Van der Pol vacuum tube triode oscillator [Minorsky, 1962, p. 423]. Example 1.4-4 illustrates how limit cycles may be produced from a linear control system driven by saturating state variable feedback control.

Example 1.4-2: Logistics Equation Consider the system

$$\begin{aligned}\dot{x}_1 &= x_2 + x_1[1 - x_1^2 - x_2^2]\\ \dot{x}_2 &= -x_1 + x_2[1 - x_1^2 - x_2^2].\end{aligned} \tag{1.4-10}$$

Converting to polar coordinates

$$r = \sqrt{x_1^2 + x_2^2} \tag{1.4-11}$$

$$\theta = \tan^{-1}\left(\frac{x_2}{x_1}\right) \tag{1.4-12}$$

and defining $N \triangleq r^2$ yields

$$\dot{N} = 2N(1 - N) \tag{1.4-13}$$

$$\dot{\theta} = -1. \tag{1.4-14}$$

The first of these transformed differential equations (1.4-13) is a classical **logistics equation**. It appears in many applications, including some World War II studies on the logistics of a two-ocean worldwide war. Today it is much more prevalent in the field of population biology, since it is one of the simplest models for the dynamics of a single-species population with a limited food supply. As we see shortly, the logistics equation also plays an important role in some very revolutionary developments underway in modern nonlinear dynamics.

For our example system, governed by (1.4-10), the limit cycle is obtained from (1.4-13) by setting $\dot{N} = 0$, which yields the two solutions $N = 0$ and $N = 1$. The latter solution corresponds to a circle, $r(t) \equiv 1$, as illustrated in Figure 1.4-2. It is the only periodic solution for the dynamic system and, in this example, all other trajectories approach the limit cycle, except for the equilibrium solution at $\mathbf{x}(t) \equiv \mathbf{0}$.

Even though all of the system trajectories (except the origin) are attracted to the limit cycle, it should be emphasized that only one trajectory passes through any point in state space. Furthermore none of the trajectories, except the limit cycle itself, ever actually reach the limit cycle. This latter behavior is not restricted to limit cycles. It is also associated with equilibrium points for either linear or nonlinear systems. Indeed limit cycles are somewhat akin to equilibrium points, in the sense that a limit cycle represents a steady condition for the system. It is not a "steady-state" (unfortunately this terminology is sometimes

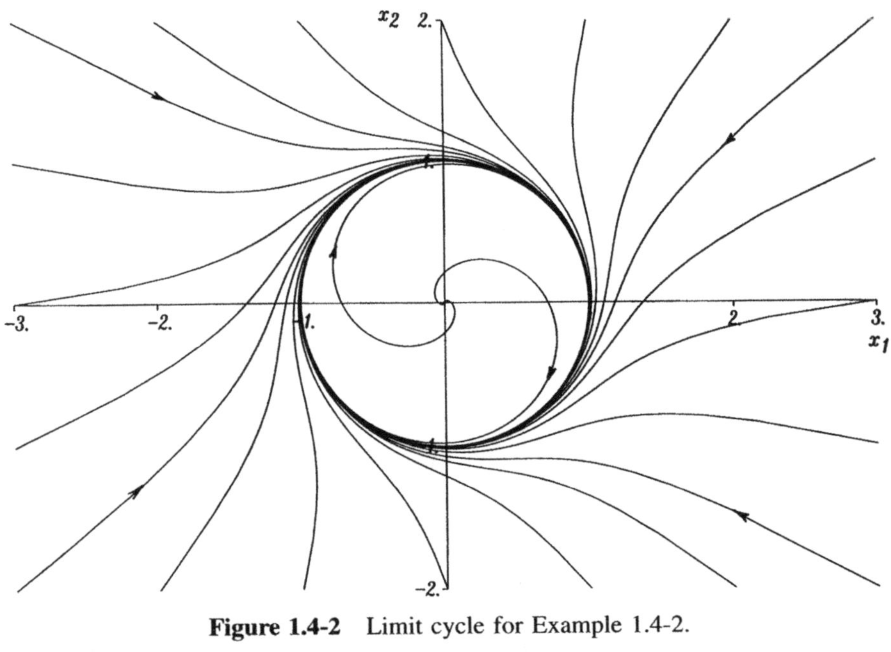

Figure 1.4-2 Limit cycle for Example 1.4-2.

used) but rather a steady oscillation. In a state-space setting "steady-state" can only have one meaning: equilibrium.

Note that the angle θ in (1.4-12) is not defined at $\mathbf{x} = \mathbf{0}$. Thus the origin is always a **singular point** for the transformation between Cartesian and polar coordinates. Because of this, the two dynamic systems are not exactly equivalent. For the system given by (1.4-10), $\mathbf{x} = \mathbf{0}$ is an equilibrium point. In fact, it is the only equilibrium point for the system. But the system given by (1.4-13) and (1.4-14) has no equilibrium solution at all. We return to this point later, when we discuss some topological analysis techniques for dynamic systems. ■

Example 1.4-3: Van der Pol Oscillator Another example of a limit cycle is given by Van der Pol's equation

$$\ddot{y} - \beta(1 - y^2)\dot{y} + y = 0. \tag{1.4-15}$$

In terms of the state variables $(x_1, x_2) = (y, \dot{y})$, Van der Pol's equation becomes

$$\begin{aligned} \dot{x}_1 &= x_2 \\ \dot{x}_2 &= -x_1 + \beta(1 - x_1^2)x_2. \end{aligned} \tag{1.4-16}$$

The system trajectories, and the limit cycle that they approach, are plotted in Figure 1.4-3, for the case where $\beta = 1$. The Van der Pol system has a limit cycle

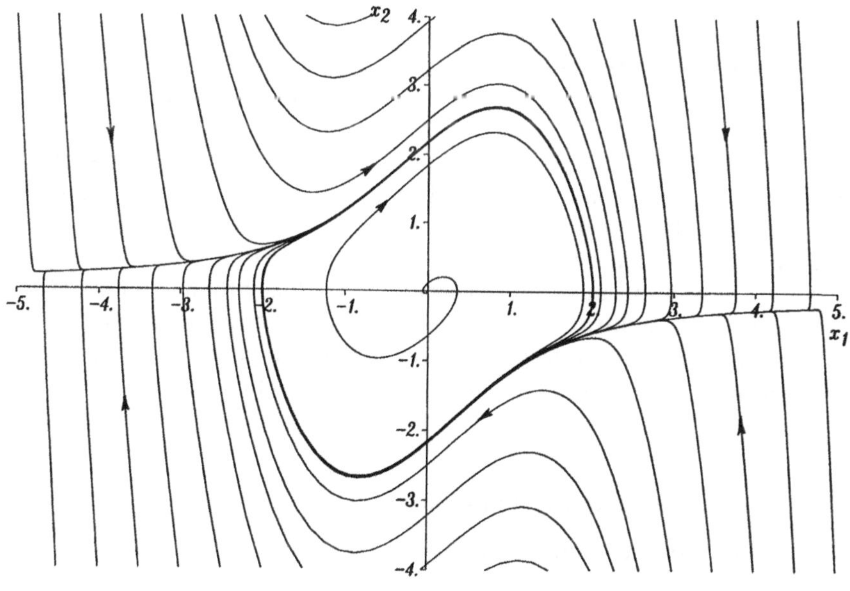

Figure 1.4-3 Van der Pol oscillator, $\beta = 1$.

for each positive value of the parameter β, but no algebraic formula is available for the limit cycle. Various estimation and asymptotic approximation techniques exist for finding limit cycles, and we return to the Van der Pol system when we discuss some of these techniques.

Example 1.4-4: Limit Cycle in a Control System Consider a control system of the form

$$\begin{aligned} \dot{x}_1 &= x_1 - 2x_2 \\ \dot{x}_2 &= x_1 + x_2 + u. \end{aligned} \tag{1.4-17}$$

With $u = 0$ this is an unstable system with complex eigenvalues given by

$$\mu = 1 \pm i\sqrt{2}.$$

However, this linear system is controllable so it can be made stable [e.g., Grantham and Vincent, 1993, p. 290]. Using a state-variable feedback controller of the form

$$u = -k_1 x_1 - k_2 x_2$$

can provide asymptotic stability to the origin for the system. For example, if we use $k_1 = -2$ and $k_2 = 4$, then the eigenvalues of the controlled system are

$$\mu = -1 \pm i\sqrt{2},$$

and the **domain of attraction** (see Section 2.2) to the equilibrium point will be the entire state space provided that there are no bounds on the magnitude of the control.

Consider now the effect of using saturating state variable feedback. Assume that the bound on the control is given by $|u| \leq 1$. The saturating state variable feedback control law for this case is given by (1.4-7) with k_1 and k_2 defined as above. As we prove later, we can find the domain of attraction to the origin for this system subject to saturating state-variable feedback by integrating (1.4-17) subject to (1.4-7) **backward** in time from any point near the neighborhood of the origin. The resultant trajectory, as illustrated in Figure 1.4-4, approaches a limit cycle with the boundary of the domain of attraction being defined by the limit cycle. ■

Stability and Controllability Boundaries

A solution $\mathbf{x}(t)$ for a dynamic system

$$\dot{\mathbf{x}} = \mathbf{f}(\mathbf{x}) \tag{1.4-18}$$

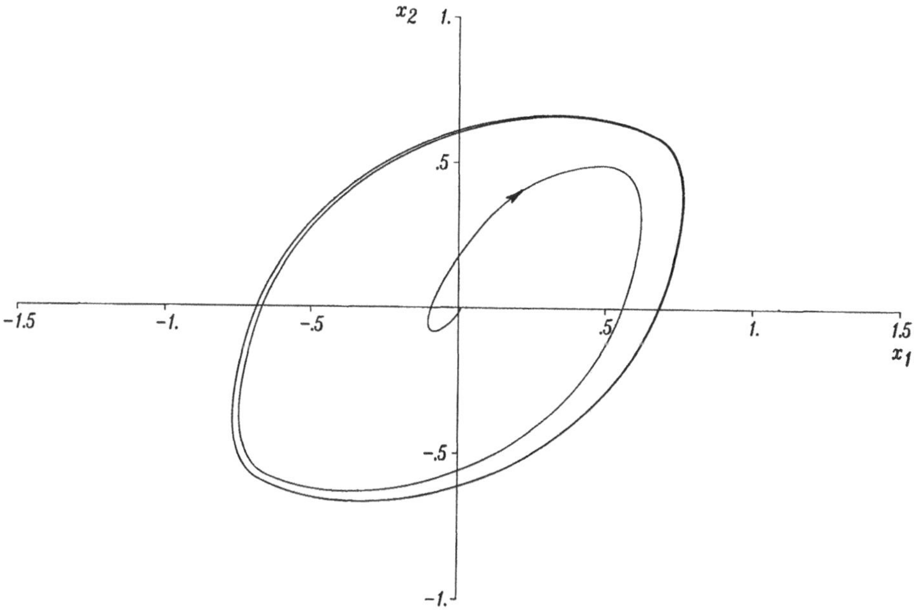

Figure 1.4-4 Retro-time limit cycle for Example 1.4-4.

is **asymptotically stable** if trajectories that start near $\mathbf{x}(t)$ remain near $\mathbf{x}(t)$ and ultimately approach $\mathbf{x}(t)$ as $t \rightarrow \infty$. Similarly, we say that a target set $\mathcal{X}$ is asymptotically stable if trajectories that start near $\mathcal{X}$ remain near $\mathcal{X}$ and ultimately approach $\mathcal{X}$ as $t \rightarrow \infty$. For a control system

$$\dot{\mathbf{x}} = \mathbf{f}(\mathbf{x}, \mathbf{u}), \tag{1.4-19}$$

an initial state is **controllable** to a target if an admissible control $\mathbf{u}(\cdot)$ exists that will transfer the state to the target.

For linear systems, stability and controllability to a target, generally the origin, are both local and global (i.e., local stability implies global stability). All state variable feedback controllers for a linear system that result in a controlled system having eigenvalues with negative real parts yield stability to the target throughout the state space. The ability to find such a control law also implies that the system is controllable throughout the state space. This is not true for nonlinear systems. Generally, not only will the **controllable set** (see Section 2.2) be some subset of the state space but the **domain of attraction** (or stability region) for a particular controller will be a subset of the controllable set.

Nonlinear control system design will usually require advanced methods in stability and controllability theory. These topics are discussed in detail in Chapters 4 and 6. For now, we simply illustrate, in the following two examples, that stability and controllability in nonlinear systems are usually not global phenomena.

Example 1.4-5: Stability Boundary Consider the system

$$\dot{x}_1 = -x_2 - x_1[1 - x_1^2 - x_2^2]$$
$$\dot{x}_2 = x_1 - x_2[1 - x_1^2 - x_2^2],$$

which is the backward-time system corresponding to (1.4-10). Thus the trajectories of the system are the same as in Figure 1.4-2, with the direction of motion reversed. The equilibrium solution $\mathbf{x}(t) \equiv \mathbf{0}$ for this retro-time system is now asymptotically stable, but its domain of attraction is only the region where $\|\mathbf{x}\| < 1$, and its **stability boundary** is the limit cycle $\|\mathbf{x}\| = 1$. ■

Example 1.4-6: Controllability Boundary Consider the scalar control system

$$\dot{x} = x + u,$$

and suppose that the control problem is to transfer the state to $x(t) = 0$. If there are no bounds on the control u, then we can make $\dot{x}$ have any value we choose, despite the current value of the state. Hence every point in the region $x \in (-\infty, \infty)$ can be transferred to the origin. But suppose the control is restricted to $|u| \leq 1$. Then $\dot{x}$ cannot have any value, and we find that the **controllable set** (see Section 2.2) to the origin is the region $x \in (-1, 1)$ and the system has the **controllability boundary** $x = \pm 1$. ■

Chaos and Strange Attractors

From our discussion in Section 1.2, the solution to a differential equation of the form

$$\dot{\mathbf{x}} = \mathbf{f}(\mathbf{x}) \tag{1.4-20}$$

is a unique differentiable function of time and of the initial conditions, since we have assumed that $\mathbf{f}(\cdot)$ and its first-order partial derivatives are continuous. Mathematically, this means that we can predict the future state exactly, given the initial state. This powerful feature forms the philosophical basis for almost all of modern science since the time of Sir Isaac Newton. In particular, the underlying objective of the "scientific method" of analysis is to develop a model for a particular process, verify the model experimentally, and then use the model to predict the future behavior of the system.

By 1903, however, the great French mathematician Jules-Henri Poincaré realized that the behavior of some otherwise very simple dynamical systems was so sensitive to changes in initial conditions, with very small changes producing wildly erratic responses, that long-term prediction would require exact knowledge of the initial conditions. Since any physical measurement of the initial conditions would necessarily contain some error, it became clear that some phenomena were simply unpredictable and that the "scientific method" would not work for these cases.

What would make a nonlinear system behave in such a fashion? Recall that with linear systems only two types of stable motion are possible: asymptotically stable equilibrium points and periodic orbits. A limit cycle represents a new type of stable motion possible with nonlinear systems. Another possibility is called **quasiperiodic motion**, as in the case of a trajectory that winds around on the surface of a torus but that, each time it returns near its initial point, is displaced infinitesimally to one side. The trajectory never repeats, but it does eventually "cover" the torus and in the limit as $t \to \infty$ it comes arbitrarily close to being periodic.

It turns out that there is one other type of stability possible with equations of the form of (1.4-20), if they are of dimension three or greater, called a **strange attractor**. Such an attractor is "strange" because it consists of trajectories that wander erratically in a bounded region, without repeating themselves, but with neighboring trajectories separating exponentially with time on the average. This sensitivity to initial conditions gives rise to a fundamental difficulty in predicting the future state of the system since its initial position can never be known exactly. A strange attractor has a structure analogous to a croissant, a pastry made by repeatedly stretching and folding dough. The "stretching" corresponds to the exponential separation of initially neighboring trajectories. The "folding," which requires at least three dimensions, comes from the fact that trajectories remain bounded but never repeat themselves. As mentioned earlier, a periodic trajectory is one-dimensional and could not cover a surface such as a sphere because trajectories for autonomous systems cannot intersect. A quasiperiodic orbit, as on a torus, does eventually cover a two-dimensional surface, like a doughnut. In the absence of periodicity or quasiperiodicity, a strange attractor must repeatedly fold back on itself, like a croissant.

Today we use the term **chaos** to denote the response of a dynamical system that is bounded, attracting, not periodic nor quasiperiodic, and so sensitive to small changes in initial conditions that its future behavior is essentially unpredictable. Such systems are extremely interesting because of a paradox: they are deterministic systems that exhibit random-like behavior. There are many examples of chaos throughout nature, such as the motion of a deflating balloon after it is released or the path that a lightning bolt will travel. Turbulence, in fluid mechanics, is chaotic motion and it remains as one of the major unsolved problems of science. Based on Poincaré's observations, it is understandable that many people feel that the problem of turbulence will never be solved.

In the midst of this gloomy outlook, some researchers have very recently discovered that order often exists within chaos [Crutchfield et al., 1986]. Indeed, many chaotic systems exhibit such a high degree of order that they may become useful paradigms for modelling unpredictable phenomena such as turbulence. We illustrate this phenomenon with two classic examples.

Example 1.4-7: Lorenz Attractor Consider the system

$$\begin{aligned} \dot{x} &= s(y - x) \\ \dot{y} &= (r - z)x - y \\ \dot{z} &= xy - bz, \end{aligned} \tag{1.4-21}$$

where b, r, and s are parameters. This system, called the **Lorenz attractor**, was developed for the study of weather prediction and it arises from a simplification of the Navier–Stokes equations of fluid mechanics applied to Rayleigh–Bènard heat convection [Lorenz, 1963; Schuster, 1988, p. 223].

Figure 1.4-5*a–c* show some three-dimensional views of a single trajectory of this system, for the case where $s = 10$, $b = 8/3$, $r = 28$. Any trajectory that starts off of the z-axis converges to the butterfly-shaped surface shown, so the surface is a strange attractor, since it is comprised of a single bounded trajectory (like the case of a limit cycle), but the trajectory is not periodic or

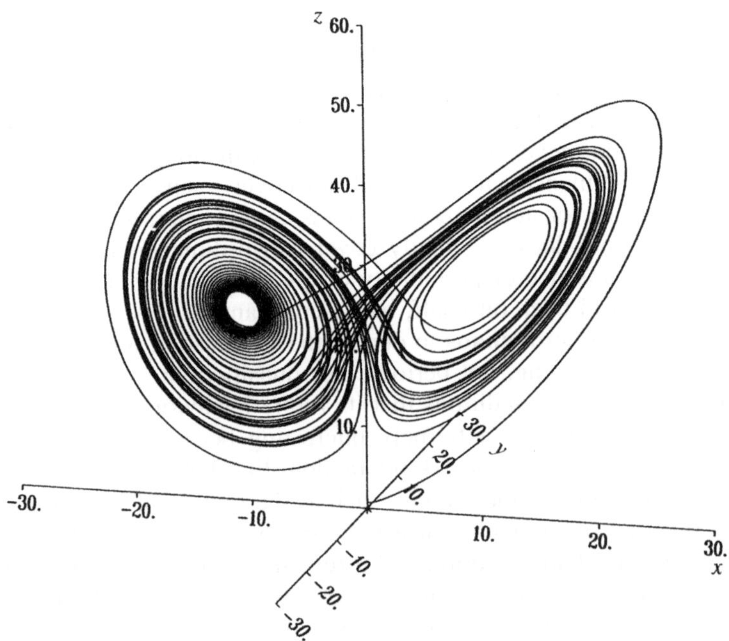

(*a*)

Figure 1.4-5 Lorenz attractor ($s = 10$, $b = 8/3$, $r = 28$). (*a*) Side view. (*b*) Edge view. (*c*) Top view. (*d*) Poincaré map for the Lorenz attractor ($y = x$ plane).

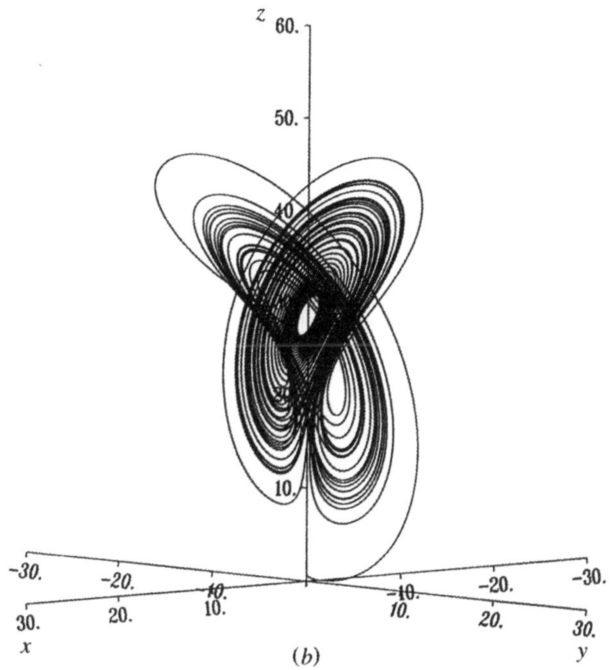

(b)

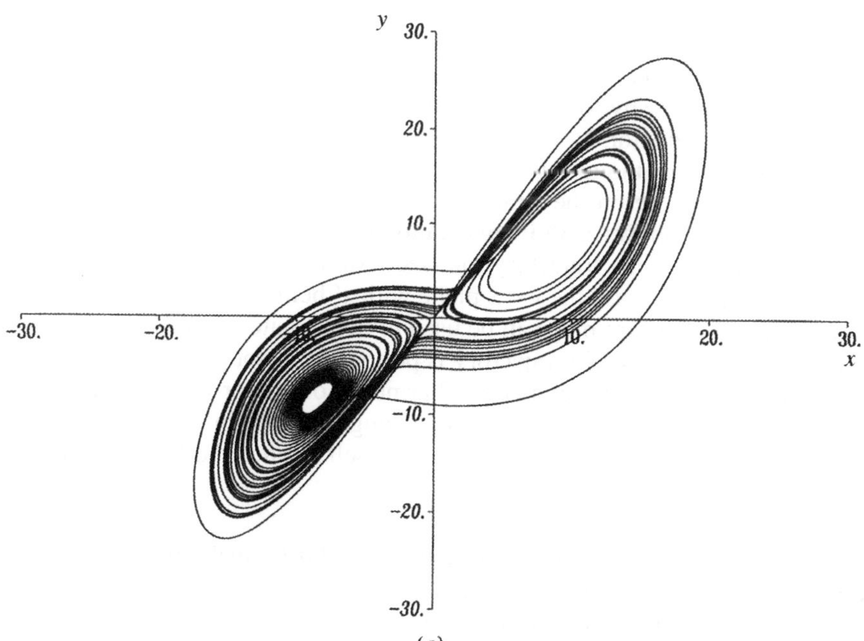

(c)

Figure 1.4-5 (*Continued*)

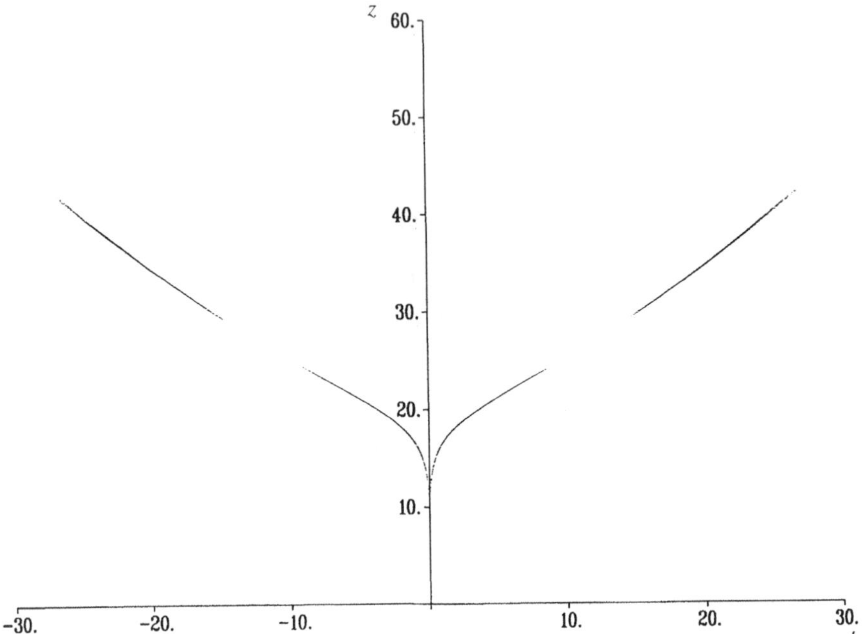

(*d*)

Figure 1.4-5 (*Continued*)

quasiperiodic. Instead, it wanders around on the surface, shifting unpredictably from one wing of the butterfly to the other. The system is chaotic, and it is impossible to make any long term prediction of the location of the state once it approaches the attractor. Nevertheless, there is clearly some structure and order to the system: even though we cannot predict the state exactly, we know that eventually it must lie on the strange attractor.

Figure 1.4-5*d* shows a **Poincaré map** for the Lorenz attractor, indicating each point at which the trajectory intersects some specified plane transverse to the trajectory, in this case the $y = x$ plane with x' being the horizontal coordinate in the plane. A Poincaré map can be used to investigate the presence or absence of periodic trajectories. If a trajectory is periodic, the dot pattern will eventually repeat itself (e.g., for the undamped harmonic oscillator only two points will occur on a Poincaré map for any line through the origin in state space). The dot pattern in Figure 1.4-5*d* never repeats itself. Hence the Lorenz attractor is not a periodic trajectory. ■

Strange attractors need not be structurally robust: small changes in the system parameters may produce distinctly different results. In the Lorenz attractor example, if we change the parameters to $s = 10$, $b = 8/3$, $r = 14$, motion is stable about one of the nonzero equilibrium points. By changing the parameters to $s = 10$, $b = 8/3$, $r = 22.4$, "preturbulent" trajectories are obtained [Sparrow,

1982]; that is, for most initial conditions, the trajectories will follow paths very much like those on the chaotic attractor, but they ultimately end up approaching one of the two nonzero equilibrium points. This sensitivity suggests that some nonlinear systems may be easily controlled to either produce chaos or to eliminate it. For example, Vincent and Yu [1991] showed that by adding a control term to the second equation of (1.4-21),

$$\dot{y} = (r - z)x - y + u,$$

and that by using the "chaotic" parameters $s = 10$, $b = 8/3$, $r = 28$, all types of motion could be obtained with very little control effort. This included both stable motion and preturbulence to one of the unstable equilibrium points of the original system.

As previously indicated, for autonomous systems of the form (1.4-20), strange attractors can only occur if the system is at least three-dimensional, since the geometry of a strange attractor involves both a stretching and a folding of trajectories [Lichtenberg and Lieberman, 1983, pp. 58–62; Crutchfield et al., 1986]. This structure is evident in the following example.

Example 1.4-8: Rössler Attractor Consider the system given by

$$\begin{aligned} \dot{x} &= -(y + z) \\ \dot{y} &= x + \tfrac{1}{5}y \\ \dot{z} &= z(x - \alpha) + \tfrac{1}{5}. \end{aligned} \tag{1.4-22}$$

Figure 1.4-6 is a plot of the trajectory for this system, after it reaches the attractor, for the case where $\alpha = 4.23$. ■

Strange attractors are closely associated with **fractals**, which are "fractional dimensional" objects [Mandelbrot, 1983]. In the usual notion of dimension, a point is zero-dimensional, a line is one-dimensional, a plane is two-dimensional, and so on. But suppose that a one-dimensional curve wiggles back and forth, somewhat like the plot of a very high frequency sine wave, in such a way that it covers a planar region. In this case the curve is said to have a (Hausdorff) "dimension" somewhere between one and two. A high frequency sine wave is not a fractal, but it illustrates the basic idea. A better example of a fractal would be the path of a lightning bolt, or a curve, such as the one constructed by Karl Weierstrass in 1872, that is so irregular that no tangent exists at any point of the curve.

Fractals are often studied in terms of an underlying **discrete-time** dynamical system, of the form

$$\mathbf{x}(k + 1) = \mathbf{f}[\mathbf{x}(k)], \tag{1.4-23}$$

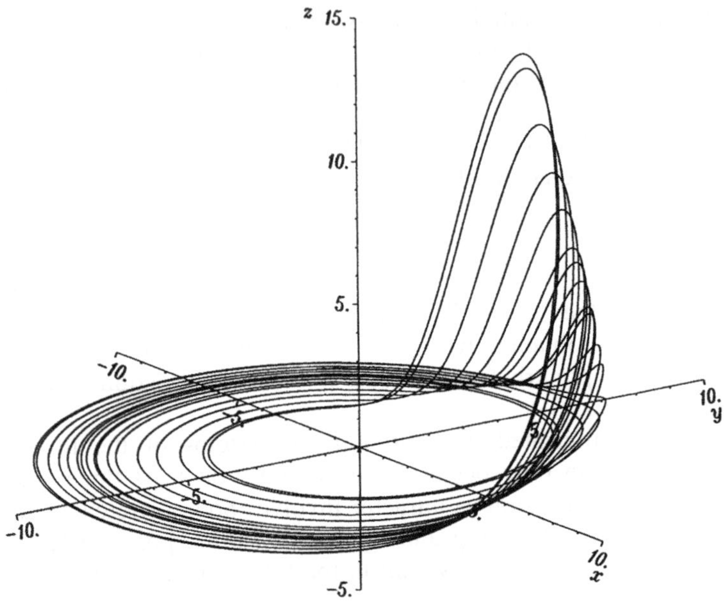

Figure 1.4-6 Rössler attractor, $\alpha = 4.23$.

where the "time" k is an integer. The difference equations of a numerical integration scheme are an example of a discrete-time system. Another example is given by Newton's method for finding a root of a system of equations. Population biologists often use difference equations to model species that only reproduce once per cycle. In nonlinear dynamics, the points and times at which a trajectory penetrates a Poincaré mapping plane correspond to a discrete-time system of the form (1.4-23).

Despite the fact that (1.4-23) is a system of algebraic equations, and therefore easier to solve than a system of differential equations, the behavior of discrete-time systems can be extremely complicated, as illustrated by the following example.

Example 1.4-9: Fractals and the Discrete Logistic Equation Consider a two-dimensional complex variable system of the form

$$z(k+1) = f[z(k)], \tag{1.4-24}$$

where

$$f(z) = \alpha z(1 - z) \tag{1.4-25}$$

with $z = x + iy$, $\alpha = \sigma + i\omega$, and $i = \sqrt{-1}$. This system is a complex-valued

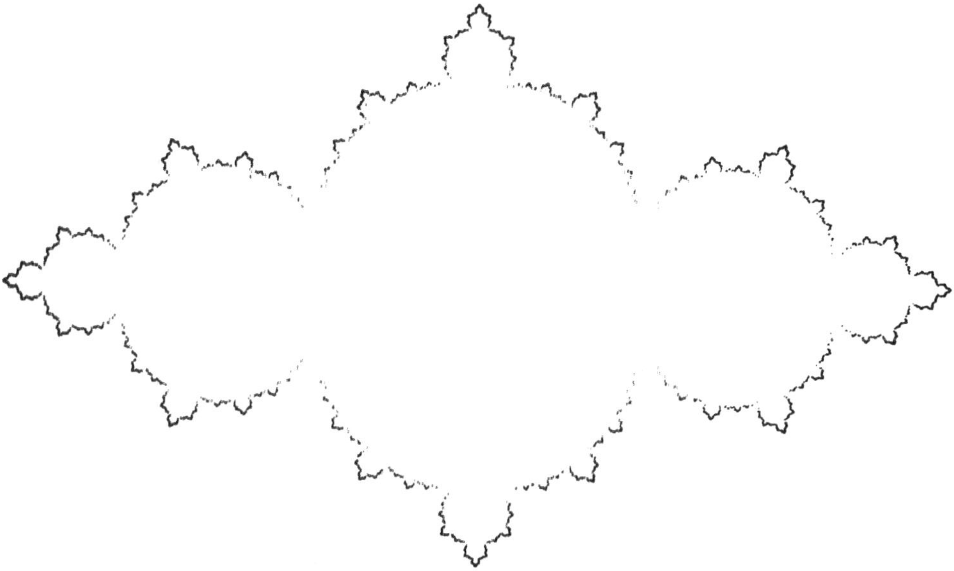

Figure 1.4-7 Fractal domain of attraction for $f(z) = 3z(1 - z)$.

discrete-time analog of the logistics equation. The **fixed points** of the system, where $f(z) = z$, are given by $z = 0$ and $z = (\alpha - 1)/\alpha$ and are analogous to the equilibrium points for a system of continuous-time differential equations.

For $\alpha = 3$, the fixed point at $z = 2/3$ is stable and the boundary of its domain of attraction is illustrated in Figure 1.4-7. The boundary points shown in the figure were constructed by propagating the state backward in time, that is, by solving (1.4-24) and (1.4-25) for $z(k)$ in terms of $z(k + 1)$. For this quadratic function there are two solutions for $z(k)$ and we randomly chose between them. The details of the algorithm are given in [Sørensen, 1984] and an alternate algorithm can be found in [Schroeder, 1986].

It can be shown that the boundary of the domain of attraction, illustrated in Figure 1.4-7, is a fractal and a strange attractor for the backward-time system [Devaney, 1986, pp. 279–291]. The state bounces around unpredictably on the boundary. Since the simulation was only performed for a finite time (200,000 iterations, or about 15 minutes on an IBM PC class computer), the boundary in Figure 1.4-7 appears to consist of discrete points. However the complete boundary is a closed connected curve that contains no smooth arcs; that is, the boundary is a fractal.

A rich variety of fractals can be generated from the logistics equation by varying the complex parameter α in (1.4-25). It is convenient to introduce a linear transformation of the complex plane, given by

$$z = aZ + b.$$

For a certain choice of the complex parameters a and b, (1.4-24) and (1.4-25) transform to

$$Z(k+1) = f[Z(k)],$$

where

$$f(Z) = Z^2 + c.$$

Figure 1.4-8 shows the fractal boundary of the domain of attraction for the fixed point $Z = \frac{1}{2} + \sqrt{\frac{1}{4} - c}$ for the case $c = i$, which corresponds to

$$\alpha = 1 + \sqrt{1 - 4i}$$

in (1.4-25). The fractal boundary in this case is reminiscent of a bolt of lightning, or a river estuary system, or a nerve fiber, which suggests a modeling approach for such systems. ■

For the remainder of this text, we focus on continuous-time rather than discrete-time systems. Also, while we are very much concerned with the problem of determining stability and controllability boundaries, we do not con-

Figure 1.4-8 Fractal domain of attraction for $f(Z) = Z^2 + i$.

sider situations where these boundaries are fractals. Instead, when we deal with curves and surfaces we make the simplifying assumption that they have a unique tangent plane "almost everywhere," except possibly at certain points that may occur when a control law $\mathbf{u}(\cdot)$ switches.

1.5 CLASSICAL ANALYSIS TECHNIQUES

In this section we present some analysis techniques for nonlinear systems. We only consider autonomous systems of the form

$$\dot{\mathbf{x}} = \mathbf{f}(\mathbf{x}), \tag{1.5-1}$$

where $\mathbf{x}$ is an n_x-dimensional state vector. Again the vector function $\mathbf{f}(\cdot)$ and its partial derivatives with respect to $\mathbf{x}$ are assumed to be continuous functions of $\mathbf{x}$. Some of the results presented here are specifically restricted to two-dimensional systems.

State Perturbation Equations

Let $\bar{\mathbf{x}}$ be an **equilibrium point** for (1.5-1), defined by $\mathbf{f}(\bar{\mathbf{x}}) = \mathbf{0}$, and let

$$\mathbf{x}(t) = \bar{\mathbf{x}} + \boldsymbol{\eta}(t) \tag{1.5-2}$$

be a solution to (1.5-1) that starts near $\bar{\mathbf{x}}$ at

$$\mathbf{x}(0) = \bar{\mathbf{x}} + \boldsymbol{\eta}(0), \tag{1.5-3}$$

where $\boldsymbol{\eta}(0) = \boldsymbol{\eta}_0$ and $\|\boldsymbol{\eta}_0\|$ is small. Substituting this solution into (1.5-1) and using Taylor's theorem to expand, to first order, the right-hand side in a neighborhood of $\bar{\mathbf{x}}$ we obtain the differential equations for $\boldsymbol{\eta}(t)$, known as the **state perturbation equations**

$$\dot{\boldsymbol{\eta}} = \mathbf{A}\boldsymbol{\eta} \tag{1.5-4}$$

where

$$\mathbf{A} = \frac{\partial \mathbf{f}(\bar{\mathbf{x}})}{\partial \mathbf{x}} \triangleq \begin{bmatrix} \dfrac{\partial f_1(\bar{\mathbf{x}})}{\partial x_1} & \cdots & \dfrac{\partial f_1(\bar{\mathbf{x}})}{\partial x_{n_x}} \\ \vdots & \ddots & \vdots \\ \dfrac{\partial f_{n_x}(\bar{\mathbf{x}})}{\partial x_1} & \cdots & \dfrac{\partial f_{n_x}(\bar{\mathbf{x}})}{\partial x_{n_x}} \end{bmatrix}. \tag{1.5-5}$$

Integrating the linear system (1.5-4) we obtain a solution as given by (1.5-2). Equation (1.5-4) is a linear representation of the nonlinear system (1.5-1) in a small neighborhood of the equilibrium point. The extent to which (1.5-2) represents a good approximation to the solution outside this neighborhood depends on the particular system at hand.

Example 1.5-1: Inverted Pendulum The damped inverted pendulum system, illustrated in Figure 1.5-1, has the equation of motion

$$ml^2\ddot{\theta} + b\dot{\theta} - mgl\sin\theta = 0, \tag{1.5-6}$$

which can be written in terms of the state variables $x_1 = \theta$ and $x_2 = \dot{\theta}$ as

$$\begin{aligned} \dot{x}_1 &= x_2 \\ \dot{x}_2 &= \omega^2 \sin x_1 - 2\zeta\omega x_2, \end{aligned} \tag{1.5-7}$$

where

$$\omega = \sqrt{\frac{g}{l}} \tag{1.5-8}$$

$$\zeta = \frac{b}{2\omega ml^2}\,. \tag{1.5-9}$$

The equilibrium points for this system are at $\bar{\mathbf{x}} = [\bar{x}_1 \quad \bar{x}_2]^T = [\pm k\pi \quad 0]^T$, where k is any integer. For k even, corresponding to an upright equilibrium of the pendulum, the linearized equations of motion are

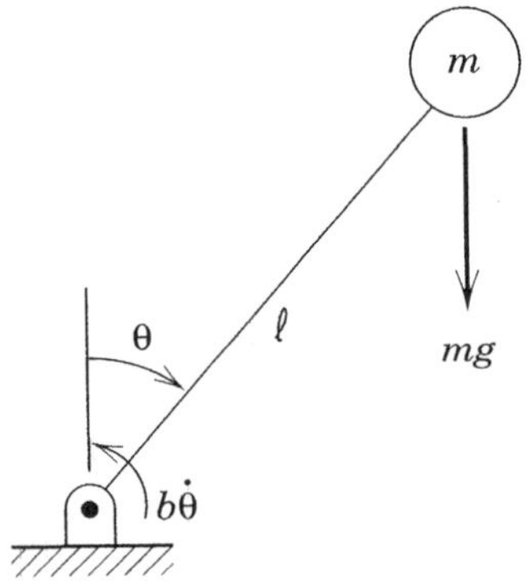

Figure 1.5-1 Inverted pendulum.

$$\begin{bmatrix} \dot{\eta}_1 \\ \dot{\eta}_2 \end{bmatrix} = \begin{bmatrix} 0 & 1 \\ \omega^2 & -2\zeta\omega \end{bmatrix} \begin{bmatrix} \eta_1 \\ \eta_2 \end{bmatrix}. \tag{1.5-10}$$

For an equilibrium at the downright position, where k is odd, the linearized equations are given by

$$\begin{bmatrix} \dot{\eta}_1 \\ \dot{\eta}_2 \end{bmatrix} = \begin{bmatrix} 0 & 1 \\ -\omega^2 & -2\zeta\omega \end{bmatrix} \begin{bmatrix} \eta_1 \\ \eta_2 \end{bmatrix}. \tag{1.5-11}$$

■

Eigenvalues

The motion of a nonlinear system (1.5-1) in a neighborhood of an equilibrium point is characterized by the **eigenvalues** μ_i and **eigenvectors** $\boldsymbol{\xi}_i$, $i = 1, \ldots, n_x$, of the matrix $\mathbf{A}$ for the corresponding linearized system (1.5-4). An eigenvector $\boldsymbol{\xi}$ is defined as a nonzero solution to the equations

$$\mathbf{A}\boldsymbol{\xi} = \mu\boldsymbol{\xi}, \tag{1.5-12}$$

which can be rewritten as

$$[\mu\mathbf{I} - \mathbf{A}]\boldsymbol{\xi} = \mathbf{0}.$$

This equation has a nonzero solution only if μ is a root of the n_x-order polynomial **characteristic equation**

$$0 = P(\mu) \triangleq |\mu\mathbf{I} - \mathbf{A}|. \tag{1.5-13}$$

If the eigenvectors of $\mathbf{A}$ are linearly independent (e.g., distinct eigenvalues), then the solution to (1.5-4) is of the form

$$\boldsymbol{\eta}(t) = \gamma_1 e^{\mu_1 t}\boldsymbol{\xi}_1 + \cdots + \gamma_{n_x} e^{\mu_{n_x} t}\boldsymbol{\xi}_{n_x}, \tag{1.5-14}$$

where the γ_i are constants determined by the initial conditions. For the more general case where the eigenvectors are not linearly independent, see Section 4.2.

Classification of Two-Dimensional Equilibrium Points

In a two-dimensional system with real-valued coefficients in the $\mathbf{A}$ matrix, the second-order equation in (1.5-13) has two roots, which are either both real or they are complex conjugates. Figure 1.5-2 illustrates the various types of motion that can occur near an equilibrium point for a two-dimensional system.

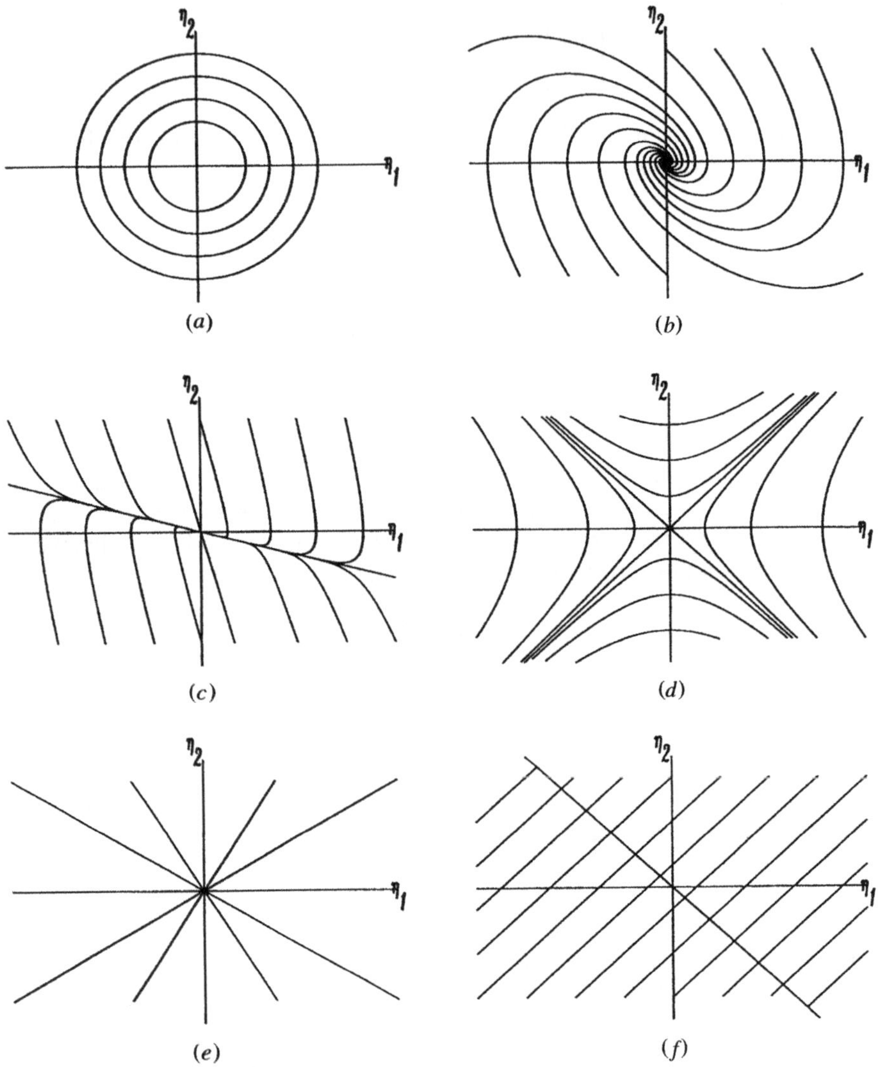

Figure 1.5-2 Motion of two-dimensional linear systems. (*a*) Center. (*b*) Focus. (*c*) Node. (*d*) Saddle. (*e*) Star. (*f*) Equilibrium line.

A **center** (Figure 1.5-2*a*) corresponds to a pair of purely imaginary conjugate roots to the characteristic equation. In general, for this case, the trajectory orbits the equilibrium point on elliptical trajectories, with the axes of the ellipses and the direction of motion being determined by the elements of the **A** matrix. An undamped oscillation is an example of this type of motion.

A **focus** (Figure 1.5-2*b*) occurs when the eigenvalues are complex and have a nonzero real part, denoted by $Re(\mu)$. The trajectories spiral around the equilibrium point. If $Re(\mu) < 0$, the trajectories approach the equilibrium point. They diverge away from it if $Re(\mu) > 0$. As an example, damped oscillation produces a focus in which the oscillation damps out as the trajectory spirals in toward the equilibrium.

The only other possibilities, typified by the cases illustrated in Figure 1.5-2*c*–*f*, all correspond to real eigenvalues. The eigenvectors are also real and lie along straight-line trajectories of the linearized system. To verify this, consider a point that lies on an eigenvector, that is, $\boldsymbol{\eta} = k\boldsymbol{\xi}$ for some k. From (1.5-4) and (1.5-12) it follows that the "velocity" vector

$$\dot{\boldsymbol{\eta}} = \mathbf{A}[k\boldsymbol{\xi}] = k\mathbf{A}\boldsymbol{\xi} = k\mu\boldsymbol{\xi}$$

must also lie on the eigenvector. We conclude that a trajectory that starts on an eigenvector stays on it, both forward and backward in time. Thus the trajectory is a straight line. We also note that, since there is only one trajectory through a point in state space, trajectories starting off of an eigenvector trajectory can never reach or intersect the eigenvector. However, nothing precludes trajectories that asymptotically approach an eigenvector. Indeed this is the usual case, as illustrated in Figure 1.5-2*c*, *d*, and *f*.

A **node** (Figure 1.5-2*c*) has real nonzero eigenvalues of the same sign. For negative eigenvalues the trajectories approach the node. For positive eigenvalues they diverge away from it. The trajectories do not spiral around the equilibrium; instead they exhibit a preferred direction of approach or departure near the equilibrium. This direction corresponds to the **slow eigenvector** associated with the eigenvalue having the smallest magnitude. Far away from the equilibrium all of the trajectories are essentially parallel to a single direction of motion. The second characteristic direction is parallel to the **fast eigenvector** corresponding to the eigenvalue with the larger magnitude.

The terms "slow" and "fast" refer to the exponential rates of change of the corresponding solutions. For instance, suppose the eigenvalues are $\mu_1 = -1$ and $\mu_2 = -10$ for a stable node. For an initial state well removed from the equilibrium, the motion is initially like that of the fast eigenvector trajectory corresponding to $\mu_2 = -10$. But this motion rapidly dies out and the trajectory quickly begins to behave like the slow eigenvector solution having eigenvalue $\mu_1 = -1$. These differences in time-scale are readily apparent in numerical integration and plotting operations and they are the underlying cause of the numerical difficulties associated with **stiff** systems, that is, systems with both large and small eigenvalues.

A **saddle-point** equilibrium (Figure 1.5-2*d*) has real nonzero eigenvalues with opposite signs, and the equilibrium is unstable. A trajectory starting on the eigenvector for the negative eigenvalue will asymptotically approach the equilibrium, but will never quite reach it, since the velocity vector $\dot{\boldsymbol{\eta}}(t)$ approaches zero as $\boldsymbol{\eta}(t)$ approaches zero. All other trajectories, except the equilibrium itself, asymptotically approach the eigenvector trajectory for the positive eigenvalue as they approach infinity.

The motion of a pendulum in the neighborhood of the upright position is an example of a saddle-point equilibrium. The pendulum can be balanced upright, but any disturbance will cause it to diverge. Also, if the pendulum is started away from upright and is given a certain critical initial velocity toward this position (corresponding to the initial state being on the eigenvector associated with the negative eigenvalue), then the pendulum will asymptotically come to rest just as it reaches the upright position. For a smaller initial velocity the pendulum will not reach upright and will fall backwards. If the initial velocity is greater than the critical velocity, the pendulum will overshoot the upright position and continue on past.

For real nonzero eigenvalues that are equal, the two characteristic eigenvector directions for a node may coalesce to a single direction (when **A** has only one eigenvector), or they may remain as two distinct directions (when **A** has two eigenvectors corresponding to the repeated root), or it may happen that every direction corresponds to a straight-line trajectory. The **star** equilibrium shown in Figure 1.5-2*e* is an illustration of the latter possibility. If $\mathbf{A} = \mu\mathbf{I}$, then $\mathbf{A}\boldsymbol{\xi} = \mu\boldsymbol{\xi}$ for every $\boldsymbol{\xi}$ and every nonzero vector is an eigenvector.

When one of the eigenvalues is zero (and the other is not) a line of equilibrium points exists through the origin. The **equilibrium line** (Figure 1.5-2*f*) contains the eigenvector associated with the zero eigenvalue. All trajectories starting off of the equilibrium line are straight-line trajectories parallel to the eigenvector for the nonzero eigenvalue. These trajectories either approach or diverge from the equilibrium line, or they are parallel to it. In either case, the directions of motion are opposite on opposite sides of the equilibrium line.

For the case of a repeated zero eigenvalue, we have either the flow depicted (if **A** has only one eigenvector) or the "static" case (when **A** has two linearly independent eigenvectors) where every state is an equilibrium point.

Separatrix and the Brouwer Fixed-Point Theorem

Knowledge of the linearized motion near equilibrium points, along with other qualitative information, can be used to estimate the behavior of a two-dimensional nonlinear system. To illustrate the process, consider the inverted pendulum of Example 1.5-1, with no damping ($\zeta = 0$) and with the time scaled so that the system has a natural frequency $\omega = 1$.

The equilibriums associated with the pendulum in the upright position are saddle points, with the shape illustrated in Figure 1.5-2*d*, and those associated with the downright position of the pendulum are circular centers, as illustrated in Figure 1.5-2*a*. The linearized motion near these equilibrium points is sketched in Figure 1.5-3*a*. Our problem is to decide what the trajectories may look like qualitatively away from these equilibrium points.

Consider an eigenvector trajectory that leaves a saddle point. Although we do not yet know where such a trajectory may go, the structure of the saddle-point flow makes it clear that a departing eigenvector trajectory must form a

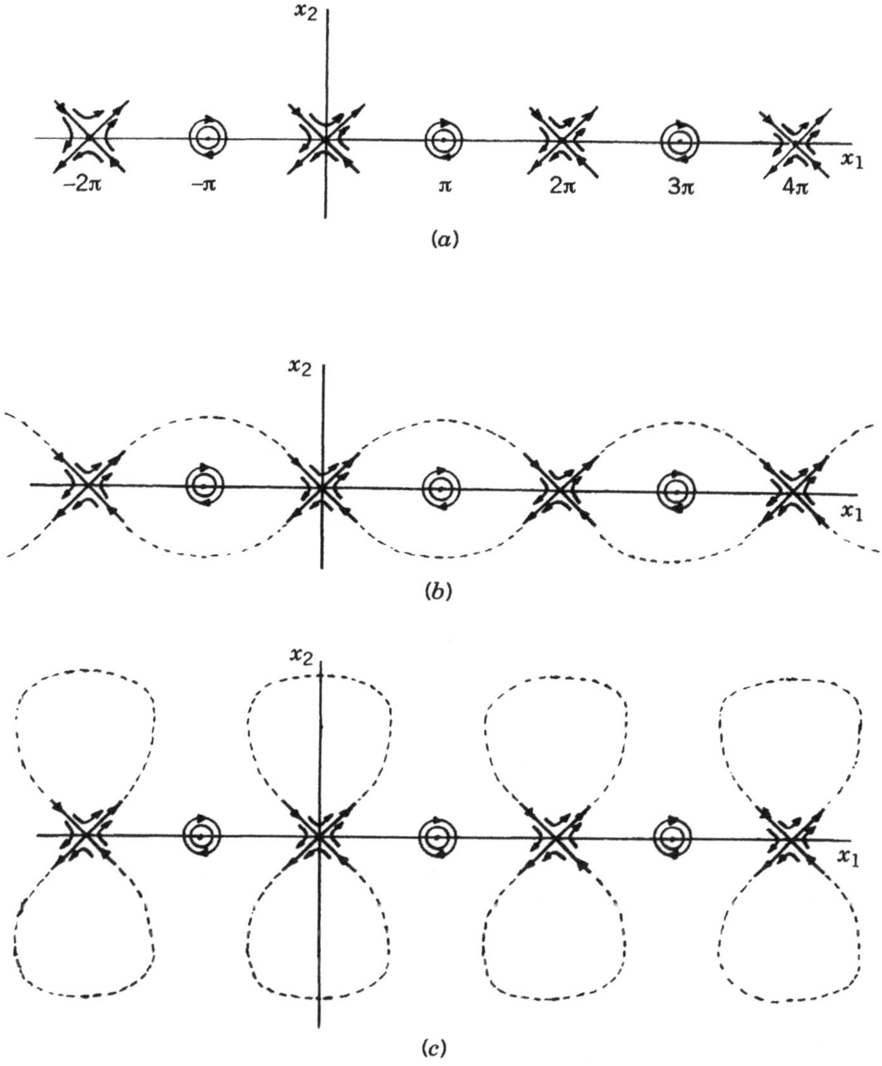

Figure 1.5-3 Separtix for the inverted pendulum. (*a*) Motion near the equilibrium points. (*b*) Separatrix. (*c*) Alternate configuration for the separatrix.

separatrix, with trajectories on either side of it having different destinations or origins.

For the linearized system, the separatrix trajectory goes to (or comes from) infinity along a straight line. From the physics of the nonlinear system it is not possible that a trajectory starting near an equilibrium can go to infinity, since there is no mechanism included for increasing the energy of the system. Thus a linearized-system eigenvector trajectory that leaves a saddle-point must be bounded for the nonlinear system. In particular, for this example, a separatrix trajectory must go somewhere in a bounded region, in a state space occupied by trajectories that cannot intersect.

Having eliminated infinity, the only other possible destination for a separatrix is one of the equilibrium points of the system. Trajectories near a center surround the equilibrium point and cannot be crossed. Thus we conclude that, for the nonlinear pendulum, an eigenvector trajectory that leaves one of the saddle-points must asymptotically approach some other (or perhaps the same) saddle-point.

Two conceivable configurations are illustrated in Figure 1.5-3*b* and *c*. Both configurations are reasonable, at this stage, but some simple topological considerations can be used to eliminate the configuration in Figure 1.5-3*c*. As a preview, we note that for the case shown in Figure 1.5-3*b* we can easily sketch a reasonable picture of the overall system trajectories. However, the configuration shown in Figure 1.5-3*c* would require the existence of some additional equilibrium points in order for us to be able to sketch its trajectories in any reasonable fashion, as the reader can verify by trying to sketch what the trajectories might look like for Figure 1.5-3*c*.

The topological property that we will use to eliminate the configuration shown in Figure 1.5-3*c* is given by the **Brouwer fixed-point theorem**, which states that every continuous transformation $\gamma(\cdot)$: $R^{n_x} \rightarrow R^{n_x}$ that has a continuous inverse and maps an n_x-dimensional closed unit ball

$$\mathcal{B} = \{\mathbf{x} \in \mathcal{R}^{n_x} \mid \|\mathbf{x}\| \leq 1\}$$

into itself has at least one **fixed point** $\hat{\mathbf{x}}$ in $\mathcal{B}$, for which $\boldsymbol{\gamma}(\hat{\mathbf{x}}) = \hat{\mathbf{x}}$. For a proof of this theorem see [Smart, 1974; Naber, 1980]. The theorem is also valid for any set $\mathcal{B}$ that is topologically equivalent to a closed unit ball, that is, for any closed set that can be stretched or compressed into the shape of an n_x-dimensional ball. For example, the football-shaped regions and either half of the figure eight regions bounded by (and including) the dotted lines in Figures 1.5-3*b* and *c* are topologically equivalent to closed two-dimensional balls, but the complete figure-eight-shaped regions are not.

The Brouwer fixed-point theorem is applicable to the general class of nonlinear dynamic systems of the form (1.5-1) since the solution $\mathbf{x}(t) = \boldsymbol{\xi}[t, \mathbf{x}(0)]$ at time t is a continuous transformation that maps the initial state $\mathbf{x}(0)$ to the state $\mathbf{x}(t)$. For $t > 0$, we may think of the solution $\boldsymbol{\xi}(t, \cdot)$ as a state-space transformation $\boldsymbol{\gamma}(\mathbf{x}) = \boldsymbol{\xi}(t, \mathbf{x})$ in the sense used in the Brouwer fixed-point theorem. In other words, for any "initial" state $\mathbf{x} = \mathbf{x}(0)$ in a given region, $\boldsymbol{\gamma}(\mathbf{x})$ maps $\mathbf{x}$ to another or the same state $\mathbf{x}(t)$ at time t. This transformation has a continuous inverse, since we can determine the initial state uniquely by integrating backwards from the state at time t.

If the solutions to the differential equations map a region of the state space that is topologically equivalent to a closed unit ball onto itself, that region must contain an equilibrium point (which is the fixed point of the transformation). Equivalently, if one can show that the state-space trajectories from every point of a state-space region topologically equivalent to a closed unit ball remain in that region for all $t > 0$, then that region must contain an equilibrium point.

Now suppose the separatrix for the nonlinear inverted pendulum had the configuration shown in Figure 1.5-3*c*, and consider the upper half of the figure-eight region at the origin. Since trajectories from the equilibrium point, combined with the equilibrium point, are not unique, we need to consider a closed region just inside the separatrix boundary, in order to exclude the saddle-point equilibrium trajectory from the ensuing discussion. Therefore, consider a trajectory passing just above the saddle-point. Since the curve consisting of the separatrix plus the saddle-point is a closed curve, continuity with respect to initial conditions implies that our neighboring trajectory must also be a closed curve. Let $\mathcal{B}$ be the closed region bounded by this trajectory. Since trajectories cannot cross each other, trajectories starting in $\mathcal{B}$ remain in $\mathcal{B}$. However, since the region $\mathcal{B}$ does not contain an equilibrium point, it violates the Brouwer fixed-point theorem. Note that the configuration in Figure 1.5-3*b* satisfies it.

Figure 1.5-4 shows the actual separatrix and system trajectories for the undamped nonlinear pendulum. For a general nonlinear system, with multiple equilibrium points, we are not usually able to obtain an algebraic expression for the separatrix. However, for the undamped pendulum we can do so by considering the total energy of the system. For the state-space form of the equations of

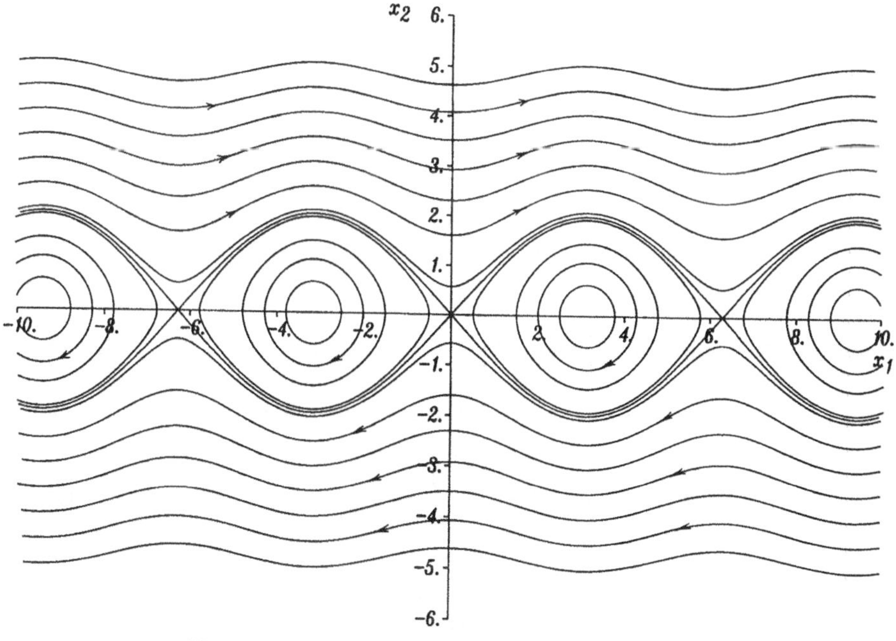

Figure 1.5-4 Inverted pendulum ($\omega = 1$, $\zeta = 0$).

motion, given by (1.5-7), the sum of the kinetic energy and the potential energy is

$$E = \tfrac{1}{2} ml^2 x_2^2 + mgl \cos x_1. \tag{1.5-15}$$

In the absence of damping the total kinetic plus potential energy is constant. Using the definition (1.5-8) we have the **first integral** (a constant of the motion) given by

$$x_2^2 + 2\omega^2 \cos x_1 = \text{ constant} \tag{1.5-16}$$

along trajectories of the system. In general, first integrals can be used to reduce the number of differential equations that need to be integrated. In this two-dimensional case, (1.5-16) can be used to plot all of the trajectories for the system (1.5-7). Figure 1.5-4 illustrates trajectories for $\omega = 1$. In particular the separatrices that pass through $x_1 = x_2 = 0$ are given by

$$x_2 = \pm\omega\sqrt{2(1 - \cos x_1)}. \tag{1.5-17}$$

Limit Cycles in Two Dimensions

Consider two-dimensional systems of the form of (1.5-1)

$$\begin{aligned} \dot{x}_1 &= f_1(x_1, x_2) \\ \dot{x}_2 &= f_2(x_1, x_2) \end{aligned} \tag{1.5-18}$$

in which f_1 and f_2 and their derivatives with respect to x_1 and x_2 are continuous. We present two useful results concerning the possible existence of limit cycles and other periodic trajectories. Both of these results are restricted to two-dimensional systems.

Bendixson's Negative Criterion This criterion [Minorsky, 1962, p. 82; Khalil, 1992, p. 349] is a sufficient condition that there be no closed trajectories in two-dimensional systems, other than equilibrium points. The criterion states that if

$$\text{div}[\mathbf{f}(\mathbf{x})] \triangleq \frac{\partial f_1}{\partial x_1} + \frac{\partial f_2}{\partial x_2}, \tag{1.5-19}$$

called the **divergence** of **f**(**x**), does not change sign (or vanish identically) in a region $\mathcal{D}$, then no bounded closed trajectory of (1.5-18) can exist in $\mathcal{D}$, except possibly an equilibrium trajectory. In terms of the "flow" of trajectories in an

n_x-dimensional space, the divergence div[$\mathbf{f}(\mathbf{x})$] is equal to the rate of change of volume per unit volume, div[$\mathbf{f}(\mathbf{x})$] $= \lim_{\mathcal{V} \to 0}[(1/\mathcal{V})d\mathcal{V}/dt]$, at the point $\mathbf{x}$, in the limit as the volume $\mathcal{V}$ about $\mathbf{x}$ approaches zero [Minorsky, 1962].

To verify Bendixson's criterion let $\gamma = \{\mathbf{x}(t)|t \geq 0\}$ be a trajectory of the two-dimensional system, and suppose that γ is a bounded closed nonequilibrium (i.e., periodic) trajectory in the region $\mathcal{D}$. Since trajectories cannot cross themselves, γ is a **Jordan curve**, topologically equivalent to a circle [Coddington and Levinson, 1955, p. 393]; that is, γ separates the plane into two regions: a bounded region $\mathcal{R}$ on and inside γ, and an unbounded region outside γ. The trajectory path γ can be determined from

$$\frac{dx_2}{dx_1} = \frac{f_2}{f_1}.$$

It follows that the path integral given by

$$\int_\gamma (f_1 dx_2 - f_2 dx_1) = 0$$

is zero. The path integral can also be written as

$$\int_0^T (\dot{x}_1 \dot{x}_2 - \dot{x}_2 \dot{x}_1) dt = 0.$$

Now since the path γ is a closed curve we can apply **Green's theorem** in the plane [Kreyszig, 1983, p. 419] to yield

$$\iint_R \left(\frac{\partial f_1}{\partial x_1} + \frac{\partial f_2}{\partial x_2} \right) dx_1 dx_2 = \int_\gamma (f_1 dx_2 - f_2 dx_1) = 0.$$

If div($\mathbf{f}(\mathbf{x})$] never changes sign, and if it is not identically zero in the region $\mathcal{D}$ containing γ, it follows that it is not possible for the surface integral in the above equation to be zero and no closed trajectory can exist.

Example 1.5-2: Linear Systems We have previously stated that limit cycles cannot exist in linear systems, since every (nonequilibrium) periodic trajectory for a linear system has periodic neighbors. Bendixson's negative criterion offers a proof of this in two dimensions. Consider a two-dimensional linear system

$$\dot{\mathbf{x}} = \mathbf{A}\mathbf{x},$$

given by

$$\dot{x}_1 = a_{11}x_1 + a_{12}x_2$$
$$\dot{x}_2 = a_{21}x_1 + a_{22}x_2.$$

The divergence is

$$\text{div}[\mathbf{f}] = \text{Tr}[\mathbf{A}] = a_{11} + a_{22} = \text{constant},$$

where Tr[**A**], the **trace** of **A**, is the sum of the diagonal elements.

From linear algebra, the trace of a square matrix is equal to the sum of the eigenvalues. Thus from the Bendixson criterion the only way that a two-dimensional linear system can have a periodic solution is that

$$\text{div}[\mathbf{f}] = \mu_1 + \mu_2 = 0.$$

Assume that this condition is satisfied. If the eigenvalues are real, then they must either have opposite signs, which yields a saddle-point flow, or they must both be zero. In either case the resulting trajectories are not periodic solutions. The remaining possibility, assuming **A** is real, is that the eigenvalues are complex conjugates, of the form

$$\mu_1 = \sigma + i\omega$$
$$\mu_2 = \sigma - i\omega.$$

In this case

$$\text{div}[\mathbf{f}] = 2\sigma = 0$$

requires that the eigenvalues be purely imaginary, and that all resulting nonzero solutions be periodic. From these requirements we conclude that two-dimensional linear systems cannot exhibit limit cycles.

The same conclusion holds true for any linear system, regardless of the dimension; however Bendixson's criterion cannot be used to show this, since it applies only to two-dimensional systems. Using an eigenvalue analysis, it can be shown that periodic solutions (all eigenvalues are imaginary) always have periodic neighbors. Thus linear systems cannot exhibit limit cycles. ■

Example 1.5-3: Van der Pol System As another example, consider the Van der Pol oscillator from Example 1.4-3, expressed as

$$\dot{x}_1 = x_2$$
$$\dot{x}_2 = -x_1 - \beta(x_1^2 - 1)x_2,$$

with β nonzero. The divergence is given by

$$\operatorname{div}(\mathbf{f}) = -\beta(x_1^2 - 1),$$

so that in each of the three regions,

$$\begin{aligned}\mathcal{D}_1 &= \{(x_1, x_2) | x_1 > 1\}\\ \mathcal{D}_2 &= \{(x_1, x_2) |\ |x_1| < 1\}\\ \mathcal{D}_3 &= \{(x_1, x_2) | x_1 < -1\},\end{aligned}$$

the divergence does not change sign, nor is it equal to zero.

Thus no limit cycle (or any other periodic solution) can exist that stays entirely within one of these regions. Note that we cannot conclude that a limit cycle does or does not exist (it does for $\beta > 0$). All we can say is that if such a trajectory exists it must cross one or both of the vertical lines $x_1 = \pm 1$. ■

As we have seen from the previous example, Bendixson's criterion is a sufficient condition for the nonexistence of closed trajectories. Unless Bendixson's criterion holds everywhere in the plane, we cannot exclude the possibility that periodic solutions and limit cycles may exist.

Poincaré–Bendixson Theorem A criterion for the existence of limit cycles, or other periodic solutions, is given by the **Poincaré–Bendixson theorem**, which states that if a nonequilibrium forward-time trajectory $\gamma = \{\mathbf{x}(t) | t \geq 0\}$ for the two-dimensional system (1.5-18) remains in a closed and bounded region of the plane, then either

(a) γ approaches an equilibrium point,
(b) γ is a closed (i.e., periodic) trajectory, or
(c) γ approaches a limit cycle.

For a proof of this result see Coddington and Levinson [1955, pp. 391–394].

Example 1.5-4: Logistics Equation In Example 1.4-2 we transformed the system

$$\begin{aligned}\dot{x}_1 &= x_2 + x_1[1 - x_1^2 - x_2^2]\\ \dot{x}_2 &= -x_1 + x_2[1 - x_1^2 - x_2^2]\end{aligned}$$

into polar coordinates (yielding a logistics equation for the squared radius $r^2 = x_1^2 + x_2^2$) in order to show that the system has a limit cycle at $r = 1$. An alternate, qualitative, approach is now possible.

To establish that a periodic solution exists, using the Poincaré–Bendixson theorem, we need to show that certain trajectories remain in a closed and bounded region and do not approach the equilibrium point at the origin. We can do this by considering the rate of change of the function

$$V(x_1, x_2) = x_1^2 + x_2^2.$$

Along a trajectory $\mathbf{x}(t)$ we have

$$\frac{dV}{dt} = \frac{\partial V}{\partial x_1}\dot{x}_1 + \frac{\partial V}{\partial x_2}\dot{x}_2 = 2(x_1^2 + x_2^2)[1 - x_1^2 - x_2^2].$$

Curves of constant V are circles centered at the origin, and we see that at any radius $r < 1$ the value of $V[\mathbf{x}(t)]$ is increasing, while at any radius $r > 1$ the value of $V[\mathbf{x}(t)]$ is decreasing. Thus the trajectories of the systems must behave qualitatively as illustrated in Figure 1.5-5, Specifically, all trajectories at the large circle in Figure 1.5-5 penetrate the circle toward a smaller radius, and at the small circle all trajectories penetrate toward a larger radius. Thus the conditions of the Poincaré–Bendixson theorem are satisfied in the annular region between the two circles, and this region must contain a periodic trajectory. Since these penetration properties persist as the two radii approach $r = 1$, we conclude that the circle at $r = 1$ is a limit cycle, since it has no neighboring periodic trajectories. ■

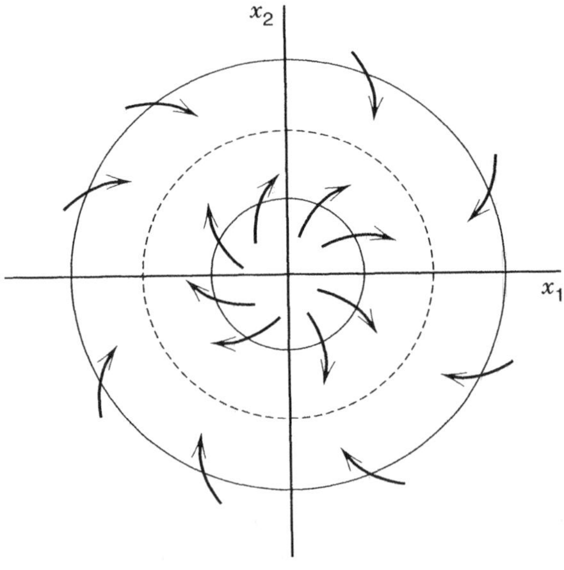

Figure 1.5-5 Poincaré–Bendixson theorem applied to Example 1.5-4.

Control System Limit Cycles

Saturating linear state-variable feedback will often produce limit cycles even in systems that are otherwise linear. For example, consider the linear single-input control system

$$\dot{\mathbf{x}} = \mathbf{A}\mathbf{x} + \mathbf{b}u, \tag{1.5-20}$$

with control constraints $|u| \leq u_{\max} > 0$. The saturating state feedback controller (1.4-7) in this case may be written in terms of the **sign function**

$$\text{sgn}(y) \triangleq \frac{y}{|y|} \qquad \text{for } y \neq 0$$

as

$$\text{sat}[u(\mathbf{x})] = \begin{cases} u(\mathbf{x}) & \text{if } |u(\mathbf{x})| \leq u_{\max} \\ u_{\max}\,\text{sgn}\,[u(\mathbf{x})] & \text{if } |u(\mathbf{x})| > u_{\max} \end{cases}. \tag{1.5-21}$$

As in the previous example consider a V function of the form

$$V = \tfrac{1}{2}\mathbf{x}^{\mathrm{T}}\mathbf{x}.$$

For V = constant, this equation defines hyperspheres in the state space. Along a trajectory we have

$$\dot{V} = \tfrac{1}{2}(\mathbf{x}^{\mathrm{T}}\dot{\mathbf{x}} + \dot{\mathbf{x}}^{\mathrm{T}}\mathbf{x}) = \mathbf{x}^{\mathrm{T}}\dot{\mathbf{x}}.$$

Thus

$$\dot{V} = \mathbf{x}^{\mathrm{T}}\mathbf{A}\mathbf{x} + \mathbf{x}^{\mathrm{T}}\mathbf{b}u.$$

If all of the eigenvalues of $\mathbf{A}$ have positive real parts, then the quadratic form $\mathbf{x}^{\mathrm{T}}\mathbf{A}\mathbf{x}$ is **positive definite**, that is, $\mathbf{x}^{\mathrm{T}}\mathbf{A}\mathbf{x} > 0$ for $\mathbf{x} \neq 0$. Since u is bounded, it follows that the first term in $\dot{V}$ must dominate for large $\mathbf{x}$. In other words, there must exist a value for V and a corresponding hypersphere beyond which it is not possible to decrease the value of V, no matter what particular saturating state-variable controller is used. This means that the domain of attraction to the origin for a particular control law, if it exists, must be bounded.

Let $\mathcal{D}$ denote a nonempty domain of attraction to the origin associated with a particular saturating state-variable feedback control law for the system (1.5-20). Assume that $\mathbf{A}$ has eigenvalues with positive real parts. Then from the above arguments, the closure of $\mathcal{D}$ is closed and bounded.

Control Domain of Attraction If the system is two-dimensional, the Poincaré–Bendixson theorem suggests how we can actually find the boundary of the domain of attraction for those cases where the closure of $\mathcal{D}$ is bounded. Consider integrating (1.5-20) backward in time from any point in $\mathcal{D}$, except the origin, using saturating state-variable feedback as given by (1.5-21). All such points must remain in $\mathcal{D}$, for otherwise, any point reached outside of $\mathcal{D}$ would be in the domain of attraction for the forward system, which in turn would violate the definition of $\mathcal{D}$. It follows, then, from the Poincaré–Bendixson theorem, that the retro-time trajectory either

(a) approaches an equilibrium point,
(b) approaches a limit cycle, or
(c) is a periodic trajectory.

Clearly alternative (c) is not possible, since such a trajectory for the forward system would not be in the domain of attraction. This leaves (a) or (b) as the only possibilities, and the trajectory must approach either a limit point on the boundary of $\mathcal{D}$ or a limit cycle that defines the boundary of $\mathcal{D}$.

Example 1.5-5: Linear System with Control Bounds We are interested in determining the domain of attraction for the controllable system

$$\begin{aligned}\dot{x}_1 &= -x_1 - x_2 + u\\ \dot{x}_2 &= 3x_1 + 2x_2 + u\\ |u| &\le \tfrac{1}{2}\end{aligned}$$

under a saturating state-variable feedback control law of the form (1.5-21) with $u_{\max} = \frac{1}{2}$. The eigenvalues of the $\mathbf{A}$ matrix,

$$\mathbf{A} = \begin{bmatrix} -1 & -1 \\ 3 & 2 \end{bmatrix},$$

are $\frac{1}{2} \pm i\frac{\sqrt{3}}{2}$. Choosing the feedback gains, $k_1 = \frac{8}{7}$, $k_2 = \frac{6}{7}$, will result in a controlled system such that whenever the control is not saturated it is equivalent to a damped-spring-mass system with a damping ratio of $\zeta = \frac{1}{2}$ and an undamped natural frequency of $\omega_n = 1$. Integrating the above system of equations under saturating state-variable feedback control backward in time starting near the origin yields the trajectory shown in Figure 1.5-6. The trajectory approaches a limit cycle that, from the above discussion, must be the boundary of the domain of attraction. ■

It should be noted that the size and shape of the domain of attraction will, in general, be a function of the feedback gains used. The backward integration method presented in the above example gives us a useful tool for comparing

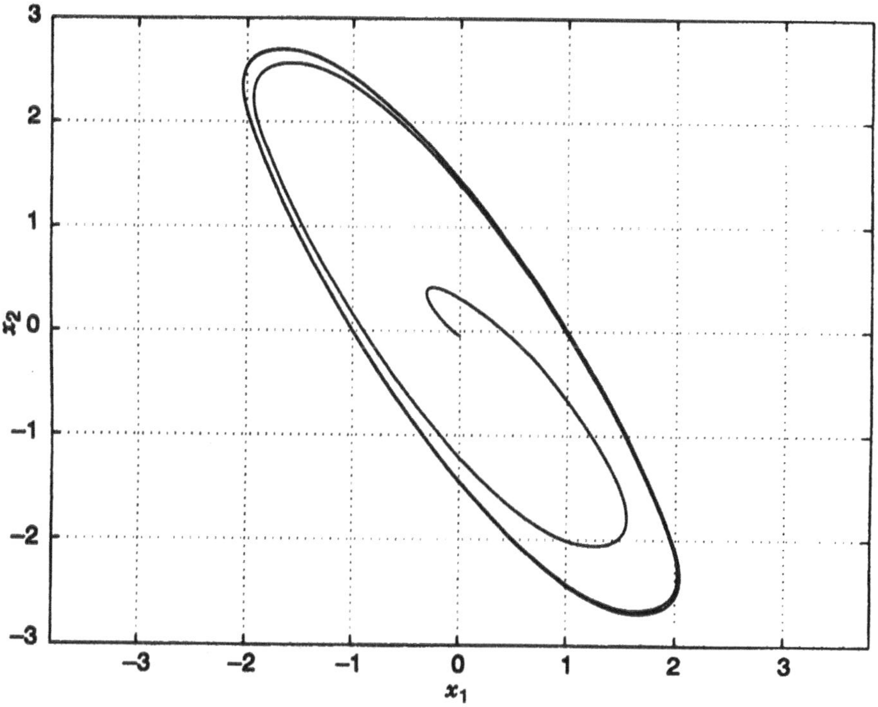

Figure 1.5-6 Domain of attraction for Example 1.5-5.

different control algorithms as the size and shape of the domain of attraction may vary considerably (see for example Exercise 1-16).

Lindstedt–Poincaré Perturbation Method

Various perturbation methods exist for quantitatively approximating a limit cycle or other periodic solutions to a nonlinear system. In general, these methods apply to nonlinear systems that consist of a linear part, having a periodic solution, plus a nonlinear part multiplied by a small parameter. The basic approach is to expand the nonlinear solution in a power series involving the small parameter.

Because of the algebra involved, the process is best applied to systems described in terms of a single n_x-order differential equation rather than to systems in state-space (i.e., vector) form. Even so, the algebra can be tedious and is subject to human mistakes. For this reason, if you plan to use these methods extensively it is highly recommended that you take advantage of one of the available symbolic algebra computer programs, such as "Mathematica" or "Maple." These programs perform operations such as differentiation, integration, and expression expansion on formulas, with the output being the resulting formula. They manipulate formulas just as powerfully as a calculator manipulates numbers, and can be a great help in eliminating human algebraic mistakes.

To illustrate perturbation methods, consider a system whose **output** $y(t)$ is a scalar quantity obeying a second-order differential equation of the form

$$\ddot{y} + \omega^2 y + \epsilon g(y) = 0, \tag{1.5-22}$$

where ϵ is a small parameter and $g(y)$ is a specified nonlinear function. The method we present also applies to various other systems, but system (1.5-22) serves to illustrate the procedure.

When ϵ is zero the system resulting from (1.5-22) is an undamped linear harmonic oscillator, analogous to a spring-mass system, and the corresponding solutions are periodic with frequency ω and period $2\pi/\omega$. For nonzero ϵ, the nonlinear term acts like a nonlinear spring, since $g(\cdot)$ is only a function of y. Thus for small values of ϵ we suspect that (1.5-22) will also have a periodic solution, such as a limit cycle, with the amplitude and period being functions of ϵ. For detailed discussions on the existence of periodic solutions for differential equations such as (1.5-22), we refer the reader to any of a number of texts on nonlinear oscillations, such as Minorsky [1962], Nayfeh and Mook [1979], and Khalil [1992].

The Lindstedt–Poincaré **perturbation method** [Minorsky, 1962, p. 224] begins by assuming that the differential equation, in this case (1.5-22), has a periodic solution $y(t)$, with an unknown period T and corresponding frequency $\Omega = 2\pi/T$. The essential feature of the Lindstedt–Poincaré procedure is that both the solution $y(t) = y(t, \epsilon)$ and its frequency $\Omega = \Omega(\epsilon)$ are expanded as power series in the small parameter ϵ. This provides the flexibility needed to eliminate **secular terms**, which are terms in the solution that would grow with time, yielding a nonperiodic response.

Fourier Series A periodic solution $y(t)$, once it has been found, will have a **Fourier series** representation of the form

$$y(t) = \sum_{n=0}^{\infty} \left(a_n \cos \frac{2n\pi t}{T} + b_n \sin \frac{2n\pi t}{T} \right), \tag{1.5-23}$$

where

$$a_0 = \int_{-T/2}^{T/2} y(t) dt \tag{1.5-24}$$

$$a_n = \int_{-T/2}^{T/2} y(t) \cos \frac{2n\pi t}{T}\, dt \tag{1.5-25}$$

$$b_n = \int_{-T/2}^{T/2} y(t) \sin \frac{2n\pi t}{T}\, dt. \tag{1.5-26}$$

To handle the unknown period T we change the time scale to

$$\tau = \Omega t + \varphi, \tag{1.5-27}$$

where φ is a parameter that we are free to choose. Then from (1.5-23) the solution in terms of the new time variable,

$$z(\tau) \triangleq y(t) = y[(\tau - \varphi)/\Omega], \tag{1.5-28}$$

has period 2π in τ. Substituting (1.5-28) into (1.5-22) yields

$$\Omega^2 \ddot{z} + \omega^2 z + \epsilon g(z) = 0, \tag{1.5-29}$$

where the differentiation is now with respect to τ.

In our search for periodic solutions, we can simplify the ensuing algebra by noting from (1.5-28) that, regardless of the resulting amplitude, that is, the unknown initial condition $z(0)$, we can always choose the parameter φ so that the initial condition on $\dot{z}(\tau)$ is

$$\dot{z}(0) = 0. \qquad (1.5\text{-}30))$$

To continue, we assume power series expansions for both the solution

$$z(\tau) = \sum_{n=0}^{\infty} \epsilon^n z_n(\tau) \tag{1.5-31}$$

and the frequency

$$\Omega^2 = \sum_{n=0}^{\infty} \epsilon^n \alpha_n, \tag{1.5-32}$$

where $z_n(\tau)$ and α_n are to be chosen so that $z(\tau)$ is a periodic solution to (1.5-29) that converges to the solution $z_0(\tau)$ for the linear case as the parameter ϵ approaches zero. We also expand the nonlinear function $g(\cdot)$ in a Taylor series about z_0, yielding

$$g(z) = g(z_0 + \epsilon z_1 + \epsilon^2 z_2 + \cdots)$$
$$= g(z_0) + \epsilon z_1 g'(z_0) + \epsilon^2 \left\{ z_2 g'(z_0) + \frac{z_1 [g''(z_0)]^2}{2!} \right\} + \cdots. \qquad (1.5\text{-}33)$$

Substituting (1.5-31)–(1.5-33) into (1.5-29) and setting the coefficient of each power of ϵ equal to zero produces the linear differential equations:

$$\begin{aligned} \alpha_0 \ddot{z}_0 + \omega^2 z_0 &= 0 \\ \alpha_0 \ddot{z}_1 + \omega^2 z_1 &= -g(z_0) - \alpha_1 \ddot{z}_0 \\ &\vdots \\ \alpha_0 \ddot{z}_{n+1} + \omega^2 z_{n+1} &= h_n(z_0, \ldots, z_n) - \alpha_{n+1} \ddot{z}_0 - \cdots - \alpha_1 \ddot{z}_n, \\ &\vdots \end{aligned} \qquad (1.5\text{-}34)$$

where $h_n(\cdot)$ is determined from the Fourier series expansion of $g(\cdot)$.

In addition to these differential equations, we require

$$\alpha_0 = \omega^2, \qquad (1.5\text{-}35)$$

so that the frequency Ω of the response, given by (1.5-33), converges to the frequency ω for the linear system as ϵ approaches zero. Also, in order for the initial condition $z(0) = 0$ on $z(\tau)$ to hold at $\tau = 0$ for all sufficiently small ϵ, we must have

$$\dot{z}_n(0) = 0 \qquad (1.5\text{-}36)$$

for all $n = 0, 1, 2, \ldots$. Furthermore we can account for any initial condition $z(0)$ by a constant of integration in the "homogeneous solution" $z_0(\tau)$, so for convenience we set the remaining constants of integration to zero, as we encounter them, in $z_n(\tau)$ for $n = 1, 2, \ldots$.

Using (1.5-35) and (1.5-36) and solving the first $N+1$ equations in (1.5-34) for $z_0, z_1, \ldots, z_N$ and $\alpha_0, \alpha_1, \ldots, \alpha_N$, we can construct the function

$$Z_N(\tau) = \sum_{n=0}^{N} \epsilon^n z_n(\tau) \qquad (1.5\text{-}37)$$

as the Nth-order approximation to the solution $z(\tau)$ of (1.5-29), yielding the corresponding approximation $Y_N(t) = Z_N(\tau)$ to the solution $y(t)$ of (1.5-22).

In view of (1.5-35) and (1.5-36), the solution to the first equation in (1.5-34) is of the form

$$z_0 = A\cos\tau. \tag{1.5-38}$$

To determine the solution to the second of equations (1.5-34), we expand the term $g(A\cos\tau)$ in a Fourier cosine series:

$$g(A\cos\tau) = \sum_{k=0}^{\infty} g_k(A)\cos k\tau$$

$$= g_0(A) + g_1(A)\cos\tau + \sum_{k=2}^{\infty} g_k(A)\cos k\tau. \tag{1.5-39}$$

The resulting differential equation is

$$\omega^2(\ddot{z}_1 + z_1) = -g_0(A) + [\alpha_1 A - g_1(A)]\cos\tau - \sum_{k=2}^{\infty} g_k(A)\cos k\tau. \tag{1.5-40}$$

Resonance and Secular Terms The $\cos\tau$ term in (1.5-40) operates at the natural frequency for the differential equation and will produce **resonance** and a **secular term** of the form $\tau\cos\tau$, yielding a nonperiodic response. To eliminate the secular term we choose

$$\alpha_1 = \frac{g_1(A)}{A}. \tag{1.5-41}$$

Then for $\dot{z}_1(0) = 0$ the resulting differential equation has the solution

$$z_1(\tau) = B\cos\tau - \frac{g_0(A)}{\omega^2} - \frac{1}{\omega^2}\sum_{k=2}^{\infty}\frac{g_k(A)\cos k\tau}{k^2-1}, \tag{1.5-42}$$

and we set $B = 0$ for convenience.

We repeat this procedure for each succeeding differential equation in (1.5-34), choosing α_n at each step to eliminate the secular term in $z_n(\tau)$, and stopping when the amplitudes of the remaining terms in $z(\tau)$ are sufficiently small.

Example 1.5-6: Perturbation Expansion In the system

$$\ddot{y} + 2y + 0.01\sin y = 0$$

let $\sin y$ be approximated for small y by

$$\sin y \approx y - \frac{y^3}{6},$$

yielding

$$\ddot{y} + 2y + 0.01\left(y - \frac{y^3}{6}\right) = 0, \tag{1.5-43}$$

which is in the form of (1.5-22) with

$$\omega^2 = 2$$
$$\epsilon = 0.01$$
$$g(y) = y - \frac{y^3}{6}.$$

Changing the time variable to

$$\tau = \Omega t + \varphi$$

and representing $z(\tau) \triangleq y(t)$ by the power series

$$z = z_0 + \epsilon z_1 + \epsilon^2 z_2 + \cdots,$$

along with

$$\Omega^2 = \omega^2 + \epsilon\alpha_1 + \epsilon^2\alpha_2 + \cdots,$$

yields

$$\Omega^2 \ddot{z} + \omega^2 z + g(z) = 0. \tag{1.5-44}$$

The expansion of $g(z) = g(z_0 + \epsilon z_1 + \epsilon^2 z_2 + \cdots)$ is given by (1.5-33) with

$$g(z_0) = z_0 - \frac{z_0^3}{6}$$

$$g'(z_0) = 1 - \frac{z_0^2}{2}$$

$$g''(z_0) = -z_0$$

$$g'''(z_0) = -1,$$

and all higher-order terms are zero. Substituting the expansion of $g(z)$ into (1.5-44), along with the series for $z(\tau)$ and Ω^2, and equating the resulting coefficients for each power of ϵ equal to zero yields the sequence of linear differential equations:

$$\ddot{z}_0 + z_0 = 0 \tag{1.5-45}$$

$$\ddot{z}_1 + z_1 = -\frac{1}{2}\left\{\alpha_1 \ddot{z}_0 + z_0 - \frac{z_0^3}{6}\right\} \tag{1.5-46}$$

$$\ddot{z}_2 + z_2 = -\frac{1}{2}\left\{\alpha_2 \ddot{z}_0 + \alpha_1 \ddot{z}_1 + z_1\left[1 - \frac{z_0^2}{2}\right]\right\} \tag{1.5-47}$$

$$\vdots$$

For $\dot{z}_0(0) = 0$ the solution to (1.5-45) is

$$z_0(\tau) = A \cos \tau, \tag{1.5-48}$$

and (1.5-46) becomes

$$\ddot{z}_1 + z_1 = -\frac{1}{2}\left\{\left[A \cos \tau - \frac{A^3 \cos^3 \tau}{6}\right] - \alpha_1 A \cos \tau\right\}. \tag{1.5-49}$$

Using the trigonometric identity

$$\cos^3 \tau = \tfrac{3}{4} \cos \tau + \tfrac{1}{4} \cos 3\tau,$$

we have

$$\ddot{z}_1 + z_1 = -\frac{1}{2}\left\{\left[1 - \frac{A^2}{8} - \alpha_1\right] A \cos \tau - \frac{A^3}{24} \cos 3\tau\right\}. \tag{1.5-50}$$

To eliminate the secular resonance term we set

$$\alpha_1 = 1 - \frac{A^2}{8}, \tag{1.5-51}$$

yielding

$$\ddot{z}_1 + z_1 = \frac{A^3}{48} \cos 3\tau. \qquad (1.5\text{-}52)$$

For $\dot{z}_1(0) = 0$ the solution to (1.5-52) is

$$z_1(\tau) = B \cos \tau - \frac{A^3}{384} \cos 3\tau.$$

Setting $B = 0$ and combining $z_1(\tau)$ with $z_0(\tau)$, we have the first-order approximation

$$z(\tau) = A \cos \tau - \epsilon \frac{A^3}{384} \cos 3\tau + \cdots,$$

and

$$\Omega^2 = \omega^2 + \epsilon \left(1 - \frac{A^2}{8}\right) + \cdots.$$

In terms of $y(t)$ and the parameter values for this example, we have

$$y(t) = A \cos(\Omega t + \varphi) - \frac{A}{38{,}400} \cos 3(\Omega t + \varphi) + \cdots,$$

where

$$\Omega^2 = 2 + 0.01 \left(1 - \frac{A^3}{8}\right) + \cdots$$

and φ is chosen to yield $y(0) = 0$. ■

Harmonic Balance and Describing Functions

The harmonic balance method [Takahashi et al., 1970, p. 523; Mees, 1981, p. 128; Khalil, 1992, p. 351] for predicting periodic trajectories does not rely on a small-parameter assumption. The basic idea is to use Fourier series representations for periodic functions. Substituting these into the differential equations converts the problem from one of solving nonlinear differential equations into one of solving systems of nonlinear algebraic equations.

Following the development in Khalil [1992, p. 351], consider a single-input single-output system

$$\dot{\mathbf{x}} = \mathbf{A}\mathbf{x} + \mathbf{b}u$$
$$y = \mathbf{c}^{\mathrm{T}}\mathbf{x} \tag{1.5-53}$$

with an output nonlinearity

$$u = \psi(y), \tag{1.5-54}$$

where the linear part (1.5-53) of the system is controllable and observable [Grantham and Vincent, 1993]. Thus the linear system Laplace transform **transfer function**

$$G(s) \triangleq \left.\frac{Y(s)}{U(s)}\right|_{\text{zero ICs}} = \mathbf{c}^{\mathrm{T}}[s\mathbf{I} - \mathbf{A}]^{-1}\mathbf{b} = \frac{Q(s)}{P(s)}$$

is strictly proper: $P(s)$ is a polynominal of higher order than the polynomial $Q(s)$. Hence $G(i\omega) \to 0$ as $\omega \to \infty$.

Now suppose $y(t)$ is periodic, with (unknown) period T and frequency $\omega = 2\pi/T$. Then $y(t)$ can be represented by a complex Fourier series

$$y(t) = \sum_{k=-\infty}^{\infty} a_k \exp(ik\omega t) \tag{1.5-55}$$

where $i = \sqrt{-1}$ and the a_k are complex with $a_{-k} = a_k^*$ (complex conjugate), given by

$$a_k = \frac{1}{T}\int_0^T y(\tau)\exp(ik\omega\tau)d\tau. \tag{1.5-56}$$

The nonlinearity $\psi[y(t)]$ is also periodic with the same frequency ω, and can be written as

$$\psi[y(t)] = \sum_{k=-\infty}^{\infty} c_k \exp(ik\omega t), \tag{1.5-57}$$

where each c_k is generally a function of all of the a_j, given by

$$c_k = \frac{1}{T}\int_0^T \psi[y(\tau)]\exp(-ik\omega\tau)d\tau. \tag{1.5-58}$$

For $y(t)$ to be a solution to (1.5-53) and (1.5-54) it must satisfy the differential equation

$$P(D)y(t) - Q(D)\psi[y(t)] = 0, \tag{1.5-59}$$

where D is the differential operator $D = d(\cdot)/dt$. Substituting (1.5-55) and (1.5-57) into (1.5-59) yields

$$\sum_{k=-\infty}^{\infty} [P(ik\omega)a_k - Q(ik\omega)c_k]\exp(ik\omega t) = 0.$$

Thus the coefficients must satisfy

$$a_k - G(ik\omega)c_k = 0 \tag{1.5-60}$$

for all integers k. However, since $a_k = a_{-k}^*$, $c_k = c_{-k}^*$, and $G(ik\omega) = G^*(-ik\omega)$, we only need to consider (1.5-60) for $k \geq 0$.

Since $G(s)$ is proper, $|G(ik\omega)| \to 0$ as $k\omega \to \infty$. Thus $a_k \to 0$ as $k \to \infty$ and we may truncate the infinite system of equations in (1.5-60) for k greater than some integer m, yielding the mth-order **harmonic balance** conditions

$$a_k - G(ik\omega)c_k = 0, \qquad k = 0, 1, \ldots, m. \tag{1.5-61}$$

The Fourier series in (1.5-55) and (1.5-57) are also truncated after the mth harmonic. For a mathematical analysis of the validity of the harmonic balance approximation, see Mees [1981, p. 128].

Describing Function The simplest approximation would be to take $m = 1$ in (1.5-61). This yields the describing function method, commonly used to handle nonlinearities in otherwise linear control systems. For a sinusoidal input, the output of a nonlinear element is approximated by the terms through the first harmonic term in the Fourier series for the nonlinearity. Specifically, recalling **Euler's formula**

$$e^{i\theta} = \cos\theta + i\sin\theta, \tag{1.5-62}$$

suppose that the input to the nonlinearity is

$$y(t) = a \sin \omega t = \frac{a}{2i} (e^{i\omega t} - e^{-i\omega t}),$$

which corresponds to

$$a_1 = \frac{a}{2i}, \qquad a_k = 0, \qquad k \neq 1$$

in (1.5-55). The output $u = \psi[y(t)]$ is periodic with the same period $T = 2\pi/\omega$ as the input and has the complex Fourier series representation (1.5-57) and (1.5-58), which can also be written in real form as

$$u(t) = \psi[y(t)] = c_0 + \sum_{k=1}^{\infty} (A_k \cos k\omega t + B_k \sin k\omega t), \tag{1.5-63}$$

where

$$\begin{aligned} c_0 &= \frac{1}{T} \int_0^T \psi[y(\tau)]\, d\tau &&= \frac{1}{2\pi} \int_0^{2\pi} \psi(a \sin \theta)\, d\theta, \qquad \theta = \omega t \\ A_k &= \frac{2}{T} \int_0^T \psi[y(\tau)] \cos k\omega\tau\, d\tau &&= \frac{1}{\pi} \int_0^{2\pi} \psi(a \sin \theta) \cos k\theta\, d\theta \\ B_k &= \frac{2}{T} \int_0^T \psi[y(\tau)] \sin k\omega\tau\, d\tau &&= \frac{1}{\pi} \int_0^{2\pi} \psi(a \sin \theta) \sin k\theta\, d\theta, \end{aligned} \tag{1.5-64}$$

which are related to the c_k by

$$c_k = \tfrac{1}{2}(A_k - iB_k). \tag{1.5-65}$$

Note, from the right-hand side of (1.5-64), that the Fourier coefficients are generally functions of a but not of ω. Also, if $\psi(y)$ is an **odd function**, $\psi(-y) = -\psi(y)$, as is frequently the case, then $c_0 = 0$ and $A_k = 0$, since the integral of an odd periodic function over one cycle is zero.

The **describing function** for the nonlinearity $u = \psi(y)$ is defined as the ratio of the output-to-input coefficients for the first harmonic terms in the complex Fourier series:

$$N(a) \triangleq \frac{c_1}{a_1} = \frac{B_1 + iA_1}{a}. \tag{1.5-66}$$

When $\psi(y)$ is an odd function $N(a)$ is real and is the ratio of the output to the input of the nonlinearity:

$$N(a) = \frac{B_1}{a} = \frac{B_1 \sin \omega t}{a \sin \omega t} = \frac{u(t)}{y(t)}.$$

Example 1.5-7: Describing Function for an On-Off Relay As illustrated in Figure 1.5-7, consider the case where $\psi(y)$ is the on-off nonlinearity

$$u = \psi(y) = \begin{cases} U & \text{if } y > 0 \\ -U & \text{if } y < 0. \end{cases}$$

If $y(t) = a \sin \omega t$, then $u(t) = \psi[y(t)]$ is the periodic square wave given by

$$u(t) = \psi[y(t)] = \begin{cases} U & \text{if } n\pi < \omega t < (n+1)\pi,\ n = \text{even} \\ -U & \text{if } n\pi < \omega t < (n+1)\pi,\ n = \text{odd}, \end{cases}$$

which has the Fourier series representation

$$u(t) = \psi[y(t)] = \sum_{k=-\infty}^{\infty} c_k \exp(ik\omega t),$$

where

$$\begin{aligned} c_k &= \frac{1}{T} \int_0^T y(\tau) \exp(ik\omega\tau) d\tau \\ &= \frac{\omega}{2\pi} \left[\int_0^{\pi/\omega} U \exp(ik\omega\tau) d\tau - \int_{\pi/\omega}^{2\pi/\omega} U \exp(ik\omega\tau) d\tau \right] \\ &= \frac{U}{ik\pi} [(-1)^k - 1] \\ &= \begin{cases} 0 & \text{if } k = \text{even} \\ -\dfrac{2U}{ik\pi} & \text{if } k = \text{odd} \end{cases} \end{aligned}$$

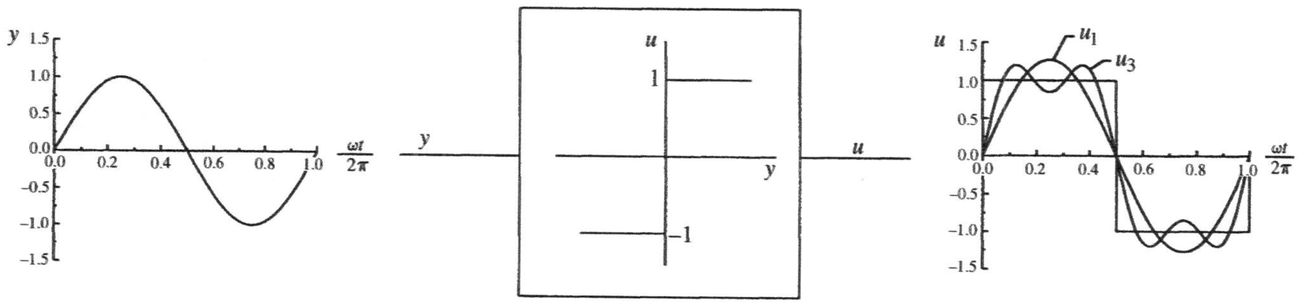

Figure 1.5-7 Describing function approximation for an on-off element.

Thus

$$
\begin{aligned}
u(t) &= \sum_{n=1}^{\infty} \frac{2U}{i\pi}\,[-\exp(ik\omega t) + \exp(-ik\omega t)] \\
&= \sum_{n=1}^{\infty} \frac{4U}{\pi(2n-1)} \sin[(2n-1)\omega t] \\
&= \frac{4U}{\pi}\left(\sin \omega t + \frac{1}{3}\sin 3\omega t + \frac{1}{5}\sin 5\omega t + \cdots\right).
\end{aligned}
$$

Truncating this series after the mth harmonic, let

$$u_m(t) = \sum_{k=1}^{m} B_k \sin k\omega t,$$

where

$$B_k = \begin{cases} 0 & \text{if } k = \text{even} \\ \dfrac{4U}{k\pi} & \text{if } k = \text{odd.} \end{cases}$$

The describing function

$$N(a) = \frac{u_1(t)}{y(t)} = \frac{B_1}{a} = \frac{4U}{\pi a}$$

corresponds to approximating the square-wave output $u(t)$ by its first harmonic. As illustrated in Figure 1.5-7, the first harmonic $u_1(t)$ does have the largest contribution to the output, with higher harmonics adding increasingly smaller contributions. ■

Now we return to the question of estimating limit cycles in a nonlinear system. Using the describing function method, with $m = 1$ in the Fourier series expansions, we seek a periodic solution of the form

$$
\begin{aligned}
y(t) &\approx a_0 + a_1 e^{i\omega t} + a_{-1} e^{-i\omega t} \\
&= a_0 + a \sin \omega t,
\end{aligned}
$$

where $a_1 = a_{-1}^* = a/2i$. For $m = 1$ the harmonic balance conditions (1.5-61) reduce to

$$a_0 - G(0)c_0 = 0$$
$$a_1 - G(i\omega)c_1 = 0. \tag{1.5-67}$$

The first of these equations allows us to determine $a_0 = a_0(c_0)$. The second equation can be written as

$$[1 - G(i\omega)N(a)]a = 0.$$

Since we are not interested in the case $a = 0$, the search for an approximate periodic solution to the original differential equations reduces to finding a pair (ω, a) satisfying the describing function conditions

$$1 - G(i\omega)N(a) = 0. \tag{1.5-68}$$

If the nonlinearity is an odd function, then the describing function $N(a)$ is real and this equation can be written as

$$1 - N(a)Re[G(i\omega)] = 0 \tag{1.5-69}$$
$$Im[G(i\omega)] = 0. \tag{1.5-70}$$

The second equation, which is independent of the nonlinearity, can be used to determine the possible frequencies of oscillation. For each of these frequencies (1.5-69) can be used to determine the amplitude a of oscillation.

Example 1.5-8: Limit Cycle in an On-Off Control System Consider the system

$$\frac{d^3y}{dt^3} + 2\frac{d^2y}{dt^2} + 4\frac{dy}{dt} = -u,$$

with $u(y) = \text{sgn}(y)$. From Example 1.5-7 the describing function for the nonlinearity is

$$N(a) = \frac{4}{\pi a}$$

and $c_0 = 0$ since the nonlinearity is an odd function. The transfer function for the linear part of the system is

$$G(s) = \frac{-1}{s(s^2 + 2s + 4)}.$$

Thus

$$G(i\omega) = \frac{2}{(4-\omega^2)^2 + 4\omega^2} + \frac{i}{\omega}\,\frac{4-\omega^2}{(4-\omega^2)^2 + 4\omega^2}.$$

From (1.5-67) $a_0 = 0$ and from (1.5-70)

$$\omega^2 - 4 = 0,$$

which has the solution $\omega = 2$. Then (1.5-69) becomes

$$1 - \frac{2}{(4-\omega^2)^2 + 4\omega^2}\,\frac{4}{\pi a} = 0,$$

which yields

$$a = \frac{8/\pi}{(4-\omega^2)^2 + 4\omega^2} = \frac{1}{2\pi}.$$

Thus we expect that the nonlinear system will oscillate at approximately

$$y(t) \approx \frac{1}{2\pi}\sin 2t. \qquad \blacksquare$$

EXERCISES

1-1. Consider the system

$$\dot{x}_1 = x_2$$
$$\dot{x}_2 = -x_1 + u$$

with $u(\mathbf{x}) = -\text{sgn}(x_2)$.

(a) Prove that state-space trajectories between switches (i.e., for constant u) are circular arcs, and find relations of the form $g(x_1, x_2, u) = 0$ for such arcs. *Hint:* Integrate dx_2/dx_1 and then complete the square.

(b) Using the results from (a), construct several state-space trajectories in a region of the origin defined by $|x_1| \le 3$, $|x_2| \le 3$. Indicate the direction of motion along the trajectories.

(c) Determine the portions of the switching surface that correspond to "dispersal," "switching," or "chatter."

(d) Determine any chattering trajectories. If they occur, determine the Filippov velocity vector and the equivalent value of u.

(e) Determine the equilibrium points, where $\mathbf{x}(t) =$ constant.

(f) Determine any initial states that do not get transferred to the origin.

1-2. Same as Exercise 1-1, but with $u(\mathbf{x}) = -\text{sgn}(x_1 + x_2)$.

1-3. Find a program (e.g., Matlab, MathCad), or write your own code, to plot state-space trajectories using either the fourth-order Runge–Kutta algorithm or (preferably) the LSODAR routine discussed in Section 1.2. Plot trajectories for the damped-spring-mass system

$$\dot{x}_1 = x_2$$
$$\dot{x}_2 = -\omega_n^2 x_1 - 2\zeta\omega x_2$$

over the range $|x_1| \leq 10$, $|x_2| \leq 10$. Make a separate plot for the cases:

(a) $\omega_n = 1$, $\zeta = 0.2$.

(b) $\omega_n = 1$, $\zeta = 1.5$.

Find any separatrices that may exist, by integrating backward in time if necessary. If you write your own program, use a separate subroutine for the right-hand sides of the state-space equations of motion, so that changing dynamic systems only requires recompiling one subroutine. Use a separate driver routine for the user interface and structure the program so that the integration procedure and its routines act as a callable subroutine that could be used in other programs.

1-4. Plot contours of constant $V(x, y)$ in the box $|x| \leq 5$, $|y| \leq 5$ for the following functions:

(a)

$$V(x, y) = [x \quad y] \begin{bmatrix} 5 & 4 \\ 4 & 5 \end{bmatrix} \begin{bmatrix} x \\ y \end{bmatrix}.$$

(b)

$$V(x, y) = y^2 + 2\cos x.$$

1-5. Modify your program from Exercise 1-3 to incorporate three-dimensional projection plots from any viewing direction. This can be done by writing a subroutine to project from three dimensions to two, and then calling this routine before any graphics call such as "move," "draw," and so on. Use your program to plot one trajectory $(x, y, z) = (x_1, x_2, t)$ for the system in Exercise 1-3a, plotting backward in time from a point near $x_1 = x_2 = 0$ at $t = 0$.

1-6. Plot state-space trajectories for the Van der Pol system

$$\dot{x}_1 = x_2$$
$$\dot{x}_2 = -x_1 + \beta(1 - x_1^2)x_2,$$

with a separate plot for each of these cases:

(a) $\beta = 0.1$.

(b) $\beta = 1$.

(c) $\beta = 5$.

1-7. Show that the **Rayleigh system** [Davis, 1962, pp. 186–187]

$$\ddot{z} + \beta(\dot{z}^2 - 1)\dot{z} + \omega^2 z = 0$$

is entirely equivalent to the Van der Pol system

$$\ddot{y} + \beta(y^2 - 1)\dot{y} + \omega^2 y = 0$$

by finding a suitable coordinate transformation $y(t) \rightarrow z(t)$, and the corresponding initial conditions for $z(0)$ and $\dot{z}(0)$ as functions of $y(0)$ and $\dot{y}(0)$.

1-8. For the Rössler system in Example 1.4-8, make (x, y, z) plots of the trajectory starting from (0, 0, 0) for each of these cases:

(a) $\alpha = 2.6$.

(b) $\alpha = 3.5$.

(c) $\alpha = 4.1$.

(d) $\alpha = 4.23$.

(e) $\alpha = 4.6$.

Use an accurate step size (or small error tolerance for LSODAR, etc.) and let the trajectories run for a long time, to investigate whether the system has a periodic solution.

1-9. For the Lorenz system and parameter values given in Example 1.4-7, construct a **Poincaré map**, by plotting a point in (x, z) space every time $y = 0$, for the trajectory starting at $(x, y, z) = (0.001, 0.001, 27)$. Use the regula falsi (or LSODAR) root-finding procedure to find these points accurately. If a periodic trajectory exists, the points should eventually repeat themselves.

1-10. Find the complex variable transformation

$$z = as + b$$

that transforms

$$z(k+1) = \alpha z(k)[1 - z(k)]$$

to

$$s(k+1) = s^2(k) + c,$$

and determine the value of α corresponding to $c = 2i$.

1-11. Consider the following nonlinear system:

$$\dot{x}_1 = 3x_1 + x_2^2 - \text{sat}(2x_2 - u)$$
$$\dot{x}_2 = 3 \sin x_1 - x_2 - u,$$

where sat is the saturation function defined by

$$\text{sat}(y) = \begin{cases} y & \text{if } |y| \leq 1 \\ \text{sgn}(y) & \text{if } |y| \geq 1. \end{cases}$$

(a) With $u = 0$, plot state-space trajectories in a neighborhood of the origin defined by $|x_1| \leq 1$, $|x_2| \leq 2$.

(b) Using (1.5-4)–(1.5-5), linearize the above system about $x_1 = x_2 = u = 0$. Show that the linearized system is unstable and design a state variable feedback controller so that the controlled linear system has eigenvalues equivalent to a damped-spring-mass system (see Exercise 1-3) with $\zeta = 0.5$, $\omega = 3$.

(c) Use the control law of part (b) to drive the original nonlinear system. Plot state-space trajectories in the neighborhood of the origin defined in part (a).

1-12. Consider the nonlinear system defined by

$$\dot{x}_1 = 3x_1 + x_2^2 - 2x_2 + u$$
$$\dot{x}_2 = 3 \sin x_1 - x_2 - u.$$

(a) With $u = 0$, plot state-space trajectories in a neighborhood of the origin defined by $|x_1| \leq 1$, $|x_2| \leq 2$.

(b) Using (1.5-4)–(1.5-5), linearize the above system about $x_1 = x_2 = u = 0$. Show that the linearized system is unstable and design a state-variable feedback controller so that the controlled linear system has eigenvalues equivalent to a damped-spring-mass system (see Exercise 1-3) with $\zeta = 0.5$, $\omega_n = 3$.

(c) Use the control law of part (b) to drive the original nonlinear system.

Plot state-space trajectories in the neighborhood of the origin defined in part (a).

1-13. Let x and y denote the populations of species I and II, respectively, where

$$\dot{x} = (a - by)x$$
$$\dot{y} = (c - dx)y,$$

with a, b, c, and d being positive constants. Without integrating the equations of motion, sketch the state-space trajectories in the region $x \geq 0$, $y \geq 0$ and estimate on the plot the region where species I will become extinct and the region where species II will become extinct. Now check your results by integrating the equations of motion.

1-14. Redo Figure 1.5-4 with $\omega = 1$ and $\zeta = 0.2$.

1-15. Discuss the possible existence of limit cycles or periodic solutions in the region $x_1 > 0$, $x_2 > 0$ for the Lotka–Volterra equations

$$\dot{x}_1 = \alpha x_1 - \beta x_1 x_2$$
$$\dot{x}_2 = -\gamma x_2 + \delta x_1 x_2$$

with positive parameters α, β, γ, and δ.

1-16. Plot the domain of attraction to the origin for the system

$$\dot{x}_1 = x_2$$
$$\dot{x}_2 = x_1 + u,$$

subject to saturating state variable feedback given by (1.4-5) and (1.4-7) with $|u| \leq 1$, using feedback gains so that the unsaturated controlled system will behave as a second-order system (see Exercise 1-3) corresponding to (a) $\zeta = 0.5$, $\omega_n = 2$, (b) $\zeta = 0.707$, $\omega_n = 20$.

1-17. The equations of motion for a DC motor-controlled inverted pendulum system are given by

$$\dot{x}_1 = x_2$$
$$\dot{x}_2 = \beta_1 \sin x_1 - \beta_2 x_2 + \beta_3 u,$$

where x_1 is the angular displacement from the vertical, x_2 is the angular rate, u is a control voltage that can produce a moment at the pivot point, and β_i are constants associated with a particular configuration. For this example suppose that $\beta_1 = 17.63/\text{sec}^2$, $\beta_2 = 0.1872/\text{sec}$, and $\beta_3 = 0.6455/\text{sec}^2$-volt.

(a) For small angles $[\sin(x_1) \approx x_1]$ and unbounded control,

(1) Design a linear state-variable feedback controller so that the controlled system will behave as a second-order system (see Exercise 1-3) with a damping ratio $\zeta = 0.5$ and undamped natural frequency $\omega_n = 1$ rad/sec.

(2) Plot several state-space trajectories using the controller of part (1).

(3) Is the Bendixson negative criterion satisfied for the controlled system?

(b) For large angles and the control bounded by $|u| \leq 15$, and using a saturating state-variable feedback controller based on part (1) above, plot the domain of attraction to the origin in a region of state space defined by $|x_1| \leq 12$, $|x_2| \leq 14$. Is the point $x_1 = \pi$, $x_2 = 0$ controllable to the origin? Note: the origin is given by $x_1 = x_2 = 0$. The points $x_1 = 2n\pi$, $n = \pm 1, \pm 2, \ldots, x_2 = 0$ are not the origin, even though the pendulum is in the upright position at these points.

1-18. Apply the Lindstedt–Poincaré perturbation method to estimate the limit cycle for the Van der Pol system in Exercise 1-6 for $\beta = 0.1$. Carry out your analysis through terms of second-order in β, and estimate the value of x_1 at the point where the limit cycle crosses the x_1-axis.

1-19. Repeat Exercise 1-18 with $\beta = 0.1$, but using the describing function method, after first finding the describing function for the nonlinearity.

2

NONLINEAR CONTROL SYSTEMS

2.1 ADMISSIBLE CONTROLS

Constrained Nonlinear Control Systems

Our general class of nonlinear control systems is of the form

$$\dot{\mathbf{x}} = \mathbf{f}(\mathbf{x}, \mathbf{u}), \tag{2.1-1}$$

where $\mathbf{x}$ is an n_x-dimensional **state** vector for the system and $\mathbf{u}$ is an n_u-dimensional **control** input, which we are free to choose. The n_x-dimensional vector function $\mathbf{f}(\cdot)$ is assumed to be C^1 (continuous and continuously differentiable) in all of its arguments. However, as we discuss in more detail shortly, we allow for the possibility that a feedback control function $\mathbf{u}(\mathbf{x})$ may be discontinuous.

We are interested in the underlying basic controllability objective of finding a feedback control function $\mathbf{u}(\mathbf{x})$ that will transfer the state to a specified **target** set $\mathcal{X} \subset R^{Nx}$, usually defined by a system of n_g equalities

$$\mathcal{X} = \{\mathbf{x} | \mathbf{g}(\mathbf{x}) = \mathbf{0}\}, \tag{2.1-2}$$

with $\mathbf{g}(\mathbf{x}) = [g_1(\mathbf{x}) \cdots g_{n_g}(\mathbf{x})]^{\mathrm{T}}$ assumed to be continuous and continuously differentiable. Furthermore, we assume that the gradient vectors $\partial g_i(\mathbf{x})/\partial \mathbf{x}$, $i = 1, \ldots, n_g$, are linearly independent on $\mathcal{X}$.

The control vector is required to satisfy a given set of constraints

$$\mathbf{u} \in \mathcal{U}, \tag{2.1-3}$$

where $\mathcal{U} \subseteq R^{n_u}$ is a specified **control constraint set**. If no control constraints are imposed, then $\mathcal{U} = R^{n_u}$. The general case allows for the possibility that $\mathcal{U}$ may depend on the current state, with the control constraint set being defined by a system of n_h inequalities

$$\mathcal{U} = \{\mathbf{u} | \mathbf{h}(\mathbf{x}, \mathbf{u}) \geq \mathbf{0}\}. \tag{2.1-4}$$

However, we usually restrict the discussion to state-independent control constraints of the form

$$\mathcal{U} = \{\mathbf{u} | \mathbf{h}(\mathbf{u}) \geq \mathbf{0}\}, \tag{2.1-5}$$

with $\mathbf{h}(\mathbf{u}) = [h_1(\mathbf{u}) \cdots h_{n_h}(\mathbf{u})]^T$ assumed to be continuous and continuously differentiable. We also require that the gradient vectors $\partial h_i(\mathbf{u})/\partial \mathbf{u}$ satisfy a **regularity** assumption. This requirement is discussed in detail in Section 3.2, where we develop first-order necessary conditions governing solutions to nonlinear constrained parameter optimization problems.

u-Vectogram

In addition to the possible presence of control constraints and controllability boundaries, consider for a moment some of the fundamental differences between dynamical systems of the form

$$\dot{\mathbf{x}} = \mathbf{f}(\mathbf{x}), \tag{2.1-6}$$

and the control systems of the form (2.1-1). At a state-space point $\mathbf{x}$ where $\mathbf{f}(\mathbf{x})$ is continuous, the dynamical system (2.1-6) has a unique **velocity vector** $\dot{\mathbf{x}}$, defined by (2.1-6). However, for the control system (2.1-1) a different velocity vector is generally obtained for each control vector $\mathbf{u}$ chosen from the control constraint set $\mathcal{U}$. Thus at every point $\mathbf{x}$ in state space, we have a set of possible velocity vectors to choose from given by

$$\mathbf{f}(\mathbf{x}, \mathcal{U}) \triangleq \{\mathbf{f}(\mathbf{x}, \mathbf{u}) | \mathbf{u} \in \mathcal{U}\}. \tag{2.1-7}$$

Isaacs (1965) coined the term **vectogram** or **u-vectogram** to refer to this set. Figure 2.1-1 illustrates the **u**-vectogram associated with a set of possible velocity vectors along a trajectory $\mathbf{x}(t)$.

If we use a feedback control law $\mathbf{u}(\mathbf{x})$, then the control system (2.1-1) becomes a dynamical system of the form (2.1-6),

$$\dot{\mathbf{x}} = \mathbf{f}(\mathbf{x}) \leftarrow \mathbf{f}[\mathbf{x}, \mathbf{u}(\mathbf{x})], \tag{2.1-8}$$

yielding a unique velocity vector at each state $\mathbf{x}$. The control design problem is

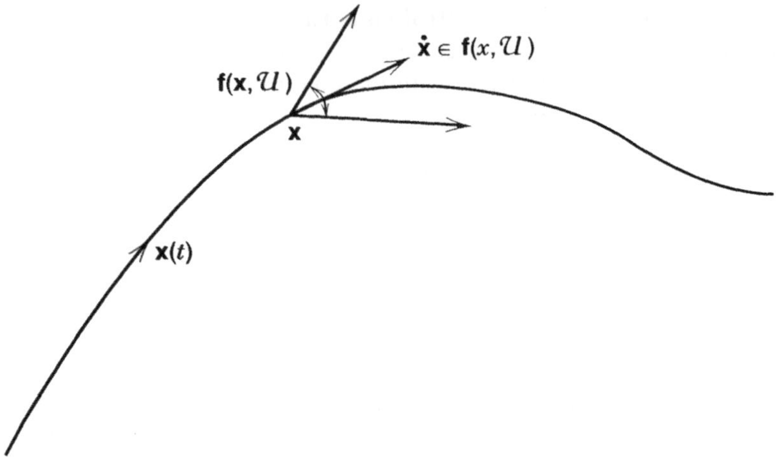

Figure 2.1-1 The set of velocity vectors at a point in state space.

to choose a feedback control law $\mathbf{u}(\mathbf{x})$ so that the resulting dynamical system (2.1-8) behaves in some desired fashion. For example, we might want to create an asymptotically state equilibrium point at $\mathbf{x} = \mathbf{0}$ such that the solutions $\mathbf{x}(t)$ to (2.1-8) reach a small neighborhood of this point in minimum time.

Corners and "Bang-Bang" Controllers

Another important distinction between the dynamical systems defined by $\mathbf{f}(\mathbf{x})$ and $\mathbf{f}(\mathbf{x}, \mathbf{u})$ is that we do not require that $\mathbf{u}(\mathbf{x})$ be a continuous function of $\mathbf{x}$. Thus $\mathbf{f}[\mathbf{x}, \mathbf{u}(\mathbf{x})]$ may be discontinuous, and along a solution $\mathbf{x}(t)$ to (2.1-8) a trajectory may have **corners** corresponding to sudden changes in the direction of the velocity vector. This can happen when some component of the control vector $\mathbf{u}(\mathbf{x})$ changes discontinuously, as in the case of a **switching**, **bang-bang**, or **on-off** control, which suddenly switches from an upper bound value to a lower bound value.

Piecewise Differentiable Feedback Controllers

While we do not require that $\mathbf{u}(\mathbf{x})$ be continuous, we do need to impose some conditions on the choice of $\mathbf{u}(\mathbf{x})$ in order to ensure that solutions $\mathbf{x}(t)$ to (2.1-8) exist and have certain properties that allow us to analyze the effects of small changes in initial conditions. The solutions that we consider are defined in the sense of Filippov, as has been discussed previously in Section 1.2.

An **admissible control** $\mathbf{u}(\mathbf{x})$ is defined as any closed-loop control law such that the following conditions hold:

(i) The control satisfies $\mathbf{u}[\mathbf{x}(t)] \in \mathcal{U}$ for each point $\mathbf{x}(t)$ along a trajectory generated by $\mathbf{u}(\mathbf{x})$.

(ii) There exists a **partition** of state space (a collection of disjoint open sets $\mathcal{S}_i \subseteq R^{n_x}$ with $\cup\{\overline{\mathcal{S}_i}\} = R^{n_x}$, where $\overline{\mathcal{S}_i}$ denotes the closure of $\mathcal{S}_i$) such that $\mathbf{u}(\cdot)$ is continuous and continuously differentiable on each region $\mathcal{S}_i$ comprising the partition.

Condition (ii) allows the control function $\mathbf{u}(\mathbf{x})$ to be piecewise continuous, with discontinuities occurring across certain **switching surfaces** in state space corresponding to the boundaries of the regions $\mathcal{S}_i$. Except at these switching surfaces, $\mathbf{u}(\mathbf{x})$ is continuously differentiable and so is $\mathbf{f}(\mathbf{x})$ in (2.1-8), so that for any initial state-space point $\mathbf{x}(0)$ inside a partition region $\mathcal{S}_i$, a unique solution $\mathbf{x}(t)$ to (2.1-8), forward in time ($t \geq 0$), exists and is a continuously differentiable function of the initial state.

Example 2.1-1: A Minimum-Time Switching Surface Consider the problem of moving an object from one point to another in minimum time. Specifically, consider rectilinear motion of an object under the action of a bounded applied force, modeled by the control system

$$\begin{aligned} \dot{x}_1 &= x_2 \\ \dot{x}_2 &= u, \end{aligned} \tag{2.1-9}$$

with

$$|u| \leq 1, \tag{2.1-10}$$

and suppose that the control objective is to transfer the state to the origin as quickly as possible. As we see in Chapter 7, because the control u appears linearly in (2.1-9) the resulting minimum-time control switches between $u = \pm 1$ each time that a trajectory $\mathbf{x}(t)$ reaches a certain switching surface. Given that $u(t)$ only takes on two possible values $u = \pm 1$, we can readily determine the minimum-time switching surface for this example system.

Regardless of the actual $u = \pm 1$ control switching sequence, the final portion of a minimum-time trajectory for this example must coincide with either the $u = +1$ or the $u = -1$ trajectory that reaches the origin. We can determine these two terminal trajectories by integrating the differential equations backward from the origin with $u =$ constant.

The retro-time equations of motion are given by

$$\frac{dx_1}{d\tau} = -x_2$$
$$\frac{dx_2}{d\tau} = -u, \tag{2.1-11}$$

where t_f is the time at which a trajectory of (2.1-9) reaches the origin and $\tau = t_f - t$ is backwards time, or the time-to-go to the target. Integrating the backwards time system (2.1-11) from the origin with $u(\tau) \equiv u =$ constant yields

$$x_1(\tau) = u\,\frac{\tau^2}{2}$$
$$x_2(\tau) = -u\tau. \tag{2.1-12}$$

From (2.1-12) it follows that the terminal portion of a forward time trajectory $\mathbf{x}(t)$ that reaches the origin under a constant control must be part of the parabola

$$x_2^2 - 2ux_1 = 0.$$

More specifically, as illustrated in Figure 2.1-2, for $u(t) = +1$ the terminal trajectory is on the half parabola

$$x_2 = -\sqrt{2x_1} \qquad x_1 \geq 0$$

in the fourth quadrant, and for $u(t) = -1$ the terminal trajectory is on the half parabola

$$x_2 = \sqrt{-2x_1} \qquad x_1 \leq 0$$

in the second quadrant.

Based on these results, and the constraints $|u(t)| \leq 1$, which lead to $u = \pm 1$, consider the switching control defined by

$$u(\mathbf{x}) = \begin{cases} -\mathrm{sgn}[\sigma(\mathbf{x})] & \text{if } \sigma(\mathbf{x}) \neq 0 \\ -\mathrm{sgn}(x_2) & \text{if } \sigma(\mathbf{x}) = 0 \text{ and } \mathbf{x} \neq \mathbf{0} \\ 0 & \text{if } \mathbf{x} = \mathbf{0}, \end{cases} \tag{2.1-13}$$

where the **switching function** $\sigma(\mathbf{x})$ is given by

$$\sigma(\mathbf{x}) = \begin{cases} x_2 + \sqrt{2x_1} & \text{if } x_1 \geq 0 \\ x_2 - \sqrt{2|x_1|} & \text{if } x_1 \leq 0. \end{cases} \tag{2.1-14}$$

Figure 2.1-2 shows a plot of the switching surface $\sigma(\mathbf{x}) = 0$ and the minimum-time trajectories for this example. The switching surface partitions the state space into two open regions, one above and one below the switching surface. The control $u(\mathbf{x})$ is constant (hence continuous and differentiable) in each of these two regions, but is discontinuous across the switching surface.

For an initial state above the switching surface the equations of motion (2.1-9) are integrated with $u = -1$, yielding a trajectory $\mathbf{x}(t)$ that reaches the switching surface at some time t_s. Switching the control to $u = 1$ and using $\mathbf{x}(t_s)$ as the new "initial" state, the equations of motion (2.1-9) are then integrated with the new control $u = 1$, until reaching this origin.

Notice from Figure 2.1-2 that if there were a slight time delay in implementing a control switch, the resulting trajectory would overshoot the switching surface and would ultimately produce a small limit cycle about the origin. Because of this, a practical application of this minimum-time controller might incorporate a stabilizing linear state feedback terminal controller after a trajectory has reached a small neighborhood of the origin. ■

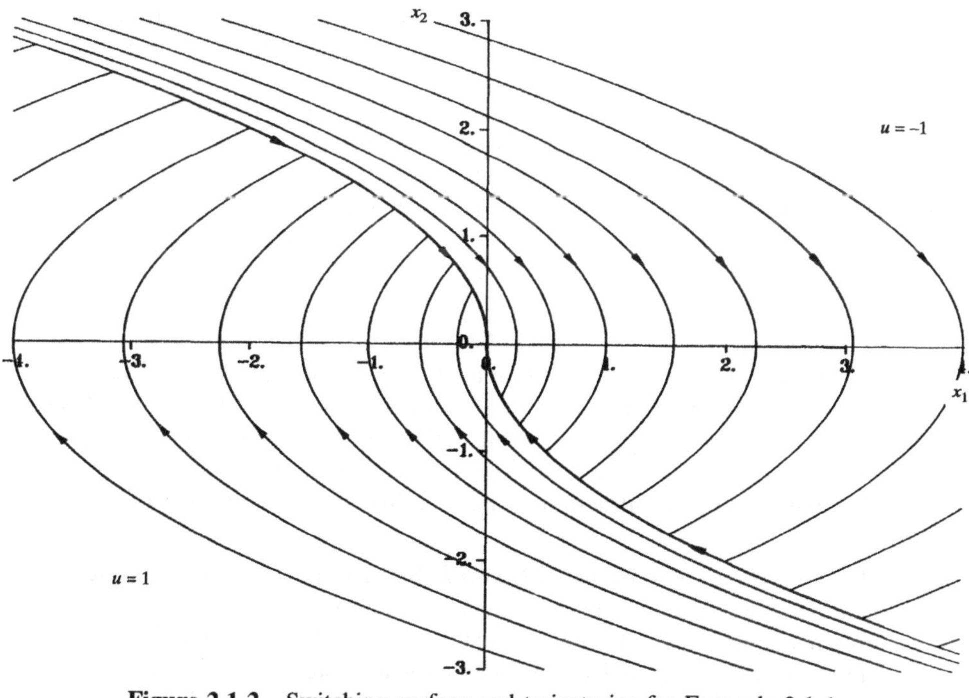

Figure 2.1-2 Switching surface and trajectories for Example 2.1-1.

2.2 CONTROLLABILITY BOUNDARIES

Controllable Set

Given a control system (2.1-1) with specified control constraints (2.1-3) and a specified target set (2.1-2), an initial state is said to be **controllable** to the target if there exists an admissible control law $\mathbf{u}(\mathbf{x})$ such that the solution to (2.2-1) transfers the initial state to the target in finite time. The set of all controllable initial states is said to be the **controllable set**. If the controllable set is the entire state space, the system is said to be **globally controllable** to the target. Not all systems are globally controllable. Example 1.5-5 is such a system. In general, the controllable set is a subset of the state space and a fundamental problem in control theory is to either find or approximate the controllable set. In Chapter 6 we consider the basic problem of determining the boundary of the controllable set.

Some control laws may be less effective than others at transferring initial states to the target. For a given admissible control law $\mathbf{u}(\mathbf{x})$, the set of initial points that actually get transferred to the target in finite time, called the **domain of effectiveness** for $\mathbf{u}(\mathbf{x})$, generally will be a subset of the controllable set. We will be interested not only in finding the domains of effectiveness for selected control laws, but also in developing control laws for which the resulting domain of effectiveness is the entire controllable set. From the definition of the controllable set, we are assured that such a control will always exist.

Note that controllability does not necessarily imply stability; trajectories may hit a target but not remain in a vicinity of the target afterwards (e.g., a rifle bullet hitting a paper target). However, for many systems we can find a $\mathbf{u}(\mathbf{x})$ that does provide stability for a target. In particular, our control design objective may be to create trajectories that asymptotically approach the target. To distinguish this situation, we need one more definition: Given an admissible control law $\mathbf{u}(\mathbf{x})$, the **domain of attraction** for $\mathbf{u}(\mathbf{x})$ is defined as the set of initial states for which the solutions to (2.1-1) yield trajectories that asymptotically approach the target.

We classify our control problems as follows:

1. *Linear Systems with Unbounded Control.* This is the classical system from linear control theory. If the system is globally controllable and if the target set is an equilibrium point, then there will always exist a state variable feedback control law such that the system is controllable from every point in state space to the target. In addition there will exist a state variable feedback control law such that the domain of attraction to the equilibrium point is the entire state space.

2. *Linear Systems with Bounded Control.* Strictly speaking, such a system is nonlinear. However, it is of interest to classify it separately from other nonlinear systems since, unless the control is against the bounds, the system will behave as in the previous case. A globally controllable linear system with unbounded controls need not be globally controllable when the control is bounded. However,

if the system is globally controllable to an equilibrium point under unbounded control, then, under bounded control, often we can find a control law which will provide either a nonempty domain of effectiveness or a nonempty domain of attraction.

3. *Nonlinear Systems.* A nonlinear system is one in which the system dynamics, as given by (2.1-1), contains nonlinear components. In general, such systems need not be globally controllable to a target, whether the control is bounded or not. While there always exists an admissible control law such that the domain of effectiveness is equal to the controllable set, there may not be any admissible control law that will provide a nonempty domain of attraction.

Linear Systems with Unbounded Control

For a linear system of the form

$$\dot{\mathbf{x}} = \mathbf{A}\mathbf{x} + \mathbf{B}\mathbf{u}, \tag{2.2-1}$$

global controllability to an equilibrium point is usually defined in terms of **complete controllability**. The state space system (2.2-1) is said to be completely controllable if there exists an unconstrained control law that can transfer any initial state to any other given point in state space. A completely controllable linear system with unbounded control will also be globally controllable to an equilibrium point.

Kalman Controllability Criterion With a constant control $\bar{\mathbf{u}}$, let $\bar{\mathbf{x}}$ be an equilibrium point for (2.2-1), satisfying

$$\mathbf{A}\bar{\mathbf{x}} + \mathbf{B}\bar{\mathbf{u}} = \mathbf{0}.$$

The **Kalman controllability criterion** [Grantham and Vincent, 1993; see also Example 6.2-1] is that the system (2.2-1) is completely controllable to the equilibrium point target $\bar{\mathbf{x}}$ if and only if

$$\operatorname{rank}[\mathcal{P}] = n_x, \tag{2.2-2}$$

where the $n_x \times n_x n_u$ matrix $\mathcal{P}$ is given by

$$\mathcal{P} = [\mathbf{B}, \mathbf{A}\mathbf{B}, \ldots, \mathbf{A}^{n_x - 1}\mathbf{B}]. \tag{2.2-3}$$

Furthermore, for a completely controllable system, there will always exist feedback gains $\mathbf{K}$ such that the system under state variable feedback of the form

$$\mathbf{u} = \bar{\mathbf{u}} - \mathbf{K}(\mathbf{x} - \bar{\mathbf{x}}) \tag{2.2-4}$$

will provide a domain of attraction to $\bar{\mathbf{x}}$ which is the entire state space, that is, $\bar{\mathbf{x}}$ is a globally asymptotically stable equilibrium point.

***The LQR Method for Determining* K** LQR stands for "linear-quadratic regulator," that is, a linear system (2.2-1) in which we wish to find an appropriate feedback gain matrix **K** for the regulator (2.2-4), which will minimize the quadratic cost

$$J = \int_0^\infty [(\mathbf{x}-\bar{\mathbf{x}})^T\mathbf{Q}(\mathbf{x}-\bar{\mathbf{x}}) + (\mathbf{u}-\bar{\mathbf{u}})^T\mathbf{R}(\mathbf{u}-\bar{\mathbf{u}})]\,dt$$

while driving the system asymptotically to the target $\bar{\mathbf{x}}$. Both **Q** and **R** are free to be chosen as long as they are constant symmetric positive definite matrices (see Section 4.3). The matrices **Q** and **R** determine the balance between the cost of not being at the origin and the cost of control to move the system to the origin.

Assume a control law of the form (2.2-4) and substitute into (2.2-1), using $\mathbf{B}\bar{\mathbf{u}} = -\mathbf{A}\bar{\mathbf{x}}$, to obtain

$$\dot{\mathbf{x}} = (\mathbf{A} - \mathbf{B}\mathbf{K})(\mathbf{x} - \bar{\mathbf{x}}). \tag{2.2-5}$$

Substituting (2.2-4) into J yields

$$J = \int_0^\infty (\mathbf{x}-\bar{\mathbf{x}})^T[\mathbf{Q} + \mathbf{K}^T\mathbf{R}\mathbf{K}](\mathbf{x}-\bar{\mathbf{x}})\,dt.$$

Consider now the quadratic form

$$V = (\mathbf{x}-\bar{\mathbf{x}})^T\mathbf{P}(\mathbf{x}-\bar{\mathbf{x}}),$$

where **P** is a constant symmetric positive definite matrix (to be chosen later). Lines of constant V are ellipsoids, in state space, centered at $\bar{\mathbf{x}}$. If we can show that $\dot{V} < 0$ along all trajectories (except at $\mathbf{x} = \bar{\mathbf{x}}$, where $\dot{V} = 0$), then all trajectories must asymptotically approach $\bar{\mathbf{x}}$ under the control law (2.2-4). This follows because V is a function that has a unique global minimum at $\bar{\mathbf{x}}$. If every trajectory decreases V, then every trajectory must eventually approach $\bar{\mathbf{x}}$. The V function introduced here is a **Lyapunov function**. Lyapunov function theory and its roll in control system design are discussed in much more detail in Chapters 4 and 5.

Taking the time derivative of V and substituting for $\dot{\mathbf{x}}$ using (2.2-5) yields

$$\dot{V} = (\mathbf{x} - \overline{\mathbf{x}})^{\mathrm{T}}[(\mathbf{A} - \mathbf{BK})^{\mathrm{T}}\mathbf{P} + \mathbf{P}(\mathbf{A} - \mathbf{BK})](\mathbf{x} - \overline{\mathbf{x}}).$$

If we now set

$$(\mathbf{A} - \mathbf{BK})^{\mathrm{T}}\mathbf{P} + \mathbf{P}(\mathbf{A} - \mathbf{BK}) = -(\mathbf{Q} + \mathbf{K}^{\mathrm{T}}\mathbf{RK}), \tag{2.2-6}$$

which will guarantee $\dot{V}$ to be negative for $(\mathbf{x} - \overline{\mathbf{x}}) \neq \mathbf{0}$ [since the integrand of J is still positive for $(\mathbf{x} - \overline{\mathbf{x}}) \neq \mathbf{0}$], it follows that, for a stable target with $\mathbf{x}(\infty) = \overline{\mathbf{x}}$,

$$J = -\int_0^\infty \dot{V}\, dt = -(\mathbf{x} - \overline{\mathbf{x}})^{\mathrm{T}}\mathbf{P}(\mathbf{x} - \overline{\mathbf{x}})\Big|_0^\infty = [\mathbf{x}(0) - \overline{\mathbf{x}}]^{\mathrm{T}}\mathbf{P}[\mathbf{x}(0) - \overline{\mathbf{x}}].$$

Since $\mathbf{P}$ is a function of $\mathbf{K}$, through (2.2-6), we can determine $\mathbf{K}$ by minimizing J with respect to $\mathbf{K}$. This operation yields [Ogata, 1990, p. 842]

$$\mathbf{K} = \mathbf{R}^{-1}\mathbf{B}^{\mathrm{T}}\mathbf{P}. \tag{2.2-7}$$

Substituting this result back into (2.2-6) results in

$$\mathbf{PA} + \mathbf{A}^{\mathrm{T}}\mathbf{P} - \mathbf{PBR}^{-1}\mathbf{B}^{\mathrm{T}}\mathbf{P} = -\mathbf{Q}. \tag{2.2-8}$$

This quadratic equation in the matrix $\mathbf{P}$ is known as the **algebraic Riccati equation**.

With full state information, the LQR design method simply involves choosing the $\mathbf{Q}$ and $\mathbf{R}$ matrices. Given $\mathbf{Q}$ and $\mathbf{R}$, $\mathbf{P}$ can be obtained from (2.2-8), and $\mathbf{K}$ is then determined from (2.2-7) for use in the feedback control law (2.2-4). Oftentimes a good control design is obtained by simply letting $\mathbf{Q}$ and $\mathbf{R}$ be identity matrices of appropriate dimension. Numerical methods are available for solving (2.2-8) (e.g., by using MATLAB's command "lqr"). This gives a distinct advantage to the LQR method as it is now possible to easily design controllers for higher-dimensional systems. An optimal control approach to LQR design is also given in Section 8.2.

Example 2.2-1: Linearized Inverted Pendulum Consider the problem of stabilizing the vertical position $\theta = 0$ of an undamped linearized inverted pendulum system whose position is described by

$$\ddot{\theta} - \theta = u.$$

The control input u corresponds to an applied torque that we will assume, in this example, to be unbounded. In terms of the state vector $\mathbf{x} = [\theta \quad \dot{\theta}]^{\mathrm{T}}$ the equations of motion can be written as

$$\dot{x}_1 = x_2$$
$$\dot{x}_2 = x_1 + u.$$

The objective is to develop a feedback control law $u(\mathbf{x})$ so that the target set

$$\mathcal{X} = \{\mathbf{0}\}$$

is asymptotically stable.

The Kalman controllability criterion (2.2-2) is satisfied for this problem. This means that every point in state space is controllable to the origin and we can find constants k_1 and k_2 such that a state variable feedback control of the form

$$u = -(k_1 x_1 + k_2 x_2)$$

will provide a globally asymptotically stable origin. For example, using the LQR method, with $\mathbf{Q}$ equal to the two-dimensional identity matrix and $\mathbf{R}$ equal to one, we obtain $k_1 = 2.4142$, $k_2 = 2.4142$. This results in a stable system with the controlled system eigenvalues $\mu_1 = -1.4142$, $\mu_2 = -1.0000$. ■

A Controllability Criterion for Second-Order Systems There is a large class of vibration problems that are described by second-order equations of the form

$$\mathbf{M}\ddot{\mathbf{y}} + \mathbf{S}\mathbf{y} = \mathbf{E}\mathbf{u}, \tag{2.2-9}$$

where $\mathbf{y}$ is an n_y-dimensional displacement vector, $\mathbf{u}$ is an n_y-dimensional force input vector, $\mathbf{M}$ is an $n_y \times n_y$ mass matrix, $\mathbf{S}$ is an $n_y \times n_y$ stiffness matrix, and $\mathbf{E}$ is an $n_y \times n_y$ force placement matrix. See, for example, Vincent et al. [1990]. The system may be put in the state space form (2.2-1) by letting, for $i = 1, \ldots, n_y$,

$$x_i = y_i$$

and, for $j = n_y + 1, \ldots, 2n_y$,

$$x_j = \dot{y}_i,$$

in which case we obtain

$$\mathbf{A} = \begin{bmatrix} \mathbf{0} & \mathbf{I} \\ -\mathbf{M}^{-1}\mathbf{S} & \mathbf{0} \end{bmatrix}$$

$$\mathbf{B} = \begin{bmatrix} \mathbf{0} \\ \mathbf{M}^{-1}\mathbf{E} \end{bmatrix}$$

If we now define

$$\overline{\mathbf{A}} = -\mathbf{M}^{-1}\mathbf{S}$$

$$\overline{\mathbf{B}} = \mathbf{M}^{-1}\mathbf{E},$$

the Kalman controllability matrix in (2.2-2) becomes

$$\mathcal{P} = \begin{bmatrix} \mathbf{0} & \overline{\mathbf{B}} & \mathbf{0} & \overline{\mathbf{A}}\overline{\mathbf{B}} & \cdots & \mathbf{0} & \overline{\mathbf{A}}^{n_y - 1}\overline{\mathbf{B}} \\ \overline{\mathbf{B}} & \mathbf{0} & \overline{\mathbf{A}}\overline{\mathbf{B}} & \mathbf{0} & \cdots & \overline{\mathbf{A}}^{n_y - 1}\overline{\mathbf{B}} & \mathbf{0} \end{bmatrix}.$$

Since each pair of column blocks of $\mathcal{P}$ is linearly independent, it is sufficient to check the rank of

$$\overline{\mathcal{P}} = [\overline{\mathbf{B}} \quad \overline{\mathbf{A}}\overline{\mathbf{B}} \quad \cdots \quad \overline{\mathbf{A}}^{n_y - 1}\overline{\mathbf{B}}]. \tag{2.2-10}$$

Example 2.2-2: A Spring-Mass System Subject to Control Consider the system illustrated in Figure 2.2-1. The system is composed of three masses m_1, m_2, and m_3 connected by four springs, as shown, each with the spring constants s_1, s_2, s_3, and s_4. Each mass has a force actuator attached. No other external forces are assumed to act on the system. In this case, $n_y = 3$, and the equation of motion is given by (2.2-9) with

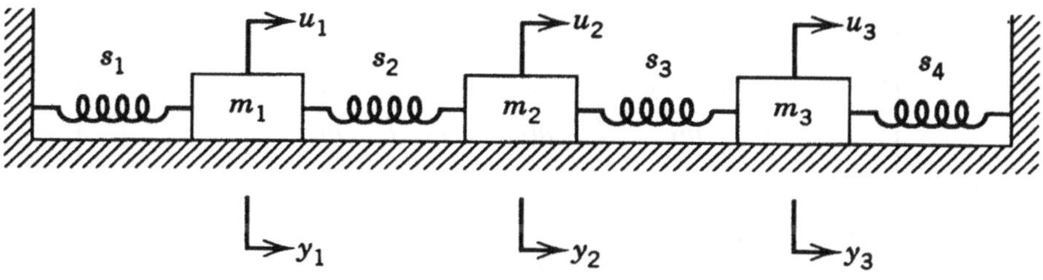

Figure 2.2-1 A spring-mass control system.

$$\mathbf{M} = \begin{bmatrix} m_1 & 0 & 0 \\ 0 & m_2 & 0 \\ 0 & 0 & m_3 \end{bmatrix}$$

$$\mathbf{S} = \begin{bmatrix} s_1 + s_2 & -s_2 & 0 \\ -s_2 & s_2 + s_3 & -s_3 \\ 0 & -s_3 & s_3 + s_4 \end{bmatrix}$$

$$\mathbf{E} = \begin{bmatrix} e_1 & 0 & 0 \\ 0 & e_2 & 0 \\ 0 & 0 & e_3 \end{bmatrix},$$

where $e_i = 1$ if that actuator is to be used for control and $e_i = 0$ otherwise. For example, if we place a force actuator only on the first mass, then $e_1 = 1$, $e_2 = 0$, $e_3 = 0$, and we have

$$\overline{\mathbf{A}} = \begin{bmatrix} \dfrac{-(s_1 + s_2)}{m_1} & \dfrac{s_2}{m_1} & 0 \\ \dfrac{s_2}{m_2} & \dfrac{-(s_2 + s_3)}{m_2} & \dfrac{s_3}{m_2} \\ 0 & \dfrac{s_3}{m_3} & \dfrac{-(s_3 + s_4)}{m_3} \end{bmatrix}$$

$$\overline{\mathbf{B}} = \begin{bmatrix} \dfrac{1}{m_1} & 0 & 0 \\ 0 & 0 & 0 \\ 0 & 0 & 0 \end{bmatrix}.$$

Thus

$$\overline{\mathbf{A}}\overline{\mathbf{B}} = \begin{bmatrix} \dfrac{-(s_1 + s_2)}{m_1} \dfrac{1}{m_1} & 0 & 0 \\ \dfrac{s_2}{m_2} \dfrac{1}{m_1} & 0 & 0 \\ 0 & 0 & 0 \end{bmatrix}$$

$$\overline{\mathbf{A}}^2\overline{\mathbf{B}} = \begin{bmatrix} \left(\dfrac{-(s_1 + s_2)}{m_1}\right)^2 \dfrac{1}{m_1} + \left(\dfrac{s_2}{m_1}\right)^2 \dfrac{1}{m_2} & 0 & 0 \\ \dfrac{s_2}{m_2} \left(\dfrac{-(s_1 + s_2)}{m_1}\right) \dfrac{1}{m_1} + \left(\dfrac{-(s_2 + s_3)}{m_2}\right) \dfrac{s_2}{m_2} \dfrac{1}{m_1} & 0 & 0 \\ \dfrac{s_3}{m_3} \dfrac{s_2}{m_2} \dfrac{1}{m_1} & 0 & 0 \end{bmatrix}.$$

From (2.2-10) the $\overline{\mathbf{P}}$ matrix for this case is

$$\overline{\mathbf{P}} = [\overline{\mathbf{B}} \quad \overline{\mathbf{A}}\overline{\mathbf{B}} \quad \overline{\mathbf{A}}^2\overline{\mathbf{B}}],$$

which clearly has rank 3. Thus the system is controllable and we may design a state-variable feedback controller that will provide a globally asymptotically stable origin. In particular, if we let $s_1 = 1$, $s_2 = 10$, $s_3 = s_4 = 1$, $m_1 = m_2 = 10$, $m_3 = 1$, then

$$\mathbf{A} = \begin{bmatrix} 0 & 0 & 0 & 1 & 0 & 0 \\ 0 & 0 & 0 & 0 & 1 & 0 \\ 0 & 0 & 0 & 0 & 0 & 1 \\ -1.1 & 1 & 0 & 0 & 0 & 0 \\ 1 & -1.1 & 0.1 & 0 & 0 & 0 \\ 0 & 1 & -2 & 0 & 0 & 0 \end{bmatrix}$$

$$\mathbf{B} = \begin{bmatrix} 0 & 0 & 0 \\ 0 & 0 & 0 \\ 0 & 0 & 0 \\ 0.1 & 0 & 0 \\ 0 & 0 & 0 \\ 0 & 0 & 0 \end{bmatrix}.$$

Using the LQR method we seek a feedback control law of the form

$$\mathbf{u} = -\mathbf{K}\mathbf{x}. \tag{2.2-11}$$

Choosing $\mathbf{Q}$ equal to 100 times the six-dimensional identity matrix and $\mathbf{R}$ equal to the three-dimensional identity matrix, the LQR method yields

$$\mathbf{K} = \begin{bmatrix} 25.7528 & -15.0930 & 3.8371 & 24.8003 & 16.6915 & -9.7498 \\ 0 & 0 & 0 & 0 & 0 & 0 \\ 0 & 0 & 0 & 0 & 0 & 0 \end{bmatrix},$$

which results in the stable system

$$\dot{\mathbf{x}} = [\mathbf{A} - \mathbf{B}\mathbf{K}]\mathbf{x}, \tag{2.2-12}$$

with the eigenvalues of $[\mathbf{A} - \mathbf{B}\mathbf{K}]$ given by

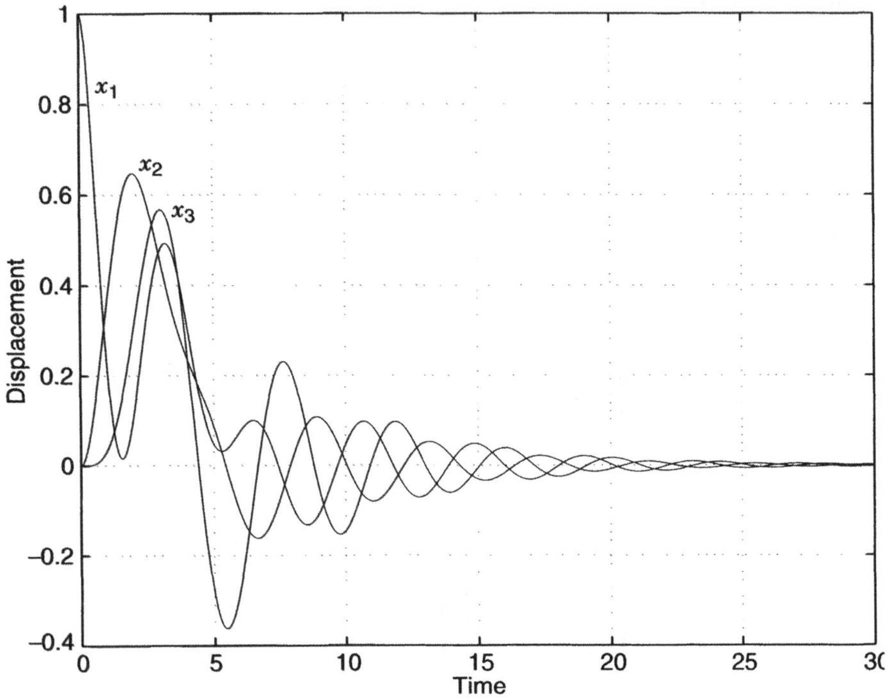

Figure 2.2-2 Response of the spring-mass system under state-variable feedback.

$$-0.6528 \pm 0.5238i$$
$$-0.3937 \pm 1.2923i$$
$$-0.1935 \pm 1.5408i.$$

Figure 2.2-2 illustrates the displacement of the three masses as a function of time as determined from (2.2-12) with the first mass given an initial displacement of 1 unit. ■

Linear Systems Under Bounded Control

Even if the Kalman condition is satisfied, as in the following two examples, the controllable set for a linear system under bounded control need not be the entire state space.

Example 2.2-3: Controllable Set as a Bounded Subset of the State Space Under unbounded control, the entire state space for the system

$$\dot{x}_1 = x_2$$
$$\dot{x}_2 = -2x_1 + 3x_2 + u \qquad (2.2\text{-}13)$$

is controllable to the origin. However, under bounded control, the controllable set is a bounded subset of the state space. For example, let the control be bounded by $|u| \leq 1$. This results in the controllable set illustrated in Figure 2.2-3. The controllability boundaries illustrated in this figure were obtained using the methods developed in Chapter 6. For this problem they reduce to simply integrating the system equations backward in time from each of the equilibrium points obtained under constant control at one of the control bounds. That is, the upper curve is obtained by starting the system at $x_1(0) = 0.5$, $x_2(0) = 0$, and integrating (2.2-13) backward in time with the control $u = -1$. This drives the system asymptotically to the equilibrium point $x_1(0) = -0.5$, $x_2(0) = 0$. The lower curve is then generated by integrating (2.2-13) backward in time from this point using the control $u = +1$. ■

Example 2.2-4: Controllable Set as an Unbounded Subset of the State Space
Consider the problem of stabilizing the vertical position $\theta = 0$ of an undamped linearized inverted pendulum system whose position is described by

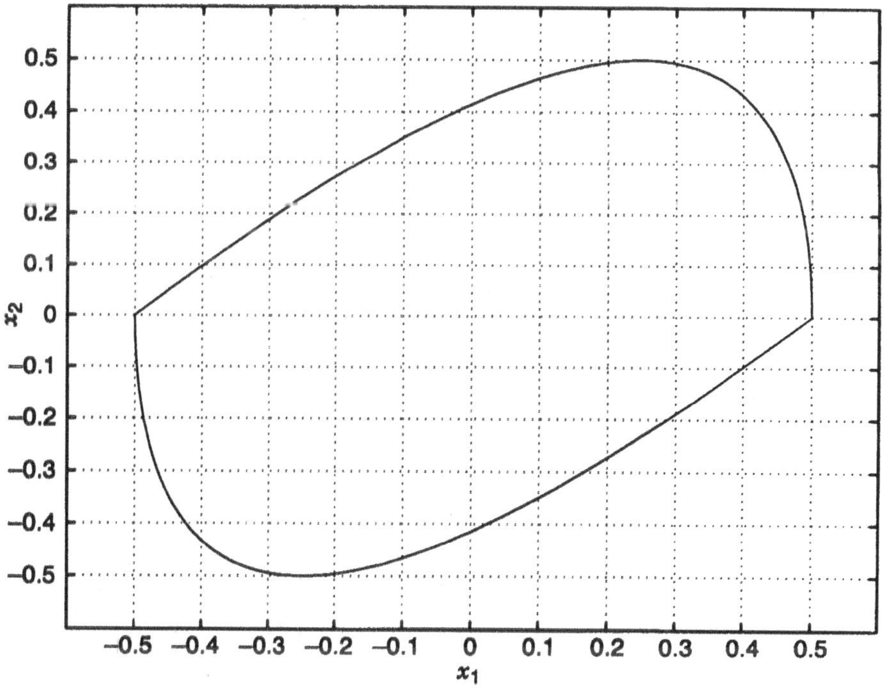

Figure 2.2-3 Controllable set for Example 2.2-3 under bounded control.

$$\ddot{\theta} - \theta = u.$$

Assume that the control input u, corresponding to an applied torque, has constraint bounds imposed upon it. Specifically, suppose that the control constraint set is given by

$$u \in \mathcal{U} = \{u| \; |u| \le 1\}.$$

In terms of the state vector $\mathbf{x} = [\theta \quad \dot{\theta}]^{\mathrm{T}}$ the equations of motion can be written as

$$\begin{aligned}\dot{x}_1 &= x_2\\ \dot{x}_2 &= x_1 + u.\end{aligned}$$

The objective is to develop a feedback control law $u(\mathbf{x})$ so that $\mathbf{x}(t) \rightarrow 0$ as $t \rightarrow \infty$. Thus the target set is

$$\mathcal{X} = \{\mathbf{0}\}.$$

To see that a controllability boundary exists for this example system, consider the constant-control trajectories shown in Figure 2.2-4. For a constant control,

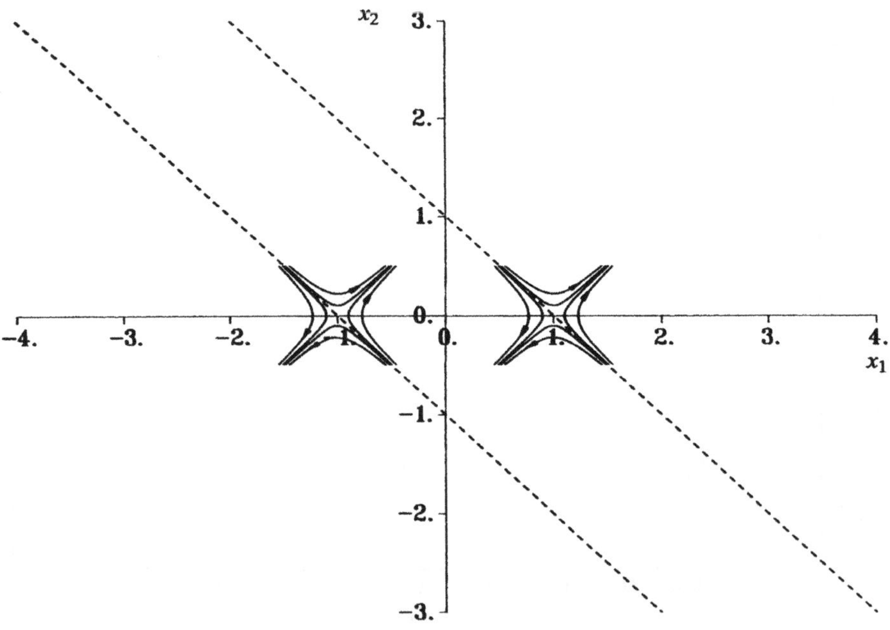

Figure 2.2-4 Controllability boundary for a linearized inverted pendulum.

the equilibrium point for the state-space system is an unstable saddle point at $\mathbf{x} = (-u, 0)$. Figure 2.2-4 illustrates the saddle-point geometry for both $u = 1$ and $u = -1$, corresponding to the upper and lower bounds imposed on the control u. For $u = 0$ the saddle-point equilibrium is located at the origin.

The eigenvalues and corresponding unit eigenvectors for the system are given by

$$\mu_1 = 1 \qquad \boldsymbol{\xi}_1 = \begin{bmatrix} \dfrac{1}{\sqrt{2}} \\ \dfrac{1}{\sqrt{2}} \end{bmatrix}$$

$$\mu_2 = -1 \qquad \boldsymbol{\xi}_2 = \begin{bmatrix} \dfrac{1}{\sqrt{2}} \\ -\dfrac{1}{\sqrt{2}} \end{bmatrix}.$$

Suppose that the initial state lies on the eigenvector $\boldsymbol{\xi}_1$ corresponding to the unstable eigenvalue $\mu_1 = 1$, at $\mathbf{x}(0) = \alpha\boldsymbol{\xi}_1$ with $\alpha > 0.707$. Then at $t = 0$,

$$\dot{\mathbf{x}}(0) = \alpha\mu_1\boldsymbol{\xi}_1 + \mathbf{b}u,$$

where $b = [0 \quad 1]^T$. The velocity projected outward along the eigenvector is given by

$$\boldsymbol{\xi}^T\dot{\mathbf{x}}(0) = \alpha + 0.707u,$$

and is positive for $\alpha > 0.707$ and $|u| \leq 1$. Thus the state moves outward toward infinity, because the control term is not large enough to counteract the effect of the outward motion associated with the unstable eigenvalue. Nearer the origin the $\mathbf{Ax}$ term in $\dot{\mathbf{x}} = \mathbf{Ax}+\mathbf{b}u$ is small and can be overcome by the $\mathbf{b}u$ term. Thus we have at least an indication that somewhere in state space there is a controllability boundary surface, beyond which the bounded control cannot drive the state to the origin.

The actual controllability boundary is given by

$$x_1 + x_2 = \pm 1$$

and is shown by the dashed lines in Figure 2.2-4. To verify this, in a more general way than in the previous eigenvalue discussion, consider the function

$$W(\mathbf{x}) = x_1 + x_2.$$

At a state $\mathbf{x}$ we have

$$\begin{aligned}\dot{W} &= x_2 + (x_1 + u) \\ &= W + u.\end{aligned}$$

Hence $\dot{W} \geq 0$ for $|u| \leq 1$ if $W(\mathbf{x}) \geq 1$ and $\dot{W} \leq 0$ for $|u| \leq 1$ if $W(\mathbf{x}) \leq -1$. Therefore, trajectories that start outside the region defined by $|W| = |x_1 + x_2| < 1$ (the controllable set to the origin) remain outside the region and can never reach the origin. Any saturating state variable feedback controller designed for this system will result in a domain of attraction that is a subset (or all) of the controllable set to the origin. ■

Example 2.2-5: Spring-Mass System Subject to Bounded Control In the spring-mass system of Example 2.2-2, the control was unbounded. For the situation depicted in Figure 2.2-2, the control magnitude is large at the first instant of time. However, it quickly reduces to $|u| \leq 2$ after about 5.7 seconds. If we now bound the control, as $|u| \leq u_{\max} = 2$, and use the saturating state variable feedback control given by (1.4-7) and (2.2-11) starting with the same initial conditions used in Figure 2.2-2, the system becomes unstable, as shown in Figure 2.2-5 (see Exercise 2-1). This illustrates the fact that the application of control bounds on an otherwise linear stable control system may destabilize the system. However, this is not to say that a controller cannot be found to stabilize the system to thc origin. In fact, the controllable set for this spring-mass system subject to bounded control is the entire state space (e.g., small damping would do). ■

Nonlinear Systems

With the following example, we make two observations: (1) The controllable set need not be the entire state space even if the control is unbounded, and (2) a nonempty controllable set to a target does not imply that a control law exists that will drive the system asymptotically to the target.

Example 2.2-6: Zermelo's Navigation Problem Consider a very wide river having a uniform flow of 2 m/sec from right to left, parallel but opposite to the x_1-axis, as illustrated in Figure 2.2-6. A swimmer in the river swims with a constant speed of 1 m/sec relative to the water, at any heading angle u that he or she chooses. The control problem of determining the swimmer's heading angle $u(\mathbf{x})$ to reach a specified target is known as **Zermelo's problem**. We take the target to be a circular island of radius one centered at the origin.

In terms of the swimmer's heading angle u from the positive x_1-axis, the equations of motion can be written as

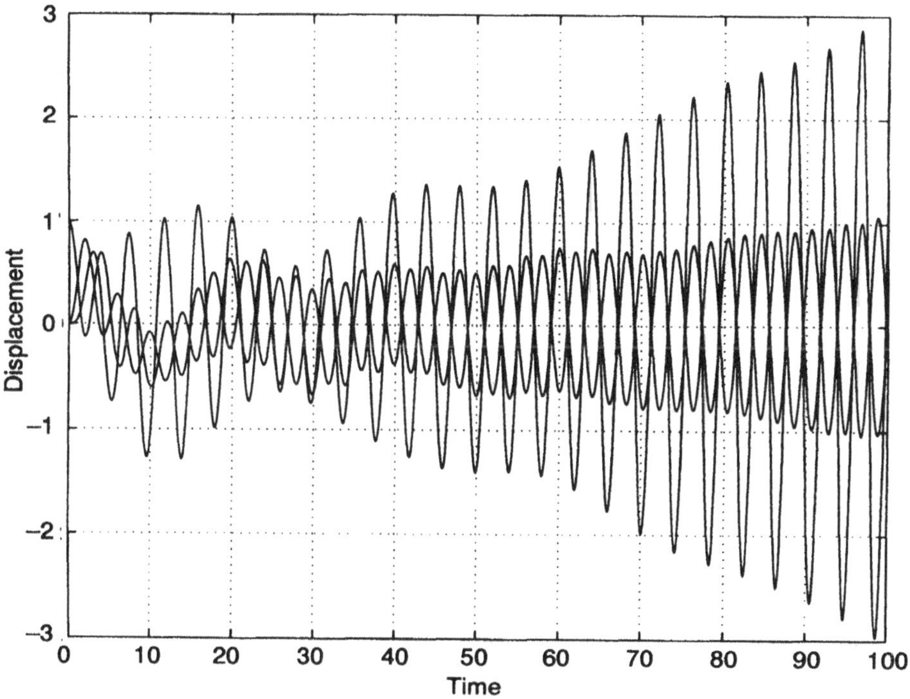

Figure 2.2-5 The spring-mass system under saturating state feedback.

$$\dot{x}_1 = -2 + \cos u$$
$$\dot{x}_2 = \sin u.$$

The target set is

$$\mathcal{X} = \{(x_1, x_2) | x_1^2 + x_2^2 = 1\},$$

and there are no constraints imposed on the heading angle u.

Since $\dot{x}_1$ is negative, regardless of the heading that the swimmer chooses, it is clear that there are some initial states from which the swimmer cannot reach the target. For example, since the swimmer cannot swim upstream, he or she cannot reach the target from points to the left of the circle in Figure 2.2-6.

The actual controllable set to the target is the cone-shaped region in Figure 2.2-6. It can be constructed, in an ad hoc fashion, by noting that the set of possible velocity vectors $\dot{\mathbf{x}}$ at a point $\mathbf{x}$ forms a circle of radius 1 centered 2 units to the left of $\mathbf{x}$. This geometry is the same at every state $\mathbf{x}$. Placing this velocity set so that it is just tangent to the target yields the controllable set shown in Figure 2.2-6. For a given control law, there will be a corresponding

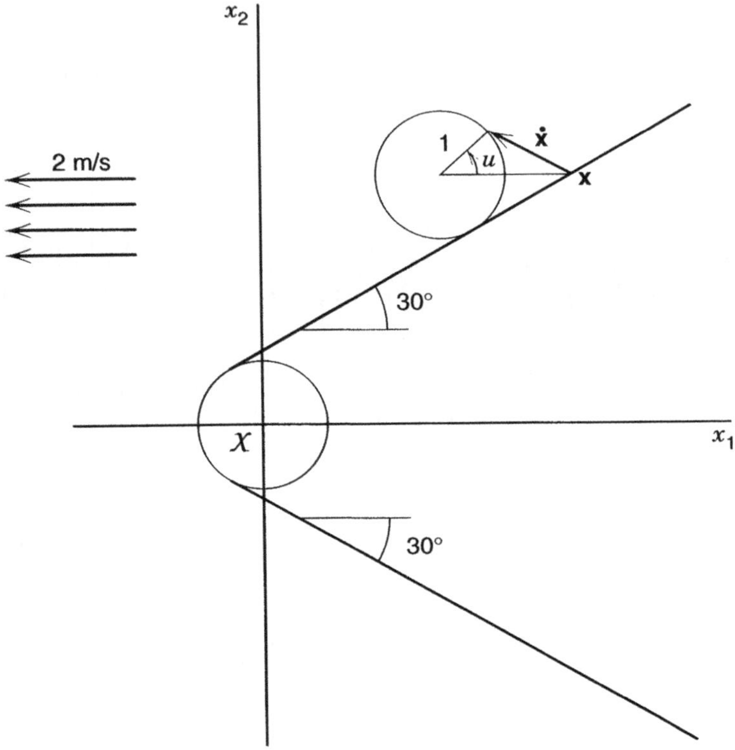

Figure 2.2-6 Controllable set for Zermelo's problem.

domain of effectiveness from which the system can be driven to the target. In general, the domain of effectiveness will be a subset of the controllable set. Hence if the controllable set can be determined, it becomes a useful tool for evaluating the performance of a given control law.

Note that at the origin, there does not exist a control that can keep the system at the origin. Thus while the system can be driven to the target there does not exist a control law that will drive the system asymptotically to the target. The domain of attraction for this problem under any control law is empty. ■

2.3 DOMAIN OF ATTRACTION BOUNDARIES

Domain of Attraction

Given the control system (2.1-1) with control constraints (2.1-3), a target set (2.1-2), and a specified control law $\mathbf{u}(\mathbf{x})$, the set of initial states such that $\mathbf{x}(t)$ asymptotically approaches the target as $t \rightarrow \infty$ is called the **domain of attraction**. The domain of attraction represents a useful performance criterion for control systems, and it is often calculated or approximated when comparing control laws. Ideally, we might like to design a stabilizing control law such that the domain of attraction is all of the controllable set. However, it should be kept in mind that doing so might result in degrading system performance in some other regard, such as the time it takes to reach the target. As mentioned in Section 1.5, it is possible to determine the boundary of the domain of attraction for two-dimensional systems by simply integrating backward from the target. This also works for one-dimensional systems. For higher-dimensional systems, this same approach may be used to give some indication of the domain of attraction. However, it is often simpler to approximate the domain of attraction using other methods. Lyapunov functions, studied in Chapter 4, are useful in this regard.

Linear Systems with Unbounded Control

Provided that the Kalman controllability criterion is satisfied, the controllable set for linear systems with unbounded control will be the entire state space. There are any number of control laws for $\mathbf{u}(\mathbf{x})$ that will result in a domain of attraction that will also be the entire state space. For example, the LQR design method can be used to find feedback gains such that a control law of the form (2.2-4) will drive the system from any point in the state space such that $\mathbf{x}(t)$ approaches the equilibrium point target $\bar{\mathbf{x}}$ as $t \to \infty$. If, however, the Kalman controllability criterion is not satisfied, then neither the controllable set nor the domain of attraction need be the entire state space.

Example 2.3-1: Global Controllability Without Kalman Criterion The following system does not satisfy the Kalman controllability criterion:

$$\dot{x}_1 = -x_1$$
$$\dot{x}_2 = u.$$

Indeed, only the $x_1 = 0$ line is controllable to the origin. However, the system is globally controllable to a neighborhood of the origin. For example, the domain of attraction to the origin under the control law $u = -x_2$ is the entire state space. Thus, under the same control law, the domain of effectiveness to a neighborhood of the origin is also the entire state space. However, we lose these global properties if the target is not on the x_2-axis. ■

Example 2.3-2: A Domain of Attraction Without Kalman Criterion The following system does not satisfy the Kalman controllability criterion:

$$\dot{x}_1 = x_1$$
$$\dot{x}_2 = u,$$

and there does not exist a control law that will drive every point in the state space to the origin. However, a nonempty domain of attraction to the origin is still possible. For example, under the control law $u = -x_2$, the domain of attraction is the line $x_1 = 0$. In this case the origin is not asymptotically stable since the domain of attraction does not contain an open neighborhood of the origin (see Section 4.1). ■

Linear Systems Under Bounded Control

As mentioned in Section 1.4, a saturating state feedback controller is often used in the control of linear systems under bounded control. Its use may either be implicit or explicit. It is implicit when certain elements of the control system saturate (a control tab hits a stop, a power amplifier saturates) unknown to the control designer. It is explicit when the control bounds have been included in the control algorithm. In the following chapters we learn how to deal explicitly with control bounds and how to include such bounds in the formulation of control algorithms. For now, we deal with control bounds simply by using saturating state variable feedback. In general, the domain of attraction under saturating state variable feedback will be a subset of the controllable set. ■

Example 2.3-3: Domain of Attraction Under Saturating State Variable Feedback Consider again the system of Example 2.2-3:

$$\begin{aligned}\dot{x}_1 &= x_2 \\ \dot{x}_2 &= -2x_1 + 3x_2 + u,\end{aligned} \tag{2.3-1}$$

with the control bounds $|u| \leq 1$. Using again the LQR method, we may design a feedback controller of the form (2.2-11) by choosing **Q** and **R** as identity matrices. The LQR method yields

$$\mathbf{K} = [0.2361 \quad 6.2361],$$

with the eigenvalues of $\mathbf{A} - \mathbf{BK}$ being $\mu_1 = -1$ and $\mu_2 = -2.2361$. Thus, without control bounds, the entire state space is controllable to the origin. In order to accommodate the control bounds, we replace the control law (2.2-11) with saturating state variable feedback as given by (1.4-7). Using the retro-time integration methods of Example 1.5-5, we find the domain of attraction under saturating state variable feedback illustrated in Figure 2.3-1. Comparing this figure with Figure 2.2-3, we see that the domain of attraction is approximately two-thirds the size of the controllable set. Of course, different saturating state variable feedback control laws will give different domains of attraction, but in no case can the domain of attraction be larger than the controllable set. ■

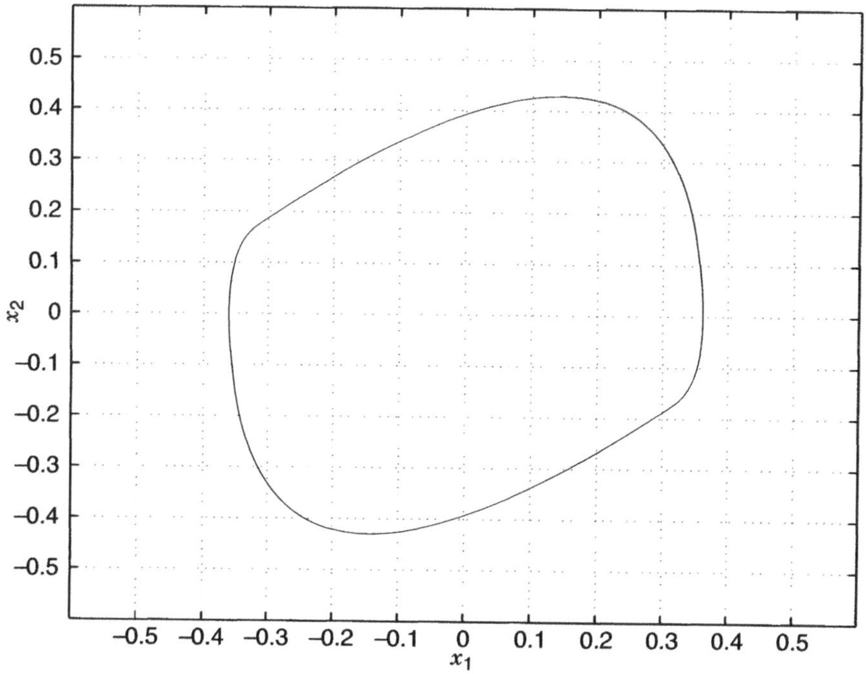

Figure 2.3-1 Domain of attraction under saturating state variable feedback.

Nonlinear Systems

Domains of attraction for two-dimensional nonlinear systems may also be determined by integrating the system equations backward in time under a specified control law $\mathbf{u}(\mathbf{x})$. Only now the domain of attraction for a particular system may be a very complex object even for a relatively simple nonlinear system.

Example 2.3-4: Domain of Attraction for a Damped Inverted Pendulum The equations of motion for a damped nonlinear inverted pendulum are given by

$$\begin{aligned} \dot{x}_1 &= x_2 \\ \dot{x}_2 &= a_1 \sin x_1 + a_2 x_2 + bu, \end{aligned} \tag{2.3-2}$$

where x_1 is the angle the pendulum makes from the upright vertical position, and a_1, a_2, and b are constants associated with the system. The term $a_1 \sin x_1$ is the torque provided by gravity, $a_2 x_2$ is the damping provided by friction and back EMF, bu is the control torque provided by a DC motor, and u is the voltage applied to the motor, with $|u| \le u_{\max}$. A particular system that we examine here has [Grantham and Vincent, 1993]

$$
\begin{aligned}
a_1 &= 17.627 \text{ rad/sec}^2 \\
a_2 &= -0.187 \text{ sec}^{-2} \\
b &= 0.6455 \text{ rad/volt-sec}^2 \\
u_{\max} &= 18 \text{ volts.}
\end{aligned}
$$

Our objective is to stabilize the inverted pendulum in the vertical upright position given by $x_1 = x_2 = 0$. If we linearize this system by setting $\sin x_1 = x_1$, then we can use LQR design with $\mathbf{Q}$ and $\mathbf{R}$ identity matrices to obtain the feedback gains

$$\mathbf{K} = [54.6333 \quad 12.7624],$$

with the eigenvalues for the linearized controlled system given by $\mu_1 = -4.5397$ and $\mu_2 = -3.8855$. We can find the domain of attraction under saturating LQR control as applied to the nonlinear system (2.3-2) by integrating backward from points very close to the equilibrium points obtained when the control is set equal to $\pm u_{\max}$. The results are illustrated in Figure 2.3-2. In order to "see"

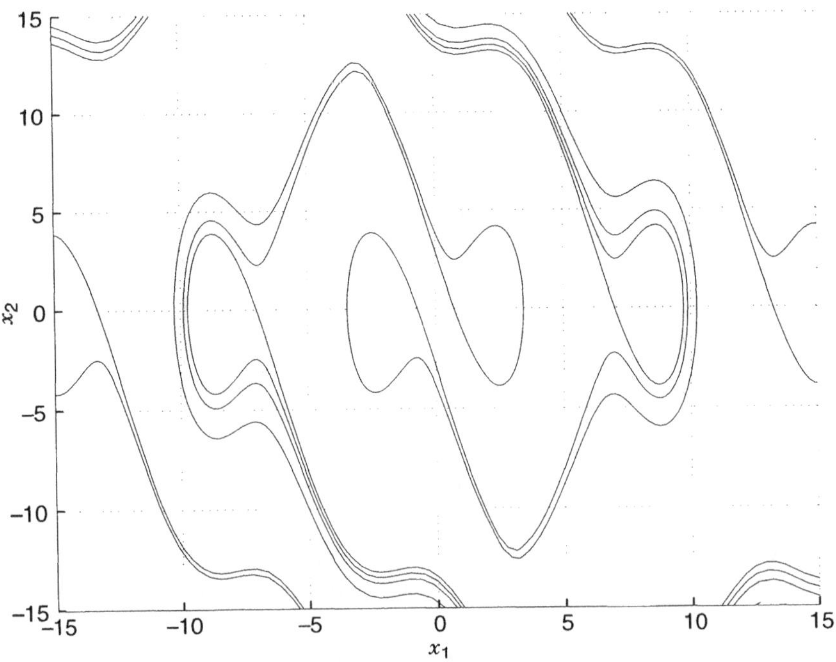

Figure 2.3-2 Domain of attraction for the damped inverted pendulum under saturating LQR control.

the domain of attraction in this figure, first pinpoint the origin. The domain of attraction to the origin is the set of all points obtained by "flooding" with a color, or by drawing all continuous curves from the origin that do not cross any of the curves contained in the figure. Observe that, in places, the domain of attraction narrows down and becomes tubular. Each of the lobe-like objects contains a different stable equilibrium point. If the system were to be started in any of the lobe-like objects, or their tubular extensions, it would remain in them, ultimately arriving at the equilibrium point contained in the lobe. Clearly, this rather interesting domain of attraction is not the entire state space. It is of interest to note, however, that for this problem, the controllable set is the entire state space. We learn later in Chapters 5 and 7 how to design controllers so that the associated domain of attraction for this problem will be the entire state space [Anderson and Grantham, 1989]. ■

2.4 CONTROL OF NONLINEAR SYSTEMS

Control Based on Linearization

The previous example introduced one way of designing controllers for nonlinear systems, a way that is often used and is often successful. The idea is to linearize the nonlinear system about some reference operating condition (e.g., an equilibrium point), design a controller for the linear system, and then apply the control to the nonlinear system. While this method will always result in a stabilizing controller in some neighborhood of the operating condition, the problem with this approach is that the domain of attraction may be difficult to determine and may also be quite small.

Any nonlinear system will behave linearly if it is sufficiently near a reference operating condition. Consider the system

$$\dot{\mathbf{X}} = \mathbf{f}(\mathbf{X}, \mathbf{U})$$

and let $\overline{\mathbf{X}}(t)$ denote a solution generated by a specific input $\overline{\mathbf{U}}(t)$. We let the perturbations $\mathbf{x}(t)$ and $\mathbf{u}(t)$ denote small changes from the reference condition $\overline{\mathbf{X}}(t)$ and $\overline{\mathbf{U}}(t)$. The corresponding solution generated by

$$\mathbf{U}(t) = \overline{\mathbf{U}}(t) + \mathbf{u}(t)$$

is of the form

$$\mathbf{X}(t) = \overline{\mathbf{X}}(t) + \mathbf{x}(t).$$

If the magnitudes of the perturbations are small for all times, then Taylor's theorem implies that $\mathbf{x}(t)$ is approximated by the solutions to the **state perturbation equations**

$$\dot{\mathbf{x}} = \left[\frac{\partial \mathbf{f}}{\partial \mathbf{X}}\right]\mathbf{x} + \left[\frac{\partial \mathbf{f}}{\partial \mathbf{U}}\right]\mathbf{u}, \tag{2.4-1}$$

where the partial derivatives matrices in (2.4-1) are evaluated along the reference conditions. If the reference conditions are all constants, as they frequently are, so that these matrices are constant, then (2.4-1) is of the form of (2.2-1) and the LQR method, or any other linear control system method, may be used to design a stable controller. As long as the system remains in a sufficiently small neighborhood of the reference condition, it will remain stable using the linear controller.

Example 2.4-1: A Linear Controller for a Double Pendulum Consider the general two-link pendulum system shown in Figure 2.4-1. The system is operating under gravity with the pivot points provided by the bearings of DC motors. These motors also provide control torques. The control torques are lumped together with any damping torques applied at the pivots into total equivalent torques of T_1 and T_2. In terms of the angles defined in Figure 2.4-1, the equations of motion are given by [Schmitt, 1992]:

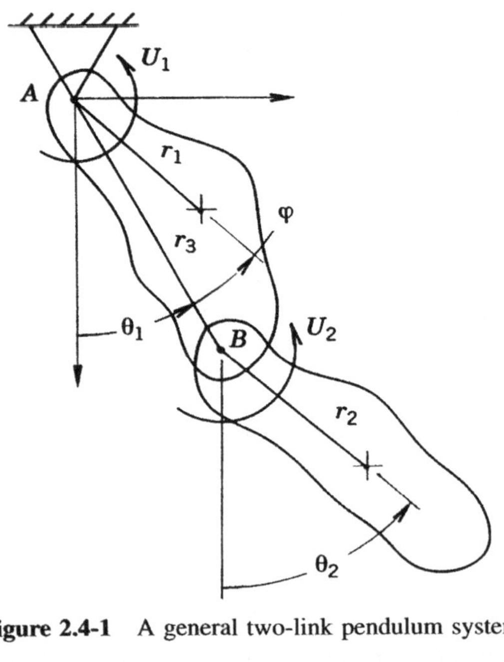

Figure 2.4-1 A general two-link pendulum system.

$$T_1 - T_2 = [(I_1 + m_2 r_3^2)\ddot{\theta}_1 + m_2 r_2 r_3 \ddot{\theta}_2 \cos(\theta_2 - \theta_1) - m_2 r_2 r_3 \dot{\theta}_2^2 \sin(\theta_2 - \theta_1)] + m_1 r_1 g \sin(\theta_1 + \varphi) + m_2 r_3 g \sin \theta_1$$

$$T_2 = I_2 \ddot{\theta}_2 + m_2 r_2 r_3 \ddot{\theta}_1 \cos(\theta_2 - \theta_1) + m_2 r_2 r_3 \dot{\theta}_1^2 \sin(\theta_2 - \theta_1) + m_2 r_2 g \sin \theta_2,$$

where

r_1 = distance to center of mass for pendulum 1
r_3 = distance to pivot point B
I_1 = moment of inertia about center of rotation for pendulum 1
m_1 = mass of pendulum 1
r_2 = distance to center of mass for pendulum 2
I_2 = moment of inertia about center of rotation for pendulum 2
m_2 = mass of pendulum 2
g = acceleration of gravity.

If we choose state variables according to

$$x_1 = \theta_1$$
$$x_2 = \dot{\theta}_1$$
$$x_3 = \theta_2 - \theta_1$$
$$x_4 = \dot{\theta}_2,$$

then, assuming that $\varphi = 0$, the equations of motion may be written as

$$A\dot{x}_2 + B\dot{x}_4 \cos x_3 - Bx_4^2 \sin x_3 + C \sin x_1 + D \sin x_1 = T_1 - T_2$$
$$E\dot{x}_4 + B\dot{x}_2 \cos x_3 + Bx_2^2 \sin x_3 + F \sin(x_3 + x_1) = T_2,$$

where

$$A = I_1 + m_2 r_3^2$$
$$B = m_2 r_2 r_3$$
$$C = m_1 r_1 g$$
$$D = m_2 r_3 g$$
$$E = I_2$$
$$F = m_2 r_2 g.$$

Using the state definitions, along with Cramer's rule, to solve for $\dot{x}_2$ and $\dot{x}_4$, we obtain the state space equations

$$\dot{x}_1 = x_2$$

$$\dot{x}_2 = \frac{[T_1 - T_2 + Bx_4^2 \sin x_3 - (C + D) \sin x_1]E - [T_2 - Bx_2^2 \sin x_3 - F \sin(x_3 + x_1)]B \cos x_3}{AE - B^2 \cos^2 x_3}$$

$$\dot{x}_3 = x_4 - x_2$$

$$\dot{x}_4 = \frac{[T_2 - Bx_2^2 \sin x_3 - F \sin(x_3 + x_1)]A - [T_1 - T_2 + Bx_4^2 \sin x_3 - (C + D) \sin x_1]B \cos x_3}{AE - B^2 \cos^2 x_3}$$

It can be shown that the denominator term will always be positive, so that there is no chance of a singularity occurring during integration.

We now consider a specific example where the torque from the first motor

$$T_1 = \frac{K_{\gamma 1}}{R_1} u_1 - \frac{K_{\gamma_1} K_{\beta_1}}{R_1} x_2$$

is obtained by supplying a control voltage u_1, subject to the limits $0 \leq u_1 \leq 19.9$. The second term is the back electromotive force (EMF) produced by rotation, and

K_{γ_1} = motor 1 torque constant

K_{β_1} = motor 1 back EMF constant

R_1 = motor 1 armature windings resistance.

The torque of the second motor,

$$T_2 = K_{\gamma_2} u_2,$$

is obtained by supplying a control current u_2 subject to the limits $0 \leq u_2 \leq 1.7$, where

K_{γ_2} = motor 2 torque constant.

The values used for all the constants are

$r_1 = 0.95833$ ft

$r_3 = 1.3542$ ft

$$
\begin{aligned}
I_1 &= 0.05 \text{ slug-ft}^2 \\
m_1 &= 1.327/g \text{ slugs} \\
r_2 &= 0.65104 \text{ ft} \\
I_2 &= 0.0035 \text{ slug-ft}^2 \\
m_2 &= 0.1684/g \text{ slugs} \\
g &= 32.17 \text{ ft/sec}^2 \\
K_{\gamma 1} &= 0.194 \text{ ft-lb/amp} \\
K_{\beta_1} &= 0.263 \text{ volt-sec}^2\text{/rad} \\
R_1 &= 7.7 \text{ ohms} \\
K_{\gamma_2} &= 0.0218 \text{ ft-lb/amp.}
\end{aligned}
$$

Assume that the objective is to control this system in a fashion such that having the first pendulum in the vertical downward position while the second pendulum is in the vertical upward position represents a stable equilibrium. Toward this end we seek to first linearize (2.4-2) about the equilibrium point given by

$$
\begin{aligned}
\overline{\mathbf{X}} &= [0 \quad 0 \quad \pi \quad 0]^{\mathrm{T}} \\
\overline{\mathbf{U}} &= [0 \quad 0]^{\mathrm{T}}.
\end{aligned}
$$

This means that the matrices in (2.4-1) must be evaluated. This is clearly a difficult chore to perform analytically due to the complicated right-hand sides of (2.4-2). However, routines are available to perform the task numerically. For example, using the MATLAB routine "linmode" in conjunction with SIMULINK, we obtain

$$
\mathbf{A} = \left.\frac{\partial \mathbf{f}}{\partial \mathbf{X}}\right|_{\overline{\mathbf{X}}, \overline{\mathbf{U}}} = \begin{bmatrix} 0 & 1 & 0 & 0 \\ -25.3238 & -0.1238 & 2.7001 & 0 \\ 0 & -1 & 0 & 1 \\ -2.0672 & -0.1633 & 34.8764 & 0 \end{bmatrix}
$$

$$
\mathbf{B} = \left.\frac{\partial \mathbf{f}}{\partial \mathbf{U}}\right|_{\overline{\mathbf{X}}, \overline{\mathbf{U}}} = \begin{bmatrix} 0 & 0 \\ 0.4708 & 0.1297 \\ 0 & 0 \\ 0.6207 & 6.3995 \end{bmatrix}.
$$

We now have a linear system of the form of (2.2-1), and any of the linear design methods may be used to design a controller. For example, if we use LQR design with

$$\mathbf{Q} = \begin{bmatrix} 1 & 0 & 0 & 0 \\ 0 & 1 & 0 & 0 \\ 0 & 0 & 10 & 0 \\ 0 & 0 & 0 & 10 \end{bmatrix}$$

and

$$\mathbf{R} = \begin{bmatrix} 1 & 0 \\ 0 & 10 \end{bmatrix},$$

the feedback gains for the two motors are given by

$$\mathbf{K} = \begin{bmatrix} 2.2222 & 1.1570 & 5.1353 & 1.1650 \\ 4.4201 & -1.0657 & 10.3299 & 2.0337 \end{bmatrix}.$$

When we apply this controller to the nonlinear system (2.4-2), we must keep in mind that the controllers will saturate, so that the actual saturating state variable feedback control will be given by (1.4-7). The above control design provides asymptotic stability for the equilibrium point $\overline{\mathbf{X}}$ in an irregular shaped domain of attraction (see Exercise 2-6). ■

As in the above example, applying saturating state variable feedback to a nonlinear system will generally result in a domain of attraction that will include some neighborhood of the target, but not necessarily all of state space. However, if the same control system can be driven chaotically, then a simple modification of the saturating state variable feedback control law can often result in a significantly larger domain of attraction.

Control of Chaotic Systems

Given a nonlinear system, assume that linearization has been used to find a saturating state variable feedback control law that has a finite nonempty domain of attraction to an equilibrium point of interest. Two essential ingredients are used in order to modify this design in a way that will generally increase the size of the domain of attraction. They are: a chaotic attractor and a controllable target. The chaotic attractor must either be "naturally present" in the system or be such that it can be created using open-loop control (e.g., sinusoidal forcing). The **controllable target** is any subset of the domain of attraction for the saturating state feedback controller that can be found with a nonempty intersection with the chaotic attractor. If we start the chaotic system at any point in the domain of attraction for the chaotic attractor, the resulting chaotic motion will ultimately drive the system to the controllable target since a chaotic trajectory is ergotic, that is, it eventually "visits" every point in the attractor [Schuster,

1988, p. 5]. The **chaotic control algorithm** simply keeps track of when the system enters the controllable target. When it does, any open-loop control that was used to create the chaos is turned off and, at the same time, the closed-loop saturating state variable feedback control is turned on. Clearly, the domain of attraction for the chaotic control algorithm will include the domain of attraction for the chaotic attractor.

A basic requirement with this approach is the ability to determine a controllable target. If we are lucky the controllable target will be the entire state space, as in Example 2.4-2 below. More generally, the controllable target will only be a small region about the equilibrium point of interest. We show in Chapter 6 how Lyapunov functions may be used to find controllable targets. For now, we must rely on more intuitive concepts.

Example 2.4-2: Lorenz Equations Consider again the Lorenz equations (1.4-21) in the following state space form with a control term u added to the second equation:

$$\begin{aligned}\dot{x}_1 &= s(x_2 - x_1)\\ \dot{x}_2 &= (r - x_3)x_1 - x_2 + u\\ \dot{x}_3 &= x_1x_2 - bx_3.\end{aligned} \tag{2.4-3}$$

This controlled form of the Lorenz equations does actually represent a physical natural convection system [Ehrhard and Müller, 1991]. We use the parameters $s = 10$, $b = \frac{8}{3}$, $r = 28$, which correspond to chaotic motion when $u = 0$. With $u = 0$, the system has three equilibrium points: the origin and the points $\bar{\mathbf{x}} = [a \quad a \quad r-1]^T$, $a = \pm\sqrt{b(r-1)}$, yielding $\bar{\mathbf{x}}_1 = [8.4853 \quad 8.4853 \quad 27]^T$ and $\bar{\mathbf{x}}_2 = [-8.4853 \quad -8.4853 \quad 27]^T$, all of which are unstable. Steady-state convection corresponds to the equilibrium points $\bar{\mathbf{x}}_1$ and $\bar{\mathbf{x}}_2$, while the state of no convection corresponds to the origin. Assume that we wish to stabilize the system at $\bar{\mathbf{x}}_1$. Linearizing (2.4-3) about $\bar{\mathbf{x}} = \bar{\mathbf{x}}_1$, $\bar{u} = 0$, we obtain

$$\begin{aligned}\dot{z}_1 &= -sz_1 + sz_2\\ \dot{z}_2 &= z_1 - z_2 - \sqrt{b(r-1)}\, z_3 + u\\ \dot{z}_3 &= (z_1 + z_2)\sqrt{b(r-1)} - bz_3,\end{aligned}$$

where

$$\begin{aligned}z_1 &= x_1 - \sqrt{b(r-1)}\\ z_2 &= x_2 - \sqrt{b(r-1)}\\ z_3 &= x_3 - (r-1).\end{aligned}$$

Note that $\mathbf{z} = \mathbf{0}$ corresponds to the equilibrium point $\overline{\mathbf{x}}_1$. The eigenvalues for the linear system with $u = 0$ are given by -13.85457, $0.09395 \pm 10.19451i$, which verifies that this equilibrium point is unstable. To maintain the linear system at this point only partial state feedback is required. Using

$$u = -kz_1$$

it is easy to show, using the Routh–Hurwitz criterion [Grantham and Vincent, 1993, p. 119], that the linear system will be stable for $k > 0.50101$. It turns out that if we apply this control, subject to the condition on k, to the nonlinear system (2.2-4), the equilibrium point $\overline{\mathbf{x}}_1$ is asymptotically stable with a domain of attraction that evidently (via simulation) is the entire state space [Vincent and Yu, 1991]. For large values of k the motion of the system to the equilibrium point is indistinguishable from any other well-behaved linear or nonlinear system. However, for values of k approaching $k = 0.50101$, the system can go though **chaotic transients** before ultimately approaching the equilibrium point in an asymptotic fashion. That is, depending on starting conditions, the motion will appear to be chaotic for possibly a long period of time before settling down to the equilibrium point. ■

Not all chaotic systems will respond as nicely to (unbounded) state variable feedback as in the previous example. Indeed, applying such a control outside of a small neighborhood of linearized equilibrium point may destabilize a nonlinear system.

Example 2.4-3: A Chaotic Controller for the Double Pendulum Consider again the double pendulum introduced in Example 2.4-1. This is a system in which chaotic motion can be maintained by driving one of the motors with a sinusoidal open-loop input. For example, with $u_2 = 0$, applying the control input

$$u_1 = 5.3 \cos 5t$$

results in chaotic motion. While this is not enough of an input for the first pendulum to swing past the vertical position, it is more than enough for the second pendulum to do so. In other words, the chaotic attractor under this control input includes states on or near the target point

$$\overline{\mathbf{X}} = [0 \quad 0 \quad \pi \quad 0]^{\mathrm{T}}.$$

If we can find a controllable target that intersects the chaotic attractor, we will then have both ingredients needed to apply the chaotic control algorithm. A controllable target for this system can be obtained for this system by finding an appropriate **Lyapunov function**. While Lyapunov stability theory is not formally introduced until Chapter 4, we simply make note here that the feedback gains **K** determined in Example 2.4-1 resulted in a stable linearized controlled system of the form

$$\dot{\mathbf{x}} = \hat{\mathbf{A}}\mathbf{x},$$

where $\hat{\mathbf{A}} = \mathbf{A} - \mathbf{BK}$. Since the controlled system is asymptotically stable, it is always possible to find a positive definite matrix **P** that is the solution to the **Lyapunov equation**

$$\mathbf{P}\hat{\mathbf{A}} + \hat{\mathbf{A}}^{\mathrm{T}}\mathbf{P} = -\mathbf{Q},$$

where **Q** is any specified positive definite matrix, such that

$$V = \mathbf{x}^{\mathrm{T}}\mathbf{P}\mathbf{x} \tag{2.4-4}$$

is a Lyapunov function for the linear system. For example, using the same **Q** matrix as in Example 2.4-1 and solving the Lyapunov equation using the MATLAB command "lyap", yields

$$\mathbf{P} = \begin{bmatrix} 61.1487 & -0.3113 & 16.2618 & 0.2801 \\ -0.3113 & 1.6989 & -0.5361 & 0.0297 \\ 16.2618 & -0.5361 & 14.6893 & 0.1617 \\ 0.2801 & 0.0297 & 0.1617 & 0.3740 \end{bmatrix}.$$

Setting $V = V_c$, where V_c is a positive constant, (2.4-4) yields an ellipsoid in state space centered about $\mathbf{x} = \mathbf{0}$. If we can demonstrate that $\dot{V} < 0$ for the nonlinear system at all states (except the equilibrium point) that lie within the level curve $V = V_c$, and if the intersection of the set of points defined by $V \leq V_c$ with the chaotic attractor is nonempty, then a controllable target is given by the set of points defined by $V \leq V_c$. Using an appropriate value for V_c and applying the chaotic control algorithm using the state feedback control of Example 2.4-1 results in a domain of attraction that is the entire state space (see Exercise 2-9). ■

EXERCISES

2-1. Reproduce the results depicted in Figures 2.2-2 and 2.2-5. Design a controller that will stabilize the origin for the spring-mass system, starting from every point in state space under the bounded control $|u| \leq 1$. *Hint:* Consider pure velocity damping.

2-2. Design a saturating state feedback controller for Example 2.2-1 and find the domain of attraction numerically.

2-3. Reexamine Example 2.3-1 with the target as any point other than the origin and with $|u| \leq 1$.

(a) Does there exist a control law with a nonempty domain of effectiveness?

(b) Does there exist a control law with a nonempty domain of attraction?

2-4. Consider the following two-dimensional system:

$$\dot{x}_1 = -x_2$$
$$\dot{x}_2 = 2x_1 + 3x_2 + u,$$

where the control must satisfy the constraints

$$|u| \leq 1.$$

Design linear state variable feedback controllers so that the unconstrained system is equivalent to a two-dimensional system with undamped natural frequency and damping ratio given by

(a) $\omega = 1, \zeta = 0.5$.

(b) $\omega = 1, \zeta = 2$.

(c) $\omega = 10, \zeta = 0.1$.

Numerically determine the domain of attraction for each case under saturating state variable feedback.

2-5. Reproduce Figure 2.3-2.

2-6. Use the saturating state variable controller design developed in Example 2.4-1 to drive the system from several points in the neighborhood of the equilibrium point. Find some starting conditions where this control law does not work. What happens if the system is started at $\overline{\mathbf{X}} = [0 \quad 0 \quad 0 \quad 0]^T$?

2-7. The control system given by

$$\dot{x}_1 = x_1 + u_1$$
$$\dot{x}_2 = x_2 + u_2$$

has a controller for driving the system to the origin given by

$$u_i = \begin{cases} -\sigma_i & \text{if } |\sigma_i| \leq 5 \\ -5 \operatorname{sgn}(\sigma_i) & \text{otherwise,} \end{cases}$$

where

$$\sigma_i = 5x_i.$$

(a) Estimate, analytically, the domain of attraction for this controller by considering the rate of change of the function

$$V = \tfrac{1}{2}(x_1^2 + x_2^2).$$

(b) Analytically determine the actual domain of attraction.

2-8. Obtain three-dimensional plots of the controlled Lorenz system given in Example 2.4-2 under the feedback control $u = -kz_1$ with $k = 1$ and $k = 10$.

2-9. Demonstrate that the chaotic controller designed in Example 2.4-3 will drive the system from the origin to the equilibrium point $\overline{\mathbf{X}} = [0 \quad 0 \quad \pi \quad 0]^{\mathrm{T}}$ using $V_c = 15$. Examine other initial conditions. Can you improve upon the design given?

3

NONLINEAR OPTIMIZATION

3.1 CONSTRAINED FUNCTION MINIMIZATION

In many of the topics to be discussed in this text we are dealing with an underlying function optimization problem of the following form:

$$\text{minimize } G(\mathbf{u}) \tag{3.1-1}$$

subject to a set of inequality constraints

$$\mathbf{u} \in \mathcal{U} = \{\mathbf{u} | \mathbf{h}(\mathbf{u}) \geq 0\}, \tag{3.1-2}$$

where $G(\mathbf{u})$ is a scalar cost function of an n_u-dimensional control vector $\mathbf{u}$, the set $\mathcal{U} \subseteq R^{n_u}$ is a specified constraint set, and $\mathbf{h}(\mathbf{u}) = [h_1(\mathbf{u}) \cdots h_{n_h}(\mathbf{u})]^{\mathrm{T}}$ is an n_h-dimensional vector of inequality constraint functions, with $G(\mathbf{u})$ and each component of $\mathbf{h}(\mathbf{u})$ being continuous and continuously differentiable in $\mathbf{u}$. The problem of maximizing a function $W(\mathbf{u})$, subject to the constraints (3.1-2), can be handled by minimizing $G(\mathbf{u}) = -W(\mathbf{u})$.

Definition of a Minimum

The function $G(\cdot)$ is said to have a **global minimum** at $\mathbf{u}^* \in \mathcal{U}$ if and only if

$$G(\mathbf{u}^*) \leq G(\mathbf{u})$$

for all $\mathbf{u} \in \mathcal{U}$. More than one point in $\mathcal{U}$ may take on a global minimum. However, it follows from the definition that the minimum cost is unique.

If $\mathcal{U}$ is a closed and bounded set, then from the theorem of Weierstrass [Vincent and Grantham, 1981] there will always exist a global minimum for G on $\mathcal{U}$. Only for linear systems will it be possible to apply necessary conditions for the determination of a global minimum. In general, the necessary conditions developed below will be used to find local minimum points. The function $G(\cdot)$ is said to have a **local minimum** at $\mathbf{u}^* \in \mathcal{U}$ if and only if there exists a ball $\mathcal{B}$ centered at $\mathbf{u}^*$ such that

$$G(\mathbf{u}^*) \leq G(\mathbf{u}^* + \Delta\mathbf{u})$$

for all $\mathbf{u}^* + \Delta\mathbf{u} \in \mathcal{B} \cap \mathcal{U}$. A ball $\mathcal{B}$ is a special type of open neighborhood defined by

$$\mathcal{B} = \{\mathbf{u}^* + \Delta\mathbf{u} \mid \|\Delta\mathbf{u}\| < \epsilon \text{ for any finite } \epsilon > 0\}.$$

It follows from these definitions that, if all local minima for $G(\cdot)$ can be found, then the global minimum of $G(\cdot)$ (if it exists) must be among them.

Constraints

The constraints (3.1-2) are in a form designed to handle a wide variety of applications. For example, suppose some component u_i of the control vector $\mathbf{u}$ has upper and lower bounds imposed on it, such as

$$a \leq u_i \leq b.$$

These constraints can be written in the form of (3.1-2) as

$$h_1(\mathbf{u}) = b - u_i \geq 0$$
$$h_2(\mathbf{u}) = u_i - a \geq 0.$$

Often an optimization problem may include equality constraints on the choice of $\mathbf{u}$, of the form

$$\boldsymbol{\psi}(\mathbf{u}) = \mathbf{0}.$$

Constraints of this type may be included directly in a more general theoretical presentation [Vincent and Grantham, 1981], or they can be included in the class of constraints given by (3.1-2) by introducing additional inequalities. For example, a scalar equality constraint can be written equivalently as two inequality constraints:

$$h_1(\mathbf{u}) = \psi(\mathbf{u}) \geq 0 \qquad h_2(\mathbf{u}) = -\psi(\mathbf{u}) \geq 0.$$

Another method for handling equality constraints is to include them in the cost function through the use of a penalty function. The penalty function approach is discussed in some detail in Section 3.5.

Parameter Optimization Problem

The function minimization problem given by (3.1-1) and (3.1-2) is called a **parameter optimization problem** (or a nonlinear programming problem, or a static optimization problem). The problem is to pick a vector $\mathbf{u} \in \mathcal{U}$ that minimizes the **cost function** $G(\mathbf{u})$. Later in this chapter we consider the simplest problem in the calculus of variations, where the problem is to pick a continuous piecewise differentiable function $\mathbf{q}(t)$ to minimize a cumulative cost integral. This problem is examined again in Chapter 7 in the context of **optimal control theory**, where the problem is to pick a feedback control function $\mathbf{u}(\mathbf{x})$ to minimize a cumulative cost integral $J[\mathbf{u}(\mathbf{x})]$. For the optimal control problem we present a very powerful result, known as Pontryagin's minimum principle, which reduces the problem of finding $\mathbf{u}(\mathbf{x})$ to a parameter optimization problem that must be solved at each point along a state space trajectory $\mathbf{x}(t)$.

3.2 TANGENT VECTORS TO A SET

To solve the minimization problem (3.1-1) subject to the constraints (3.1-2), we develop a set of first-order necessary conditions that a minimizing control $\mathbf{u}$ must satisfy. To do so we examine the effects of small perturbations in the control, of the form $\mathbf{u} + \Delta\mathbf{u}$, with the requirement that $\mathbf{u} + \Delta\mathbf{u}$ must always lie in the constraint set $\mathcal{U}$ during the limiting process as $\|\Delta\mathbf{u}\| \to 0$, where $\|\cdot\|$ denotes the Euclidean norm (length) of a vector.

To handle this constraint requirement mathematically, we need the notion of a **tangent vector** to a set, which is an extension of the familiar concept of a tangent vector to a curve. Let $\mathbf{u} \in \mathcal{U}$ be any point in the set $\mathcal{U}$. We say that a vector $\mathbf{e} \in R^{n_u}$ is **tangent** to $\mathcal{U}$ at $\mathbf{u}$ if and only if a continuous vector-valued function $\delta\mathbf{u}(\alpha)$ exists such that for all scalar values α in some interval $(0, \beta)$ with $\beta > 0$, the following conditions hold:

(i) $\mathbf{u} + \alpha\,\delta\mathbf{u}(\alpha) \in \mathcal{U}$;

(ii) $\delta\mathbf{u}(\alpha) \to \mathbf{e}$ as $\alpha \to 0$.

In terms of this notation we set the perturbation $\Delta\mathbf{u} = \alpha\,\delta\mathbf{u}(\alpha)$.

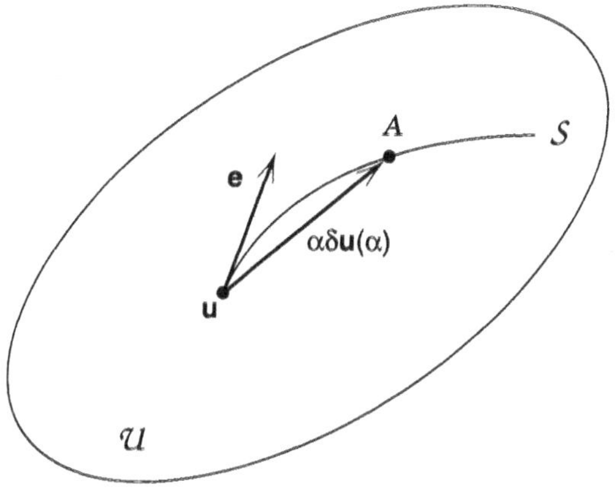

Figure 3.2-1 Tangent vector to a set.

Figure 3.2-1 illustrates how a tangent vector can be obtained from this definition. Let $\mathcal{S}$ be any smooth curve that lies in the set $\mathcal{U}$ and starts at the point $\mathbf{u} \in \mathcal{U}$. Let $\delta\mathbf{u}(\alpha)$ be a vector directed from $\mathbf{u}$ toward any moving point A on $\mathcal{S}$. To visualize the geometry it may help to think of $\delta\mathbf{u}(\alpha)$ as a unit vector, although this is not a requirement. The parameter α is used to adjust the direction of the vector $\delta\mathbf{u}(\alpha)$ and as a multiplier for $\alpha\,\delta\mathbf{u}(\alpha)$, so that the tip of the vector $\mathbf{u} + \alpha\,\delta\mathbf{u}(\alpha)$ always lies on the moving point A. As the point A is moved to $\mathbf{u}$ by decreasing α, the direction of the vector $\delta\mathbf{u}(\alpha)$ converges to the direction of a vector tangent to the curve $\mathcal{S}$ and the set $\mathcal{U}$. Many such tangent vectors exist for the set $\mathcal{U}$, corresponding to any family of smooth curves $\mathcal{S}$ in $\mathcal{U}$ that pass through $\mathbf{u}$ and to any nonnegative multiples of vectors tangent to the curves.

The set of all vectors tangent to $\mathcal{U}$ at $\mathbf{u}$ defines the **tangent set** $\mathcal{T}$ to $\mathcal{U}$ at a point $\mathbf{u} \in \mathcal{U}$. Figure 3.2-2 illustrates the tangent set at an interior point and at a boundary point of an example set $\mathcal{U}$. At the interior point of $\mathcal{U}$ every vector is a tangent vector. At the boundary point the tangent set is a cone of vectors.

To develop formulas that define the tangent set, let $\mathbf{e}$ be any tangent vector to $\mathcal{U}$ at a point $\mathbf{u} \in \mathcal{U}$. Then, from Taylor's theorem and the definition of a tangent vector, there is a corresponding function $\delta\mathbf{u}(\alpha)$ such that, for each inequality constraint function $h_i(\mathbf{u})$ evaluated at $\mathbf{u} + \alpha\,\delta\mathbf{u}(\alpha) \in \mathcal{U}$, we have

$$0 \le h_i[\mathbf{u} + \alpha\,\delta\mathbf{u}(\alpha)] = h_i(\mathbf{u}) + \frac{\partial h_i(\mathbf{u})}{\partial \mathbf{u}}\,\alpha\,\delta\mathbf{u}(\alpha) + R_i(\alpha) \tag{3.2-1}$$

for all sufficiently small $\alpha > 0$, where $\delta\mathbf{u}(\alpha) \to \mathbf{e}$ and $R_i(\alpha)/\alpha \to 0$ as $\alpha \to 0$. Note that, when an argument of a gradient is specified as in $\partial h_i(\mathbf{u})/\partial\mathbf{u}$, it means that the gradient is to be evaluated at the argument value. It has exactly the same meaning as the alternate notation $\partial h_i/\partial\mathbf{u}|_{\mathbf{u}}$.

We say that a constraint $h_i(\mathbf{u}) \ge 0$ is **active** at $\mathbf{u}$ if $h_i(\mathbf{u}) = 0$, and **inactive** if

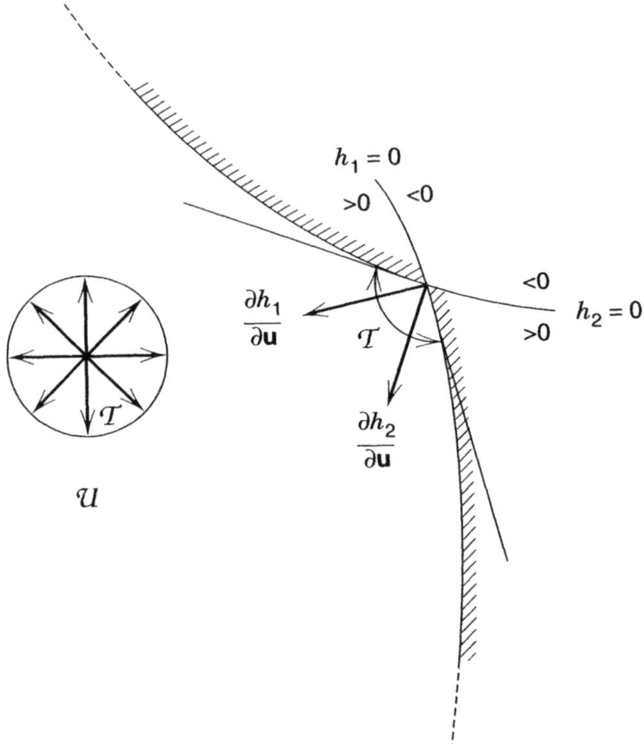

Figure 3.2-2 Tangent sets at interior points and boundary points of the set $\mathcal{U}$.

$h_i(\mathbf{u}) > 0$. For inactive constraints (3.2-1) is satisfied automatically since $h_i(\mathbf{u}) > 0$ dominates the other terms on the right-hand side of (3.2-1) for sufficiently small α. For active constraints, with $h_i(\mathbf{u}) = 0$, dividing the right-hand side of (3.2-1) by $\alpha > 0$ and taking the limit as $\alpha \to 0$ yields

$$\frac{\partial h_i(\mathbf{u})}{\partial \mathbf{u}} \mathbf{e} \geq 0 \tag{3.2-2}$$

for each active constraint, that is, for each i such that $h_i(\mathbf{u}) = 0$.

As an example, consider a point $\mathbf{u}$ in the boundary of $\mathcal{U}$ at which only one inequality constraint $h_i(\mathbf{u})$ is active. The equation $h_i(\mathbf{u}) = 0$ defines an $(n_u - 1)$-dimensional surface and the constraint gradient vector $\partial h_i(\mathbf{u})/\partial \mathbf{u}$ is perpendicular to this surface and points to the inside of the constraint region $\mathcal{U}$. The tangent set to the boundary of $\mathcal{U}$ consists of an $(n_u - 1)$-dimensional plane made up of all of the vectors $\mathbf{e}$ that are orthogonal to $\partial h_i(\mathbf{u})/\partial \mathbf{u}$. The tangent set to $\mathcal{U}$ itself consists of all of the vectors that satisfy (3.2-2). This tangent set is a "half space" in R^{n_u}, with the plane perpendicular to $\partial h_i(\mathbf{u})/\partial \mathbf{u}$ as its boundary and with $\partial h_i(\mathbf{u})/\partial \mathbf{u}$ pointing inside of $\mathcal{U}$.

Every vector **e** tangent to $\mathcal{U}$ at **u** must satisfy (3.2-2) for each constraint $h_i(\mathbf{u})$ that is active at **u**. This is a necessary condition for a tangent vector, but it is not a sufficient condition. In particular, there may be vectors that satisfy (3.2-2) but that are not tangent to $\mathcal{U}$. This can happen, for example, when **u** is a boundary point of $\mathcal{U}$ and the active constraints meet tangentially and form a **cusp** at **u**. This cusp geometry seldom occurs in real applications, however, and we now exclude it from further consideration.

We say that $\mathbf{u} \in \mathcal{U}$ is a **regular** point of $\mathcal{U}$ if the tangent set at **u** consists of exactly those vectors that satisfy (3.2-2) for each active inequality constraint $h_i(\mathbf{u}) = 0$. This condition is almost always satisfied in practical applications, and we assume that this condition holds for the remainder of this text. In parameter optimization texts this regularity assumption is known historically as the Kuhn–Tucker **constraint qualification** [Vincent and Grantham, 1981, p. 39]. A more stringent condition, which implies regularity, would be to assume that the gradient vectors $\partial h_i(\mathbf{u})/\partial \mathbf{u}$ for the active constraints are linearly independent of each other. This is very often the case, but the more general and less stringent regularity assumption is adequate for our purposes.

3.3 NECESSARY CONDITIONS FOR A LOCAL MINIMUM

Basic Necessary Condition for Tangent Vectors

For a control vector $\mathbf{u}^* \in \mathcal{U}$ to provide a local minimum for the cost function $G(\mathbf{u})$, we must have

$$G(\mathbf{u}^*) \le G(\mathbf{u}^* + \Delta\mathbf{u}) \tag{3.3-1}$$

for all arbitrarily small control perturbations $\Delta\mathbf{u}$ satisfying the constraints $\mathbf{u}^* + \Delta\mathbf{u} \in \mathcal{U}$.

In terms of first-order variations in **u**, the requirement in (3.3-1), that no allowable small variation $\Delta\mathbf{u}$ in **u** can decrease the value of $G(\mathbf{u})$, can be written as

$$0 \le \Delta G = \frac{\partial G(\mathbf{u}^*)}{\partial \mathbf{u}} \Delta\mathbf{u}.$$

To be precise about an "allowable $\Delta\mathbf{u}$," let **e** be any tangent vector to $\mathcal{U}$ at $\mathbf{u}^*$. Geometrically, it may help to think of **e** as the unit vector $\Delta\mathbf{u}/\|\Delta\mathbf{u}\|$, in the limit as $\|\Delta\mathbf{u}\| \to 0$. From the definition of a tangent vector to $\mathcal{U}$, there is a function $\delta\mathbf{u}(\alpha)$ that converges to **e** as $\alpha > 0$ approaches zero, while satisfying $\mathbf{u}^* + \alpha\,\delta\mathbf{u}(\alpha) \in \mathcal{U}$ for all sufficiently small $\alpha > 0$. Using $\Delta\mathbf{u} = \alpha\,\delta\mathbf{u}(\alpha)$ and applying Taylor's theorem to the right-hand side of (3.3-1) yields

$$G(\mathbf{u}^*) \le G(\mathbf{u}^*) + \frac{\partial G(\mathbf{u}^*)}{\partial \mathbf{u}} \alpha\, \delta\mathbf{u}(\alpha) + R(\alpha),$$

where $R(\alpha)/\alpha \to 0$ as $\alpha \to 0$. Canceling $G(\mathbf{u}^*)$, dividing by $\alpha > 0$, and then taking the limit as $\alpha \to 0$, we get

$$\frac{\partial G(\mathbf{u}^*)}{\partial \mathbf{u}} \mathbf{e} \ge 0 \tag{3.3-2}$$

for all $\mathbf{e}$ tangent to $\mathcal{U}$.

Condition (3.3-2) is the **basic necessary condition** for a local minimum of $G(\mathbf{u})$ at $\mathbf{u} = \mathbf{u}^*$. It is the first-order equivalent to the requirement that, at a minimizing point $\mathbf{u}^*$, no allowable small control perturbation $\Delta\mathbf{u}$ (in a direction $\mathbf{e}$) can decrease $G(\mathbf{u}^* + \Delta\mathbf{u})$ below $G(\mathbf{u}^*)$.

If $\mathbf{u}^*$ is an interior point of the control constraint set $\mathcal{U}$, then every vector $\mathbf{e} \in R^{n_u}$ is tangent to $\mathcal{U}$, as illustrated in Figure 3.2-2. In that case, both $\mathbf{e}$ and $-\mathbf{e}$ are tangent vectors and (3.3-2) reduces to the familiar necessary condition

$$\frac{\partial G(\mathbf{u}^*)}{\partial \mathbf{u}} = \mathbf{0}^{\mathrm{T}}. \tag{3.3-3}$$

Farkas' Lemma

In the general case where $\mathbf{u}^*$ may be either an interior point or a boundary point of $\mathcal{U}$, we assume that $\mathbf{u}^*$ is a regular point. Then the inequality condition in (3.3-2) must hold for all $\mathbf{e}$ satisfying (3.2-2). Another way of stating this is that for any vector $\Delta\mathbf{u}$ with $[\partial G(\mathbf{u}^*)/\partial\mathbf{u}]\Delta\mathbf{u} < 0$, which would decrease $G(\mathbf{u}^* + \Delta\mathbf{u})$, the point $\mathbf{u}^* + \Delta\mathbf{u}$ must violate one of the contraints. Specifically, from **Farkas' lemma** [Vincent and Grantham, 1981, p. 22], condition (3.3-2) requires that the cost gradient vector $\partial G(\mathbf{u}^*)/\partial\mathbf{u}$, which points in the direction of the steepest increase in $G(\mathbf{u})$, must lie in the **constraint cone** generated by the active constraint gradients $\partial h_i(\mathbf{u}^*)/\partial\mathbf{u}$, which point inside of the constraint region. That is, $\partial G(\mathbf{u}^*)/\partial\mathbf{u}$ must be a convex linear combination of the active constraint gradients $\partial h_i(\mathbf{u}^*)/\partial\mathbf{u}$ at $\mathbf{u}^*$:

$$\frac{\partial G(\mathbf{u}^*)}{\partial \mathbf{u}} = \sum \hat{\gamma}_i \frac{\partial h_i(\mathbf{u}^*)}{\partial \mathbf{u}}, \tag{3.3-4}$$

where the coefficients $\hat{\gamma}_i \ge 0$ are nonnegative and the sum in (3.3-4) is over the active constraints $h_i(\mathbf{u}^*) = 0$. If none of the constraints are active at $\mathbf{u}^*$, then the right-hand side of (3.3-4) is the zero vector and (3.3-4) reduces to the previous result in (3.3-3).

Figure 3.3-1a illustrates the geometry of condition (3.3-4). If $\partial G(\mathbf{u}^*)/\partial\mathbf{u}$ is

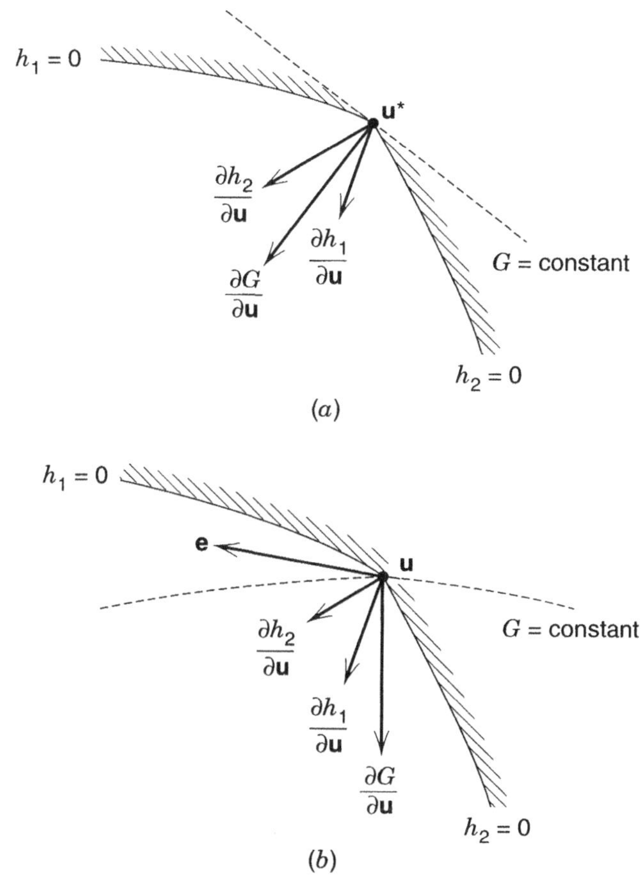

Figure 3.3-1 Cost and constraint gradient geometry for minimizing $G(\mathbf{u})$. (*a*) At a minimizing point $\mathbf{u}*$. (*b*) At a nonminimizing point $\mathbf{u}$.

not in the active constraint cone, as in Figure 3.3-1*b*, then there would exist an admissible change in **u** (in the direction of a tangent vector **e**) that would decrease $G(\mathbf{u})$, which contradicts the definition of a local minimum given in (3.3-1). For a complete proof see Vincent and Grantham 1981, p. 52].

General Necessary Conditions

The general **necessary conditions** for a local minimum, given by (3.3-4), can be written compactly as follows:

Minimization Necessary Conditions: *If* $\mathbf{u}^* \in \mathcal{U}$ *is a regular local minimizing point for* $G(\mathbf{u})$ *subject to the constraints (3.1-2), then there exists a vector* $\boldsymbol{\gamma} = [\gamma_1 \quad \cdots \quad \gamma_{n_h}]^{\mathrm{T}}$ *such that*

$$\mathbf{0}^{\mathrm{T}} = \frac{\partial L(\mathbf{u}^*, \boldsymbol{\gamma})}{\partial \mathbf{u}} \tag{3.3-5}$$

$$\mathbf{0} \leq \mathbf{h}(\mathbf{u}^*) \tag{3.3-6}$$

$$\mathbf{0} \leq \boldsymbol{\gamma} \tag{3.3-7}$$

$$0 = \boldsymbol{\gamma}^{\mathrm{T}}\mathbf{h}(\mathbf{u}^*), \tag{3.3-8}$$

where $\boldsymbol{\gamma}$ is called a Lagrange multiplier vector and the Lagrangian function $L(\cdot)$ in (3.3-5) is defined as

$$L(\mathbf{u}, \boldsymbol{\gamma}) \triangleq G(\mathbf{u}) - \boldsymbol{\gamma}^{\mathrm{T}}\mathbf{h}(\mathbf{u}). \tag{3.3-9}$$

These necessary conditions incorporate all of the inequality constraint functions, both active and inactive. Note that (3.3-6)–(3.3-8) imply that $\gamma_i = 0$ if $h_i(\mathbf{u}^*) > 0$, so that inactive constraints will not play a role.

As previously noted, an equality constraint, $\psi(\mathbf{u}) = 0$, can be handled using the above necessary conditions, by introducing the constraint as two inequality constraints, say as $h_i = \psi \geq 0$ and $h_{i+1} = -\psi \geq 0$. The L function would then contain the term $(\gamma_i - \gamma_{i+1})\psi$, in which the coefficient of ψ could be positive, negative, or zero. We can simply treat that coefficient as a single Lagrange multiplier whose sign is unrestricted. In other words, suppose that some components of $\mathbf{h}(\mathbf{u})$ are equality constraints, rather than inequality constraints. In forming the L function, we could still use just one γ_i for each constraint (rather than two for equality constraints) and then require that only the γ_i associated with inequality constraints must satisfy $\gamma_i \geq 0$ with $\gamma_i = 0$ if $h_i(\mathbf{u}^*) > 0$. The γ_i associated with equality constraints would have no sign restrictions imposed on them by the necessary conditions.

Example 3.3-1: A Tangency Condition In two dimensions, with $\mathbf{u} = [u_1 \quad u_2]^{\mathrm{T}}$, consider the problem of minimizing

$$G(\mathbf{u}) = u_1^2 + u_2^2$$

subject to

$$0 \leq h(\mathbf{u}) = u_1 + u_2 - 1.$$

The Lagrangian function is

$$L(\mathbf{u}, \gamma) = u_1^2 + u_2^2 - \gamma(u_1 + u_2 - 1)$$

and the necessary conditions (3.3-5)–(3.3-8) yield

$$0 = \frac{\partial L}{\partial u_1} = 2u_1 - \gamma \tag{3.3-10}$$

$$0 = \frac{\partial L}{\partial u_2} = 2u_2 - \gamma \tag{3.3-11}$$

$$0 \le u_1 + u_2 - 1 \tag{3.3-12}$$

$$0 \le \gamma \tag{3.3-13}$$

$$0 = \gamma(u_1 + u_2 - 1). \tag{3.3-14}$$

There are two possible cases, depending on whether the constraint (3.3-12) is active or not. Suppose $0 < u_1 + u_2 - 1$. Then (3.3-13) and (3.3-14) would imply $\gamma = 0$ and (3.3-10) and (3.3-11) would yield $u_1 = u_2 = 0$, which would not satisfy (3.3-12). Thus we conclude that the inequality constraint is active, with

$$0 = u_1 + u_2 - 1.$$

Furthermore, after eliminating γ from (3.3-10) and (3.3-11), we also have

$$0 = u_1 - u_2.$$

Solving these two equations yields

$$u_1^* = u_2^* = \tfrac{1}{2}, \tag{3.3-15}$$

with $\gamma = 1$ following from either (3.3-10) or (3.3-11). The corresponding minimum cost is $G(\mathbf{u}^*) = \frac{1}{2}$.

Figure 3.3-2 shows the geometry for this example, in which the geometric necessary condition (3.3-4), or the equivalent condition (3.3-5), reduces to a requirement that the curves of constant $G(\mathbf{u})$ must be tangent to the $h(\mathbf{u}) = 0$ boundary of the constraint region at the point $\mathbf{u}^*$. ■

Example 3.3-2: Minimum Integral-Square Error Consider again the undamped inverted pendulum of Example 2.2-1. The linearized equation of motion

$$\ddot{\theta} - \theta - u,$$

where θ is the angle of the pendulum measured from the vertical, has the following equivalent state space representation:

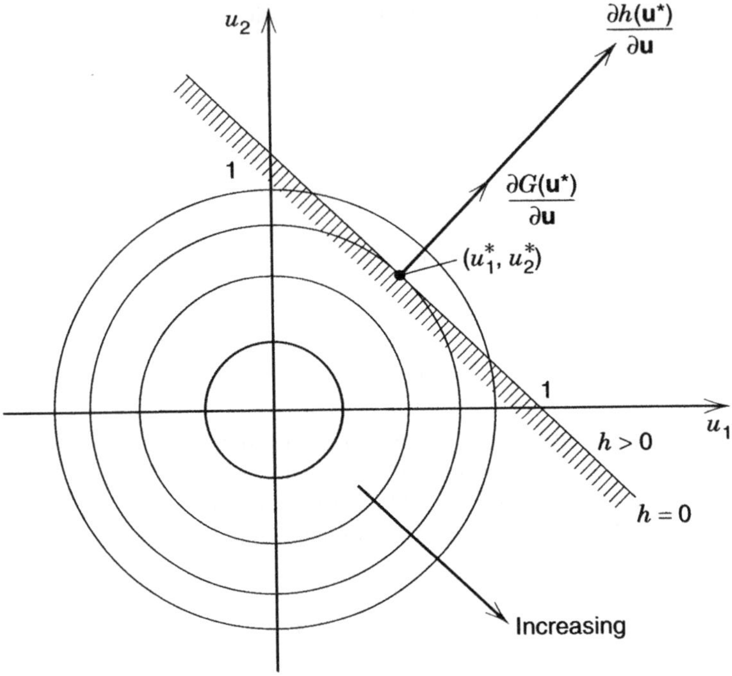

Figure 3.3-2 Geometry for Example 3.3-1.

$$\begin{bmatrix} \dot{x}_1 \\ \dot{x}_2 \end{bmatrix} = \begin{bmatrix} 0 & 1 \\ 1 & 0 \end{bmatrix} \begin{bmatrix} x_1 \\ x_2 \end{bmatrix} + \begin{bmatrix} 0 \\ 1 \end{bmatrix} u \tag{3.3-16}$$

where $\mathbf{x} = [\theta \quad \dot{\theta}]^T$.

The origin, $\mathbf{x} = \mathbf{0}$, is unstable and we wish to design a linear state-feedback controller

$$u = r - k_1 x_1 - k_2 x_2 = r - k_1 \theta - k_2 \dot{\theta} \tag{3.3-17}$$

so that a step command input $r(t) = \hat{\theta}$ yields an asymptotically stable equilibrium at $\theta(t) \to \hat{\theta}$ for the closed-loop system

$$\ddot{\theta} + 2\zeta\omega_n \dot{\theta} + \omega_n^2 \theta = \hat{\theta}, \tag{3.3-18}$$

where

$$\omega_n^2 = k_1 - 1$$
$$2\zeta\omega_n = k_2.$$

Also, we want the step response for (3.3-18) to be such that the **integral-square error**, defined by

$$J = \int_0^\infty [\hat{\theta} - \theta(t)]^2 dt,$$

for a trajectory starting at $\mathbf{x}(0) = \mathbf{0}$, is minimized. By integrating the equation of motion it can be shown that [Grantham and Vincent, 1993, p. 140]

$$J = \left[\frac{\hat{\theta}}{\omega_n}\right]^2 \frac{1}{\omega_n}\left[\zeta + \frac{1}{4\zeta}\right]. \qquad (3.3\text{-}19)$$

For $\theta(t) \to \hat{\theta}$ we need $\omega_n = 1$ and $\zeta > 0$. To minimize $J(\zeta)$ the necessary conditions (3.3-5)–(3.3-9) reduce to

$$0 = \frac{\partial J(\zeta)}{\partial \zeta} = \left[\frac{\hat{\theta}}{\omega_n}\right]^2 \frac{1}{\omega_n}\left[1 - \frac{1}{4\zeta^2}\right], \qquad (3.3\text{-}20)$$

which yields $\zeta = \frac{1}{2}$. These results produce the minimum integral-square error state-feedback controller

$$u = \hat{\theta} - 2x_1 - x_2 \qquad (3.3\text{-}21)$$

for the system (3.3-16).

The closed-loop system

$$\begin{bmatrix} \dot{x}_1 \\ \dot{x}_2 \end{bmatrix} = \begin{bmatrix} 0 & 1 \\ -1 & -1 \end{bmatrix} \begin{bmatrix} x_1 \\ x_2 \end{bmatrix} + \begin{bmatrix} 0 \\ 1 \end{bmatrix} \hat{\theta}$$

has a unique equilibrium point at $x_1 = \hat{\theta}$, $x_2 = 0$. The characteristic equation (1.5-13) for the closed-loop system is given by

$$0 = \mu^2 + \mu + 1,$$

which yields the eigenvalues

$$\mu = -\tfrac{1}{2} \pm i\,\frac{\sqrt{3}}{2}.$$

Therefore, the equilibrium is globally asymptotically stable, as desired. ■

Linear-Quadratic Parameter Optimization

In this section we present a preview of Lyapunov stability methods and use the results to simplify a very large class of parameter optimization problems similar to the linear-quadratic regulator problem discussed in Section 2.2. Consider a linear system

$$\dot{\mathbf{x}} = \mathbf{A}\mathbf{x}, \tag{3.3-22}$$

where the system matrix $\mathbf{A} = \mathbf{A}(\mathbf{u})$ depends on a vector of parameters $\mathbf{u}$ to be chosen to make the origin asymptotically stable and minimize a quadratic cost function

$$J = \int_0^\infty \mathbf{x}^{\mathrm{T}}\mathbf{Q}\mathbf{x}\,dt, \tag{3.3-23}$$

where $\mathbf{Q}$ is a given symmetric positive-definite (or, possibly, only positive-semidefinite) matrix (see Section 4.3).

As we see in Chapter 4, the origin for the linear system (3.3-22) will be asymptotically stable $[\mathbf{x}(t) \to \mathbf{0}]$ if and only if there exists a symmetric positive-definite matrix $\mathbf{P}$, with a corresponding **quadratic Lyapunov function**

$$V(\mathbf{x}) = \mathbf{x}^{\mathrm{T}}\mathbf{P}\mathbf{x}, \tag{3.3-24}$$

such that $dV/dt < 0$ (or ≤ 0, but not identically zero) along any nonzero trajectory $\mathbf{x}(t)$. Let us choose $\mathbf{P}$ such that

$$\frac{dV}{dt} = -\mathbf{x}^{\mathrm{T}}\mathbf{Q}\mathbf{x} \leq 0,$$

so that $V[\mathbf{x}(t)] \to 0$ as $t \to \infty$; hence $\mathbf{x}(t) \to \mathbf{0}$. Thus we require

$$\begin{aligned} -\mathbf{x}^{\mathrm{T}}\mathbf{Q}\mathbf{x} &= \dot{\mathbf{x}}^{\mathrm{T}}\mathbf{P}\mathbf{x} + \mathbf{x}^{\mathrm{T}}\mathbf{P}\dot{\mathbf{x}} \\ &= \mathbf{x}^{\mathrm{T}}\mathbf{A}^{\mathrm{T}}\mathbf{P}\mathbf{x} + \mathbf{x}^{\mathrm{T}}\mathbf{P}\mathbf{A}\mathbf{x} \\ &= \mathbf{x}^{\mathrm{T}}(\mathbf{A}^{\mathrm{T}}\mathbf{P} + \mathbf{P}\mathbf{A})\mathbf{x}. \end{aligned}$$

Hence $\mathbf{P}$ must satisfy the **matrix Lyapunov equation**

$$\mathbf{A}^{\mathrm{T}}\mathbf{P} + \mathbf{P}\mathbf{A} = -\mathbf{Q}. \tag{3.3-25}$$

Having determined $\mathbf{P}$ (as a function of the parameters $\mathbf{u}$ in $\mathbf{A}$) we can integrate (3.3-23), yielding

$$J = -\int_0^{\infty} \frac{dV}{dt}\, dt = -\mathbf{x}^{\mathrm{T}}\mathbf{P}\mathbf{x}\Big|_0^{\infty} = -\mathbf{x}^{\mathrm{T}}(\infty)\mathbf{P}\mathbf{x}(\infty) + \mathbf{x}^{\mathrm{T}}(0)\mathbf{P}\mathbf{x}(0).$$

Since the parameters $\mathbf{u}$ are chosen such that $\mathbf{x}(\infty) = \mathbf{0}$, we have

$$J = \mathbf{x}^{\mathrm{T}}(0)\mathbf{P}\mathbf{x}(0). \tag{3.3-26}$$

Thus the cost function J can be evaluated without having to directly integrate the equations of motion (3.3-22), and we can now choose $\mathbf{u}$ by minimizing J with respect to $\mathbf{u}$.

Example 3.3-3: Minimum Integral-Square Error Revisited Consider, as in Example 3.3-2, the problem of determining the damping ratio $\zeta > 0$ to minimize the integral-square error

$$J = \int_0^{\infty} [\hat{\theta} - \theta(t)]^2\, dt$$

for the step response of the system

$$\ddot{\theta} + 2\zeta\dot{\theta} + \theta = \hat{\theta}$$

with initial conditions $\theta(0) = \dot{\theta}(0) = 0$. In terms of the output error $e = \hat{\theta} - \theta$ and the state vector $\mathbf{x} = [e \quad \dot{e}]^{\mathrm{T}}$, the problem can be written in the form of (3.3-22) and (3.3-23) as follows: Minimize

$$J = \int_0^{\infty} e^2\, dt = \int_0^{\infty} [x_1 \quad x_2]^{\mathrm{T}} \begin{bmatrix} 1 & 0 \\ 0 & 0 \end{bmatrix} \begin{bmatrix} x_1 \\ x_2 \end{bmatrix} dt$$

for the system

$$\begin{bmatrix} \dot{x}_1 \\ \dot{x}_2 \end{bmatrix} = \begin{bmatrix} 0 & 1 \\ -1 & -2\zeta \end{bmatrix} \begin{bmatrix} x_1 \\ x_2 \end{bmatrix},$$

with initial conditions $\mathbf{x}(0) = [\hat{\theta} \quad 0]^{\mathrm{T}}$.

Instead of integrating the equations of motion for $\mathbf{x}(t)$ and then using the result to evaluate J directly, we achieve the same result indirectly, using the Lyapunov equation (3.3-25) for $\mathbf{P}$:

$$\begin{bmatrix} 0 & -1 \\ 1 & -2\zeta \end{bmatrix}\begin{bmatrix} p_{11} & p_{12} \\ p_{12} & p_{22} \end{bmatrix}+\begin{bmatrix} p_{11} & p_{12} \\ p_{12} & p_{22} \end{bmatrix}\begin{bmatrix} 0 & 1 \\ -1 & -2\zeta \end{bmatrix}=-\begin{bmatrix} 1 & 0 \\ 0 & 0 \end{bmatrix},$$

which yields

$$\begin{aligned} -2p_{12} &= -1 \\ p_{11} - p_{22} - 2\zeta p_{12} &= 0 \\ -4\zeta p_{22} + 2p_{12} &= 0. \end{aligned}$$

Thus

$$\mathbf{P} = \begin{bmatrix} \zeta + \dfrac{1}{4\zeta} & \dfrac{1}{2} \\ \dfrac{1}{2} & \dfrac{1}{4\zeta} \end{bmatrix}$$

and, from (3.3-26),

$$J = \mathbf{x}^T(0)\mathbf{P}\mathbf{x}(0) = [\hat{\theta} \quad 0]\begin{bmatrix} \zeta + \dfrac{1}{4\zeta} & \dfrac{1}{2} \\ \dfrac{1}{2} & \dfrac{1}{4\zeta} \end{bmatrix}\begin{bmatrix} \hat{\theta} \\ 0 \end{bmatrix} = \hat{\theta}^2\left(\zeta + \frac{1}{4\zeta}\right).$$

This is the same result as (3.3-19) in Example 3.3-2. Setting $dJ/d\zeta = 0$ again yields the optimal damping ratio $\zeta = \frac{1}{2}$. ■

3.4 MIN-MAX PARAMETRIC GAMES

We also have occasion to examine optimization problems involving cost functions that depend on the control choices of two independent agencies, with each agent having a different objective. Such optimization problems are called **games** and the separate control agents are called **players**.

Two-Player Zero-Sum Games

We consider a particular class of games in which there are two players, one choosing $\mathbf{u} \in \mathcal{U}$ and the other choosing $\mathbf{v} \in \mathcal{V}$. The $\mathbf{u}$ player wishes to minimize a cost function $G(\mathbf{u}, \mathbf{v})$, while the $\mathbf{v}$ player has the opposite objective of maximizing the **u** player's cost. Such a game is called a **two-player zero-sum game** because what one player wins the other player loses.

To solve the game we must find controls $\mathbf{u}^* \in \mathcal{U}$ and $\mathbf{v}^* \in \mathcal{V}$ that yield

$$\min_{\mathbf{u}\in\mathcal{U}} \max_{\mathbf{v}\in\mathcal{V}} G(\mathbf{u},\mathbf{v}).$$

More precisely, $\mathbf{u}^*$ and $\mathbf{v}^*$ taken together must satisfy what is known as a game-theoretic **saddle-point** condition [Isaacs, 1965]

$$G(\mathbf{u}^*,\mathbf{v}) \le G(\mathbf{u}^*,\mathbf{v}^*) \le G(\mathbf{u},\mathbf{v}^*) \qquad (3.4\text{-}1)$$

for all $\mathbf{u} \in \mathcal{U}$ and $\mathbf{v} \in \mathcal{V}$. That is,

$$G(\mathbf{u}^*,\mathbf{v}^*) = \min_{\mathbf{u}\in\mathcal{U}} \max_{\mathbf{v}\in\mathcal{V}} G(\mathbf{u},\mathbf{v}) = \max_{\mathbf{v}\in\mathcal{V}} \min_{\mathbf{u}\in\mathcal{U}} G(\mathbf{u},\mathbf{v}). \qquad (3.4\text{-}2)$$

Necessary conditions for solving this problem are available from **parametric min-max game** theory [Vincent and Grantham, 1981].

Before we present the necessary conditions, consider the question of whether a solution for the game saddle-point condition even exists. With respect to the saddle condition as stated in (3.4-1), it should be noted that when u and v are scalars, a game saddle is a special case of a mathematical saddle that is aligned with the u- and v-axes. In general a mathematical saddle point has a local maximum along one direction and a local minimum along another direction. As we see when we reach Example 3.4-1, it is not enough that u^* and v^* generate a mathematical saddle; they must produce a game saddle, in which the minimum occurs along the u-direction and the maximum occurs along the v-direction. In other words, if a mathematical saddle exists for the function $G(u,v)$, but the axes of the saddle do not align with the u and v axes, then the game saddle-point condition may not have a solution.

While the game saddle condition as stated in (3.4-2) requires that the max-min and the min-max of G be equal, this need not be true in general. Indeed it is not difficult to prove that for a function $G(\mathbf{u},\mathbf{v})$

$$\min_{\mathbf{u}\in\mathcal{U}} \{\max_{\mathbf{v}\in\mathcal{V}} G(\mathbf{u},\mathbf{v})\} \ge \max_{\mathbf{v}\in\mathcal{V}} \{\min_{\mathbf{u}\in\mathcal{U}} G(\mathbf{u},\mathbf{v})\}. \qquad (3.4\text{-}3)$$

Fortunately, in many applications the cost function is **separable** in **u** and **v**, of the form

$$G(\mathbf{u},\mathbf{v}) = \mathbf{g}(\mathbf{u}) + \bar{\mathbf{g}}(\mathbf{v}),$$

with neither of the constraint sets $\mathcal{U}$ and $\mathcal{V}$ depending on the other player's control. For such separable systems **Weierstrass' theorem** [Vincent and Grantham,

1981, p. 17] ensures that the game saddle-point condition will have a solution if:

(i) The function $G(\cdot)$ is continuous in **u** and **v**, as we have assumed,

(ii) It is separable in **u** and **v**, and

(iii) The control constraint sets $\mathcal{U}$ and $\mathcal{V}$ are compact (closed and bounded) and decoupled.

These conditions are sufficient but not mandatory for a game saddle-point solution to exist.

Henceforth we simply assume that a game saddle-point solution exists, in order to develop necessary conditions for finding such a solution. Furthermore, to simplify the notation, we refer to the game saddle-point condition simply as a min-max condition.

Parametric Min-Max Necessary Conditions

As in the parametric optimization case in Section 3.3, in order to apply parametric min-max game theory the sets $\mathcal{U}$ and $\mathcal{V}$ must be defined in terms of algebraic relations. In general, $\mathcal{U}$ and $\mathcal{V}$ may be functions of the current state **x** of a control system, as well as the controls **u** and **v**. However, if $\mathcal{U}$ and $\mathcal{V}$ are just fixed, constant, decoupled sets, then considerable simplification of the necessary conditions occurs. Henceforth we assume that $\mathcal{U}$ and $\mathcal{V}$ are defined in terms of inequality constraints of the form

$$\begin{aligned} \mathcal{U} &= \{\mathbf{u} \in E^{n_u} | \mathbf{h}(\mathbf{u}) \geq \mathbf{0}\} \\ \mathcal{V} &= \{\mathbf{v} \in E^{n_v} | \bar{\mathbf{h}}(\mathbf{v}) \geq \mathbf{0}\}, \end{aligned} \tag{3.4-4}$$

where $\mathbf{h} = [h_1 \ \cdots \ h_{n_h}]^T$ is an n_h-dimensional vector function of the control vector **u** and $\bar{\mathbf{h}} = [\bar{h}_1 \ \cdots \ \bar{h}_{n_{\bar{h}}}]^T$ is an $n_{\bar{h}}$-dimensional vector function of the control vector **v**. Every component of the **h** and $\bar{\mathbf{h}}$ functions is assumed to be continuous in its arguments and to have continuous partial derivatives as well. The functions are further restricted to that class that will produce **regular** sets for $\mathcal{U}$ and $\mathcal{V}$, as defined in Section 3.2.

Under these assumptions we have the following **min-max necessary conditions** [Vincent and Grantham, 1981]:

Min-Max Necessary Conditions: *Let* $\mathbf{u}^*$ *be a regular point of* $\mathcal{U}$ *and let* $\mathbf{v}^*$ *be a regular point of* $\mathcal{V}$. *If* $G(\mathbf{u}, \mathbf{v})$ *takes on a min-max with respect to* **u** *at* $\mathbf{u}^*$ *and* **v** *at* $\mathbf{v}^*$, *then Lagrange multiplier vectors* $\boldsymbol{\gamma} = [\gamma_1 \ \cdots \ \gamma_{n_h}]^T$ *and* $\bar{\boldsymbol{\gamma}} = [\bar{\gamma}_1 \ \cdots \ \bar{\gamma}_{n_{\bar{h}}}]^T$ *exist such that*

$$\mathbf{0}^{\mathrm{T}} = \frac{\partial L}{\partial \mathbf{u}}$$
$$\mathbf{0}^{\mathrm{T}} = \frac{\partial L}{\partial \mathbf{v}} \tag{3.4.5}$$

$$\mathbf{0} \le \mathbf{h}(\mathbf{u}^*)$$
$$\mathbf{0} \le \overline{\mathbf{h}}(\mathbf{v}^*) \tag{3.4.6}$$

$$0 = \boldsymbol{\gamma}^{\mathrm{T}}\mathbf{h}(\mathbf{u}^*)$$
$$0 = \boldsymbol{\gamma}^{\mathrm{T}}\overline{\mathbf{h}}(\mathbf{v}^*) \tag{3.4.7}$$

$$\mathbf{0} \le \boldsymbol{\gamma}$$
$$\mathbf{0} \ge \overline{\boldsymbol{\gamma}}, \tag{3.4.8}$$

where

$$L(\mathbf{u}, \mathbf{v}, \boldsymbol{\gamma}, \overline{\boldsymbol{\gamma}}) \triangleq G(\mathbf{u}, \mathbf{v}) - \boldsymbol{\gamma}^{\mathrm{T}}\mathbf{h}(\mathbf{u}) - \overline{\boldsymbol{\gamma}}^{\mathrm{T}}\overline{\mathbf{h}}(\mathbf{v}) \tag{3.4-9}$$

and the partial derivatives in (3.4-5) are evaluated at $\mathbf{u} = \mathbf{u}^*$, $\mathbf{v} = \mathbf{v}^*$.

Example 3.4-1: Game-Theoretic Saddle Point Consider a two-player zero-sum game with cost function

$$G(u, v) = u^2 - v^2$$

and with no constraints on the scalar controls u and v. For this simple problem the min-max solution $u^* = v^* = 0$ can be found by inspection. This solution does satisfy the necessary conditions (3.4-5)–(3.4-9), which for $L = G$ reduce to $0 = \partial G/\partial u = \partial G/\partial v$. Figure 3.4-1$a$ shows a three-dimensional plot of G and Figure 3.4-1b is a plot of lines of constant $G(u, v)$.

Now suppose we introduce a change of coordinates given by

$$\begin{bmatrix} u \\ v \end{bmatrix} = \begin{bmatrix} \cos\theta & -\sin\theta \\ \sin\theta & \cos\theta \end{bmatrix} \begin{bmatrix} u' \\ v' \end{bmatrix},$$

which represents a rotation of the (u, v) coordinate axes clockwise by an angle θ. The geometry is illustrated in Figure 3.4-1b for $\theta = \pi/3$, yielding

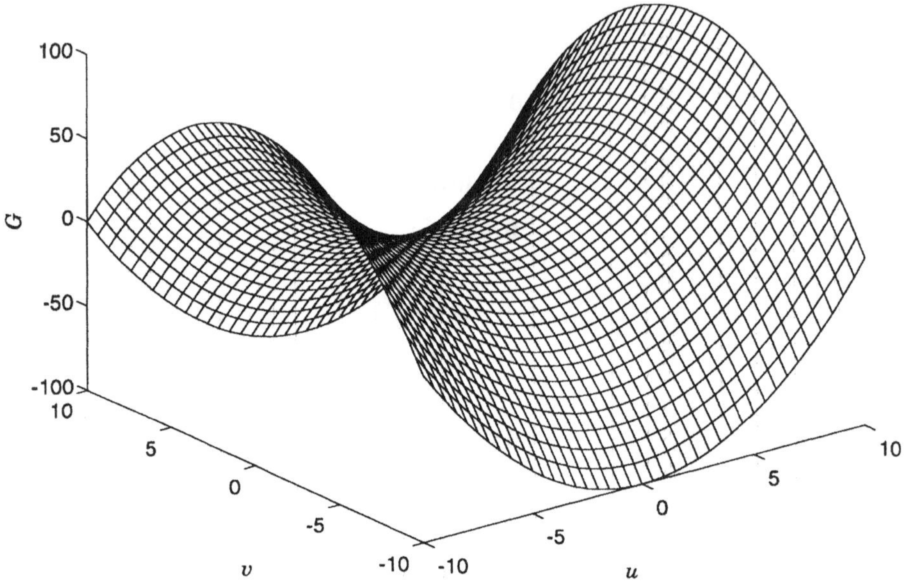

(a)

Figure 3.4-1 Geometry of Example 3.4-1. (*a*) Saddle-point surface. (*b*) $G(u, v)$ contours.

$$u = \tfrac{1}{2}(u' - \sqrt{3}\,v')$$
$$v = \tfrac{1}{2}(\sqrt{3}\,u' + v').$$

If we now take u' and v' as the control variables, the cost function becomes

$$\begin{aligned} G'(u', v') \triangleq G(u, v) &= \tfrac{1}{4}[(u' - \sqrt{3}\,v')^2 - (\sqrt{3}\,u' + v')^2] \\ &= \tfrac{1}{2}[-u'^2 - 2\sqrt{3}\,u'v' + v'^2]. \end{aligned}$$

The necessary conditions

$$0 = \frac{\partial G'}{\partial u'} = -u' - \sqrt{3}v'$$
$$0 = \frac{\partial G'}{\partial v'} = -\sqrt{3}u' + v'$$

again yield the origin, $u' = v' = 0$. However, while this point is a mathematical

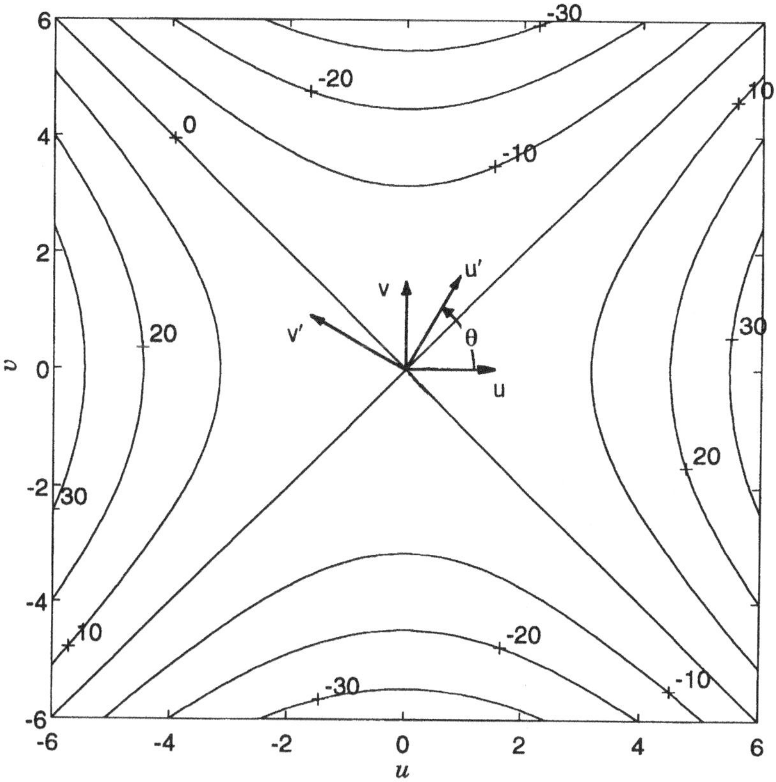

(*b*)

Figure 3.4-1 (*Continued*)

saddle point for $G'(u', v')$, it is not a game-theoretic saddle point, despite the fact that it satisfies the first-order necessary conditions for a game-theoretic saddle point. In fact, *no game saddle point exists* for $G'(u', v')$. To see this, suppose the minimizing player chooses $u' = 0$. By inspection of $G'(u', v')$, the maximizing player will pick $|v'|$ as large as possible and certainly would not pick $v' = 0$. A similar argument holds for the minimizing player's choice of u' given $v' = 0$. ■

3.5 NUMERICAL OPTIMIZATION METHODS

There are a number of numerical methods that have been used to solve parameter optimization problems. Some of the better known methods include: the gradient method [Hostetter et al., 1991], Newton's method [Hostetter et al., 1991], variable metric methods [Press et al., 1986], and sequential quadratic programming [Gill et al., 1985]. The gradient method relies on the availability of a good one-dimensional search algorithm, such as golden section [Press et al., 1986], to make a number of one-dimensional searches to minimize $G(\mathbf{u})$ in the direction of $-\partial G/\partial \mathbf{u}$. At the end of each search the gradient is recalculated and a new one-dimensional search is made. This process is continued until a convergence criterion is satisfied. The advantage of the method is that, at each iteration, the cost must decrease. Its disadvantage is that the rate of convergence may be slow. Newton's method approximates $G(\mathbf{u})$ by a quadratic function, finds $\mathbf{u}$ to minimize the quadratic, and repeats the process. Variable metric methods are similar to Newton's method except that the Hessian $\mathbf{H}(\mathbf{u}) = [\partial^2 G/\partial \mathbf{u}^2]^{-1}$ is not evaluated directly. Instead rank one updates are used, based on $\partial G(\mathbf{u})/\partial \mathbf{u}$ and $\partial G(\mathbf{u} + \Delta \mathbf{u})/\partial \mathbf{u}$. Sequential quadratic programming can be used with problems having constraints and works best in conjunction with Hestenes' method of multipliers for constrained problems. Powell's augmented cost function, $W = G(\mathbf{u}) + |\boldsymbol{\gamma}^T \mathbf{h}(\mathbf{u})|$, generally gives the best results [Baker, 1992]. The Lagrange multipliers are determined by solving a quadratic subproblem, in which the cost function is approximated by a quadratic and is subjected to linearized constraints. Numerically stable implementations employ an efficient Cholesky decomposition of the Hessian, to ensure positive definiteness. Many of these methods have been programmed in commercial software such as MATLAB [Grace, 1992].

Sequential Quadratic Programming

For reference, we present a basic version of the sequential quadratic programming algorithm for unconstrained optimization. We are concerned with the problem of

$$\min_{\mathbf{u}} G(\mathbf{u})$$

where $\mathbf{u} = [u_1 \quad \cdots \quad u_{n_u}]^T$. The basic approach is to approximate $G(\mathbf{u})$ by a quadratic, move toward the point that minimizes the quadratic approximation, and then repeat the process.

(a) Initialize: $\mathbf{u}_0$, $\mathbf{B}_0 = \mathbf{I}$. Compute $\partial G(\mathbf{u}_0)/\partial \mathbf{u}$.

(b) At $\mathbf{u} = \mathbf{u}_k$, $\mathbf{B} = \mathbf{B}_k$ (which hopefully converges to $\partial^2 G/\partial \mathbf{u}^2 > 0$), solve the quadratic subproblem for a search direction $\mathbf{s}$ (corresponding to a $\Delta \mathbf{u}$):

$$\min_{\mathbf{s}} \Phi(\mathbf{s}),$$

where

$$\Phi(\mathbf{s}) = \tfrac{1}{2}\mathbf{s}^{\mathrm{T}}\mathbf{B}\mathbf{s} + \frac{\partial G(\mathbf{u})}{\partial \mathbf{u}}\,\mathbf{s}$$

$$= \tfrac{1}{2}[s_1 \quad \cdots \quad s_{n_u}]\begin{bmatrix} B_{11} & \cdots & B_{1n_u} \\ \vdots & \ddots & \vdots \\ B_{1n_u} & \cdots & B_{n_u n_u} \end{bmatrix}\begin{bmatrix} s_1 \\ \vdots \\ s_{n_u} \end{bmatrix}$$

$$+ \left[\frac{\partial G}{\partial u_1} \quad \cdots \quad \frac{\partial G}{\partial u_{n_u}}\right]\begin{bmatrix} s_1 \\ \vdots \\ s_{n_u} \end{bmatrix}.$$

(c) Solve for α^* to yield a "sufficient decrease" in $G(\mathbf{u} + \alpha\mathbf{s})$.

(d) $\mathbf{u}_{k+1} = \mathbf{u}_k + \alpha^*\mathbf{s}$.

(e) Determine $\partial G(\mathbf{u}_{k+1})/\partial \mathbf{u}$.

(f) Check stopping conditions:

$$\left\| \frac{\partial G(\mathbf{u}_{k+1})}{\partial \mathbf{u}} \right\| \le \epsilon_1$$

or

$$\|\mathbf{u}_{k+1} - \mathbf{u}_k\| \le \epsilon_2$$

or

$$k \ge k_{\max}.$$

(g) Update $\mathbf{B}_k \rightarrow \mathbf{B}_{k+1}$.

(h) $k \rightarrow k + 1$. Go to (b).

Gradient Computation Each element of $\partial G(\mathbf{u})/\partial \mathbf{u}$ can be approximated by forward differences:

$$\frac{\partial G(\mathbf{u})}{\partial u_i} \approx \frac{G(\mathbf{u} + \Delta\mathbf{u}_i) - G(\mathbf{u})}{h}$$

where

$$\Delta \mathbf{u}_i = \begin{bmatrix} 0 \\ \vdots \\ h \\ \vdots \\ 0 \end{bmatrix}$$

is zero except at the ith position, and h is a small step, for example, $h \approx 10^{-7}$, 10^{-8}.

Quadratic Subproblem Using Gauss elimination, for example, find a point $\mathbf{s}$ satisfying $\partial \Phi(\mathbf{s})/\partial \mathbf{s} = \mathbf{0}^{\mathrm{T}}$, that is,

$$\mathbf{Bs} = -\nabla G \triangleq -\left[\frac{\partial G(\mathbf{u})}{\partial \mathbf{u}}\right]^{\mathrm{T}}.$$

Sufficient Decrease

(a) For $\Psi(\alpha) = G(\mathbf{u} + \alpha \mathbf{s})$ set

$$\Psi(0) = G(\mathbf{u})$$

$$\Psi'(0) = \frac{\partial G(\mathbf{u})}{\partial \mathbf{u}} \mathbf{s} = \begin{bmatrix} \dfrac{\partial G}{\partial u_1} & \cdots & \dfrac{\partial G}{\partial u_{n_u}} \end{bmatrix} \begin{bmatrix} s_1 \\ \vdots \\ s_{n_u} \end{bmatrix}$$

$$\alpha_1 = 1$$

$$\gamma = 0.1.$$

(b) Compute $\Psi(\alpha_i)$. If

$$\Psi(\alpha_i) \le \Psi(0) + \gamma \Psi'(0)\alpha_i,$$

then exit ($\alpha^* = \alpha_i$). Otherwise go to (c).

(c) Determine α_{i+1} to minimize the (two point + slope) quadratic approximation to $\Psi(\alpha)$:

$$\psi(\alpha) = \tfrac{1}{2} C_2 \alpha^2 + C_1 \alpha + C_0,$$

where

$$C_0 = \Psi(0), \qquad C_1 = \Psi'(0), \qquad C_2 = 2\,\frac{\Psi(\alpha_i) - \Psi(0) - \Psi'(0)\alpha_i}{\alpha_i^2}.$$

Then $\psi'(\alpha) = 0$ yields

$$\alpha_{i+1} = \frac{-\frac{1}{2}\Psi'(0)\alpha_i^2}{\Psi(\alpha_i) - \Psi(0) - \Psi'(0)\alpha_i}.$$

$i \rightarrow i + 1$. Go to (b)

Update B The Broyden–Fletcher–Goldfarb–Shanno (BFGS) update algorithm [Gill et al., 1985] is given by

$$\mathbf{B}_{k+1} = \begin{cases} \mathbf{B} - \dfrac{1}{\beta}\,(\mathbf{Bp})(\mathbf{Bp})^{\mathrm{T}} + \dfrac{1}{\sigma}\,\mathbf{x}\mathbf{x}^{\mathrm{T}} & \text{if } \sigma > 0 \\ \mathbf{B} & \text{if } \sigma \leq 0, \end{cases}$$

where

$$\begin{aligned} \mathbf{p} &= \mathbf{u}_{k+1} - \mathbf{u}_k \\ \mathbf{x} &= \nabla G(\mathbf{u}_{k+1}) - \nabla G(\mathbf{u}_k) \\ \beta &= \mathbf{p}^{\mathrm{T}}\mathbf{B}\mathbf{p} \\ \sigma &= \mathbf{p}^{\mathrm{T}}\mathbf{x}. \end{aligned}$$

Trajectory Following

All of the numerical methods previously discussed require iteratively solving discrete algebraic equations. We present here a method [Kelley, 1962; Arrow and Hurwicz, 1977] based on solving continuous differential equations, whose equilibrium solutions satisfy the necessary conditions for a minimum, maximum, or min-max, depending on the problem to be solved. This method has the advantage that no special programming skill is required to generate solutions and the differential equation approach fits much more naturally with the other themes of this book. The method is "trajectory following" in the sense that we move from an initial guess for the solution along a trajectory generated by a set of differential equations to a solution point. It is also quite robust, in that many problems, well known to be difficult to solve by other methods, can be easily solved. The same assumptions on G and $\mathcal{U}$ will be made as in Section 3.1.

Minimizing G Subject to No Constraints The basic idea behind the trajectory following method is to form a set of differential equations from the gradient of the cost function. Consider first the problem of minimizing $G(\mathbf{u})$ subject to no constraints. Suppose that we use the local minimal point $\mathbf{u}^*$ as an initial condition for integrating the differential equations,

$$\dot{\mathbf{u}} = \mathbf{f}(\mathbf{u}), \tag{3.5-1}$$

where $\mathbf{f}(\cdot)$ is a function at our disposal, to be determined shortly. Suppose also that we calculate $G(\mathbf{u})$ along the trajectory generated by the solution to (3.5-1). Since $\mathbf{u}^*$ is a local minimal point, it follows from our basic necessary condition for minimizing $G(\mathbf{u})$ as given by (3.3-2) and the fact that $\mathbf{f}(\mathbf{u}^*)$ is a tangent vector to the trajectory generated by (3.5-1) that

$$\left.\frac{dG}{dt}\right|_{\mathbf{u}^*} = \left.\frac{\partial G}{\partial \mathbf{u}}\right|_{\mathbf{u}^*} \mathbf{f}(\mathbf{u}^*) \geq 0. \tag{3.5-2}$$

That is, as we move away from a minimal point, along any feasible trajectory in a neighborhood of the minimal point, the time derivative of the cost must be nonnegative. Since we are interested in a trajectory that will search for a minimum, the above observation suggests that we integrate (3.5-1) by choosing

$$\mathbf{f}(\mathbf{u}) = -\left[\frac{\partial G}{\partial \mathbf{u}}\right]^{\mathrm{T}}.$$

To see why this is a good choice, suppose that $G(\cdot)$ has a proper local global minimum at $\mathbf{u}^*$. This means that $G(\mathbf{u}) > G(\mathbf{u}^*)$ for all $\mathbf{u} \neq \mathbf{u}^*$ in some open neighborhood containing $\mathbf{u}^*$. Along a trajectory generated by the **unconstrained trajectory following equations**

$$\dot{\mathbf{u}} = -\left[\frac{\partial G}{\partial \mathbf{u}}\right]^{\mathrm{T}}, \tag{3.5-3}$$

Starting from any point in the open neighborhood, we have

$$\frac{dG}{dt} = -\frac{\partial G}{\partial \mathbf{u}}\left[\frac{\partial G}{\partial \mathbf{u}}\right]^{\mathrm{T}} < 0$$

for all points along the trajectory except at $\mathbf{u} = \mathbf{u}^*$, where $dG/dt = 0$. Thus, in order to solve the problem of minimizing $G(\cdot)$ in an unconstrained control space, one need only choose an appropriate initial condition for $\mathbf{u}$ and integrate (3.5-3) until the equilibrium solution is approached. Unfortunately, since we only have asymptotic stability, the equilibrium point will never be obtained in finite time and we must stop the integration process when some accuracy criterion has been satisfied. We will not get an exact solution, but we can come as close to an exact solution as desired by specifying that the integration be stopped only when, for example, the norm of the gradient is less than a prespecified amount. Note that in using (3.5-3) to search for a minimum it is usually more convenient to calculate the gradient using a numerical method, such as central difference, then to use an analytical gradient.

While the trajectory following method has some similarities with the gradient method mentioned previously, it is not the same, since the gradient vector under trajectory following is continuously updated. Because of this, we avoid the convergence problems associated with the numerical gradient method. The numerical gradient method has difficulties converging for certain problems, which can best be described in a two-dimensional control space. Suppose that G plotted versus $\mathbf{u}$ forms a "canyon" with steep sides and a relatively flat bottom that curves a lot. Anyone who has hiked Arizona's Paria Canyon would have a good mental image of this geometry. Using trajectory following with such a G function results in a **stiff system** of differential equations. There is a fast time scale associated with integrating down the sides of the canyon and a slow time scale associated with integrating along the bottom of the canyon. In order to solve such a problem, using trajectory following, the integration routine must be able to handle stiff systems.

Example 3.5-1: Rosenbrock's Banana Function One of the more famous examples that has a canyon-like G function described above is Rosenbrock's banana function [Grace, 1992], given by

$$G = 100(u_1^2 - u_2)^2 + (1 - u_1)^2.$$

This function has lines of constant G as illustrated in Figure 3.5-1. Using trajectory following, starting at various initial points so that we can get an idea of the flow in the neighborhood of the minimal point, we obtain the trajectories illustrated in Figure 3.5-2. A Gear integration routine using a numerical central difference method for calculating the gradient was used to generate this plot. Note that the algorithm first moves the system down the sides of the canyon and then moves the system along the bottom of the canyon to the minimal point. All trajectories ultimately converge to the minimal point given by $u_1 = u_2 = 1$. This problem is much more difficult to solve using traditional numerical methods if the constant 100 is increased in value. Considerably larger values of this constant do not present difficulties using the trajectory following method. See Exercise 3-4. ■

Minimizing G Subject to Simple Inequality Constraints Consider now the problem of minimizing $G(\mathbf{u})$ subject to **simple inequality constraints**, of the

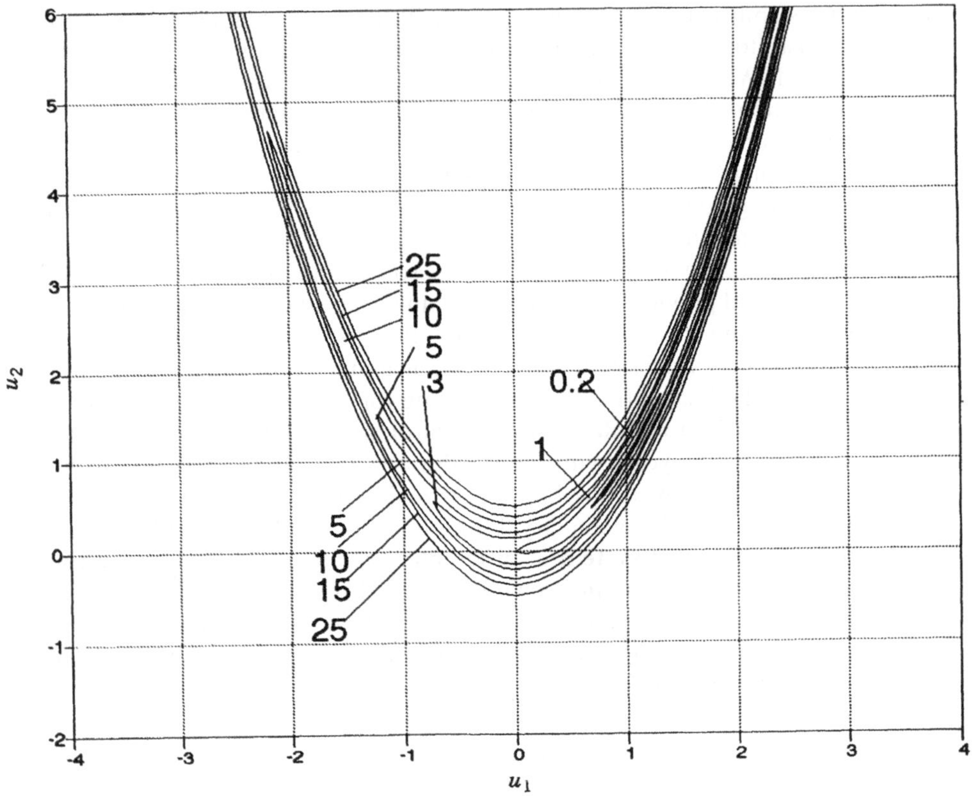

Figure 3.5-1 Lines of constant cost for the banana function.

form

$$\mathbf{u}_{\min} \leq \mathbf{u} \leq \mathbf{u}_{\max},$$

where every component of the vector $\mathbf{u}_{\min}$ is less than every component of the vector $\mathbf{u}_{\max}$. One advantage of introducing simple inequality constraints, even for a problem that can be formulated without any constraints, is that we are now guaranteed that a global minimum (and a global maximum) will exist. The necessary conditions for a minimum reduce to a straightforward set of equations for this case. We then use this result to formulate a trajectory following algorithm.

From (3.3-9)

$$L = G(\mathbf{u}) - \boldsymbol{\gamma}_{\max}^{\mathrm{T}}(\mathbf{u}_{\max} - \mathbf{u}) - \boldsymbol{\gamma}_{\min}^{\mathrm{T}}(\mathbf{u} - \mathbf{u}_{\min}),$$

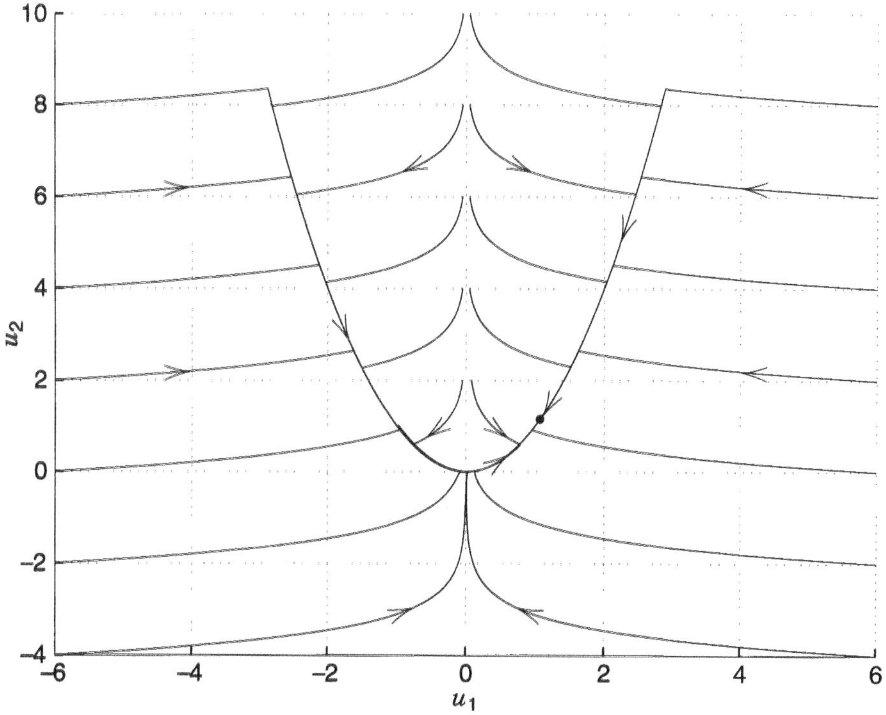

Figure 3.5-2 Trajectory to the minimal point resulting from using the unconstrained trajectory following equations.

so that the necessary conditions for $\mathbf{u}^*$ to be a minimum become

$$\mathbf{0} = \left.\frac{\partial G}{\partial \mathbf{u}}\right|_{\mathbf{u}=\mathbf{u}^*} + \boldsymbol{\gamma}_{\max}^{\mathrm{T}} - \boldsymbol{\gamma}_{\min}^{\mathrm{T}}$$

$$\mathbf{0} \le \mathbf{u}_{\max} - \mathbf{u}^*$$

$$\mathbf{0} \le \mathbf{u}^* - \mathbf{u}_{\min}$$

$$\mathbf{0} \le \boldsymbol{\gamma}_{\max}$$

$$\mathbf{0} \le \boldsymbol{\gamma}_{\min}$$

$$0 = \boldsymbol{\gamma}_{\max}^{\mathrm{T}}(\mathbf{u}_{\max} - \mathbf{u}^*) + \boldsymbol{\gamma}_{\min}^{\mathrm{T}}(\mathbf{u}^* - \mathbf{u}_{\min}).$$

It is easy to show that this set of necessary conditions is equivalent to the following conditions:

$$\begin{aligned}
&\text{if } u_i^* = u_{i_{\max}} && \text{then } \left.\frac{\partial G}{\partial u_i}\right|_{u_i = u_i^*} \le 0 \\
&\text{if } u_i^* = u_{i_{\min}} && \text{then } \left.\frac{\partial G}{\partial u_i}\right|_{u_i = u_i^*} \ge 0 \\
&\text{if } u_{i_{\min}} < u_i^* < u_{i_{\max}} && \text{then } \left.\frac{\partial G}{\partial u_i}\right|_{u_i = u_i^*} = 0.
\end{aligned} \tag{3.5-4}$$

Consider now the following trajectory following algorithm to be used for all $\mathbf{u} \in \mathcal{U}$:

$$\dot{u}_i = \begin{cases} 0 & \text{if } u_i = u_{i_{\max}} \text{ and } \dfrac{\partial G}{\partial u_i} \le 0 \\ 0 & \text{if } u_i = u_{i_{\min}} \text{ and } \dfrac{\partial G}{\partial u_i} \ge 0 \\ -\dfrac{\partial G}{\partial u_i} & \text{otherwise,} \end{cases} \tag{3.5-5}$$

where $i = 1, \ldots, n_u$. Note that any equilibrium solution generated by the trajectory following algorithm will satisfy (3.5-4) and hence will represent a local minimum candidate. This algorithm can handle initial conditions $\mathbf{u} \notin \mathcal{U}$, as well as any problems associated with integrating past a boundary if the integration routine includes a test for the inequality conditions being satisfied and setting u_i to the appropriate boundary value for any inequality not satisfied. In this way the first two conditions of (3.5-5) will be automatically satisfied.

It is well worth the extra coding effort to take care of situations where $\mathbf{u} \notin \mathcal{U}$ for any integration time step. If MATLAB is being used to solve these problems, this is easily done using the limiting integrator in SIMULINK. The advantage gained is that if a minimal solution is on the boundary of $\mathcal{U}$, then *exact* values can be obtained for all components of $\mathbf{u}$ that lie on the boundary of $\mathcal{U}$.

If it is known that G has several local minimal points, the above algorithm can be modified to search for all of them. In modifying the algorithm to search for multiple minimal points, it turns out that this process is facilitated by searching for all local minimal points and all maximal points at the same time. To start the process an initial point is chosen and (3.5-5) is integrated forward in time to find the first local minimal candidate. A new initial point is then randomly chosen in the neighborhood of the first local minimal candidate and the system (3.5-5) is then integrated backward in time to find the first local maximal candidate. A new initial point is randomly chosen in the neighborhood of this maximal point with forward integration again used. This process continues, using forward-backward integration, until some stopping criterion is satisfied. Clearly, there is no guarantee that this process will find all local minimal and maximal points, but it will find a lot of them, some points many times over. In fact, if after a long run, this process always finds the same minimal point and the same maximal point, one would have evidence that the minimal and maximal points are unique.

Example 3.5-2: Quadratic Cost A two-dimensional quadratic cost function subject to simple inequality constraints will either have one global minimum point with one or more local maximal points or one global maximal point with one or more local minimal points. Consider the problem of minimizing

$$G = u_1^2 - 2u_1u_2 + 4u_2^2$$

subject to the constraints

$$-10 \leq u_i \leq 10.$$

Suppose that we wish to find all local minimal points using trajectory following. Since this problem could have more than one local minimal point, we attempt to find all local minimal and maximal points. Rather than automating this process and just printing out the solutions obtained, the iterative process has been done by hand in order to illustrate the trajectories obtained as shown in Figure 3.5-3. The first forward trajectory finds a local minimum at the origin. The first backward trajectory finds a local maximum at $u_1 = 10$, $u_2 = -10$, and so on. ■

Example 3.5-3: Multiple Maxima and Minima A simple one-dimensional example illustrates the efficiency of using trajectory following for the determination of multiple minima and maxima. Consider the problem of finding all minima and maxima for the function

$$G = e^{-u}\sin(10u)$$

subject to the simple inequality constraint

$$-\pi \leq u \leq \pi.$$

This problem has 11 local maxima points, including one at π, and 11 local minimal points, including one at $-\pi$, as illustrated in Figure 3.5-4. Table 3.5-1 shows the results of using trajectory following modified to look for all local minima and all local maxima.

In this case the algorithm was instructed to do 100 searches (50 backward integrations and 50 forward integrations) and all 22 points were found. Obviously many of the points were found several times as the algorithm converged 78 other times. The code was written to remove redundant solutions and to put the results in the order of increasing cost. Only a few minutes on a modern PC is required to produce this table. It is possible to obtained the same results with

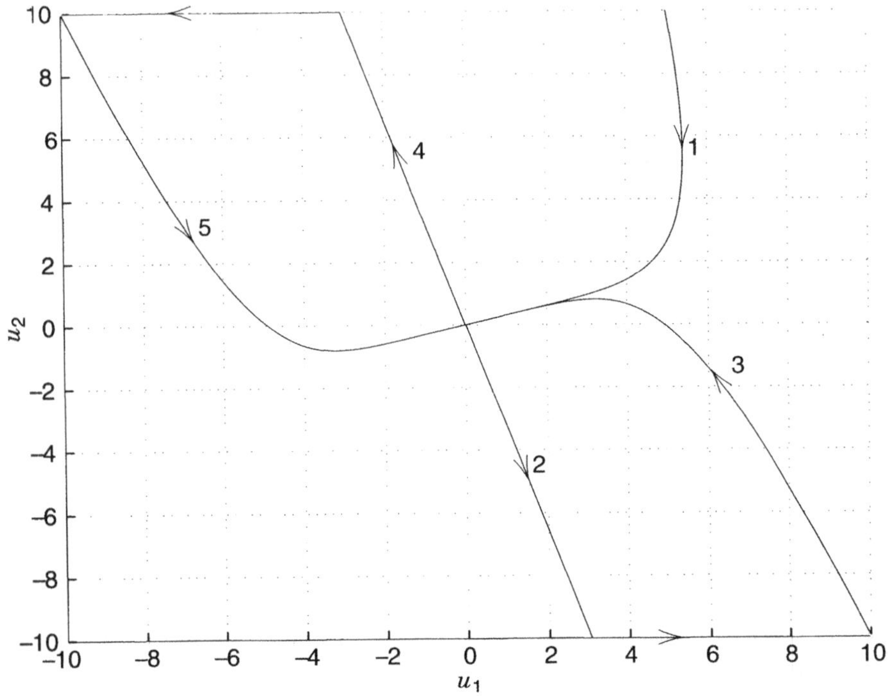

Figure 3.5-3 Using the trajectory following method to search for all minimal and maximal points.

fewer runs (if we are lucky, as few as 22); however even at 100 runs not all points need be found due to the random resetting of initial conditions. ■

Minimizing G Subject to Constraints The trajectory following method classifies all inequality constraints as being either "simple" or "nonsimple." Since the simple constraints (i.e., upper and lower bounds) are handled by the algorithm itself, in this section we use the notation $\mathbf{h}(\mathbf{u}) \geq \mathbf{0}$ to denote any nonsimple constraints that are included in the constraint set. In addition we include the possibility of equality constraints of the form $\boldsymbol{\psi}(\mathbf{u}) = \mathbf{0}$. Both of these constraints can be included in the trajectory following algorithm through the use of a penalty function. In particular, we form a new function

$$W = G + \beta_h \sum_{i=1}^{n_h} \gamma_i h_i^2 + \beta_\psi \sum_{i=1}^{n_\psi} \psi_i^2,$$

where β_h and β_ψ are large positive numbers and γ_i is chosen such that $\gamma_i = 0$ if $h_i \geq 0$ and $\gamma_i = 1$ if $h_i < 0$. Since each term added to G adds a positive

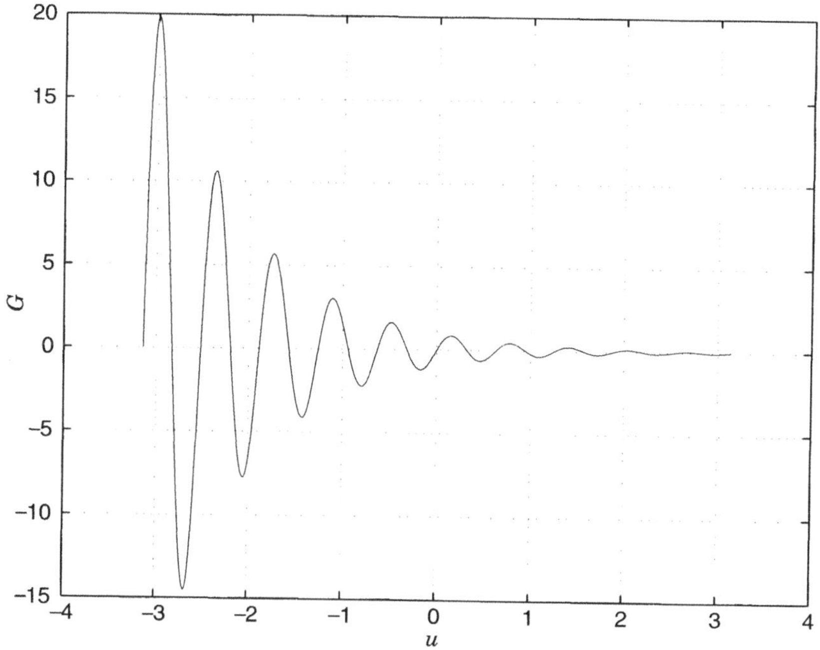

Figure 3.5-4 The cost function for Example 3.5-3.

penalty, the minimum value of W should be very close to the minimum value of G provided that the penalty parameters are taken sufficiently large.

Trajectory following is used to find the minimum of W by replacing G in (3.5-5) with W. It should be noted that even if G by itself would not make (3.3-5) a stiff system, large penalty values certainly will. In fact, most numerical

TABLE 3.5-1 Minima and Maxima for Example 3.5-3

G_{min}	u	G_{max}	u
−14.5174	−2.6803	0.0000	3.1416
−7.7448	−2.0520	0.0696	2.6604
−4.1318	−1.4237	0.1304	2.0321
−2.2043	−0.7954	0.2445	1.4037
−1.1759	−0.1670	0.4582	0.7754
−0.6274	0.4613	0.8589	0.1471
−0.3347	1.0896	1.6100	−0.4812
−0.1786	1.7179	3.0179	−1.1095
−0.0953	2.3462	5.6569	−1.7378
−0.0508	2.9745	10.6035	−2.3662
0.0000	−3.1416	19.8758	−2.9945

procedures that use penalty functions for minimizing G recommend using what is called **sequential nonlinear programming**. This is a procedure in which W is first minimized using small values for the penalty parameters. The solution obtained is used as a starting condition with larger values of the penalty parameters until some convergence criterion is obtained. Such a sequential procedure is often not needed to minimize W using trajectory following, provided that we use an integration routine that can handle stiff systems, such as Gear's routine. In fact, very large values for the penalty functions can often be used at the onset, and if the solution obtained satisifies prespecified convergence criteria, only one run need be made. Note that convergence criteria should include the amount that the equality constraints are allowed to deviate from zero and the amount that the nonsimple inequality constraints are allowed to be negative.

Example 3.5-4: A Two-Dimensional Example Using trajectory following to find all minimal and maximal points for [Reklaitis et al., 1983, p. 621]

$$G = u_1^2 u_2^4,$$

subject to the constraints

$$\begin{aligned} 0.25 - (u_1)^{3/4} u_2 &= 0 \\ u_1 u_2^4 + u_2^2 \sqrt{u_1} - 1 &\geq 0 \end{aligned}$$

and the bounds

$$\begin{aligned} 0 \leq u_1 \leq 10 \\ 0 \leq u_2 \leq 10, \end{aligned}$$

we obtain

$$\begin{aligned} G_{\min} &= 0.386062 \\ u_1 &= 0.100959 \\ u_2 &= 1.394652 \end{aligned}$$

and

$$\begin{aligned} G_{\max} &= 0.542615 \\ u_1 &= 0.007416 \\ u_2 &= 10.00000. \end{aligned}$$

Note that since u_2 is on the boundary for the maximal solution, its value is exact. ∎

Example 3.5-5: Higher-Dimensional Example The problem of minimizing

$$G = u_1^2 + u_2^2 + 2u_3^2 + u_4^2 - 5u_1 - 5u_2 - 21u_3 + 7u_4$$

subject to the inequality constraints

$$\begin{aligned} 8 - u_1^2 - u_2^2 - u_3^2 - u_4^2 - u_1 + u_2 - u_3 + u_4 &\geq 0 \\ 10 - u_1^2 - 2u_2^2 - u_3^2 - 2u_4^2 + u_1 + u_4 &\geq 0 \\ 5 - 2u_1^2 - u_2^2 - u_3^2 - 2u_1 + u_2 + u_4 &\geq 0 \end{aligned}$$

is an example problem used by Rosen and Suzuki [1965] to illustrate their procedure for generating problems with known solutions. In this case, the solution point is $\mathbf{u} = [0 \quad 1 \quad 2 \quad -1]^{\mathrm{T}}$, with $G_{\min} = -44$. Using trajectory following with $\beta_h = 1000$, we obtain

$$\begin{aligned} G_{\min} &= -44.0025 \\ u_1 &= 0.0003 \\ u_2 &= 0.9972 \\ u_3 &= 2.0009 \\ u_4 &= -0.9995. \end{aligned}$$

Of course, knowing the answer, a more accurate solution could have been obtained by increasing β_h and perhaps adjusting other parameters such as those used to calculate the numerical gradient, stop the integration, and so on. Less than a minute of PC computer time was required to obtain the above solution, starting from $\mathbf{u} = \mathbf{0}$, using MATLAB with the Gear integrator. ■

Min-Max via Trajectory Following A similar approach can be used to develop a trajectory following algorithm for min-max optimization [Vincent et al., 1992]. Without going into details, the algorithm for min-max $G(\mathbf{u}, \mathbf{v})$ subject to simple inequality constraints is given by

$$\dot{u}_i = \begin{cases} 0 & \text{if } u_i = u_{i_{\max}} \text{ and } \dfrac{\partial G}{\partial u_i} \leq 0 \\ 0 & \text{if } u_i = u_{i_{\min}} \text{ and } \dfrac{\partial G}{\partial u_i} \geq 0 \\ -\dfrac{\partial G}{\partial u_i} & \text{otherwise} \end{cases} \tag{3.5-6}$$

and

$$\dot{v}_i = \begin{cases} 0 & \text{if } v_i = v_{i_{\max}} \text{ and } \dfrac{\partial G}{\partial v_i} \geq 0 \\ 0 & \text{if } v_i = v_{i_{\min}} \text{ and } \dfrac{\partial G}{\partial v_i} \leq 0 \\ \dfrac{\partial G}{\partial v_i} & \text{otherwise.} \end{cases} \tag{3.5-7}$$

Example 3.5-6: Advertising Model Here we will seek a numerical solution to a problem originally proposed by Leitmann and Schmitendorf [1978]. The problem involves two firms selling substitutable products, with each firm attempting to maximize its own profit. It is assumed that a firm's share of the market is influenced only by its own advertising and the advertising of its competitor. The original problem was posed as a dynamical model. However, a steady-state version of the model has been previously investigated by Vincent and Grantham [1981], and that is the version we use here.

The equilibrium fractions of the market, x_1 and x_2, that each firm receives are given by

$$\begin{aligned} -3x_1 + u - u^2 - x_1 v &= 0 \\ -2x_2 + v - v^2 - 2x_2 u &= 0, \end{aligned} \tag{3.5-8}$$

where $u \geq 0$ and $v \geq 0$ are the advertising expenditure rates for firms 1 and 2, respectively. We also have the constraints $x_1 \geq 0$, $x_2 \geq 0$, and $x_1 + x_2 \leq 1$. The steady-state profits of firms 1 and 2 are taken, respectively, as

$$\begin{aligned} H_1 &= 5x_1 - u \\ H_2 &= 3x_2 - v. \end{aligned}$$

Of course, either firm might temporarily choose to ignore its own profit and, in an effort to drive the other firm out of the market, seek to minimize the other's profits. Thus a min-max solution that would test the vulnerability of each firm is of interest. In order to set up the problem so that it can be solved using the above min-max trajectory following algorithm involving just simple constraints, we first formulate cost functions as the negative of profits,

$$\begin{aligned} G_1 &= u - 5x_1 \\ G_2 &= v - 3x_2. \end{aligned}$$

We then use (3.5-8) to solve for x_1 and x_2 in terms of u and v and substitute into the cost functions to obtain

$$G_1 = u + \frac{5(u^2 - u)}{3 + v}$$

$$G_2 = v + \frac{3(v^2 - v)}{2 + 2u}.$$

It follows from (3.5-8) that $x_1 \geq 0$ implies $u \leq 1$ and $x_2 \geq 0$ implies $v \leq 1$. Thus the bounds on u and v are given by

$$0 \leq u \leq 1$$
$$0 \leq v \leq 1.$$

Under these bounds, the state constraint, $x_1 + x_2 \leq 1$, can never be violated and may be eliminated from the analysis.

Using G_1 and applying the min-max trajectory following algorithm from various points in (u, v) space, with u as the minimizer and v as the maximizer, results in trajectories illustrated in Figure 3.5-5. All the trajectories converge to the min-max point $u = 0.1$ and $v = 1$. Similarly, we obtain the trajectories of Figure 3.5-6 for the min-max of G_2 with v as the minimizer and u as the

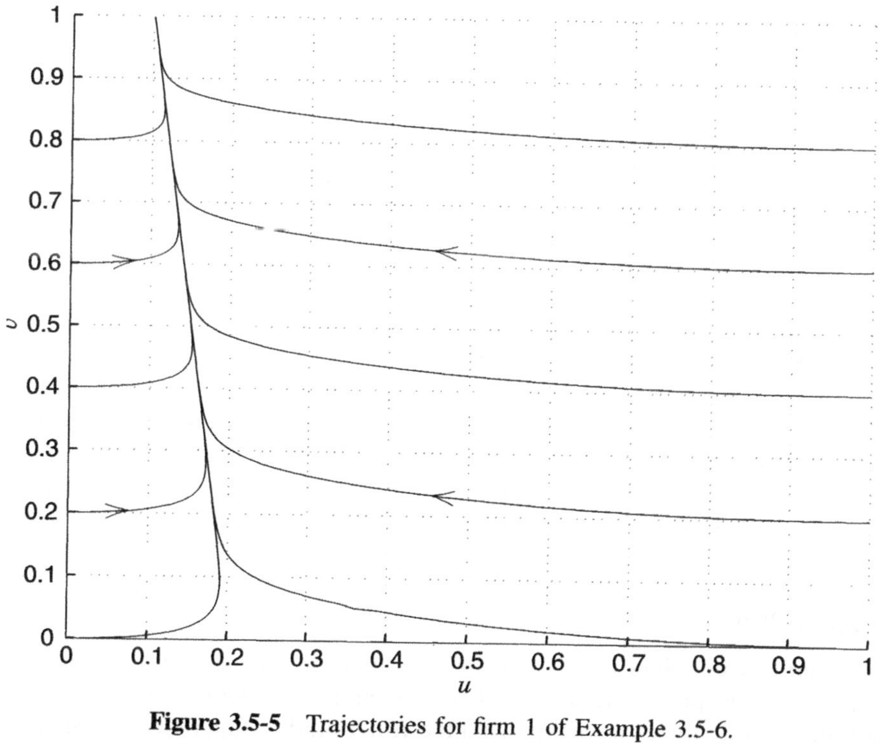

Figure 3.5-5 Trajectories for firm 1 of Example 3.5-6.

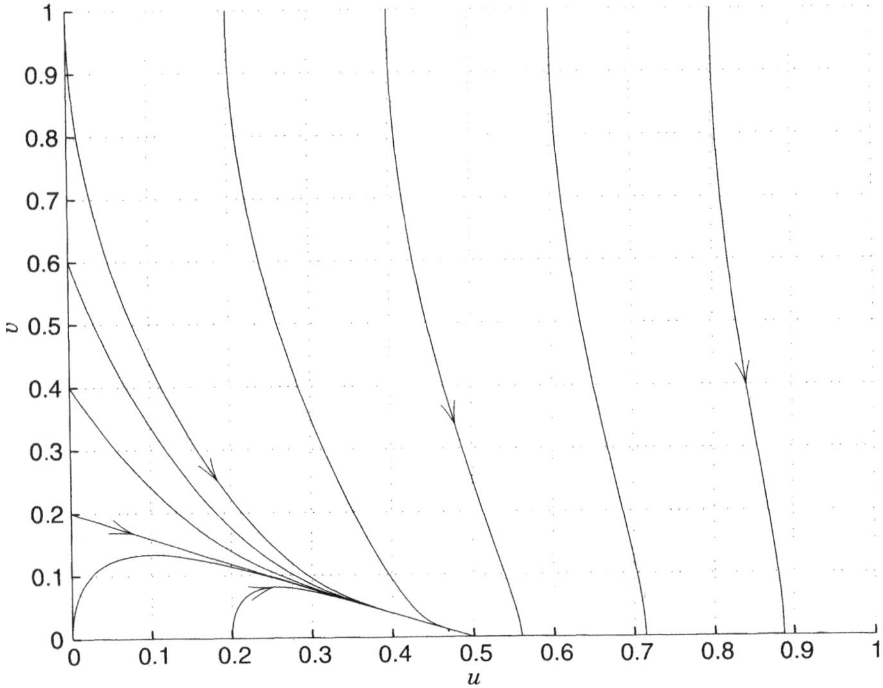

Figure 3.5-6 Trajectories for firm 2 of Example 3.5-6.

maximizer (the algorithm can be used unchanged in this case if retro integration is used instead of forward integration). In this case we obtain $u \in [0.5, 1]$ and $v = 0$. Both of these results agree with the analytical solution (see Exercise 3-11). ■

3.6 CALCULUS OF VARIATIONS

The parameter optimization problems that we have considered are concerned with finding a parameter vector to minimize a function. The calculus of variations is concerned with finding a function of time, such as a trajectory path, to minimize a path-dependent cumulative cost integral.

Simplest Problem

The **simplest problem** in the calculus of variations [Bliss, 1946] is: to find a continuous and piecewise differentiable function $\mathbf{q}(t)$, defined on $[t_0, t_f]$, that extremizes (minimizes or maximizes) the **functional** (a function of a function)

$$J[\mathbf{q}(\cdot)] = \int_{t_0}^{t_f} \phi(\mathbf{q}, \dot{\mathbf{q}}, t)dt, \tag{3.6-1}$$

where $\dot{\mathbf{q}} = d\mathbf{q}/dt$, the scalars t_0 and t_f, and the n-vectors $\mathbf{q}(t_0) = \mathbf{q}_0$ and $\mathbf{q}(t_f) = \mathbf{q}_f$ are specified and fixed, and $\phi(\cdot)$ is a specified continuous scalar-valued function having continuous partial derivatives of at least second-order in some neighborhood of $\mathbf{q}(t)$, except possibly at the endpoints of $\mathbf{q}(t)$.

Suppose $\mathbf{q}^*(t)$ extremizes $J[\mathbf{q}(\cdot)]$. Let $\boldsymbol{\eta}(t)$, an n-vector, be any continuous and differentiable function on $[t_0, t_f]$ satisfying

$$\boldsymbol{\eta}(t_0) = \boldsymbol{\eta}(t_f) = \mathbf{0}, \tag{3.6-2}$$

and, for arbitrarily small α, let

$$\mathbf{q}(t) = \mathbf{q}^*(t) + \alpha\boldsymbol{\eta}(t) \tag{3.6-3}$$

denote a **variation** in $\mathbf{q}^*(t)$. The situation is illustrated in Figure 3.6-1.

The difference in the value of the integral (3.6-1), evaluated along $\mathbf{q}(t)$ and $\mathbf{q}^*(t)$, is given by

$$\Delta J(\alpha) \triangleq J[\mathbf{q}^*(\cdot) + \alpha\boldsymbol{\eta}(\cdot)] - J[\mathbf{q}^*(\cdot)],$$

yielding

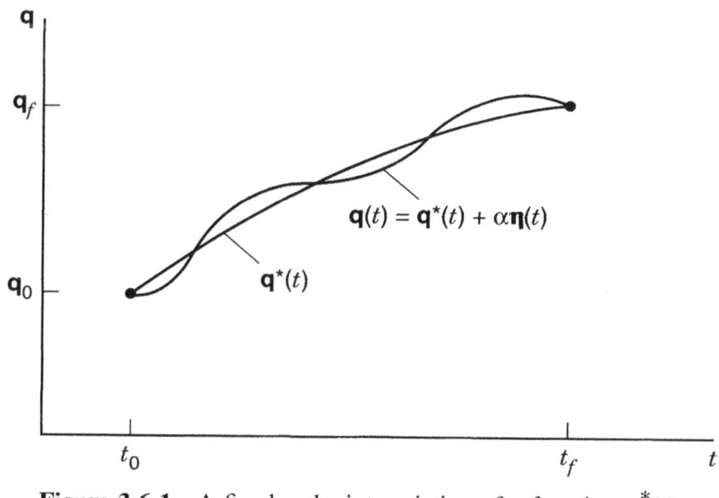

Figure 3.6-1 A fixed-endpoint variation of a function $\mathbf{q}^*(t)$.

$$\Delta J(\alpha) = \int_{t_0}^{t_f} \phi[\mathbf{q}^*(t) + \alpha\boldsymbol{\eta}(t), \dot{\mathbf{q}}^*(t) + \alpha\dot{\boldsymbol{\eta}}(t), t]dt - \int_{t_0}^{t_f} \phi[\mathbf{q}^*(t), \dot{\mathbf{q}}^*(t), t]\,dt.$$

Expanding $\phi(\cdot)$ in a Taylor series near $\mathbf{q}^*(t)$, we have

$$\Delta J(\alpha) = \alpha \int_{t_0}^{t_f} \left[\frac{\partial\phi}{\partial\mathbf{q}}\,\boldsymbol{\eta} + \frac{\partial\phi}{\partial\dot{\mathbf{q}}}\,\dot{\boldsymbol{\eta}} \right] dt + O(\alpha^2),$$

where $O(\alpha^2)/\alpha \to 0$ and $\alpha \to 0$. Dividing by α and then taking the limit as $\alpha \to 0$ produces a first-order approximation for $\Delta J(\alpha)$, called the **first variation**:

$$\delta J \triangleq \lim_{\alpha \to 0} \frac{\Delta J(\alpha)}{\alpha} = \int_{t_0}^{t_f} \left[\frac{\partial\phi}{\partial\mathbf{q}}\,\boldsymbol{\eta} + \frac{\partial\phi}{\partial\dot{\mathbf{q}}}\,\dot{\boldsymbol{\eta}} \right] dt, \tag{3.6-4}$$

where the partial derivative terms in the integral are evaluated along $\mathbf{q}^*(t)$. If $\mathbf{q}^*(t)$ extremizes $J[\mathbf{q}(\cdot)]$ then, to first-order, it is necessary that

$$\delta J = 0. \tag{3.6-5}$$

That is, the integral $J[\mathbf{q}(\cdot)]$ in (3.6-1) must be **stationary**.

Now, consider the second term of the integral in (3.6-4). Integrating this term by parts yields

$$0 = \delta J = \frac{\partial\phi}{\partial\dot{\mathbf{q}}}\,\boldsymbol{\eta}\Big|_{t_0}^{t_f} + \int_{t_0}^{t_f} \left\{ \frac{\partial\phi}{\partial\mathbf{q}} - \frac{d}{dt}\left[\frac{\partial\phi}{\partial\dot{\mathbf{q}}}\right] \right\} \boldsymbol{\eta}\,dt.$$

Since $\boldsymbol{\eta}(t_0) = \boldsymbol{\eta}(t_f) = \mathbf{0}$, this reduces to

$$0 = \delta J = \int_{t_0}^{t_f} \left\{ \frac{\partial\phi}{\partial\mathbf{q}} - \frac{d}{dt}\left[\frac{\partial\phi}{\partial\dot{\mathbf{q}}}\right] \right\} \boldsymbol{\eta}\,dt. \tag{3.6-6}$$

Euler's Equations

A **fundamental lemma** in the calculus of variations [Gelfand and Fomin, 1963] states that, if $\boldsymbol{\beta}(t) = [\beta_1(t) \cdots \beta_n(t)]^T$ is continuous on $[t_0, t_f]$ and satisfies the condition

$$0 = \int_{t_0}^{t_f} [\boldsymbol{\beta}(t)]^{\mathrm{T}} \boldsymbol{\eta}(t)dt \tag{3.6-7}$$

for all $\boldsymbol{\eta}(t) = [\eta_1(t) \cdots \eta_n(t)]^{\mathrm{T}}$ that are continuous on $[t_0, t_f]$ and satisfy $\boldsymbol{\eta}(t_0) = \boldsymbol{\eta}(t_f) = \mathbf{0}$, then $\boldsymbol{\beta}(t) \equiv \mathbf{0}$ on $[t_0, t_f]$.

To verify this result, by way of contradiction, suppose some component of $\boldsymbol{\beta}(t)$ is nonzero at a time $t \in (t_0, t_f)$, say $\beta_i(t) > 0$. Then since $\boldsymbol{\beta}(t)$ is continuous, we have $\beta_i(t) > 0$ in some neighborhood of t, that is, $\beta_i(t) > 0$ for all $t \in (t_1, t_2) \subseteq (t_0, t_f)$. Choose $\boldsymbol{\eta}(t)$ such that all of its components are zero except for the ith component, for example

$$\eta_i(t) = \begin{cases} (t - t_1)(t_2 - t) > 0 & \text{if } t \in (t_1, t_2) \\ 0 & \text{if } t \notin (t_1, t_2). \end{cases}$$

Then

$$\int_{t_0}^{t_f} [\boldsymbol{\beta}(t)]^{\mathrm{T}} \boldsymbol{\eta}(t)dt = \int_{t_1}^{t_2} \beta_i(t)\eta_i(t)dt > 0,$$

which contradicts (3.6-7).

In view of the fundamental lemma, the necessary condition (3.6-6) implies that if $\mathbf{q}^*(t)$ extremizes the integral (3.6-1), with fixed endpoints $\mathbf{q}^*(t_0)$ and $\mathbf{q}^*(t_f)$, then $\mathbf{q}^*(t)$ must satisfy the **Euler equations**:

$$\frac{d}{dt}\left[\frac{\partial \phi}{\partial \dot{\mathbf{q}}}\right] - \frac{\partial \phi}{\partial \mathbf{q}} = \mathbf{0}^{\mathrm{T}}, \tag{3.6-8}$$

which may be written in component form as

$$\frac{d}{dt}\left[\frac{\partial \phi}{\partial \dot{q}_k}\right] - \frac{\partial \phi}{\partial q_k} = 0, \qquad k = 1, \ldots, n. \tag{3.6-9}$$

Any function $\mathbf{q}(t)$ that satisfies Euler's equations (3.6-8) is called an **extremal** for the corresponding calculus of variations problem. Note that an extremal need not satisfy the boundary conditions on $\mathbf{q}(t)$, and may not minimize (or maximize) $J[\mathbf{q}(\cdot)]$. That is, if the problem is to minimize $J[\mathbf{q}(\cdot)]$, a particular extremal $\mathbf{q}(t)$ may actually maximize $J[\mathbf{q}(\cdot)]$ or yield an inflection or saddle point, since Euler's equations are only first-order necessary conditions. However, the actual optimizing solution $\mathbf{q}^*(t)$ must be one from the **family of extremals** satisfying Euler's equations, which, along with the boundary conditions, greatly limits the search for $\mathbf{q}^*(t)$. There are available various additional (e.g., second-order) necessary conditions and even some sufficient conditions [e.g., Gelfand and Fomin, 1963; Leitmann, 1981a], that we do not present here.

Example 3.6-1: Hamilton's Principle and Lagrangian Dynamics Consider a dynamical system whose configuration is defined by n independent coordinates $\mathbf{q} = [q_1 \quad \cdots \quad q_n]^T$ and in which all the applied forces are derivable from the **gradient** of a scalar potential energy function, that is, $\mathbf{F} = -\nabla_\mathbf{q} V \triangleq -[\partial V(\mathbf{q}, t)/\partial \mathbf{q}]^T$. **Hamilton's principle** in dynamics [Greenwood, 1977, p. 158] states that for such systems the actual path $\mathbf{q}(t)$ followed by the system during a fixed time interval t_0 to t_1 is such that the integral

$$J[\mathbf{q}(\cdot)] = \int_{t_0}^{t_1} \mathcal{L}(\mathbf{q}, \dot{\mathbf{q}}, t)\, dt$$

is stationary with respect to path variations that vanish at the endpoints, where

$$\mathcal{L}(\mathbf{q}, \dot{\mathbf{q}}, t) \triangleq T(\mathbf{q}, \dot{\mathbf{q}}, t) - V(\mathbf{q}, t)$$

is the difference between the kinetic energy T and the potential energy V.

For the integral to be stationary, Euler's equations apply, yielding

$$\frac{d}{dt}\left[\frac{\partial \mathcal{L}}{\partial \dot{\mathbf{q}}}\right] - \frac{\partial \mathcal{L}}{\partial \mathbf{q}} = \mathbf{0}^T,$$

which are known as **Lagrange's equations** of motion. They are one of the principle results of analytical dynamics, which examines dynamics in term of scalar-valued functions (work, kinetic energy, potential energy, etc.), in contrast to Newtonian mechanics, which deals with vector quantities (force, acceleration). ■

Example 3.6-2: Brachistochrone A particle slides frictionlessly along a planar curve $y(x)$ between two fixed points in a uniform gravitational field, as illustrated in Figure 3.6-2. The particle has mass m and starts from rest at the origin. The problem is to determine the curve $y(x)$ so that the particle travels from its initial point at $y(0) = 0$ to a specified final point $y(a) = b > 0$ in minimum time.

This classical problem is called the **brachistochrone** problem, and its name is derived from the Greek words "$\beta\rho\alpha\chi\iota\sigma\tau o\varsigma$," meaning "shortest," and "$\chi\rho o\nu o\varsigma$," meaning "time." As noted in Bliss, [1946], the brachistochrone problem was originally discussed by Galileo Galilei in 1630 and later solved by the Swiss mathematicians Johann Bernoulli and his older brother, and professor and

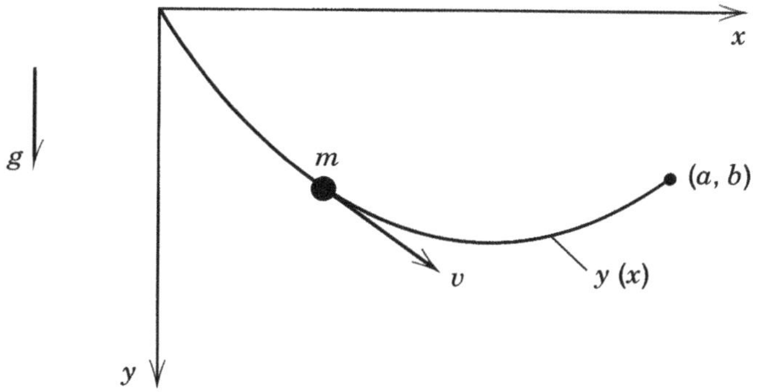

Figure 3.6-2 Geometry of the Brachristochrone problem.

predecessor at Basel, Jakob Bernoulli around 1696. (Another student of Jakob Bernoulli, his nephew Niklaus Bernoulli, made contributions to probability theory, and Johann Bernoulli's son and student, Daniel Bernoulli, is known for his basic research in fluid flow.) These solutions to the brachistochrone problem led to the development of the theory of the calculus of variations in 1744 by Leonhard Euler, who studied under Johann Bernoulli at Basel, with later contributions by Adrien Marie Legendre (1786), Comte Joseph Louis Lagrange (1787), Sir William R. Hamilton (1835), Carl Gustav Jacob Jacobi (1837), Karl Weierstrass (1879), and many others. For more on the theory of the calculus of variations and its history, see Bliss [1946].

The brachistochrone problem can be formulated as a calculus of variations problem by noting that the differential change in time along $y(x)$ can be written in terms of the speed v and the differential arc length ds as

$$\begin{aligned} dt &= \frac{1}{v}\, ds \\ &= \frac{1}{v}\sqrt{(dx)^2 + (dy)^2} \\ &= \frac{1}{v}\sqrt{1 + (y')^2}\, dx, \end{aligned}$$

where $y'(x) = dy(x)/dx$ and the speed $v(y)$ is determined from the conservation of energy condition

$$\tfrac{1}{2}mv^2 - mgy = 0,$$

which implies that

$$v = \sqrt{2gy}.$$

Thus the brachistochrone is equivalent to the problem of minimizing

$$J[y(\cdot)] = \int_0^a \phi(y, y')\,dx, \tag{3.6-10}$$

where

$$\phi(y, y') = \sqrt{\frac{1 + y'^2}{y}}. \tag{3.6-11}$$

The Euler equation for problems of the form (3.6-10) is given by

$$\frac{d}{dx}\left[\frac{\partial \phi}{\partial y'}\right] - \frac{\partial \phi}{\partial y} = 0. \tag{3.6-12}$$

For problems in which the integrand $\phi(y, y')$ does not explicitly contain the independent variable x, the Euler equation (3.6-12) has a **first integral**

$$y' \frac{\partial \phi}{\partial y'} - \phi = \text{constant}. \tag{3.6-13}$$

For the brachistochrone problem, the first integral (3.6-13) yields

$$y(1 + y'^2) = 2k = \text{constant} \tag{3.6-14}$$

and the extremals through $y(0) = 0$ are cycloids, given parametrically by

$$\begin{aligned} x &= k(\theta - \sin \theta) \\ y &= k(1 - \cos \theta). \end{aligned} \tag{3.6-15}$$

There are an infinite number of extremals for $b > 0$. Additional results from the calculus of variations are required to decide which extremal is optimal (i.e., using second-order conditions, conjugate points, etc.), and we refer the reader to the texts by Bliss [1946] and Hestenes [1966]. For example, the conjugate point condition requires that the cycloid not have a cusp beyond the initial one. Among the class of extremals that satisfy this condition, there will only be one cycloid between the initial point and the final point and it is the optimal solution. ■

EXERCISES

3-1. Using the necessary conditions for a minimum, find all candidates for minimizing

$$G = u_1 + 3u_2 - 6u_3^2$$

subject to

$$1 - (u_1^2 + u_2^2 + u_3^2) \geq 0.$$

3-2. A lifeguard sees a girl cry for help in the water at point A. He is on the sand at point B. Both of them are equal distances from the shore line. Consider B to be the origin of an (x, y) coordinate system with the x-axis parallel to the shore line. The coordinates of A are given by $(b, 2a)$. The lifeguard can run in the sand with speed V_s and swim in the water with speed V_w. Determine a relationship between his heading angles in the sand and water to minimize the time of rescue. In optics, the corresponding result is known as Snell's law.

3-3. Maximum range for a jet aircraft: The equations of motion for a jet aircraft flying in a vertical plane over a flat earth in constant gravity are given by

$$\begin{aligned} m\dot{V} &= T - D - W \sin \gamma \\ mV\dot{\gamma} &= L - W \cos \gamma \\ \dot{x} &= V \cos \gamma \\ \dot{h} &= V \sin \gamma \\ \dot{W} &= -cT, \end{aligned}$$

where x denotes horizontal distance, h is altitude, V is the speed of the aircraft, γ is the flight path climb angle, T is thrust, and $W = mg$ is the weight. The lift L and drag D are given by

$$\begin{aligned} L &= C_L \tfrac{1}{2} \rho S V^2 \\ D &= C_D \tfrac{1}{2} \rho S V^2, \end{aligned}$$

where ρ is density, S is the wing planform area, and the lift and drag coefficients are related by

$$C_D = C_{D_0} + KC_L^2,$$

with

$$0 \le C_L \le C_{L_{\max}}.$$

We are interested in determining the lift coefficient C_L to achieve maximum range for a given fuel load W_F. To simplify matters we assume quasi-steady flight ($\dot{V} \approx 0$, $\dot{\gamma} \approx 0$) at a constant altitude ($\gamma \approx 0$). For an initial weight $W(0) = W_0 > W_F$ and a final weight $W(t_f) = W_0 - W_F$, the range can be computed as

$$R = \int_0^{t_f} \dot{x}\, dt = \int_0^{t_f} V\, dt.$$

Using $dW = -cT\, dt$, we have

$$R = \int_{W_0}^{W(t_f)} -\frac{V}{cT}\, dW = \int_{W_0 - W_F}^{W_0} \frac{V}{cT}\, dW.$$

From $T = D$ we get

$$R = \int_{W_0 - W_F}^{W_0} \left(\frac{2}{cC_D \rho S V} \right) dW.$$

From $L = W$,

$$V = \sqrt{\frac{2L}{C_L \rho S}} = \sqrt{\frac{2W}{C_L \rho S}}.$$

Thus

$$\begin{aligned} R &= \int_{W_0 - W_F}^{W_0} \frac{2}{c} \frac{\sqrt{C_L}}{C_D} \frac{1}{\sqrt{2\rho S}} \frac{dW}{\sqrt{W}} \\ &= \frac{2}{c} \sqrt{\frac{2}{\rho S}} \frac{\sqrt{C_L}}{C_D} (\sqrt{W_0} - \sqrt{W_0 - W_F}). \end{aligned}$$

Hence the maximum range problem reduces to maximizing

$$\frac{\sqrt{C_L}}{C_D}$$

subject to the constraints

$$C_D = C_{D0} + KC_L^2$$

and

$$0 \leq C_L \leq C_{L\max}.$$

Solve the problem analytically for the case where $C_{D0} = 0.1$, $K = 0.08$, $C_{L\max} = 1.5$.

3-4. Resolve Example 3.5-1 using a constant of 100,000 instead of 100.

3-5. Consider the problem of minimizing

$$G = 8u_1^2 + 10u_2^2$$

subject to

$$3 - 4u_1 - 6u_2 = 0.$$

Using a penalty function, convert this problem to an unconstrained minimization problem and analytically solve this problem using a penalty parameter of 1 and 10.

3-6. Solve Exercise 3-5 using the trajectory following method.

3-7. Use the trajectory following algorithm to find all local minimal and maximal points for

$$G = u_1^2 + (u_2 - u_1)^2 + (u_3 - u_2)^2 + (1 - u_3)^2$$

subject to the constraints

$$0 \leq u_1 \leq 1$$
$$0 \leq u_2 \leq 1$$
$$0 \leq u_3 \leq 1.$$

3-8. Solve Exercise 3-1 using the trajectory following method.

3-9. Use trajectory following to find all minimal points for Powell's badly scaled function

$$G = f_1^2 + f_2^2,$$

where

$$f_1 = 10^4 u_1 u_2 - 1$$
$$f_2 = e^{-u_1} + e^{-u_2} - 1.0001$$

and

$$|u_1| \leq 10$$
$$|u_2| \leq 10.$$

3-10. The circular-to-circular orbit satellite transfer problem is to minimize the sum of the squares of two impulsive velocity changes Δv_1 and Δv_2 required to transfer the satellite from a circular orbit of radius r_1 to a circular orbit of radius r_2. An initial impulse Δv_1 is given to the satellite at r_1 to put it on a transfer orbit, and a final impulse Δv_2 is given at r_2 to put it on the final circular orbit. Let γ_1 and γ_2 be the angles of thrust measured from a tangent to each circle; v_{c_1} and v_{c_2} be the circular orbit velocity at r_1 and r_2; r_p and r_a be the radius of perigee and apogee of the transfer orbit; a, e, and h be the semimajor axis, eccentricity, and angular momentum of the transfer orbit. In terms of these variables, the optimization problem becomes: Minimize

$$G = \Delta v_1^2 + \Delta v_2^2$$

subject to

$$\Delta v_1 = \sqrt{v_1^2 + v_{c_1}^2 - v_1 v_{c_1} \cos \gamma_1}$$

$$\Delta v_2 = \sqrt{v_2^2 + v_{c_2}^2 - v_2 v_{c_2} \cos \gamma_2}$$

$$v_{c_1} = \sqrt{\frac{1}{r_1}}$$

$$v_{c_2} = \sqrt{\frac{1}{r_2}}$$

$$a = \frac{r_p + r_a}{2}$$

$$e = \frac{r_p - r_a}{2a}$$

$$h = \sqrt{a(1 - e^2)}$$

$$v_1 = \sqrt{\frac{2}{r_1} - \frac{1}{a}}$$

$$v_2 = \sqrt{\frac{2}{r_2} - \frac{1}{a}}$$

$$h = r_1 v_1 \cos \gamma_1$$

$$h = r_2 v_2 \cos \gamma_2$$

$$r_p \leq r_1$$

$$r_a \geq r_2.$$

Show that the optimal solution results in what is called a Hohmann transfer, where $r_p = r_1$ and $r_a = r_2$ (which implies that $\gamma_1 = \gamma_2 = 0$). The initial radius and final radius are fixed. Assume that $r_1 = 1$ and $r_2 = 2$. *Hint:* use the trajectory following method to solve this problem.

3-11. Use the min-max necessary conditions (3.4-5)–(3.4-9) to solve Example 3.5-6.

3-12. Show that the curve of shortest distance (geodesic) between two points on a spherical Earth is a great circle.

3-13. A uniform string of constant weight w per unit length hangs between two fixed points in a uniform gravitational field g with a shape that corresponds to minimum total potential energy. Show that the shape is an arc of a catenary.

3-14. Verify that for $\phi = \phi(y, y')$ the function $y(x)$ satisfying the first integral (3.6-13) is an extremal for the integral

$$J = \int_{x_0}^{x_1} \phi(y, y')\, dx.$$

In other words, show that $y(x)$ satisfies Euler's equation (3.6-12).

3-15. Verify that the cycloids (3.6-15) are solutions to (3.6-14) through $y(0) = 0$.

3-16. Find a continuously differentiable function $y(x) > 0$, passing through two fixed points $(x_1, y_1) = (0, 1)$ and $(x_2, y_2) = (1, \frac{1}{2})$, that minimizes

$$J[y(\cdot)] = \int_0^1 \sqrt{y[1 + (y')^2]}\, dx,$$

where $y' = dy/dx$.

3-17. A continuously differentiable function $y(x) > 0$, with fixed endpoints $(x_1, y_1) = (0, 1)$ and $(x_2, y_2) = (2, 1)$, is revolved about the x-axis to create a surface of revolution. The resulting "container" is hollow with open ends. Determine $y(x)$ so that the surface area is minimized.

3-18. Consider the problem of finding a function $x(t)$ to extremize the integral

$$J = \int_{t_0}^{t_1} g(x, \dot{x}, \ddot{x}, t)dt$$

subject to the fixed endpoint conditions $x(t_0) = x_0$, $x(t_1) = x_1$, where x_0 and x_1 are given. Convert this problem to the form of a "simplest problem" in the calculus of variations.

4

LYAPUNOV STABILITY

4.1 STABILITY CONCEPTS

As we show in Section 4.2, stability for linear systems of the form $\dot{\mathbf{x}} = \mathbf{Ax}$ usually deals with asymptotic stability and is determined by the real part of the eigenvalues of the matrix **A**. For nonlinear systems there are analogous quantities, called Lyapunov exponents, that measure the rate at which neighboring trajectories diverge. But to discuss stability in general for nonlinear systems we need to introduce new concepts, beyond local asymptotic stability. In the process we arrive at a very elegant approach to testing for stability. As a preview of Lyapunov stability methods, we first examine gradient-seeking systems.

Gradient Systems

A dynamical system is a **gradient system** if its equations of motion are of the form

$$\dot{\mathbf{x}} = -\nabla\phi(\mathbf{x}) \triangleq -\left[\frac{\partial\phi(\mathbf{x})}{\partial\mathbf{x}}\right]^{\mathrm{T}}, \qquad (4.1\text{-}1)$$

where $\mathbf{x}$ is an n_x-dimensional vector and $\phi(\mathbf{x})$ is a **velocity potential**, which can be any differentiable scalar-valued function, analogous to a potential energy function.

Properties of Gradient Systems The trajectories of a gradient system are everywhere orthogonal to the surfaces $\phi(\mathbf{x})$ = constant, since $\partial\phi(\mathbf{x})/\partial\mathbf{x}$ is normal to the surface. For a gradient system, equilibrium solutions $\mathbf{x}(t) \equiv \hat{\mathbf{x}} = \text{constant}$ occur only where

$$\frac{\partial \phi(\hat{\mathbf{x}})}{\partial \mathbf{x}} = \mathbf{0}^{\mathrm{T}}. \tag{4.1-2}$$

Elsewhere, the (row) gradient vector $\partial \phi[\mathbf{x}(t)]/\partial \mathbf{x}$ is not zero and we have

$$\begin{aligned} \frac{d\phi[\mathbf{x}(t)]}{dt} &= \frac{\partial \phi[\mathbf{x}(t)]}{\partial \mathbf{x}} \dot{\mathbf{x}}(t) \\ &= -\left\| \frac{\partial \phi[\mathbf{x}(t)]}{\partial \mathbf{x}} \right\|^2 \\ &< 0. \end{aligned} \tag{4.1-3}$$

Thus the nonequilibrium trajectories $\mathbf{x}(t)$ of a gradient system, forward in time, always decrease the value of the potential function $\phi[\mathbf{x}(t)]$. Since (4.1-2) is a necessary condition for a minimum or maximum of $\phi(\mathbf{x})$, any stable equilibrium points occur at the local minima of the function $\phi(\cdot)$, and the unstable equilibrium points occur where $\phi(\cdot)$ has a local maximum or a saddle (inflection) point. The trajectory following method introduced in Chapter 3 used this property. By choosing $\dot{\mathbf{u}} = -[\partial G/\partial \mathbf{u}]^{\mathrm{T}}$, a gradient system was created.

Note that we can conclude that a gradient system will always move toward a local minimal point of ϕ without having to integrate the equations of motion. This is characteristic of the type of conclusions that can be made using the Lyapunov analysis techniques that are developed later in this chapter.

Gradient Control of Linear Systems Given the linear control system

$$\dot{\mathbf{x}} = \mathbf{A}\mathbf{x} + \mathbf{B}\mathbf{u}, \tag{4.1-4}$$

where $\mathbf{x}$ is an n_x-dimensional state vector and $\mathbf{u}$ is an unbounded n_u-dimensional control vector, consider the problem of designing a feedback controller $\mathbf{u}(\mathbf{x})$ that will transfer $\mathbf{x}(t) \to \mathbf{0}$, with the resulting closed-loop system being a linear gradient system. In other words, we seek a feedback control $\mathbf{u}(\mathbf{x})$ that will transform (4.1-4) to the gradient form (4.1-1), using a potential function

$$\phi(\mathbf{x}) = \tfrac{1}{2}\mathbf{x}^{\mathrm{T}}\mathbf{P}\mathbf{x}, \tag{4.1-5}$$

where $\mathbf{P}$ is a symmetric matrix ($\mathbf{P}^{\mathrm{T}} = \mathbf{P}$) chosen so that $\phi(\mathbf{x}) > 0$ for $\mathbf{x} \neq \mathbf{0}$. The resulting gradient system, if it can be constructed via some $\mathbf{u}(\mathbf{x})$, will be a linear system given by

$$\dot{\mathbf{x}} = -\left[\frac{\partial\phi(\mathbf{x})}{\partial\mathbf{x}}\right]^{\mathrm{T}} = -\mathbf{P}\mathbf{x}. \tag{4.1-6}$$

For a specified linear control system (4.1-4), it may not be possible to create a desired gradient system. For example, consider the system

$$\begin{bmatrix} \dot{x}_1 \\ \dot{x}_2 \end{bmatrix} = \begin{bmatrix} 0 & 1 \\ 1 & 0 \end{bmatrix} \begin{bmatrix} x_1 \\ x_2 \end{bmatrix} + \begin{bmatrix} 0 \\ 1 \end{bmatrix} u,$$

and suppose that we wanted to create a "star" equilibrium, in which the trajectories are straight lines converging to the origin, as illustrated in Figure 1.5-2*e*, and orthogonal to lines of constant ϕ, chosen as circles centered at the origin. We cannot do so, in this case, since the control u does not directly affect $\dot{x}_1$.

For the linear control system (4.1-4), suppose $\mathbf{B}^{-1}$ exists (which in part requires $\mathbf{B}$ to be $n_x \times n_x$). Then the state of the system is certainly controllable from any point to any other point, since the **Kalman controllability condition**

$$\mathrm{rank}[\mathbf{B}, \mathbf{AB}, \ \ldots \ , \mathbf{A}^{n_x}\mathbf{B}] = n_x \tag{4.1-7}$$

is satisfied by $\mathrm{rank}[\mathbf{B}] = n_x$. Therefore from linear control systems theory we know that we can achieve any desired set of closed-loop eigenvalues by using a linear state feedback controller of the form

$$\mathbf{u} = -\mathbf{K}\mathbf{x}. \tag{4.1-8}$$

Furthermore, in the current situation, $\mathbf{K}$ is a square $n_x \times n_x$ matrix. Hence it contains enough elements to allow us to specify not only all of the eigenvalues but also all of the eigenvectors for the resulting closed-loop system

$$\dot{\mathbf{x}} = [\mathbf{A} - \mathbf{BK}]\mathbf{x}. \tag{4.1-9}$$

We can create the linear gradient system (4.1-6) by choosing

$$\mathbf{K} = \mathbf{B}^{-1}[\mathbf{A} + \mathbf{P}]. \tag{4.1-10}$$

Example 4.1-1: Gradient Stabilization of a Linear System Consider the system

$$\begin{bmatrix} \dot{x}_1 \\ \dot{x}_2 \end{bmatrix} = \begin{bmatrix} 0 & 1 \\ 1 & 0 \end{bmatrix} \begin{bmatrix} x_1 \\ x_2 \end{bmatrix} + \begin{bmatrix} 1 & 0 \\ 0 & 1 \end{bmatrix} \begin{bmatrix} u_1 \\ u_2 \end{bmatrix}.$$

For $\mathbf{u} = \mathbf{0}$ this system has an unstable saddle-point equilibrium at $\mathbf{x} = \mathbf{0}$. Suppose we want to keep the same eigenvectors as the saddle point and turn the closed-loop system into a gradient system. We can do this by choosing a $\mathbf{P}$ matrix such as

$$\mathbf{P} = \begin{bmatrix} 3 & 1 \\ 1 & 3 \end{bmatrix},$$

yielding a function $\phi(\mathbf{x})$ in (4.1-5) with elliptical contours whose semimajor and semiminor axes are on the eigenvectors of $\mathbf{A}$ (and $\mathbf{P}$). The corresponding feedback matrix for the control law (4.1-8) is computed from (4.1-10) as

$$\mathbf{K} = \begin{bmatrix} 3 & 2 \\ 2 & 3 \end{bmatrix}.$$

The resulting system trajectories, which are orthogonal to the $\phi(\mathbf{x})$ = constant contours, are plotted in Figure 4.1-1. ■

Lyapunov Stability

Because of the rich variety of dynamical behavior that nonlinear systems can exhibit, we need to expand the concept of asymptotic stability and introduce

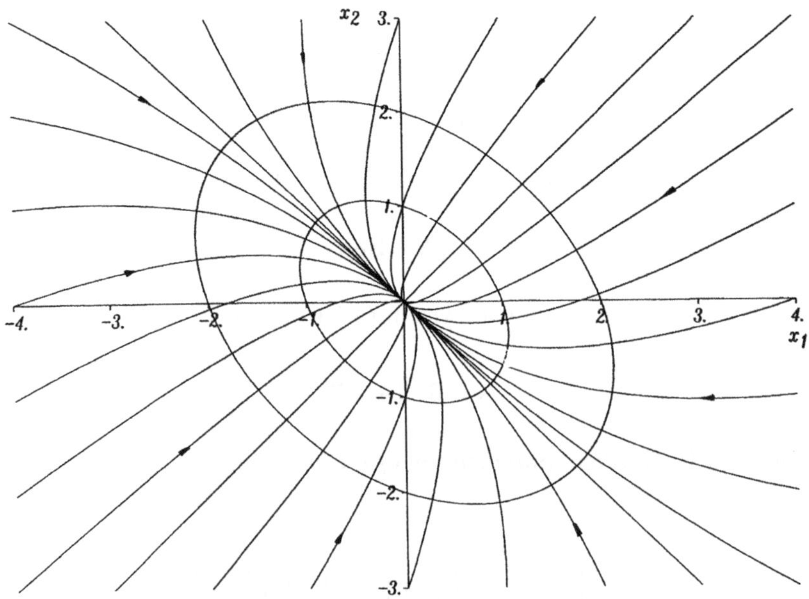

Figure 4.1-1 Gradient motion in a linear system.

several new stability concepts. Given a feedback controller $\mathbf{u}(\mathbf{x})$, the nonlinear control system (1.1-1) becomes a nonlinear dynamical system of the form

$$\dot{\mathbf{x}} = \mathbf{f}(\mathbf{x}) \leftarrow \mathbf{f}[\mathbf{x}, \mathbf{u}(\mathbf{x})]. \tag{4.1-11}$$

A **tracking** problem is one in which the control design objective is to develop a feedback control $\mathbf{u}(\mathbf{x})$ so that solutions $\mathbf{x}(t)$ of (4.1-11) approach a reference solution $\overline{\mathbf{x}}(t)$ and remain there. In other words, we want to choose $\mathbf{u}(\mathbf{x})$ so that $\overline{\mathbf{x}}(t)$ is a stable solution to (4.1-11). A special case of tracking is the **regulator** problem, where the reference trajectory is an equilibrium (constant state) solution $\overline{\mathbf{x}}(t) \equiv \hat{\mathbf{x}}$. If $\mathbf{f}(\mathbf{x})$ is continuous at $\mathbf{x}$, then the dynamical system (4.1-11) will have an **equilibrium** (constant state) at $\overline{\mathbf{x}}(t) \equiv \hat{\mathbf{x}}$ if and only if

$$\mathbf{f}(\hat{\mathbf{x}}) = \mathbf{0}. \tag{4.1-12}$$

We consider the problem of determining whether the design objective (stability) for the tracking problem has been achieved, and to what extent. To simplify the discussion of stability we assume that the function $\mathbf{f}(\cdot)$ is not only continuous but also has continuous partial derivatives of any order required by the ensuing analysis. We do this, for the moment, to avoid mathematical difficulties that can result from discontinuous systems of the form (4.1-11), caused by discontinuous feedback control laws $\mathbf{u}(\mathbf{x})$. One such difficulty, previously discussed in Section 1.2 in connection with Filippov's solution concept, is a "chattering" or "sliding state" trajectory $\mathbf{x}(t)$. For these trajectories, it is possible for the state to asymptotically approach a limit point $\hat{\mathbf{x}}$, but with $\mathbf{f}(\hat{\mathbf{x}}) \neq \mathbf{0}$.

A solution $\overline{\mathbf{x}}(t)$ to (4.1-11) is Lyapunov stable if solutions that start near $\overline{\mathbf{x}}(0)$ remain near $\overline{\mathbf{x}}(t)$ for all $t \geq 0$. More precisely, as depicted in Figure 4.1-2, a solution $\overline{\mathbf{x}}(t)$ to (4.1-11) is **Lyapunov stable** if for any $\varepsilon > 0$, no matter how small, there exists a $\delta = \delta(\varepsilon) > 0$ such that for every $\mathbf{x}(0)$ in which

$$\|\mathbf{x}(0) - \overline{\mathbf{x}}(0)\| < \delta$$

the solution $\mathbf{x}(t)$ satisfies

$$\|\mathbf{x}(t) - \overline{\mathbf{x}}(t)\| < \varepsilon \qquad \text{for all } t > 0.$$

If, in addition,

$$\|\mathbf{x}(t) - \overline{\mathbf{x}}(t)\| \to 0 \qquad \text{as } t \to \infty, \tag{4.1-13}$$

then $\overline{\mathbf{x}}(t)$ is **asymptotically stable**. If the solution $\overline{\mathbf{x}}(t)$ is stable and (4.1-13) holds for every solution $\mathbf{x}(t)$ to (4.1-11), then $\overline{\mathbf{x}}(t)$ is **globally asymptotically stable**. It is common practice to use the terminology "Lyapunov stable," (or

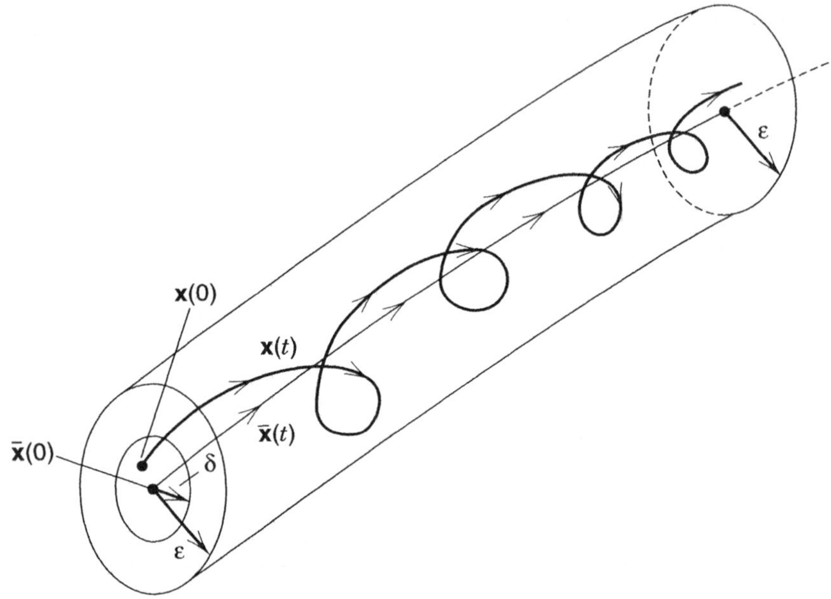

Figure 4.1-2 Lyapunov stability.

"stable in the sense of Lyapunov") and "stable" interchangeably. We will do likewise. Stable will always mean Lyapunov stable.

In these definitions, $\|\cdot\|$ denotes any suitable vector norm. However, unless otherwise noted, we always use $\|\cdot\|$ to mean the length of the vector, defined by the Euclidean norm

$$\|\mathbf{x}\| = \sqrt{\mathbf{x}^{\mathrm{T}}\mathbf{x}} = \sqrt{x_1^2 + \cdots + x_{n_x}^2}.$$

More generally, if the elements of $\mathbf{x} = \boldsymbol{\sigma} + i\boldsymbol{\omega}$ are complex numbers, with $i = \sqrt{-1}$, we use

$$\|\mathbf{x}\| = \sqrt{\mathbf{x}^{*\mathrm{T}}\mathbf{x}} = \sqrt{\boldsymbol{\sigma}^{\mathrm{T}}\boldsymbol{\sigma} + \boldsymbol{\omega}^{\mathrm{T}}\boldsymbol{\omega}},$$

where $\mathbf{x}^* = \boldsymbol{\sigma} - i\boldsymbol{\omega}$ is the complex conjugate of $\mathbf{x}$. For an $n_x \times n_x$ matrix $\mathbf{A}$, possibly complex-valued, we define a matrix norm by

$$\|\mathbf{A}\| = \max_{\|\mathbf{x}\|=1} \|\mathbf{A}\mathbf{x}\|.$$

For the Euclidean vector norm the corresponding matrix norm [Faddeev and Faddeeva, 1963, p. 109] is given by

$$\|\mathbf{A}\| = \sqrt{\mu_{\max}}, \tag{4.1-14}$$

where $\mu_{\max}$ is the largest eigenvalue of the matrix $\mathbf{A}^{*\mathrm{T}}\mathbf{A}$.

As an example of the difference between stability and asymptotic stability, consider the undamped harmonic oscillator, with output

$$\ddot{y} + \omega^2 y = 0.$$

In a **phase plane** state space, where $(x_1, x_2) = (y, \dot{y})$, the trajectories are concentric ellipses centered at the origin and the flow near the equilibrium is a center, as illustrated in Figure 1.5-2a for $\omega = 1$. The equilibrium solution $\mathbf{x}(t) \equiv \mathbf{0}$ is stable, since trajectories that start near the origin stay nearby. But it is not asymptotically stable, since the trajectories do not approach the origin as $t \to \infty$.

Example 4.1-2: Lotka–Volterra Prey-Predator System Consider a two-species population system, consisting of a prey species, having no limits on its food supply and no crowding, and a predator species, which eats only the prey species. This type of system has been modeled by the **Lotka–Volterra equations** [Clark, 1976; Luenberger, 1979; Goh, 1980]:

$$\begin{aligned} \dot{x}_1 &= \alpha x_1 - \beta x_1 x_2 \\ \dot{x}_2 &= -\gamma x_2 + \delta x_1 x_2, \end{aligned} \tag{4.1-15}$$

where x_1 is the prey population, x_2 is the predator population, and α, β, γ, and δ are positive parameters. The first term in $\dot{x}_1$ corresponds to a Malthusian exponential growth of the prey population in the absence of predators. Similarly, the first term in $\dot{x}_2$ corresponds to an exponential decay of the predator population in the absence of prey. The second terms in these equations model the rate at which the predators find and eat the prey, assumed to be proportional to both the prey and predator populations.

The nonzero equilibrium populations are given by

$$\hat{x}_1 = \frac{\delta}{\gamma}, \qquad \hat{x}_2 = \frac{\alpha}{\beta}. \tag{4.1-16}$$

Figure 4.1-3 shows the trajectories for this system, which are closed periodic solutions given by $V(\mathbf{x})$ = constant, where

$$V(x_1, x_2) = \gamma\left[\frac{x_1}{\hat{x}_1} - \ln\left(\frac{x_1}{\hat{x}_1}\right)\right] + \alpha\left[\frac{x_2}{\hat{x}_2} - \ln\left(\frac{x_2}{\hat{x}_2}\right)\right]. \tag{4.1-17}$$

It follows that the equilibrium at $\hat{\mathbf{x}}$ is stable, but not asymptotically stable. ■

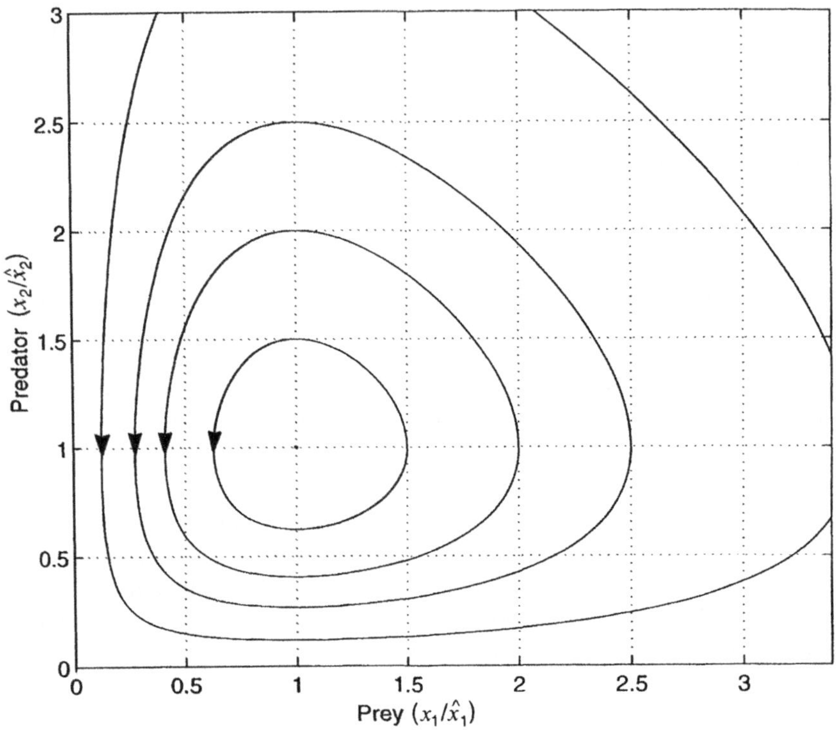

Figure 4.1-3 Trajectories for a prey-predator system.

Example 4.1-3: An Unstable Quadratic System Consider the homogeneous quadratic system [Newton, 1978] given by

$$\dot{x}_1 = (a+3)x_1^2 + (b-3)x_1x_2 + x_2^2$$
$$\dot{x}_2 = x_1^2 + ax_1x_2 + bx_2^2.$$

Trajectories are plotted in Figure 4.1-4 for the case where $a = b = -2$. Every trajectory, except the $x_1 = x_2$ line in the third quadrant, approaches the origin as $t \to \infty$, but $\mathbf{x}(t) \equiv \mathbf{0}$ is not stable, since there are trajectories starting arbitrarily close to the origin that cannot be kept close to the origin for all time. Since the origin is not stable, it cannot be asymptotically stable, even though (almost all) of the trajectories converge to the origin. ■

A solution $\overline{\mathbf{x}}(t)$ of (4.1-11) is a **periodic solution** if $\overline{\mathbf{x}}(t+T) = \overline{\mathbf{x}}(t)$ for some **period** $T > 0$. An equilibrium solution satisfies this definition, but is not what we usually mean by a periodic solution: $\overline{\mathbf{x}}(t+T) = \overline{\mathbf{x}}(t)$ and $\overline{\mathbf{x}}(\tau) \neq \overline{\mathbf{x}}(t)$ for

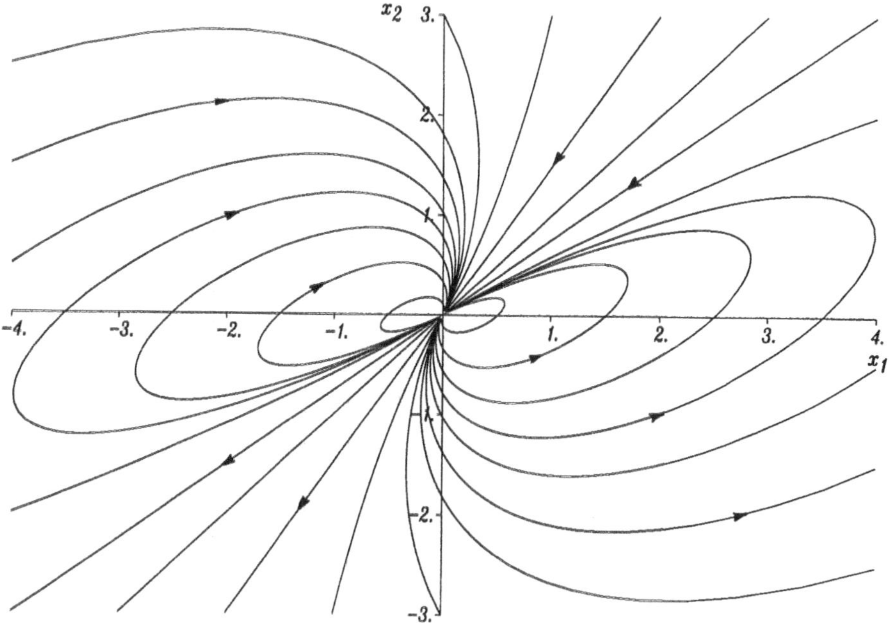

Figure 4.1-4 An unstable quadratic system.

$t < \tau < t+T$. Henceforth we usually exclude equilibrium points when we discuss periodic solutions. For the autonomous system (4.1-11) a periodic trajectory $\overline{\mathbf{x}}(t)$ (other than an equilibrium point) *cannot* be asymptotically stable [Coddington and Levinson, 1955, p. 322]. It can be stable, but not asymptotically stable. To see this, let $\mathbf{x}(t) = \boldsymbol{\xi}(t, \mathbf{x}_0)$ denote the solution to (4.1-11) starting from $\mathbf{x}(0) = \boldsymbol{\xi}(0, \mathbf{x}_0) = \mathbf{x}_0$ at $t = 0$. Let $\mathbf{x}(t)$ start from $\mathbf{x}_0 = \overline{\mathbf{x}}(\Delta t) \neq \overline{\mathbf{x}}(0)$, that is, from a slightly later point on $\overline{\mathbf{x}}(t) = \boldsymbol{\xi}[t, \overline{\mathbf{x}}(0)]$. Then even though $\mathbf{x}(t)$ and $\overline{\mathbf{x}}(t)$ are on the same trajectory curve, with a delay Δt, their displacement $\|\mathbf{x}(t) - \overline{\mathbf{x}}(t)\|$ does not asymptotically approach zero, since it periodically repeats its initial nonzero value. Note that this argument only applies for autonomous systems. Periodic trajectories for nonautonomous systems, of the form $\dot{\mathbf{x}} = \mathbf{f}(\mathbf{x}, t)$, may be asymptotically stable, uniformly for all starting times.

In nonlinear systems, unlike linear systems, it often happens that a solution $\mathbf{x}(t)$ is asymptotically stable, but not globally asymptotically stable. Let $\mathbf{x}(t) = \boldsymbol{\xi}(t, \mathbf{x}_0)$ denote the solution to (4.1-11) with $\mathbf{x}(0) = \mathbf{x}_0$. The **domain of attraction** $\mathcal{D}$ for an asymptotically stable solution $\overline{\mathbf{x}}(t)$ is the set of initial states $\mathbf{x}_0$ from which solutions $\mathbf{x}(t)$ converge to $\overline{\mathbf{x}}(t)$ as $t \to \infty$.

Another useful concept is that of an invariant set. A set X in state space is a **positively invariant set** if for every $\mathbf{x}(0)$ in X the corresponding trajectory $\mathbf{x}(t)$ remains in X for all future times $t \geq 0$. A set X is an **invariant set** if $\mathbf{x}(0) \in X$ implies $\mathbf{x}(t) \in X$ for all time, $-\infty < t < \infty$.

Example 4.1-4: Domain of Attraction and Positive Invariant Sets Consider a (backward-time) logistics-type system given in polar coordinates by

$$\dot{r} = -r(1 - r)$$
$$\dot{\theta} = 0.$$

The trajectories of this system (see Exercise 4-1) in Cartesian coordinates $(x_1, x_2) = (r\cos\theta, r\sin\theta)$ are all radial straight lines, which intersect at the origin as illustrated in Figure 4.1-5. The origin is an equilibrium point, which is asymptotically stable, and each of the equilibrium points on the circle at $r = 1$ is unstable. The domain of attraction for the equilibrium at the origin is the region $0 \leq r < 1$. Each region of the form $0 \leq r \leq R$, for $0 \leq R \leq 1$, is a positively invariant set, as are the regions $r = 1$ and $r > 1$. Of course each trajectory $\mathbf{x}(t)$, $t \geq 0$, is also a positively invariant set. ■

The domain of attraction for asymptotically stable control systems can provide a qualitative measure of the effectiveness of a particular controller. Con-

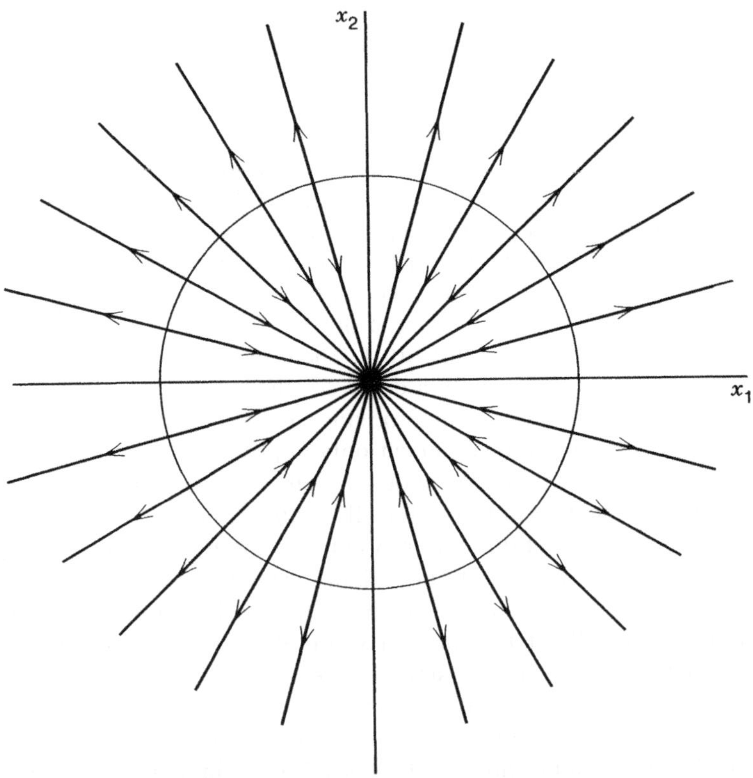

Figure 4.1-5 Trajectories for Example 4.1-4.

sider a **regulator** control design problem, in which the objective is to choose a feedback control function $\mathbf{u}(\mathbf{x})$ to produce an asymptotically stable equilibrium for (4.1-11) at a specified point $\hat{\mathbf{x}}$. Frequently, with two-dimensional nonlinear systems, the domain of attraction for an asymptotically stable equilibrium is bounded by a limit cycle. One way to decide qualitatively if one control function is better than another is to compare the size and shape of the resulting stability regions. Later in this chapter we present some methods for estimating the domain of attraction.

In general, the controllable set C to a target (see Section 2.2) depends on the equations of motion, the target, and the control constraint set $\mathcal{U}$. It is not defined in terms of any particular control choice, but instead reflects the capabilities associated with all possible control choices $\mathbf{u}(\mathbf{x}) \in \mathcal{U}$. In other words, the controllable set is related to the question of *controllability*, before any control choice has been made. After a control choice $\mathbf{u}(\mathbf{x})$ has been made, the corresponding domain of effectiveness or domain of attraction may or may not equal the controllable set. Some controls are going to be less effective than others and may not transfer every state in the controllable set C to the target. However, from the definition of the controllable set to a target, there will always exist a control law $\bar{\mathbf{u}}(\mathbf{x})$ such that its domain of effectiveness to the target is equal to the controllable set and there may exist a control $\hat{\mathbf{u}}(\mathbf{x})$ for which the resulting domain of attraction to the target will be all of the controllable set, that is, $\mathcal{D} = C$. Ideally for a point target $\hat{\mathbf{x}}$, we not only want $\mathcal{D} = C$, but also asymptotic stability for $\hat{\mathbf{x}}$ (see Example 2.3-2 for a case where $\mathcal{D} = C$, but asymptotic stability for the equilibrium point can not be achieved). In Chapter 7 we will learn how to design controllers so that the domain of effectiveness is equal to the controllable set. For now we consider the control $\mathbf{u}(\mathbf{x})$ to be given and we will focus on the stability of the equilibrium points, limit cycles, and so on, that result.

While Lyapunov stability is what we will always mean when we use the word "stable," there are so many other stability concepts in the literature that, outside of this book, one needs to approach with caution the phrase, "the system is stable." It could refer to asymptotic stability, Lyapunov stability, or any one of the following stability concepts which have been found to be useful in various applications. Furthermore, stability generally does not refer to the system but to a particular trajectory (e.g., an equilibrium point) generated by the system. A "system" may exhibit both stable and unstable trajectories.

Lagrange Stability

For a nonlinear system it may happen that a particular equilibrium solution $\mathbf{x}(t) \equiv \hat{\mathbf{x}}$ to (4.1-11) is unstable, yet solutions starting near $\hat{\mathbf{x}}$ remain in some bounded neighborhood of $\hat{\mathbf{x}}$. For this case, where neighboring solutions are **ultimately bounded**, the equilibrium is called **Lagrange stable**. For example, consider the Van der Pol oscillator illustrated in Figure 1.4-3. The equilibrium at the origin is Lyapunov unstable, but trajectories starting near it remain inside a bounded region defined by the limit cycle, so the origin is Lagrange stable. As another example, a geared servomotor positioning device, because of "backlash" due to unavoidable clearances between gear teeth, may create a small limit cycle oscillation about the desired servo position instead of creating asymptotic stability.

Orbital Stability

Let $\gamma = \{\bar{\mathbf{x}}(t) | -\infty < t < \infty\}$ be a **trajectory**, or **orbit**, corresponding to a solution $\bar{\mathbf{x}}(t)$ of (4.1-11). Usually by "orbit" we mean a closed periodic trajectory, but this is not required. A trajectory γ is "orbitally stable" if solutions that start near γ remain nearby, and it is "asymptotically orbitally stable" if, in addition, neighboring solutions approach γ as $t \to \infty$. More specifically, let $\mathbf{x}(t)$ be a solution that starts near γ and let $\rho(t)$ denote the distance between $\mathbf{x}(t)$ and γ at time t, that is, the "closest approach" distance

$$\rho(t) = \min_{\tau} \|\mathbf{x}(t) - \bar{\mathbf{x}}(\tau)\|. \tag{4.1-18}$$

Then γ is **orbitally stable** if for any $\varepsilon > 0$ there exists a $\delta(\varepsilon) > 0$ such that $\rho(0) < \delta$ implies $\rho(t) < \varepsilon$ for all $t \geq 0$. The trajectory γ is **asymptotically orbitally stable** if it is orbitally stable and $\rho(t) \to 0$ as $t \to \infty$.

The basic difference between Lyapunov stability and orbital stability is that with Lyapunov stability the two points $\mathbf{x}(t)$ and $\bar{\mathbf{x}}(t)$ must stay close together at every time $t \geq 0$, while for orbital stability all we require is that the two trajectory curves $\{\mathbf{x}(t) | t \geq 0\}$ and $\{\bar{\mathbf{x}}(t) | t \geq 0\}$ remain close to each other, but the two points $\mathbf{x}(t)$ and $\bar{\mathbf{x}}(t)$ may drift apart, as illustrated in the following example.

Example 4.1-5: Vortex Flow To contrast orbital stability with the notion of Lyapunov stability, consider a gradient system for a **vortex flow** centered at the origin in an (x_1, x_2) Cartesian state space, as shown in Figure 4.1-6.

In polar coordinates

$$x_1 = r \cos \theta$$

$$x_2 = r \sin \theta$$

$$r = \sqrt{x_1^2 + x_2^2}$$

$$\theta = \tan^{-1}\left(\frac{x_2}{x_1}\right),$$

the velocity potential function is

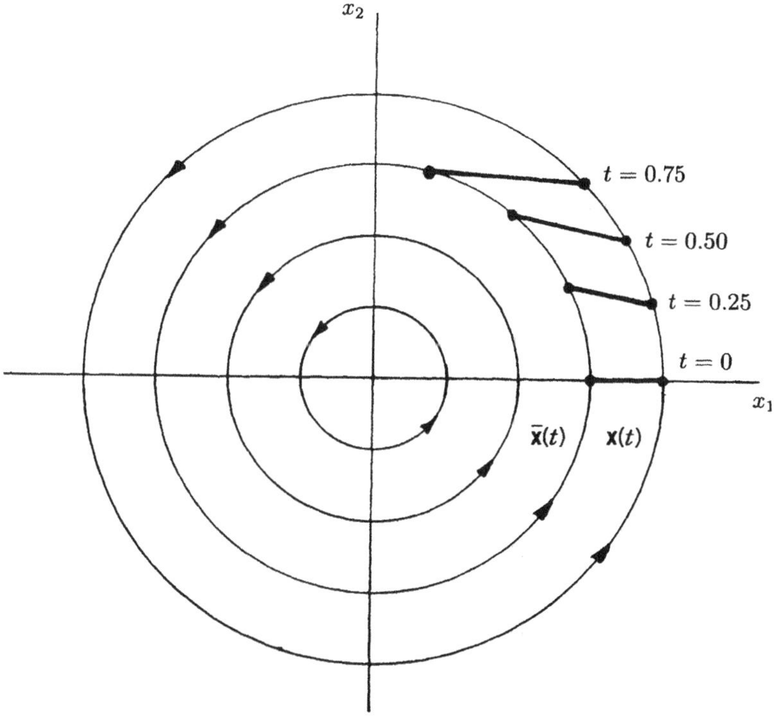

Figure 4.1-6 A vortex flow.

$$\phi(r, \theta) = -k\theta$$

and the motion of the gradient system

$$\dot{\mathbf{x}} = -\nabla\phi$$

is given in polar coordinates by

$$v_r = \dot{r} = -\frac{\partial\phi}{\partial r} = 0$$

$$v_\theta = r\dot{\theta} = -\frac{1}{r}\frac{\partial\phi}{\partial\theta} = \frac{k}{r},$$

where v_r and v_θ are the components of the velocity vector $\dot{\mathbf{x}}$ in the r and θ directions, respectively. Thus the trajectories are concentric circles centered at the origin, given by

$$r(t) = R$$

$$\theta(t) = \frac{k}{R^2} t + \alpha,$$

where $R > 0$ and α are the initial conditions on r and θ, respectively. Along a trajectory, the speed is constant:

$$v = \|\dot{\mathbf{x}}\| = v_\theta = \frac{k}{r}.$$

Now consider an arbitrary reference solution $\overline{\mathbf{x}}(t)$, starting at $\overline{r}(0) = R$ and $\overline{\theta}(0) = 0$, given by

$$\overline{r}(t) = R$$

$$\overline{\theta}(t) = \frac{k}{R^2} t,$$

and a neighboring solution $\mathbf{x}(t)$ starting at $r(0) = R + \Delta R$ and $\theta(0) = 0$, given by

$$r(t) = R + \Delta R$$

$$\theta(t) = \frac{k}{(R + \Delta R)^2} t.$$

The trajectory $\gamma = \{\overline{\mathbf{x}}(t) | -\infty < t < \infty\}$ is orbitally stable since $\mathbf{x}(t)$ is at a constant distance $|\Delta R|$ from γ, which we can make as small as desired by choosing ΔR sufficiently small. However, because of the speed difference along the two trajectories, the solution $\overline{\mathbf{x}}(t)$ is not Lyapunov stable. Even if we take $|\Delta R| > 0$ arbitrarily small, the points $\overline{\mathbf{x}}(t)$ and $\mathbf{x}(t)$ do not stay close to each other because the two solutions, which are periodic in Cartesian coordinates, have different periods. In fact, for two solutions that start close together but at slightly different radii, there will be an infinite sequence of times at which the two solutions will be 180° out of phase, that is, on opposite sides of the origin. ■

Structural Stability

The concept of a **robust system**, which maintains its qualitative behavior under small perturbations to the equations of motion, was first introduced in Andronov and Pontryagin [1937]. Structural stability [Hirsch and Smale, 1974, p. 313; Mees, 1981, p. 40; Guckenheimer and Holmes, 1983, p. 38] deals with this concept.

Consider a C^1 system

$$\dot{\mathbf{x}} = \mathbf{f}(\mathbf{x}) \tag{4.1-19}$$

and a perturbed system

$$\dot{\mathbf{x}} = \mathbf{f}(\mathbf{x}) + \varepsilon\mathbf{g}(\mathbf{x}), \tag{4.1-20}$$

where ε is a small parameter and $\mathbf{g}(\cdot)$ is C^1. The system (4.1-19) is **structurally stable** if small C^1 changes in the equations of motion yield topologically equivalent trajectory flows, that is, if the resulting solution trajectories for (4.1-19) and (4.1-20) can be continuously deformed into one another, possibly with a distortion in time but without changing its sign.

As an example consider the linear system

$$\begin{aligned} \dot{x}_1 &= x_2 \\ \dot{x}_2 &= -x_1 + \varepsilon x_2. \end{aligned}$$

For $\varepsilon = 0$ the system is structurally unstable. In particular, the equilibrium at the origin changes from a stable focus (a "sink") for small $\varepsilon < 0$, to a center for $\varepsilon = 0$, and then becomes an unstable focus (a "source") for small $\varepsilon > 0$. More generally, a linear system $\dot{\mathbf{x}} = \mathbf{A}\mathbf{x}$ is structurally unstable if any of the eigenvalues of $\mathbf{A}$ have zero real parts. In order for a linear system to be structurally stable, it is necessary that none of the eigenvalues have zero real parts. However, this condition alone is not sufficient for structural stability, except in R^2 [Guckenheimer and Holmes, 1983, p. 41].

4.2 LOCAL STABILITY: LYAPUNOV'S FIRST METHOD

In the remainder of this chapter we present two approaches to Lyapunov stability analysis. The first approach, presented in this section, is a local one involving linearization of the equations of motion. The second method, presented in Section 4.3, uses an elegant technique that is valid for determining Lyapunov stability (even global stability if it exists) for nonlinear systems.

Linearization and the State Transition Matrix

We can often learn about the Lyapunov stability of a solution to a system of nonlinear equations, at least locally, by examining the **linearized equations** of motion. This approach is known as **Lyapunov's first (or indirect) method**, and it can provide local stability properties in many applications. There are, however, some exceptions that we note as we come to them.

Let $\overline{\mathbf{X}}(t)$ be a reference solution to the nonlinear system

$$\dot{\mathbf{X}} = \mathbf{f}(\mathbf{X}), \tag{4.2-1}$$

starting at time t_0, and let $\mathbf{X}(t) = \overline{\mathbf{X}}(t) + \mathbf{x}(t)$ be the solution corresponding to a small perturbation $\mathbf{x}(t_0)$ in initial conditions. Then from Taylor's theorem the (initially small) perturbation $\mathbf{x}(t)$ obeys a system of linear **state perturbation** differential equations

$$\dot{\mathbf{x}} = \mathbf{A}(t)\mathbf{x}, \tag{4.2-2}$$

where $\mathbf{A}(t)$ is the time-varying $n_x \times n_x$ matrix of partial derivatives,

$$\mathbf{A}(t) = \frac{\partial \mathbf{f}[\overline{\mathbf{X}}(t)]}{\partial \mathbf{X}} = \begin{bmatrix} \dfrac{\partial f_1}{\partial X_1} & \cdots & \dfrac{\partial f_1}{\partial X_{n_x}} \\ \vdots & \ddots & \vdots \\ \dfrac{\partial f_{n_x}}{\partial X_1} & \cdots & \dfrac{\partial f_{n_x}}{\partial X_{n_x}} \end{bmatrix}_{\overline{\mathbf{X}}(t)}, \tag{4.2-3}$$

evaluated along the reference solution $\overline{\mathbf{X}}(t)$. Thus we can investigate the local stability of $\overline{\mathbf{X}}(t)$ by examining the stability of the equilibrium solution $\mathbf{x}(t) \equiv 0$ for the linearized system (4.2-2).

Time-Varying Linear Systems One approach to solving (4.2-2) is to first consider the time-varying transformation

$$\mathbf{x}(t) = \boldsymbol{\Gamma}(t)\boldsymbol{\xi}(t), \tag{4.2-4}$$

where $\boldsymbol{\Gamma}(t)$ is any nonsingular transformation matrix satisfying

$$\dot{\boldsymbol{\Gamma}}(t) = \mathbf{A}(t)\boldsymbol{\Gamma}(t), \qquad |\boldsymbol{\Gamma}(t_o)| \neq 0, \tag{4.2-5}$$

where $|\cdot|$ denotes the determinant of a square matrix. It follows that

$$\dot{\mathbf{x}} = \dot{\boldsymbol{\Gamma}}(t)\boldsymbol{\xi} + \boldsymbol{\Gamma}(t)\dot{\boldsymbol{\xi}}.$$

Solving for

$$\dot{\boldsymbol{\xi}} = \boldsymbol{\Gamma}^{-1}(t)[\dot{\mathbf{x}} - \dot{\boldsymbol{\Gamma}}(t)\boldsymbol{\xi}],$$

and substituting from (4.2-2) and (4.2-5) yields

$$\dot{\boldsymbol{\xi}} = \boldsymbol{\Gamma}^{-1}(t)[\mathbf{A}(t)\mathbf{x} - \mathbf{A}(t)\boldsymbol{\Gamma}(t)\boldsymbol{\xi}].$$

It then follows from (4.2-4) that $\dot{\boldsymbol{\xi}} \equiv \mathbf{0}$, which implies that $\boldsymbol{\xi}(t) \equiv \boldsymbol{\xi}(t_0)$. Substituting this result into (4.2-4) evaluated at t_0,

$$\mathbf{x}(t_0) = \boldsymbol{\Gamma}(t_0)\boldsymbol{\xi}(t),$$

allows us to solve for $\boldsymbol{\xi}(t)$, which is then substituted back into (4.2-4) to yield

$$\mathbf{x}(t) = \boldsymbol{\Gamma}(t)\boldsymbol{\Gamma}^{-1}(t_0)\mathbf{x}(t_0). \tag{4.2-6}$$

A matrix $\boldsymbol{\Gamma}(t)$ satisfying (4.2-5) is called a **fundamental matrix** and the transformation (4.2-4) is called a **Lyapunov transformation**. One way to compute a fundamental matrix is to take

$$\boldsymbol{\Gamma}(t) = [\mathbf{x}_1(t) \quad \cdots \quad \mathbf{x}_{n_x}(t)], \tag{4.2-7}$$

where the columns $\mathbf{x}_j(t)$, $j = 1, \ldots, n_x$, are any linearly independent solutions of $\dot{\mathbf{x}} = \mathbf{A}(t)\mathbf{x}$, for example, resulting from a linearly independent set of initial states $\mathbf{x}_j(t_0)$.

Defining the **state transition matrix**

$$\boldsymbol{\Phi}(t,\tau) \triangleq \boldsymbol{\Gamma}(t)\boldsymbol{\Gamma}^{-1}(\tau), \tag{4.2-8}$$

we see that the general solution to (4.2-2), starting from any initial time $\tau = t_0$, is given by

$$\mathbf{x}(t) = \boldsymbol{\Phi}(t, t_0)\mathbf{x}(t_0), \tag{4.2-9}$$

where $\boldsymbol{\Phi}(t, t_0)$ satisfies the matrix differential equation

$$\dot{\boldsymbol{\Phi}}(t, t_0) \triangleq \frac{\partial \boldsymbol{\Phi}(t, t_0)}{\partial t} = \mathbf{A}(t)\boldsymbol{\Phi}(t, t_0), \qquad \boldsymbol{\Phi}(t_0, t_0) = \mathbf{I}. \tag{4.2-10}$$

Note that a fundamental matrix $\boldsymbol{\Gamma}(t)$ and the state transition matrix $\boldsymbol{\Phi}(t, t_0)$ both satisfy the same differential equations, but do not necessarily have the same initial conditions. In particular, the fundamental matrices from (4.2-5) are not unique. But the state transition matrix is unique, since the solution $\mathbf{x}(t)$ to (4.2-2) given by (4.2-9) is a unique function of the initial conditions t_0 and $\mathbf{x}(t_0)$. Also note that, by construction, the state transition matrix is nonsingular, with

$$\boldsymbol{\Phi}^{-1}(t, t_0) = \boldsymbol{\Phi}(t_0, t), \tag{4.2-11}$$

and, from (4.2-9), has the property that

$$\mathbf{\Phi}(t_3, t_1) = \mathbf{\Phi}(t_3, t_2)\mathbf{\Phi}(t_2, t_1) \qquad \text{for all } t_1, t_2, t_3. \tag{4.2-12}$$

Constant-Coefficient Linear Systems If the reference solution is an equilibrium point, $\overline{\mathbf{X}}(t) \equiv \hat{\mathbf{X}} = \text{constant}$, the state perturbation equations (4.2-2) become

$$\dot{\mathbf{x}} = \mathbf{A}\mathbf{x}, \tag{4.2-13}$$

where $\mathbf{A}$ is a constant matrix. In this case, the solutions depend only on the time interval $t - t_0$ and, for convenience, we may set $t_0 = 0$. Also, we denote the state transition matrix by $\mathbf{\Phi}(t) \leftarrow \mathbf{\Phi}(t, 0)$.

Stability of the origin for the system (4.2-13) is determined by the **eigenvalues** μ_j, $j = 1, \ldots, n_x$, of $\mathbf{A}$, which are the solutions to the n_x-order polynomial **characteristic equation**

$$0 = |\mu \mathbf{I} - \mathbf{A}| = P(\mu) = \mu^{n_x} + p_{n_x - 1}\mu^{n_x - 1} + \cdots + p_1\mu + p_0. \tag{4.2-14}$$

The **eigenvectors** $\boldsymbol{\xi}_j$ of $\mathbf{A}$ are the corresponding solutions to

$$\mathbf{A}\boldsymbol{\xi} = \mu\boldsymbol{\xi}. \tag{4.2-15}$$

We now need to relate the eigenvalues and eigenvectors of $\mathbf{A}$ to the general solution of (4.2-13). The solution $\mathbf{x}(t)$ can be written as

$$\mathbf{x}(t) = \mathbf{\Phi}(t)\mathbf{x}(0), \tag{4.2-16}$$

where the state transition matrix is the **matrix exponential**

$$\mathbf{\Phi}(t) = e^{\mathbf{A}t} \triangleq \mathbf{I} + \mathbf{A}t + \mathbf{A}^2\frac{t^2}{2!} + \mathbf{A}^3\frac{t^3}{3!} + \cdots. \tag{4.2-17}$$

Note that $e^{\mathbf{A}t} \neq [e^{a_{ij}t}]$ unless $\mathbf{A} = [a_{ij}]$ is a diagonal matrix. The fact that $e^{\mathbf{A}t}$ is the state transition matrix follows from (4.2-10), since $e^{\mathbf{A}0} = \mathbf{I}$ and

$$\frac{de^{\mathbf{A}t}}{dt} = \mathbf{A}e^{\mathbf{A}t}. \tag{4.2-18}$$

Also note that as a consequence of the **Cayley–Hamilton theorem**, which states that every square matrix $\mathbf{A}$ satisfies its own characteristic equation, $P(\mathbf{A}) = \mathbf{0}$, the infinite series in (4.2-17) has an equivalent finite series representation [Grantham and Vincent, 1993]

$$\mathbf{\Phi}(t) = e^{\mathbf{A}t} = \alpha_0(t)\mathbf{I} + \alpha_1(t)\mathbf{A} + \alpha_2(t)\mathbf{A}^2 + \cdots + \alpha_{n_x - 1}(t)\mathbf{A}^{n_x - 1}. \tag{4.2-19}$$

Assume for now that the eigenvectors $\{\boldsymbol{\xi}_j\}$ are linearly independent. Then $\mathbf{M} \triangleq [\boldsymbol{\xi}_1 \cdots \boldsymbol{\xi}_{n_x}]$ has an inverse. Using a coordinate transformation $\mathbf{z} = \mathbf{M}^{-1}\mathbf{x}$ to diagonalize (4.2-13) yields

$$\dot{\mathbf{z}} = \boldsymbol{\Lambda}\mathbf{z},$$

where

$$\boldsymbol{\Lambda} = \mathbf{M}^{-1}\mathbf{A}\mathbf{M} = \text{diag}[\mu_1, \ldots, \mu_{n_x}] \triangleq \begin{bmatrix} \mu_1 & & \mathbf{0} \\ & \ddots & \\ \mathbf{0} & & \mu_{n_x} \end{bmatrix}.$$

Then

$$\mathbf{z}(t) = \hat{\boldsymbol{\Phi}}(t)\mathbf{z}(0),$$

where, for a diagonal matrix $\boldsymbol{\Lambda}$,

$$\hat{\boldsymbol{\Phi}}(t) = e^{\boldsymbol{\Lambda}t} = \text{diag}[e^{\mu_1 t}, \ldots, e^{\mu_{n_x} t}] = \begin{bmatrix} e^{\mu_1 t} & & \mathbf{0} \\ & \ddots & \\ \mathbf{0} & & e^{\mu_{n_x} t} \end{bmatrix}. \tag{4.2-20}$$

Transforming back to the $\mathbf{x}$ coordinate system, we see that the state transition matrix in (4.2-16) can be computed as

$$\boldsymbol{\Phi}(t) = \mathbf{M}\, e^{\boldsymbol{\Lambda}t}\mathbf{M}^{-1}. \tag{4.2-21}$$

Thus the solution to (4.2-13) can be written in terms of the eigenvalues and eigenvectors of $\mathbf{A}$ as

$$\mathbf{x}(t) = \mathbf{M}\, e^{\boldsymbol{\Lambda}t}\mathbf{M}^{-1}\mathbf{x}(0) = \gamma_1\, e^{\mu_1 t}\boldsymbol{\xi}_1 + \cdots + \gamma_{n_x} e^{\mu_{n_x} t}\boldsymbol{\xi}_{n_x}, \tag{4.2-22}$$

where $\boldsymbol{\gamma} = [\gamma_1 \cdots \gamma_{n_x}]^{\mathrm{T}}$ is determined from the initial conditions as

$$\boldsymbol{\gamma} = \mathbf{M}^{-1}\mathbf{x}(0). \tag{4.2-23}$$

If the eigenvalues of μ_j of $\mathbf{A}$ are distinct ($\mu_j \neq \mu_k$), then the eigenvectors $\{\boldsymbol{\xi}_j\}$ are linearly independent. If the n_x eigenvectors of $\mathbf{A}$ are not linearly independent, then it is not possible to transform $\mathbf{A}$ to a diagonal matrix as was done here and a different approach is required.

Jordan Block Diagonal Form It is always possible to transform $\mathbf{A}$ to an $n_x \times n_x$ block diagonal **Jordan matrix**

$$\mathbf{J} = \text{diag}[\mathbf{J}_1, \mathbf{J}_2, \ldots, \mathbf{J}_N], \tag{4.2-24}$$

where N is the number of linearly independent eigenvectors and each Jordan block $\mathbf{J}_j$, associated with an eigenvalue μ_j of $\mathbf{A}$, is a square submatrix in upper triangular form with μ_j on the diagonal, ones above and adjacent to the diagonal, and zeroes elsewhere:

$$\mathbf{J}_j = \begin{bmatrix} \mu_j & 1 & & & \\ & \mu_j & 1 & \mathbf{0} & \\ & & \mu_j & \ddots & \\ & \mathbf{0} & & \ddots & 1 \\ & & & & \mu_j \end{bmatrix}. \tag{4.2-25}$$

For any distinct eigenvalue μ_j, the corresponding Jordan block is the scalar 1×1 matrix $\mathbf{J}_j = [\mu_j]$. For a repeated eigenvalue μ_j of multiplicity m, there will be $p \leq m$ linearly independent eigenvectors, where $p = n_x - \text{rank}[\mu_j\mathbf{I} - \mathbf{A}]$ is called the geometric multiplicity of μ_j. For the general case $1 \leq p \leq m$, there will be p Jordan blocks of size $n_r \times n_r$, where $r = 1, \ldots, p$ and $\Sigma n_r = m$. For example, an eigenvalue with multiplicity $m = 3$ and $p = 2$ linearly independent eigenvectors will yield $p = 2$ Jordan blocks, with one 2×2 and the other 1×1. For $m = 4$ and $p = 2$, there will be $p = 2$ Jordan blocks, with one 3×3 and the other 1×1, or with both 2×2. However, except for the sequence of the Jordan blocks, the Jordan matrix $\mathbf{J}$ is a unique representation of the dynamical system (4.2-13) [Lee and Marcus, 1967, p. 86].

The development of the matrix $\mathbf{M}$, so that $\mathbf{z} = \mathbf{M}^{-1}\mathbf{x}$ transforms $\mathbf{A}$ to Jordan block diagonal form, is essentially the same as in the case of complete diagonalization. The difference is that in order for the matrix $\mathbf{M}$ to have an inverse, any linearly dependent eigenvectors of $\mathbf{A}$ are replaced by "generalized eigenvectors" to form an $\mathbf{M}$ matrix with n_x linearly independent columns. A nonzero vector $\boldsymbol{\xi}$ is a **generalized eigenvector** of rank $k > 1$, associated with an eigenvalue μ of $\mathbf{A}$, if

$$\begin{aligned} [\mu\mathbf{I} - \mathbf{A}]^k\boldsymbol{\xi} &= \mathbf{0} \\ [\mu\mathbf{I} - \mathbf{A}]^{k-1}\boldsymbol{\xi} &\neq \mathbf{0} \\ &\vdots \\ [\mu\mathbf{I} - \mathbf{A}]\boldsymbol{\xi} &\neq \mathbf{0} \end{aligned} \tag{4.2-26}$$

The procedure for computing the generalized eigenvectors begins by first finding the linearly independent eigenvectors of $\mathbf{A}$. For each eigenvalue μ_j there will be a set $E(\mu_j)$ of $n = n_x - \text{rank}[\mu_j\mathbf{I} - \mathbf{A}] \geq 1$ linearly independent eigenvectors. If an eigenvalue has multiplicity $m > n$, we find an additional

$m - n$ generalized eigenvectors as follows: Pick an eigenvector $\boldsymbol{\xi}_j$ in $E(\mu_j)$, let $G(\mu_j)$ denote the set of generalized eigenvectors generated so far for μ_j, and set $\boldsymbol{\xi} = \boldsymbol{\xi}_j$. Then find a generalized eigenvector $\hat{\boldsymbol{\xi}}$ as a nonzero solution to the equation

$$[\mu_j \mathbf{I} - \mathbf{A}]\hat{\boldsymbol{\xi}} = -\boldsymbol{\xi}. \tag{4.2-27}$$

If a solution is found it will be linearly independent of the vectors in $E(\mu_j)$ or $G(\mu_j)$ [Brogan, 1982, p. 143]; add it to $G(\mu_j)$, set $\boldsymbol{\xi} = \hat{\boldsymbol{\xi}}$, and find the next generalized eigenvector by using (4.2-27) again. Otherwise, repeat the process for the next eigenvector in $E(\mu_j)$, if any remain. If not, repeat the entire procedure for the next repeated eigenvalue, until a set of n_x linearly independent eigenvectors or generalized eigenvectors has been found.

Let $\mu_1, \ldots, \mu_{n_x}$ and $\boldsymbol{\xi}_1, \ldots, \boldsymbol{\xi}_{n_x}$ be the eigenvalues and the associated linearly independent eigenvectors or generalized eigenvectors. Order them such that each generalized eigenvector $\boldsymbol{\xi}_j$ is generated from $\boldsymbol{\xi}_{j-1}$ by

$$[\mu_j \mathbf{I} - \mathbf{A}]\boldsymbol{\xi}_j = -\boldsymbol{\xi}_{j-1}, \tag{4.2-28}$$

while each eigenvector $\boldsymbol{\xi}_j$ satisfies

$$[\mu_j \mathbf{I} - \mathbf{A}]\boldsymbol{\xi}_j = \mathbf{0}. \tag{4.2-29}$$

Then the transformation $\mathbf{z} = \mathbf{M}^{-1}\mathbf{x}$, with $\mathbf{M} = [\boldsymbol{\xi}_1 \cdots \boldsymbol{\xi}_{n_x}]$, converts $\mathbf{A}$ to a Jordan block diagonal form $\mathbf{J} = \mathbf{M}^{-1}\mathbf{A}\mathbf{M}$, since $\mathbf{MJ} = \mathbf{AM}$.

For a system $\dot{\mathbf{z}} = \mathbf{Jz}$ in Jordan block diagonal form the corresponding state transition matrix will be of the form

$$\hat{\boldsymbol{\Phi}}(t) = \begin{bmatrix} e^{\mathbf{J}_1 t} & & \mathbf{0} \\ & \ddots & \\ \mathbf{0} & & e^{\mathbf{J}_N t} \end{bmatrix}, \tag{4.2-30}$$

where, if $\mathbf{J}_j$ is $n \times n$ with corresponding eigenvalue μ_j, then

$$e^{\mathbf{J}_j t} = e^{\mu_j t} \begin{bmatrix} 1 & t & \dfrac{t^2}{2!} & \cdots & \dfrac{t^{n-1}}{(n-1)!} \\ & 1 & t & \cdots & \dfrac{t^{n-2}}{(n-2)!} \\ & & \ddots & \ddots & \vdots \\ & & & 1 & t \\ \mathbf{0} & & & & 1 \end{bmatrix}. \tag{4.2-31}$$

This result follows directly from integrating, in reverse order, the equations $\dot{\mathbf{z}}_j = \mathbf{J}_j \mathbf{z}_j$ for the jth block-decoupled subsystem.

Example 4.2-1: Block Diagonalization To illustrate the process of transforming to Jordan block diagonal form, consider a system $\dot{\mathbf{x}} = \mathbf{A}\mathbf{x}$, with

$$\mathbf{A} = \begin{bmatrix} 1 & 1 & 2 \\ 1 & 2 & 4 \\ 4 & -\frac{5}{2} & 3 \end{bmatrix}.$$

The characteristic equation

$$0 = |\mu \mathbf{I} - \mathbf{A}| = \mu^3 - 6\mu^2 + 12\mu - 8$$

has the triple root $\mu = \mu_1 = \mu_2 = \mu_3 = 2$ and a single eigenvector $\boldsymbol{\xi}_1 = [-4 \quad -6 \quad 1]^T$ is obtained as the solution to

$$[\mu \mathbf{I} - \mathbf{A}]\boldsymbol{\xi}_1 = \mathbf{0}.$$

The first generalized eigenvector $\boldsymbol{\xi}_2 = [-10 \quad -16 \quad 1]^T$ is obtained from $\boldsymbol{\xi}_1$ by solving

$$[\mu \mathbf{I} - \mathbf{A}]\boldsymbol{\xi}_2 = -\boldsymbol{\xi}_1,$$

and then the second generalized eigenvector $\boldsymbol{\xi}_3 = [-20 \quad -32 \quad 1]^T$ is obtained from $\boldsymbol{\xi}_2$ by solving

$$[\mu \mathbf{I} - \mathbf{A}]\boldsymbol{\xi}_3 = -\boldsymbol{\xi}_2.$$

The transformation $\mathbf{z} = \mathbf{M}^{-1}\mathbf{x}$, where $\boldsymbol{\xi}_j$ is the jth column of

$$\mathbf{M} = \begin{bmatrix} -4 & -10 & -20 \\ -6 & -16 & -32 \\ 1 & 1 & 1 \end{bmatrix},$$

converts the system $\dot{\mathbf{x}} = \mathbf{A}\mathbf{x}$ to Jordan form $\dot{\mathbf{z}} = \mathbf{J}\mathbf{z}$, where

$$\mathbf{J} = \mathbf{M}^{-1}\mathbf{A}\mathbf{M} = \begin{bmatrix} 2 & 1 & 0 \\ 0 & 2 & 1 \\ 0 & 0 & 2 \end{bmatrix}.$$

■

In terms of the eigenvalues and (generalized) eigenvectors of $\mathbf{A}$, the general solution to (4.2-13) can be written as

$$\mathbf{x}(t) = \gamma_1(t)e^{\mu_1 t}\boldsymbol{\xi}_1 + \cdots + \gamma_{n_x}(t)e^{\mu_{n_x} t}\boldsymbol{\xi}_{n_x}, \tag{4.2-32}$$

where the coefficients γ_j are at worst polynomials in t.

A Matrix Perturbation Technique It turns out that we can always work in terms of distinct eigenvalues if we make a small perturbation ε in the elements of $\mathbf{A}$ that cause repeated eigenvalues. Then the solution $\mathbf{x}(t)$ follows from (4.2-16) and (4.2-21) in the limit as $\varepsilon \to 0$. The principle result of interest is that for a repeated eigenvalue the corresponding coefficients of the exponential $e^{\mu_j t}$ in the state transition matrix (4.2-21), or in the solution (4.2-32), will be polynomials in t if the corresponding eigenvectors are not linearly independent. Because of the availability of this technique, we usually assume that the eigenvalues of $\mathbf{A}$ are distinct, so that complete diagonalization is possible.

Example 4.2-2: Repeated Eigenvalue For any (real or imaginary) number ε, consider the system

$$\begin{bmatrix} \dot{x}_1 \\ \dot{x}_2 \end{bmatrix} = \begin{bmatrix} 0 & 1 \\ -\varepsilon^2 & 0 \end{bmatrix} \begin{bmatrix} x_1 \\ x_2 \end{bmatrix},$$

with eigenvalues and eigenvectors

$$\mu_1 = \varepsilon, \qquad \boldsymbol{\xi}_1 = \begin{bmatrix} 1 \\ \varepsilon \end{bmatrix}$$

$$\mu_2 = -\varepsilon, \qquad \boldsymbol{\xi}_2 = \begin{bmatrix} 1 \\ -\varepsilon \end{bmatrix}.$$

For $\varepsilon \neq 0$ the eigenvectors are linearly independent and, from (4.2-16) and (4.2-21), the solution $\mathbf{x}(t)$ is given by

$$\begin{bmatrix} x_1(t) \\ x_2(t) \end{bmatrix} = \frac{1}{2} \begin{bmatrix} e^{\varepsilon t} + e^{-\varepsilon t} & \dfrac{e^{\varepsilon t} - e^{-\varepsilon t}}{\varepsilon} \\ \varepsilon e^{\varepsilon t} - \varepsilon e^{-\varepsilon t} & e^{\varepsilon t} + e^{-\varepsilon t} \end{bmatrix} \begin{bmatrix} x_1(0) \\ x_2(0) \end{bmatrix}. \tag{4.2-33}$$

In the limit as $\varepsilon \to 0$, we have $\mu_1 = \mu_2 = 0$ and $\boldsymbol{\xi}_1 = \boldsymbol{\xi}_2 = [1 \quad 0]^{\mathrm{T}}$. Recall **L'Hôpital's rule** for the limit of an indefinite ratio: if $f(\varepsilon)$ and $g(\varepsilon)$ are differentiable, with $f(0) = 0$ or ∞ and $g(0) = 0$ or ∞, then

$$\lim_{\varepsilon \to 0} \frac{f(\varepsilon)}{g(\varepsilon)} = \frac{\lim_{\varepsilon \to 0} [df(\varepsilon)/d\varepsilon]}{\lim_{\varepsilon \to 0} [dg(\varepsilon)/d\varepsilon]},$$

provided both limits exist. Applying L'Hôpital's rule to $(e^{\varepsilon t} - e^{-\varepsilon t})/\varepsilon$ in (4.2-33) in the limit as $\varepsilon \rightarrow 0$ yields

$$\begin{bmatrix} x_1(t) \\ x_2(t) \end{bmatrix} = \begin{bmatrix} 1 & t \\ 0 & 1 \end{bmatrix} \begin{bmatrix} x_1(0) \\ x_2(0) \end{bmatrix}. \qquad \blacksquare$$

Stability of an Equilibrium Point

The stability of the origin for the linearized equations (4.2-13) can be determined in terms of the eigenvalues and (generalized) eigenvectors of **A**, using (4.2-32) for the solution $\mathbf{x}(t)$ to (4.2-13). The **Cauchy–Schwartz inequality** states that, for any pair of n_x-dimensional vectors **y** and **z** (and any "norm" $\|\cdot\|$),

$$\|\mathbf{y} + \mathbf{z}\| \leq \|\mathbf{y}\| + \|\mathbf{z}\|,$$

where equality holds only if $\mathbf{y} = \mathbf{z}$. Thus for unit eigenvectors, with $\|\boldsymbol{\xi}_j\| = 1$ in (4.2-32), we have

$$\|\mathbf{x}(t)\| \leq |\gamma_1 \, e^{\mu_1 t}| + \cdots + |\gamma_{n_x} \, e^{\mu_{n_x} t}|.$$

Allowing for possibly complex conjugate eigenvalues $\mu = \sigma \pm i\omega$ and using **Euler's formula**, we find

$$e^{(\sigma \pm i\omega)t} = e^{\sigma t}(\cos\, \omega t \pm i \sin\, \omega t), \tag{4.2-34}$$

which yields

$$\|\mathbf{x}(t)\| \leq |\gamma_1| e^{\sigma_1 t} + \cdots + |\gamma_{n_x}| e^{\sigma_{n_x} t}.$$

Alternatively [Brogan, 1982, p. 265], we can consider

$$\|\mathbf{x}(t)\| = \|\boldsymbol{\Phi}(t)\mathbf{x}(0)\| \leq \|\boldsymbol{\Phi}(t)\| \, \|\mathbf{x}(0)\|,$$

where from (4.1-14)

$$\|\boldsymbol{\Phi}(t)\|^2 = \mu_{\max}\{\boldsymbol{\Phi}^{*T}(t)\boldsymbol{\Phi}(t)\} = \max_j \, \{e^{\mu_j^* t} \, e^{\mu_j t}\} = \max_j \, e^{2\sigma_j t}. \tag{4.2-35}$$

Therefore

$$\|\mathbf{x}(t)\| \leq \|\mathbf{x}(0)\| \, \max_j \, e^{\sigma_j t}. \tag{4.2-36}$$

Based on either result, we find that in order for the origin to be stable for the constant-coefficient linear system (4.2-13), it is necessary that all of the eigenvalues of **A** have $Re(\mu_j) \leq 0$, where $Re(\cdot)$ denotes real part. If any of the eigenvalues has a positive real part, then there is at least one solution $\mathbf{x}(t)$ to (4.2-13), starting arbitrarily near $\mathbf{x} = \mathbf{0}$, for which $\|\mathbf{x}(t)\| \rightarrow \infty$ as $t \rightarrow \infty$, which implies that the origin is unstable. Therefore we conclude: *for the constant-coefficient linear system (4.2-13) the origin is globally asymptotically stable if and only if all of the eigenvalues of the matrix* **A** *have negative real parts.*

A critical case occurs if some of the eigenvalues have zero real parts and the others have negative real parts. If the eigenvalues with zero real parts are not repeated, then the origin for the linear system (4.2-13) is stable, but not asymptotically stable, since the eigenvalues with zero real parts correspond to terms in the solution having a constant amplitude. They do not grow with time nor do they decay to zero. If there are repeated eigenvalues with zero real parts and the corresponding eigenvectors are not linearly independent, as in Examples 4.2-2, then some of the coefficients in the state transition matrix are polynomials in time with no counteracting exponential decay factor, and the origin is unstable for the linear system.

We can draw conclusions about the **local stability** or instability of an equilibrium point for a nonlinear system (4.2-1), based on the stability or instability of the linearized system (4.2-13), provided none of the eigenvalues of the linearized system have zero real parts. The principle difficulty with the case where some of the eigenvalues of **A** have zero real parts is that the linearized system is structurally unstable. Even though they are small, the neglected nonlinear terms may (or may not) make the equilibrium point unstable for the nonlinear system. Conversely, the equilibrium for a nonlinear system may be stable even if the linearized system is unstable because of repeated eigenvalues with zero real parts, with the other eigenvalues having negative real parts.

Example 4.2-3: Automobile Speed Consider the speed $S(t) > 0$ of an automobile moving on level ground through still air, with the engine developing a constant force F on the vehicle in the direction of motion. Accounting for aerodynamic drag, and letting F denote all other forces on the vehicle, Newton's equation of motion yields

$$\dot{S} = -\frac{1}{2m} C_D \rho A S^2 + \frac{F}{m} \triangleq f(S), \tag{4.2-37}$$

where C_D is the drag coefficient, ρ is the air density, m is the mass of the vehicle, and A is the frontal cross-sectional area of the vehicle.

For a specified constant accelerating force F, the equilibrium speed is

$$\hat{S} = \sqrt{\frac{2F}{C_D \rho A}}.$$

Let $S(t) = \hat{S} + s(t)$, where $s(t)$ is a small perturbation from the equilibrium speed. Differentiating the right-hand side of (4.2-37) with respect to S and evaluating the result at $\hat{S}$, we can write the linearized equations of motion (4.2-13) as

$$\dot{s} = as,$$

where

$$a = \frac{\partial f(\hat{S})}{\partial S} = -\frac{1}{m}\sqrt{2FC_D\rho A}.$$

The eigenvalue for the linearized system is $\mu = a < 0$, so we conclude that the equilibrium speed $S(t) = \hat{S}$ for the nonlinear system (4.2-36) is asymptotically stable, at least locally. ■

Example 4.2-4: Van der Pol Oscillator The Van der Pol oscillator from Example 1.4-3 is given by

$$\dot{x}_1 = x_2$$
$$\dot{x}_2 = -x_1 + \beta(1 - x_1^2)x_2.$$

This system has a single equilibrium at the origin. To investigate its stability we write the linearized equations near $\mathbf{x} = \mathbf{0}$ as

$$\dot{\mathbf{x}} = \mathbf{A}\mathbf{x},$$

where

$$\mathbf{A} = \frac{\partial \mathbf{f}(\mathbf{0})}{\partial \mathbf{x}} = \begin{bmatrix} 0 & 1 \\ -1 & \beta \end{bmatrix}.$$

The characteristic equation for $\mathbf{A}$ is

$$0 = |\mu\mathbf{I} - \mathbf{A}| = \mu^2 - \beta\mu + 1,$$

and the eigenvalues are

$$\mu = \frac{1}{2}\left(\beta \pm \sqrt{\beta^2 - 4}\right).$$

If $\beta > 0$, the eigenvalues have a positive real part and the origin is an unstable equilibrium for the nonlinear system. For $\beta < 0$ the equilibrium is asymptotically stable, at least locally. For $\beta = 0$ the eigenvalues $\mu = \pm i$ are imaginary and we need to examine the full nonlinear system in order to determine the stability of the equilibrium at the origin. However for $\beta = 0$ the nonlinear Van der Pol system is simply an undamped linear harmonic oscillator, which is stable but not asymptotically stable. In summary, we have found that the equilibrium point for the nonlinear Van der Pol system is unstable for $\beta > 0$, stable for $\beta \leq 0$, and at least locally asymptotically stable for $\beta < 0$. ■

Center Manifold Theory

When some of the eigenvalues for the linearized system have zero real parts and the others have negative real parts, we cannot make definite conclusions about the stability or instability of the nonlinear system based on a linearized analysis. Instead, we must examine the basic nonlinear system directly.

Example 4.2-5: Imaginary Eigenvalues Consider a spring-mass-damper system

$$m\ddot{y} + b\dot{y}^n + ky = 0$$

consisting of a mass m, a linear spring of stiffness k, and a nonlinear damper that develops a force proportional to some integer power $n \geq 1$ of the velocity $\dot{y}$ ("velocity" instead of "speed" since we allow $\dot{y} < 0$). In terms of phase-plane state variables $(x_1, x_2) = (y, \dot{y})$, the state equations are given by

$$\begin{aligned} \dot{x}_1 &= x_2 \\ \dot{x}_2 &= -\omega^2 x_1 - 2\zeta\omega x_2^n, \end{aligned}$$

where

$$\omega = \sqrt{\frac{k}{m}} > 0, \qquad \zeta = \frac{b}{2\sqrt{km}}.$$

This system, which is a variation of Rayleigh's system for $n = 3$ (see Exercise 1-7), has a single equilibrium at the origin. For $n > 1$ the linearized equations

$$\dot{x}_1 = x_2$$
$$\dot{x}_2 = -\omega^2 x_1$$

are those of an undamped harmonic oscillator. The eigenvalues for the linearized system are purely imaginary and we cannot conclude anything from these results about the stability of the equilibrium for the nonlinear system. We must somehow account directly for the nonlinearity. This fact is emphasized by noting that regardless of the value that we pick for the integer $n > 1$, we get exactly the same system of linearized equations, with no hint of the changes made to the nonlinear system. In Example 4.3-4 we see that for $\zeta > 0$ the equilibrium for the nonlinear system is globally asymptotically stable if n is odd. ■

For the case where some eigenvalues have zero real parts there is an important result, called the center manifold theorem [Guckenheimer and Holmes, 1983, p. 123; Khalil, 1992, p. 217], which allows us to reduce the dimension of the nonlinear system that needs to be studied.

Center Manifold Theorem: *For the system*

$$\dot{\mathbf{X}} = \mathbf{f}(\mathbf{X}), \qquad \mathbf{f}(\hat{\mathbf{X}}) = \mathbf{0}, \tag{4.2-38}$$

let $\mathbf{f}(\cdot)$ *be* C^k *(continuous and with continuous partial derivatives through order* k*), with* $k \geq 2$*, and let* $\mathbf{A} = \partial\mathbf{f}(\hat{\mathbf{X}})/\partial\mathbf{X}$. *Divide the eigenvalue spectrum of* $\mathbf{A}$ *into three parts,* σ_s, σ_c, σ_u, *with*

$$Re(\mu)\begin{cases} < 0 & \text{if } \mu \in \sigma_s \\ = 0 & \text{if } \mu \in \sigma_c \\ > 0 & \text{if } \mu \in \sigma_u. \end{cases}$$

Let the (generalized) eigenspaces of σ_s, σ_c, *and* σ_u *be* E^s, E^c, *and* E^u, *respectively. Then there exist* C^k *stable and unstable invariant manifolds (i.e., surfaces)* W^s *and* W^u *tangent to* E^s *and* E^u *at* $\hat{\mathbf{X}}$ *and a* C^{k-1} *center manifold* W^c *tangent to* E^c *at* $\hat{\mathbf{X}}$. *The manifolds* W^s, W^c, *and* W^u *are all invariant sets for the trajectories of* $\dot{\mathbf{X}} = \mathbf{f}(\mathbf{X})$. *The stable and unstable manifolds are unique, but* W^c *need not be.*

Suppose linearization fails to resolve the stability of the equilibrium $\hat{\mathbf{X}}$, because $\mathbf{A}$ has $n_c > 0$ eigenvalues with $Re(\mu) = 0$, while the remaining $n_s = n_x - n_c$ eigenvalues have $Re(\mu) < 0$. The center manifold theorem implies that the stability of the equilibrium can be determined by considering only the flow on the center manifold W^c, a surface of reduced dimension n_c. In other words, we can study the stability of the equilibrium for the nonlinear system by examining a related system whose dimension is equal to the number of eigenvalues with $Re(\mu) = 0$.

By translating the origin if necessary, let $\hat{\mathbf{X}} = \mathbf{0}$. Order the eigenvalues μ_j and (generalized) eigenvectors $\boldsymbol{\xi}_j$ of $\mathbf{A} = \partial\mathbf{f}(\mathbf{0})/\partial\mathbf{X}$ so that $Re(\mu_j) = 0$, $j = 1, \ldots, n_c$, with $Re(\mu_j) < 0$, $j = n_c + 1, \ldots, n_x$, and let $\mathbf{M} = [\boldsymbol{\xi}_1 \quad \cdots \quad \boldsymbol{\xi}_{n_x}]$. Then the coordinate transformation

$$\begin{bmatrix} \mathbf{x} \\ \mathbf{y} \end{bmatrix} = \mathbf{M}^{-1}\mathbf{X}, \qquad \mathbf{x} \in R^{n_c},\ \mathbf{y} \in R^{n_s} \tag{4.2-39}$$

transforms

$$\begin{aligned} \dot{\mathbf{X}} = \mathbf{f}(\mathbf{X}) &= \mathbf{A}\mathbf{X} + \left\{ \mathbf{f}(\mathbf{X}) - \frac{\partial\mathbf{f}(\mathbf{0})}{\partial\mathbf{X}}\,\mathbf{X} \right\} \\ &= \mathbf{A}\mathbf{X} + \tilde{\mathbf{f}}(\mathbf{X}) \end{aligned} \tag{4.2-40}$$

into Jordan block diagonal form

$$\begin{aligned} \dot{\mathbf{x}} &= \mathbf{C}\mathbf{x} + \mathbf{g}(\mathbf{x}, \mathbf{y}) \\ \dot{\mathbf{y}} &= \mathbf{S}\mathbf{y} + \mathbf{h}(\mathbf{x}, \mathbf{y}), \end{aligned} \tag{4.2-41}$$

where $\mathbf{C}$ and $\mathbf{S}$ are $n_c \times n_c$ and $n_s \times n_s$ constant matrices, respectively, and $\mathbf{g}(\cdot)$ and $\mathbf{h}(\cdot)$ vanish, along with their first partial derivatives, at the origin.

From the center manifold theorem, if $\mathbf{f}(\cdot)$ is C^k, $k \geq 2$, then there exists an n_s-vector function $\boldsymbol{\psi}(\mathbf{x})$, which is C^{k-1}, such that in some neighborhood of the origin the center manifold is of the form

$$W^c = \{(\mathbf{x}, \mathbf{y}) | \mathbf{y} = \boldsymbol{\psi}(\mathbf{x})\}, \qquad \boldsymbol{\psi}(\mathbf{0}) = \mathbf{0}, \quad \frac{\partial\boldsymbol{\psi}(\mathbf{0})}{\partial\mathbf{x}} = \mathbf{0} \tag{4.2-42}$$

and is an invariant manifold for the n_c-dimensional system

$$\dot{\mathbf{x}} = \mathbf{C}\mathbf{x} + \mathbf{g}[\mathbf{x}, \boldsymbol{\psi}(\mathbf{x})]. \tag{4.2-43}$$

For this system we have the following result [Guckenheimer and Holmes, 1983, p. 131]: *If the origin* $\mathbf{x} = \mathbf{0}$ *of (4.2-43) is locally asymptotically stable (unstable), then the origin of (4.2-38) is locally asymptotically stable (unstable).*

To determine the center manifold, substitute $\mathbf{y} = \boldsymbol{\psi}(\mathbf{x})$ into (4.2-41), yielding

$$\dot{\mathbf{y}} = \frac{\partial\boldsymbol{\psi}(\mathbf{x})}{\partial\mathbf{x}}\,\dot{\mathbf{x}} = \frac{\partial\boldsymbol{\psi}(\mathbf{x})}{\partial\mathbf{x}}\,\{\mathbf{C}\mathbf{x} + \mathbf{g}[\mathbf{x},\ \boldsymbol{\psi}(\mathbf{x})]\} = \mathbf{S}\boldsymbol{\psi}(\mathbf{x}) + \mathbf{h}[\mathbf{x}, \boldsymbol{\psi}(\mathbf{x})],$$

or

$$\mathcal{N}[\boldsymbol{\psi}] \triangleq \frac{\partial \boldsymbol{\psi}}{\partial \mathbf{x}} \{\mathbf{Cx} + \mathbf{g}[\mathbf{x}, \boldsymbol{\psi}]\} - \mathbf{S}\boldsymbol{\psi} - \mathbf{h}[\mathbf{x}, \boldsymbol{\psi}] = \mathbf{0}, \tag{4.2-44}$$

with boundary conditions

$$\boldsymbol{\psi}(\mathbf{0}) = \mathbf{0}, \qquad \frac{\partial \boldsymbol{\psi}(\mathbf{0})}{\partial \mathbf{x}} = \mathbf{0}. \tag{4.2-45}$$

In general, this partial differential equation cannot be solved exactly for $\boldsymbol{\psi}(\mathbf{x})$, but the solution can be approximated arbitrarily closely as a Taylor series near $\mathbf{x} = \mathbf{0}$.

Example 4.2-6: Lorenz System Consider the stability of the origin for the Lorenz system

$$\begin{aligned} \dot{X}_1 &= s(X_2 - X_1) \\ \dot{X}_2 &= (r - X_3)X_1 - X_2 \\ \dot{X}_3 &= X_1 X_2 - bX_3. \end{aligned} \tag{4.2-46}$$

The linearized system has

$$\mathbf{A} = \begin{bmatrix} -s & s & 0 \\ r & -1 & 0 \\ 0 & 0 & -b \end{bmatrix}.$$

For $r = 1$ the eigenvalues are $\mu_1 = 0$, $\mu_2 = -(s+1)$, $\mu_3 = -b$, with corresponding eigenvectors $\boldsymbol{\xi}_1 = [1 \quad 1 \quad 0]^T$, $\boldsymbol{\xi}_2 = [s \quad -1 \quad 0]^T$, $\boldsymbol{\xi}_3 = [0 \quad 0 \quad 1]^T$. With $s = r = b = 1$ and

$$\mathbf{M} = \begin{bmatrix} 1 & 1 & 0 \\ 1 & -1 & 0 \\ 0 & 0 & 1 \end{bmatrix},$$

the coordinate transformation (4.2-39) yields

$$\begin{aligned} x &= \tfrac{1}{2}(X_1 + X_2) \qquad & X_1 &= x + y_1 \\ y_1 &= \tfrac{1}{2}(X_1 - X_2) \qquad & X_2 &= x - y_1 \\ y_2 &= X_3 \qquad & X_3 &= y_2, \end{aligned} \tag{4.2-47}$$

and (4.2-46), written as (4.2-40)

$$\begin{bmatrix} \dot{X}_1 \\ \dot{X}_2 \\ \dot{X}_3 \end{bmatrix} = \begin{bmatrix} -1 & 1 & 0 \\ 1 & -1 & 0 \\ 0 & 0 & -1 \end{bmatrix} \begin{bmatrix} X_1 \\ X_2 \\ X_3 \end{bmatrix} + \begin{bmatrix} 0 \\ -X_1X_3 \\ X_1X_2 \end{bmatrix},$$

transforms to

$$\begin{bmatrix} \dot{x} \\ \dot{y}_1 \\ \dot{y}_2 \end{bmatrix} = \begin{bmatrix} 0 & 0 & 0 \\ 0 & -2 & 0 \\ 0 & 0 & -1 \end{bmatrix} \begin{bmatrix} x \\ y_1 \\ y_2 \end{bmatrix} + \begin{bmatrix} \frac{1}{2} & \frac{1}{2} & 0 \\ \frac{1}{2} & -\frac{1}{2} & 0 \\ 0 & 0 & 1 \end{bmatrix} \begin{bmatrix} 0 \\ -(x+y_1)y_2 \\ x^2 - y_1^2 \end{bmatrix},$$

which is equivalent to (4.2-41) with

$$\mathbf{C} = [0], \qquad \mathbf{S} = \begin{bmatrix} -2 & 0 \\ 0 & -1 \end{bmatrix}, \qquad g(x, \mathbf{y}) = -\tfrac{1}{2}(x+y_1)y_2, \qquad \mathbf{h}(x, \mathbf{y}) = \begin{bmatrix} \frac{1}{2}(x+y_1)y_2 \\ x^2 - y_1^2 \end{bmatrix}.$$

Substituting $\mathbf{y} = \boldsymbol{\psi}(x) = [\psi_1(x) \quad \psi_2(x)]^{\mathrm{T}}$, equation (4.2-43) becomes

$$\dot{x} = -\tfrac{1}{2}[x + \psi_1(x)]\psi_2(x). \tag{4.2-48}$$

Then the partial differential equation (4.2-44) for $\boldsymbol{\psi}(x)$ can be written as

$$-\mathcal{N}[\boldsymbol{\psi}] = \tfrac{1}{2}(x+\psi_1)\psi_2 \begin{bmatrix} \dfrac{\partial \psi_1}{\partial x} \\ \dfrac{\partial \psi_2}{\partial x} \end{bmatrix} + \begin{bmatrix} -2 & 0 \\ 0 & -1 \end{bmatrix} \begin{bmatrix} \psi_1 \\ \psi_2 \end{bmatrix} + \begin{bmatrix} \frac{1}{2}(x+\psi_1)\psi_2 \\ x^2 - \psi_1^2 \end{bmatrix} = \begin{bmatrix} 0 \\ 0 \end{bmatrix},$$

which is equivalent to

$$0 = (x+\psi_1)\psi_2 \frac{\partial \psi_1}{\partial x} - 4\psi_1 + (x+\psi_1)\psi_2 \tag{4.2-49}$$

$$0 = (x+\psi_1)\psi_2 \frac{\partial \psi_2}{\partial x} - 2\psi_2 + 2(x^2 - \psi_1^2), \tag{4.2-50}$$

with boundary conditions

$$\psi_1(0) = \frac{\partial\psi_1(0)}{\partial x} = 0$$
$$\psi_2(0) = \frac{\partial\psi_2(0)}{\partial x} = 0. \qquad (4.2\text{-}51)$$

To solve equations (4.2-49)–(4.2-51) we take

$$\psi_1(x) = a_1x^2 + b_1x^3 + \cdots$$
$$\psi_2(x) = a_2x^2 + b_2x^3 + \cdots$$

and substitute these expressions into (4.2-49)–(4.2-50) to find the coefficients a_j, b_j. From this we obtain the center manifold equations

$$y_1 = \psi_1(x) = \tfrac{1}{4}x^3 + O(x^4)$$
$$y_2 = \psi_2(x) = x^2 + O(x^4),$$

where $O(x^n)$ denotes terms of order x^n or higher. Then (4.2-48) becomes

$$\dot{x} = -\tfrac{1}{2}(x + \tfrac{1}{4}x^3 + \cdots)(x^2 + \cdots).$$

Thus on the center manifold $\mathbf{y} = \boldsymbol{\psi}(x)$ we have the one-dimensional motion

$$\dot{x} = -\tfrac{1}{2}x^3 + O(x^5) \approx -\tfrac{1}{2}x^3, \qquad (4.2\text{-}52)$$

corresponding to a single eigenvalue with zero real part in the original system. For small $x > 0$ we have $\dot{x} < 0$, and for small $x < 0$ we have $\dot{x} > 0$. Thus the origin for this equation is (at least locally) asymptotically stable, and we conclude that the origin for the original Lorenz system (4.2-46), with $s = r = b = 1$, is also (at least locally) asymptotically stable. ■

Orbital Stability and Floquet Theory

So far, we have used linearization to examine local stability of an equilibrium state for the nonlinear system (4.2-1) based on the resulting constant-coefficient linearized equations (4.2-13). However, linearization is not restricted to equilibrium points and if we linearize about a nonequilibrium solution we obtain a **time-varying linear system** of the form (4.2-2). Unfortunately, the eigenvalue stability criteria for the constant-coefficient case is not applicable for time-varying systems. For a time-varying linear system of the form (4.2-2), the eigenvalues $\mu(t)$ of $\mathbf{A}(t)$ have no direct relationship to the stability of the equilibrium for the linear system (4.2-2), much less for the nonlinear system (4.2-1). In particular, it is *not* sufficient for asymptotic stability of a time-varying linear system that the eigenvalues $\mu(t)$ of $\mathbf{A}(t)$ all have negative real parts. This is so even if the eigenvalues are all negative constants, as is illustrated by the following counterexample due to Vinograd [see Hahn, 1967, p. 307].

Example 4.2-7: Eigenvalues in a Time-Varying System Consider the linear system

$$\begin{bmatrix} \dot{x}_1 \\ \dot{x}_2 \end{bmatrix} = \begin{bmatrix} -1 - 9\cos^2 6t + 6\sin 12t & 12\cos^2 6t + \frac{9}{2}\sin 12t \\ -12\sin^2 6t + \frac{9}{2}\sin 12t & -1 - 9\sin^2 6t - 6\sin 12t \end{bmatrix} \begin{bmatrix} x_1 \\ x_2 \end{bmatrix}. \tag{4.2-53}$$

Despite being a time-varying system, the corresponding characteristic equation

$$0 = |\mu(t)\mathbf{I} - \mathbf{A}(t)| = \mu^2 + 11\mu + 10$$

has constant coefficients and yields constant eigenvalues of $\mu_1 = -1$ and $\mu_2 = -10$. But the following general solution to the equations of motion,

$$\begin{bmatrix} x_1(t) \\ x_2(t) \end{bmatrix} = ae^{2t} \begin{bmatrix} \cos 6t + 2\sin 6t \\ 2\cos 6t - \sin 6t \end{bmatrix} + be^{-13t} \begin{bmatrix} \sin 6t - 2\cos 6t \\ 2\sin 6t + \cos 6t \end{bmatrix}, \tag{4.2-54}$$

demonstrates that the equilibrium solution $\mathbf{x}(t) \equiv 0$ is unstable. This example also illustrates two other facts: (1) it is possible to have equilibrium solutions for time-varying systems, for example, $\dot{\mathbf{x}} = \mathbf{A}(t)\mathbf{x}$, and (2) the eigenvalues of $\mathbf{A}(t)$ may have very little to do with the form of the solution. ■

When the reference solution for (4.2-1) is periodic with period T, then so is the matrix $\mathbf{A}(t)$ in the linearized perturbation equations (4.2-2). In this case, certain stability conclusions can be deduced using **Floquet theory** [Coddington and Levinson, 1955, pp. 78, 323; Hartman, 1964, p. 60; Skelton, 1988, p. 173].

Consider the system, starting from $t_0 = 0$,

$$\dot{\mathbf{x}} = \mathbf{A}(t)\mathbf{x}, \qquad \mathbf{A}(t+T) = \mathbf{A}(t), \tag{4.2-55}$$

where $\mathbf{A}(t)$ is periodic with a period $T \neq 0$. In terms of the state transition matrix, define a constant matrix $\mathbf{R}$ by

$$e^{\mathbf{R}T} \triangleq \mathbf{\Phi}(T, 0) \tag{4.2-56}$$

and a nonsingular matrix $\mathbf{P}(t)$ by

$$\mathbf{P}(t) \triangleq \mathbf{\Phi}(t,0)e^{-\mathbf{R}t}. \tag{4.2-57}$$

It can be shown that [Coddington and Levinson, 1955, p. 78; Skelton, 1988, p. 173]

$$\mathbf{\Phi}(t,\tau) = \mathbf{P}(t)e^{\mathbf{R}(t-\tau)}\mathbf{P}^{-1}(\tau), \qquad \mathbf{P}(t+T) = \mathbf{P}(t). \tag{4.2-58}$$

The coordinate transformation

$$\mathbf{z}(t) = \mathbf{P}^{-1}(t)\mathbf{x}(t), \tag{4.2-59}$$

along with (4.2-10) and (4.2-18), yields

$$\begin{aligned} \dot{\mathbf{x}} &= \dot{\mathbf{P}}\mathbf{z} + \mathbf{P}\dot{\mathbf{z}} \\ \mathbf{A}\mathbf{x} &= [\dot{\mathbf{\Phi}}e^{-\mathbf{R}t} - \mathbf{\Phi}e^{-\mathbf{R}t}\mathbf{R}]\mathbf{z} + \mathbf{\Phi}e^{-\mathbf{R}t}\dot{\mathbf{z}} \\ \mathbf{A}\mathbf{\Phi}e^{-\mathbf{R}t}\mathbf{z} &= [\mathbf{A}\mathbf{\Phi}e^{-\mathbf{R}t} - \mathbf{\Phi}e^{-\mathbf{R}t}\mathbf{R}]\mathbf{z} + \mathbf{\Phi}e^{-\mathbf{R}t}\dot{\mathbf{z}} \end{aligned}$$

and transforms (4.2-55) to the equivalent constant-coefficient system

$$\dot{\mathbf{z}} = \mathbf{R}\mathbf{z}. \tag{4.2-60}$$

The eigenvalues $\mu_1, \ldots, \mu_{n_x}$ of $\mathbf{R}$ are called **characteristic exponents** associated with $\mathbf{A}(t)$. They determinc the stability of the origin in (4.2-55). However, they are generally not unique, since the matrices $\mathbf{R}$ and $\mathbf{P}$ are not unique. But the eigenvalues $\varphi_1, \ldots, \varphi_{n_x}$ of $e^{\mathbf{R}T} = \mathbf{\Phi}(T,0)$ are unique and are called the **characteristic (Floquet) multipliers** associated with $\mathbf{A}(t)$.

Let $\overline{\mathbf{X}}(t)$ be a reference periodic trajectory for (4.2-1), with period $T \neq 0$. Then

$$\mathbf{A}(t) = \frac{\partial \mathbf{f}[\overline{\mathbf{X}}(t)]}{\partial \mathbf{X}}$$

is periodic with period T. If the characteristic exponents μ_j all had negative real parts, then the origin for (4.2-60) and (4.2-55) would be asymptotically stable. However, we have already seen that a reference periodic solution cannot be asymptotically stable for an autonomous system (4.2-1). Furthermore, the argument now establishes that, for a nonequilibrium periodic trajectory $\overline{\mathbf{X}}(t)$, at least one of the characteristic exponents μ_i must zero, with an eigenvector corresponding to $\dot{\mathbf{x}} = \mathbf{0}$ along $\overline{\mathbf{X}}(t)$.

As to the stability of $\overline{\mathbf{X}}(t)$, we have the following result [Coddington and Levinson, 1955, p. 323]: If $n_x - 1$ characteristic exponents of (4.2-2) have negative real parts, then the reference periodic solution $\overline{\mathbf{X}}(t)$ of (4.2-1) is asymptotically orbitally stable. This result can be stated more conveniently in terms of the Floquet multipliers φ_j, one of which will be 1, corresponding to a zero characteristic exponent. Since

$$\mathbf{x}(t_k + T) = e^{\mathbf{R}T}\mathbf{x}(t_k), \quad t_k = kT, \quad k = 0, 1, \ldots$$

is a **discrete-time map** (a Poincaré first-return map), *the reference periodic solution $\overline{\mathbf{X}}(t)$ to (4.2-1) will be asymptotically orbitally stable if all except one of the eigenvalues of $e^{\mathbf{R}T} = \mathbf{\Phi}(T, 0)$* satisfy $|\varphi_j| < 1$.

Example 4.2-8: Limit Cycle Stability In Example 1.4-2 the system

$$\begin{aligned}\dot{X}_1 &= X_2 + X_1[1 - X_1^2 - X_2^2]\\ \dot{X}_2 &= -X_1 + X_2[1 - X_1^2 - X_2^2]\end{aligned}$$

was shown to have a limit cycle at $X_1^2 + X_2^2 = 1$. The corresponding periodic solution from $\mathbf{X}(0) = [1 \quad 0]^{\mathrm{T}}$,

$$\overline{\mathbf{X}}(t) = \begin{bmatrix} \cos t \\ -\sin t \end{bmatrix},$$

has period $T = 2\pi$. The question is whether or not the limit cycle is orbitally stable.

For $\mathbf{X} = \overline{\mathbf{X}}(t) + \mathbf{x}$, the linearized equations of motion $\dot{\mathbf{x}} = \mathbf{A}(t)\mathbf{x}$ near $\overline{\mathbf{X}}(t)$ are

$$\begin{aligned}\dot{\mathbf{x}} &= \begin{bmatrix} -2\overline{X}_1^2 & 1 - 2\overline{X}_1\overline{X}_2 \\ -1 - 2\overline{X}_1\overline{X}_2 & -2\overline{X}_2^2 \end{bmatrix}_{\overline{\mathbf{X}}(t)} \mathbf{x}\\ &= \begin{bmatrix} -2\cos^2 t & 1 + 2\cos t \sin t \\ -1 + 2\cos t \sin t & -2\sin^2 t \end{bmatrix} \mathbf{x}\\ &= \begin{bmatrix} -1 - \cos 2t & 1 + \sin 2t \\ -1 + \sin 2t & -1 + \cos 2t \end{bmatrix} \mathbf{x}.\end{aligned}$$

Solving (4.2-10) numerically over one period $T = 2\pi$ for the reference limit cycle yields

$$\mathbf{\Phi}(T, 0) = \begin{bmatrix} 3.4950 \times 10^{-6} & -1.7642 \times 10^{-7} \\ -2.9979 \times 10^{-7} & 1.0000 \end{bmatrix},$$

with eigenvalues $\varphi_1 = 1$, $\varphi_2 = 3.495 \times 10^{-6}$. Since $|\varphi_j| < 1$ for all φ_j except one, the limit cycle $\overline{\mathbf{X}}(t)$ is asymptotically orbitally stable. ■

Example 4.2-9: Floquet Multipliers for Periodic Matrices In calculating characteristic multipliers, the state transition matrix usually has to be determined numerically. However, for the system in Example 4.2-7 the state transition matrix can be determined analytically.

The matrix $\mathbf{A}(t)$ in (4.2-53) has (minimum) period $T = \pi/3$. Letting $\mathbf{x}(0)$ be $[1 \quad 0]^{\mathrm{T}}$ and $[0 \quad 1]^{\mathrm{T}}$ in turn and using the solutions (4.2-54) as columns yields the fundamental matrix

$$\boldsymbol{\Gamma}(t) = \begin{bmatrix} e^{2t}(\cos 6t + 2\sin 6t) & e^{-13t}(\sin 6t - 2\cos 6t) \\ e^{2t}(2\cos 6t - \sin 6t) & e^{-13t}(2\sin 6t + \cos 6t) \end{bmatrix}.$$

From the definition (4.2-8) of the state transition matrix, we have

$$\boldsymbol{\Phi}(T,0) = \boldsymbol{\Gamma}(T)\boldsymbol{\Gamma}^{-1}(0) = \begin{bmatrix} e^{2T} & -2e^{-13T} \\ 2e^{2T} & e^{-13T} \end{bmatrix} \begin{bmatrix} 1 & -2 \\ 2 & 1 \end{bmatrix}^{-1},$$

which yields

$$\boldsymbol{\Phi}(T,0) = \frac{1}{5} \begin{bmatrix} e^{2T} + 4e^{-13T} & 2e^{2T} - 2e^{-13T} \\ 2e^{2T} - 2e^{-13T} & 4e^{2T} + e^{-13T} \end{bmatrix}.$$

For $T = \pi/3$ the eigenvalues of $\boldsymbol{\Phi}(T,0)$ are $\varphi_1 = 8.1205$, $\varphi_2 = 1.228 \times 10^{-6}$. Thus the origin for (4.2-53) is unstable.

Recall that, for asymptotic stability, all of the Floquet multipliers must satisfy $|\varphi_j| < 1$ except for one that must equal 1. Since $\boldsymbol{\Phi}(T,0)$ does not have a $\varphi = 1$ eigenvalue, it follows that (4.2-53) cannot correspond to linearization about a (nonequilibrium) periodic orbit of period $T = \pi/3$ for an autonomous system of the form (4.2-1). To explore other possibilities, consider other values for T, such as $T = k\pi/3$, $k = 1, 2, \ldots$. The eigenvalue of $\boldsymbol{\Phi}(T,0)$ are

$$\varphi_1(T) = \tfrac{1}{2}\left\{e^{2T} + e^{-13T} + \sqrt{e^{4T} - 2e^{-11T} + e^{-26T}}\right\}$$
$$\varphi_2(T) = \tfrac{1}{2}\left\{e^{2T} + e^{-13T} - \sqrt{e^{4T} - 2e^{-11T} + e^{-26T}}\right\}.$$

Both are real and positive, with

$$\varphi_1(T) > \varphi_1(0) = 1$$

and

$$\varphi_2(T) < \varphi_2(0) = 1$$

for all $T > 0$. Since no $\varphi = 1$ Floquet multiplier exists for any $T > 0$, the system (4.2-53) does not correspond to linearization about a nonequilibrium periodic orbit for any autonomous system. Still, the characteristic Floquet multipliers do allow us to determine the stability of the origin for the time-varying linear system itself. ■

Lyapunov Exponents and Chaotic Systems

In this section we present some quantitative measures that not only extend the notion of eigenvalues, to encompass nonlinear systems, but also have proven useful in connection with chaotic systems.

Lyapunov Exponents Lyapunov exponents are generalizations of eigenvalues and characteristic (Floquet) multipliers that provide information about the (average) rates at which neighboring trajectories converge or diverge in a nonlinear system. Let $\mathbf{x}(t)$ and $\tilde{\mathbf{x}}(t)$ be solutions to

$$\dot{\mathbf{x}} = \mathbf{f}(\mathbf{x}), \tag{4.2-61}$$

starting from neighboring initial conditions, and let $\rho(t) = \|\tilde{\mathbf{x}}(t) - \mathbf{x}(t)\|$ be the distance between the trajectory $\mathbf{x}(t)$ and the perturbed trajectory $\tilde{\mathbf{x}}(t)$ at time t. If $\rho(0)$ is arbitrarily small and

$$\rho(t) \longrightarrow \rho(0)e^{\sigma t} \qquad \text{as } t \longrightarrow \infty, \tag{4.2-62}$$

then σ is called a **Lyapunov exponent** for the reference trajectory $\mathbf{x}(t)$. The distance between the trajectory points $\mathbf{x}(t)$ and $\tilde{\mathbf{x}}(t)$ grows, shrinks, or remains constant for $\sigma > 0$, $\sigma < 0$, or $\sigma = 0$, respectively. In an n_x-dimensional state space there are n_x real Lyapunov exponents, $\sigma_1 \geq \sigma_2 \geq \cdots \geq \sigma_{n_x}$, corresponding to exponential growth rates in n_x orthogonal directions. For a given trajectory $\mathbf{x}(t)$ the Lyapunov exponents are unique, but are functions of the initial state. Arbitrarily close initial states (e.g., on and to either side of a separatrix) may yield trajectories with different Lyapunov exponents, corresponding to different behaviors as $t \longrightarrow \infty$.

If $\mathbf{f}(\cdot)$ is continuous and continuously differentiable the Lyapunov exponents can be calculated in terms of the state perturbation equations

$$\dot{\boldsymbol{\eta}} = \mathbf{A}(t)\boldsymbol{\eta}, \tag{4.2-63}$$

where

$$\mathbf{A}(t) = \frac{\partial \mathbf{f}[\mathbf{x}(t)]}{\partial \mathbf{x}}$$

is evaluated along a trajectory $\mathbf{x}(t)$ and, for small α, $\tilde{\mathbf{x}}(t) = \mathbf{x}(t) + \alpha\boldsymbol{\eta}(t) + \mathbf{O}(\alpha^2)$ is an initially neighboring trajectory.

Let $\boldsymbol{\Phi}(t, \tau)$ be the state transition matrix for (4.2-63). The Lyapunov exponents σ_j are related to the eigenvalues $\varphi_j(t)$ of $\boldsymbol{\Phi}(t, 0)$ by [Parker and Chua, 1989, p. 67]

$$\sigma_j = \lim_{t \to \infty} \left\{ \frac{1}{t} \ln |\varphi_j(t)| \right\}. \tag{4.2-64}$$

For the special case of an equilibrium $\mathbf{x}(t)$ = constant, so that $\mathbf{A}$ is constant and $\boldsymbol{\Phi}(t) = e^{\mathbf{A}t}$, the Lyapunov exponents σ_j are the real parts of the eigenvalues μ_j of $\mathbf{A}$. This is most easily shown for the case where the eigenvectors of $\mathbf{A}$ are linearly independent. The (similarity) transformation to eigenvector coordinates preserves the eigenvalues $\varphi_j(t) = e^{\mu_j t}$ of the state transition matrix. Let $\mu_j = a + ib$. Then

$$|\varphi_j(t)| = \sqrt{e^{\mu_j^* t} e^{\mu_j t}} = \sqrt{e^{(\mu_j^* + \mu_j)t}} = \sqrt{e^{2at}} = e^{at}$$

and (4.2-64) yields $\sigma_j = Re(\mu_j)$.

One way to determine Lyapunov exponents numerically is to integrate the equations of motion (4.2-61), along with n_x copies of the perturbation equations (4.2-63), one for each of n_x initially orthogonal unit perturbations $\boldsymbol{\eta}_j(0)$, corresponding to the semi-axes of an initially spherical n_x-dimensional ellipsoid in state space. This is equivalent to integrating the matrix differential equation (4.2-10) for $\boldsymbol{\Phi}(t, 0)$. As the trajectory $\mathbf{x}(t)$ moves through state space, the perturbation vectors will all tend to align with the major axis of the ellipse, corresponding to the largest Lyapunov exponent, in a manner analogous to the power method for generating the dominant eigenvalue and eigenvector of a matrix.

Since some of the Lyapunov exponents may be positive, particularly in chaotic systems, the algorithm we present for computing Lyapunov exponents incorporates a periodic discontinuous rescaling of the perturbation vectors, to avoid numerical overflow. The algorithm, using a Gramm–Schmidt orthonormalization procedure [Wolf et al., 1985; Parker and Chua, 1989, pp. 73–81], is given by:

$$
\begin{aligned}
\boldsymbol{\psi}_1 &= \boldsymbol{\eta}_1 \\
\boldsymbol{\rho}_1 &= \|\boldsymbol{\psi}_1\| \\
\boldsymbol{\eta}_1 &= \frac{\boldsymbol{\psi}_1}{\|\boldsymbol{\psi}_1\|} \\
&\vdots \\
\boldsymbol{\psi}_m &= \boldsymbol{\eta}_m - \sum_{j=1}^{m-1} (\boldsymbol{\eta}^{\mathrm{T}}\boldsymbol{\eta}_j)\boldsymbol{\eta}_j, \qquad m = 2, \ldots, n_x \\
\rho_m &= \|\boldsymbol{\psi}_m\| \\
\boldsymbol{\eta}_m &= \frac{\boldsymbol{\psi}_m}{\|\boldsymbol{\psi}_m\|},
\end{aligned}
\tag{4.2-65}
$$

where all quantities are evaluated at time $t_k = k\,\Delta t$, $k = 1, 2, \ldots$. At time t_k the lengths $\rho_j(t_k) = \|\boldsymbol{\psi}_j(t_k)\|$ of the orthogonalized vectors, prior to rescaling, are the lengths of the transformed ellipsoidal semi-axes after a time interval Δt, starting from a unit sphere at time t_{k-1}. From these results the Lyapunov exponents can be calculated as

$$
\sigma_j = \lim_{K \to \infty} \left\{ \frac{1}{K\,\Delta t} \sum_{k=1}^{K} \ln\,[\rho_j(k\,\Delta t)] \right\}. \tag{4.2-66}
$$

The largest Lyapunov exponent dominates the others [Schuster, 1988] and can be calculated as

$$
\sigma_1 = \lim_{t \to \infty} \left\{ \frac{1}{t} \ln \left[\frac{\|\boldsymbol{\eta}(t)\|}{\|\boldsymbol{\eta}(\mathbf{0})\|} \right] \right\}, \tag{4.2-67}
$$

where $\boldsymbol{\eta}(t)$ is the solution to (4.2-63) starting from an (essentially) arbitrary $\boldsymbol{\eta}(0)$.

Divergence From Lichtenberg and Lieberman [1983], at a point $\mathbf{x}$ in state space, the instantaneous rate of change of volume $\mathcal{V}(t)$ of points moving with the flow defined by (4.2-61) is given, in terms of the **divergence** of $\mathbf{f}$, by

$$
\delta(\mathbf{x}) = \lim_{\mathcal{V} \to 0} \left[\frac{1}{\mathcal{V}} \frac{d\mathcal{V}}{dt} \right] = \operatorname{div}[\mathbf{f}(\mathbf{x})] \triangleq \sum_{j=1}^{n_x} \frac{\partial f_j(\mathbf{x})}{\partial x_j}. \tag{4.2-68}
$$

At various points, $\delta(\mathbf{x})$ may be positive (expanding) or negative (contracting). The average contraction rate along a trajectory is given by

$$\bar{\delta} \triangleq \lim_{t \to \infty} \left\{ \frac{1}{t} \lim_{\mathcal{V}(0) \to 0} \ln \left[\frac{\mathcal{V}(t)}{\mathcal{V}(0)} \right] \right\} = \sum_{j=1}^{n_x} \sigma_j. \tag{4.2-69}$$

Chaotic Systems For **dissipative systems**, defined by $\sum_{j=1}^{n_x}\{\sigma_j\} < 0$, volumes in state space contract. For such systems the associated trajectories must approach attractors, of dimension lower than that of the underlying state space. **Chaotic motion** (see Section 1.4) is associated with bounded nonperiodic motion exhibiting (on the average) exponential separation of nearby trajectories, which requires at least one positive Lyapunov exponent, for example, $\sigma_1 > 0$. A dissipative system with bounded nonperiodic trajectories and a positive Lyapunov exponent gives rise to **strange attractors**, in which trajectories remain in a bounded region but wander around in an erratic fashion that covers more than a one-dimensional region usually associated with a one-parameter trajectory curve $\mathbf{x}(t)$. From the Poincaré–Bendixson theorem, chaotic strange attractors, in systems defined by differential equations, can only occur in state spaces of dimension three or greater. In addition, the "stretching" and "folding" associated with bounded chaotic strange attractors implies that they can only occur in nonlinear systems.

Consider a three-dimensional dissipative system with Lyapunov exponents $\boldsymbol{\sigma} = (\sigma_1, \sigma_2, \sigma_3)$. At least one of the Lyapunov exponents must be negative in order for the sum of Lyapunov exponents to be negative. In addition, one of the Lyapunov exponents must be zero if the attractor does not contain an equilibrium point [Haken, 1983]. Thus the possible combinations of Lyapunov exponents and corresponding attractors for a dissipative three-dimensional system are: $\boldsymbol{\sigma} = (-, -, -)$ for an equilibrium point, $\boldsymbol{\sigma} = (0, -, -)$ for a limit cycle, $\boldsymbol{\sigma} = (0, 0, -)$ for a torus or quasiperiodic orbit, and $\boldsymbol{\sigma} = (+, 0, -)$ for a chaotic strange attractor. For more detailed information on chaotic systems, strange attractors, and their applications, see the texts by [Guckenheimer and Holmes, 1983; Lichtenberg and Lieberman, 1983; Schuster, 1988].

Fractal Dimension Strange attractors are **fractals**, that is, they have fractional dimension. Of the many possible definitions of fractional dimension, the simplest is the Kolmogorov **capacity dimension**, which is a simplification of the Hausdorff dimension. The capacity dimension d_C of a set is defined as

$$d_C \triangleq \lim_{\varepsilon \to 0} \frac{\ln N(\varepsilon)}{\ln (1/\varepsilon)}, \tag{4.2-70}$$

where $N(\varepsilon)$ is the minimum number of n_x-dimensional cubes of side ε needed to cover the set. For standard Euclidean objects the capacity dimension is an integer. For a point, $N(\varepsilon) = 1$ and $d_C = 0$. For a curve of length l, $N(\varepsilon) \to l/\varepsilon$ as $\varepsilon \to 0$. Thus $d_C = 1$ for a curve. For an area A, $N(\varepsilon) \to A/\varepsilon^2$ as $\varepsilon \to 0$, and $d_C = 2$. But in general the capacity dimension may be fractional. For example, the Cantor middle-thirds set, obtained by successively deleting the middle third of a line segment and each resulting segment, has a capacity dimension $d_C = \ln(2)/\ln(3) = 0.6309\ldots$.

The **Lyapunov dimension** [Farmer et al., 1983] of a trajectory is defined as

$$d_L \triangleq k + \frac{\sum_{j=1}^{k} \{\sigma_j\}}{|\sigma_{k+1}|}, \tag{4.2-71}$$

where $\sigma_1 \geq \sigma_2 \geq \cdots \geq \sigma_{n_x}$ are the Lyapunov exponents and k is the largest integer for which $\sigma_1 \geq \sigma_2 \geq \cdots \geq \sigma_k \geq 0$. If $\sigma_j \geq 0$ for all $j = 1, \ldots, n_x$, then $d_L = n_x$. If $\sigma_j < 0$ for all $j = 1, \ldots, n_x$, then $d_L = 0$.

The Hausdorff and capacity dimensions are usually (but not always) equal [Farmer et al., 1983], and the common value is called the **fractal dimension** d_F. The Lyapunov dimension defined by (4.2-71) was introduced by Kaplan and Yorke [1979] as a lower bound on the fractal dimension, $d_L \leq d_F$, along with a conjecture that $d_L = d_F$ for most attractors in dissipative systems. This conjecture has not been proven in general, but several results [Farmer et al., 1983] tend to support its validity.

4.3 LYAPUNOV'S SECOND METHOD

In this section we present a qualitative approach to stability analysis, valid for both linear and nonlinear systems, known as **Lyapunov's second (or direct) method**. This approach was developed by A. M. Lyapunov in Russia in 1892. His original paper was translated into French in 1907, but the approach did not receive much attention in the West until around 1960, about the same time that another more-recent Russian development, that of Pontryagin's minimum principle for optimal control also become known. Since that time, both Lyaponov's second method and Pontryagin's minimum principle (Chapter 7) have been important research topics in control theory.

Lyapunov's second method is closely related to the idea that the energy function for a damped-spring-mass system with no external force input will decrease along any trajectory of the system. The method requires finding a function of the state (such as energy for some systems) called a Lyapunov function, which has some very special properties. The process of finding such functions is still very much an art, relying heavily on the analyst's experience and intuition. We present the fundamental results of Lyapunov's approach, and their rationale, but we leave out most of the mathematical details of the various proofs. For these we refer the reader to two very good introductions: the paper by Kalman and Bertram [1960] and the introductory text by LaSalle and Lefschetz [1961], and to advanced texts such as Hahn [1967].

Consider the autonomous system

$$\dot{\mathbf{x}} = \mathbf{f}(\mathbf{x}), \tag{4.3-1}$$

where $\mathbf{f}(\cdot)$ is C^1 (continuous and has continuous first partial derivatives) and $\mathbf{f}(\bar{\mathbf{x}}) = \mathbf{0}$, so that $\mathbf{x} = \bar{\mathbf{x}}$ is an equilibrium for the system. We present the analysis in terms of the zero equilibrium state. The stability of any other equilibrium point can be obtained by simply translating the coordinate system so that the equilibrium of interest is at the origin in the new coordinates.

Lyapunov Functions

Suppose we could find a function $V(\mathbf{x})$ analogous to energy in dissipative mechanical systems, or simply "distance to the origin" in others, whose value $V[\mathbf{x}(t)]$ always decreases along trajectories $\mathbf{x}(t)$ of the system (4.3-1), as illustrated in Figure 4.3-1. It is easy to show that the origin is asymptotically stable. The steps leading to this conclusion are illustrated in the following example.

Example 4.3-1: A Damped Linear Oscillator For the system

$$\dot{x}_1 = x_2$$
$$\dot{x}_2 = -x_1 - x_2,$$

consider the distance-squared function

$$V(\mathbf{x}) = x_1^2 + x_2^2.$$

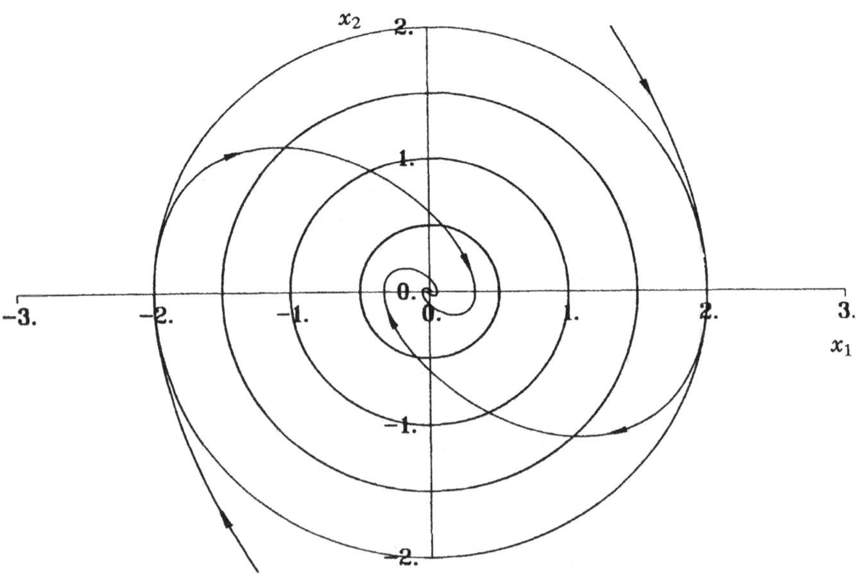

Figure 4.3-1 An example of Lyapunov stability analysis.

As illustrated in Figure 4.3-1 the function $V(\mathbf{x})$ has a unique global minimum at $\mathbf{x} = \mathbf{0}$ and its constant value curves, $V(\mathbf{x}) = c$, surround the origin in concentric circles centered at the origin.

Along a trajectory $\mathbf{x}(t)$ we have

$$\frac{dV[\mathbf{x}(t)]}{dt} = \frac{\partial V}{\partial x_1}\dot{x}_1 + \frac{\partial V}{\partial x_2}\dot{x}_2 = -2x_2^2 \leq 0.$$

Since $V[\mathbf{x}(t)]$ never increases along trajectories, any region of the form $\{\mathbf{x}|V(\mathbf{x}) \leq c\}$, with $c \geq 0$, is a positively invariant set; trajectories starting in such a region remain in the region for all $t \geq 0$. Since this result holds for arbitrarily small $c \geq 0$, the origin is stable.

Furthermore, $dV[\mathbf{x}(t)]/dt < 0$, except on the x_1-axis, where $dV[\mathbf{x}(t)]/dt = 0$. Since the equations of motion guarantee that no trajectory (except the equilibrium solution $\mathbf{x}(t) \equiv \mathbf{0}$) remains on the x_1-axis, it follows that $\mathbf{x}(t)$ penetrates the contours of constant $V(\mathbf{x})$, from higher values of $V(\mathbf{x})$ to lower values. We conclude that $V[\mathbf{x}(t)] \to 0$ as $t \to \infty$, which implies that $\mathbf{x}(t) \to 0$, since the minimum value of $V(\mathbf{x})$ occurs at $\mathbf{x} = \mathbf{0}$. These results hold for all trajectories $\mathbf{x}(t)$, so the origin is globally asymptotically stable. ■

The analysis in the previous example is typical of Lyapunov's direct method, where the objective is to find a certain type of function $V(\mathbf{x})$, and then draw stability conclusions based on $V[\mathbf{x}(t)]$, without having to integrate the equations of motion to determine $\mathbf{x}(t)$. The particular choice of $V(\cdot)$ function used in Example 4.3-1 cannot be used for all systems of the form (4.3-1). However, the procedure is easily generalized to include a much broader, more flexible class of functions.

The essential properties that we want a $V(\cdot)$ function to have, at least in some neighborhood of the origin, are the following:

(i) $V(\mathbf{x})$ has a minimum value at $\mathbf{x} = \mathbf{0}$;

(ii) Surfaces of $V(\mathbf{x}) = c =$ constant surround the origin, with the regions $\{\mathbf{x}|V(\mathbf{x}) \leq c\}$ being bounded and nested: $\{\mathbf{x}|V(\mathbf{x}) \leq c_1\} \subset \{\mathbf{x}|V(\mathbf{x}) \leq c_2\}$ for $c_1 < c_2$;

(iii) $V[\mathbf{x}(t)]$ decreases along trajectories $\mathbf{x}(t)$ of (4.3-1).

In order to establish that these conditions hold for a particular function $V(\mathbf{x})$, we introduce some definitions and auxiliary results. Also, we require that $V(\mathbf{x})$ be C^1, so that we can compute $dV[\mathbf{x}(t)]/dt$.

Positive Definite Functions A continuous function $V(\mathbf{x})$ is **positive definite**, in an open region X of state space containing $\mathbf{x} = \mathbf{0}$, if the following conditions hold:

(i) $V(\mathbf{0}) = 0$.

(ii) $V(\mathbf{x}) > 0$ for all nonzero $\mathbf{x} \in X$.

A function is **positive semidefinite** if $V(\mathbf{0}) = 0$ and $V(\mathbf{x}) \geq 0$ for all nonzero $\mathbf{x} \in X$. A continuous function $W(\mathbf{x})$ is **negative definite**, in an open region X containing $\mathbf{x} = \mathbf{0}$, if $W(\mathbf{0}) = 0$ and $W(\mathbf{x}) < 0$ for all nonzero $\mathbf{x} \in X$. The function is **negative semidefinite** if $W(\mathbf{0}) = 0$ and $W(\mathbf{x}) \leq 0$ for all nonzero $\mathbf{x} \in X$. An **indefinite** function is one that is neither positive semidefinite nor negative semidefinite. Such a function has both positive and negative values near $\mathbf{x} = \mathbf{0}$, while definite and semidefinite functions do not change sign. We are primarily concerned with positive definite $V(\mathbf{x})$ functions.

A positive definite function $V(\mathbf{x})$ has an isolated global minimum with respect to $\mathbf{x} \in X$ at $\mathbf{x} = \mathbf{0}$. Furthermore, for sufficiently small $c > 0$, regions of the form $\{\mathbf{x} | V(\mathbf{x}) \leq c\}$ are bounded, nested, and contain the origin $\mathbf{x} = \mathbf{0}$ as an interior point. A proof of these statements can be seen by noting that $V(\mathbf{x}) = 0$ at $\mathbf{x} = \mathbf{0}$ and that if we take a small step away from the origin in any direction we arrive at points $\mathbf{x} \in X$ where $V(\mathbf{x}) > 0$. Thus, since $V(\cdot)$ is continuous, for sufficiently small $c > 0$ the surface $V(\mathbf{x}) = c$ encloses the origin and the region $V(\mathbf{x}) \leq c$ is bounded. The nesting property also follows from continuity. For a detailed proof see Barbishin and Krasovskiĭ[1953] and Barbishin [1970]. For a less stringent requirement than positive definiteness see Chamberland and Lewis [1994].

In two dimensions the function

$$V(x_1, x_2) = x_1^2 + x_2^2$$

is an example of a positive definite function. However, a function such as

$$V(x_1, x_2) = x_1^2$$

is only positive semidefinite, since it has a value of zero at places other than at $\mathbf{x} = \mathbf{0}$. For $c > 0$ the regions $V(\mathbf{x}) \leq c$ are unbounded vertical bands, between the lines $x_1 = \pm\sqrt{c}$. As another example, the function

$$V(x_1, x_2) = x_1 x_2$$

is indefinite and the $V(\mathbf{x}) =$ constant contours are hyperbolas, as illustrated in Figure 1.2-2. Note that for the latter two example functions, the regions $V(\mathbf{x}) \leq c$ are not closed and bounded regions surrounding the origin.

These three examples are all special cases of a **quadratic form**

$$V(\mathbf{x}) = \mathbf{x}^{\mathrm{T}}\mathbf{P}\mathbf{x} = [x_1 \quad \cdots \quad x_{n_x}]\begin{bmatrix} p_{11} & \cdots & p_{1n_x} \\ \vdots & \ddots & \vdots \\ p_{n_x 1} & \cdots & p_{n_x n_x} \end{bmatrix}\begin{bmatrix} x_1 \\ \vdots \\ x_{n_x} \end{bmatrix}$$

$$= \sum_{i=1}^{n_x} \sum_{j=1}^{n_x} p_{ij} x_i x_j, \tag{4.3-2}$$

where $\mathbf{P} = [p_{ij}]$ is a constant $n_x \times n_x$ matrix. In two dimensions the $V(\mathbf{x}) =$ constant contours for a positive definite quadratic form are ellipses. In three dimensions the contours are ellipsoidal surfaces. If $\mathbf{P}$ is **symmetric** ($\mathbf{P}^{\mathrm{T}} = \mathbf{P}$, i.e., $p_{ij} = p_{ji}$), then the axes of the ellipse are aligned with the eigenvectors of the matrix $\mathbf{P}$.

From **Sylvester's theorem** [Hahn, 1967, p. 100], the quadratic form (4.3-2) is positive definite if and only if the determinants of the principle minors are all positive:

$$|p_{11}| > 0, \qquad \begin{vmatrix} p_{11} & p_{12} \\ p_{21} & p_{22} \end{vmatrix} > 0, \qquad \ldots, \qquad \begin{vmatrix} p_{11} & \cdots & p_{1n_x} \\ \vdots & \ddots & \vdots \\ p_{n_x 1} & \cdots & p_{n_x n_x} \end{vmatrix} > 0. \tag{4.3-3}$$

As an alternate test, suppose that the matrix $\mathbf{P}$ is symmetric. Then (a) the eigenvalues of $\mathbf{P}$ are all real and (b) the quadratic form (4.3-2) is positive definite if and only if the eigenvalues of $\mathbf{P}$ are all positive [Kailath, 1980, pp. 663–667].

Occasionally we consider functions of the form

$$V(\mathbf{x}) = \mathbf{x}^{\mathrm{T}}[\mathbf{P}(\mathbf{x})]\mathbf{x},$$

where the elements of the matrix $\mathbf{P}$ are themselves functions of $\mathbf{x}$. In this case, a sufficient condition for $V(\mathbf{x})$ to be positive definite is that Sylvester's conditions (4.3-3) hold for all $\mathbf{x}$. Similarly, if all of the eigenvalues of $\mathbf{P}(\mathbf{x})$ are positive for all $\mathbf{x}$, then $V(\mathbf{x})$ is positive definite. However, it should be emphasized that when $\mathbf{P} = \mathbf{P}(\mathbf{x})$, these conditions are only sufficient conditions, whereas they are necessary and sufficient conditions when $\mathbf{P}$ is a constant matrix (Exercise 4-7).

The assumption that $\mathbf{P}$, or $\mathbf{P}(\mathbf{x})$, is symmetric is not restrictive, since any function of the form $V(\mathbf{x}) = \mathbf{x}^{\mathrm{T}}\mathbf{P}\mathbf{x}$ can be written equivalently in terms of a symmetric matrix. To do so for any matrix $\mathbf{P}$, construct the two matrices

$$\mathbf{P}_s = \tfrac{1}{2}[\mathbf{P} + \mathbf{P}^{\mathrm{T}}]$$
$$\mathbf{P}_{ss} = \tfrac{1}{2}[\mathbf{P} - \mathbf{P}^{\mathrm{T}}],$$

where $\mathbf{P}_s$ is symmetric and $\mathbf{P}_{ss}$ is **skew-symmetric** ($\mathbf{P}_{ss}^{\mathrm{T}} = -\mathbf{P}_{ss}$). Then $V(\mathbf{x}) = \mathbf{x}^{\mathrm{T}}\mathbf{P}\mathbf{x}$ becomes

$$V(\mathbf{x}) = \mathbf{x}^{\mathrm{T}}[\mathbf{P}_s + \mathbf{P}_{ss}]\mathbf{x} = \mathbf{x}^{\mathrm{T}}\mathbf{P}_s\mathbf{x}, \tag{4.3-4}$$

since the term $\mathbf{x}^{\mathrm{T}}\mathbf{P}_{ss}\mathbf{x}$ is identically zero for any skew-symmetric matrix $\mathbf{P}_{ss}$. To verify this, consider the scalar-valued function

$$\psi(\mathbf{x}) = \mathbf{x}^{\mathrm{T}}\mathbf{P}_{ss}\mathbf{x}.$$

Since $\psi(\mathbf{x})$ is a scalar, equal to the transpose of itself, we also have

$$\psi(\mathbf{x}) = \mathbf{x}^{\mathrm{T}}\mathbf{P}_{ss}^{\mathrm{T}}\mathbf{x} = -\mathbf{x}^{\mathrm{T}}\mathbf{P}_{ss}\mathbf{x}.$$

Adding these two results, we conclude that $\psi(\mathbf{x}) \equiv 0$.

From now on, when discussing quadratic forms, we always assume that the matrix $\mathbf{P}$ in (4.3-2) is symmetric. We use the notations $\mathbf{P} > 0$ (positive definite) and $\mathbf{P} \geq 0$ (positive semidefinite) to indicate the positive definiteness or positive semidefiniteness, respectively, of the quadratic form (4.3-2) corresponding to the matrix $\mathbf{P}$. Similar notations also apply for negative definite ($\mathbf{P} < 0$) and negative semidefinite ($\mathbf{P} \leq 0$) quadratic forms.

Example 4.3-2: Positive Definite Quadratic Form In three dimensions, consider the function

$$\begin{aligned} V(\mathbf{x}) &= 2x_1^2 + 2x_1x_2 + 4x_2^2 - 4x_1x_3 + 5x_3^2 \\ &= [x_1 \quad x_2 \quad x_3]\begin{bmatrix} 2 & 1 & -2 \\ 1 & 4 & 0 \\ -2 & 0 & 5 \end{bmatrix}\begin{bmatrix} x_1 \\ x_2 \\ x_3 \end{bmatrix} \\ &= \mathbf{x}^{\mathrm{T}}\mathbf{P}\mathbf{x}. \end{aligned}$$

In its expanded form it is not at all clear that $V(\mathbf{x})$ is positive definite, because of the x_1x_2 and x_1x_3 terms. But in matrix form we can apply Sylvester's theorem, yielding principle minor determinants

$$|2| = 2, \qquad \begin{vmatrix} 2 & 1 \\ 1 & 4 \end{vmatrix} = 7, \qquad \begin{vmatrix} 2 & 1 & -2 \\ 1 & 4 & 0 \\ -2 & 0 & 5 \end{vmatrix} = 19,$$

which are all positive. Thus $V(\mathbf{x})$ is positive definite. Alternatively, the fact that the eigenvalues $(\mu_1, \mu_2, \mu_3) = (6.1028, 4.1464, 0.7509)$ of $\mathbf{P}$ of all positive implies that the quadratic form $V(\mathbf{x})$ is positive definite. ■

Bounded V(x) Contours For a positive definite quadratic form, the ellipsoidal regions $V(\mathbf{x}) \leq c$ are all bounded regions containing the origin, and the regions are nested for increasing values of c. In a more general setting, where $V(\mathbf{x})$ is not a quadratic form, positive definiteness alone is not sufficient for the regions $V(\mathbf{x}) \leq c$ to be bounded. This boundedness condition plays an important role in Lyapunov stability analysis, which we illustrate shortly with an example. First, we present conditions that ensure that the $V(\mathbf{x}) \leq c$ regions are nested and bounded.

In Barbishin and Krasovskiĭ[1953] and Barbishin [1970, p. 22; see also Letov, 1961, pp. 8–11] it is shown that if $V(\mathbf{x})$ is positive definite and $V(\mathbf{x}) \rightarrow \infty$ as $\|\mathbf{x}\| \rightarrow \infty$ (in any direction), then for any $c > 0$, $V(\mathbf{x}) = c$ defines a **simple closed surface** enclosing the origin, and $\{\mathbf{x} | V(\mathbf{x}) \leq c_1\} \subset \{\mathbf{x} | V(\mathbf{x}) \leq c_2\}$ for all c_1, c_2 with $c_1 < c_2$. The term "simple" means that the surface does not intersect itself. Figure "8" regions, as in Figure 1.5-3, are examples of curves that are not "simple." The term "closed" means that the surface (enclosing the origin) intersects every ray from the origin to infinity. If each ray only intersects the surface once, then a simple closed surface is topologically equivalent to an n_x-dimensional sphere.

A positive definite quadratic form

$$V(\mathbf{x}) = \mathbf{x}^{\mathrm{T}}\mathbf{P}\mathbf{x}, \qquad \mathbf{P} > 0$$

is an example of a positive definite function that satisfies the condition $V(\mathbf{x}) \rightarrow \infty$ as $\|\mathbf{x}\| \rightarrow \infty$. The surfaces $V(\mathbf{x}) = c > 0$ are concentric ellipsoids centered at $\mathbf{x} = \mathbf{0}$, and they can be stretched into the shape of n_x-dimensional spheres by changing the scales of the axes.

A more stringent condition for bounded and nested $V(\mathbf{x}) \leq c$ regions is given by $V(\mathbf{0}) = 0$ with $V(\mathbf{x}) > 0$ and $\partial V(\mathbf{x})/\partial \mathbf{x} \neq \mathbf{0}^{\mathrm{T}}$ except at $\mathbf{x} = \mathbf{0}$. These conditions imply that $V(\mathbf{x})$ is positive definite and $V(\mathbf{x}) \rightarrow \infty$ as $\|\mathbf{x}\| \rightarrow \infty$, with the $V(\mathbf{x}) = c$ surfaces being topologically equivalent to n_x-dimensional spheres. However, situations in which $\partial V(\mathbf{x})/\partial \mathbf{x}$ is zero away from the origin can occur in many applications, as the following example illustrates.

Example 4.3-3: Nested Disconnected V(x) Regions In two dimensions consider the function given by

$$V(\mathbf{x}) = 6r^4 - 22r^3 + 21r^2,$$

where

$$r^2 = x_1^2 + x_2^2.$$

This function is positive definite (see Exercise 4-6) and $V(\mathbf{x}) \rightarrow \infty$ as $\|\mathbf{x}\| \rightarrow \infty$.

However, the gradient vector $\partial V(\mathbf{x})/\partial \mathbf{x}$, which is perpendicular to the $V(\mathbf{x}) = c$ surfaces, is zero at places other than the origin, that is, along the $V(\mathbf{x})$ contours corresponding to the local-maximum ridge and the local-minimum valley in Figure 4.3-2*a*. However, as indicated in Figure 4.3-2*b*, the $V(\mathbf{x}) = c$ surfaces do define nested bounded regions, but not all of the regions are spherical. For example the shaded $V(\mathbf{x}) \leq 4$ region in Figure 4.3-2*b* consists of a spherical ball inside a spherical shell. The $V(\mathbf{x}) \leq 3$ region has a similar structure, contained inside the $V(\mathbf{x}) \leq 4$ region. ■

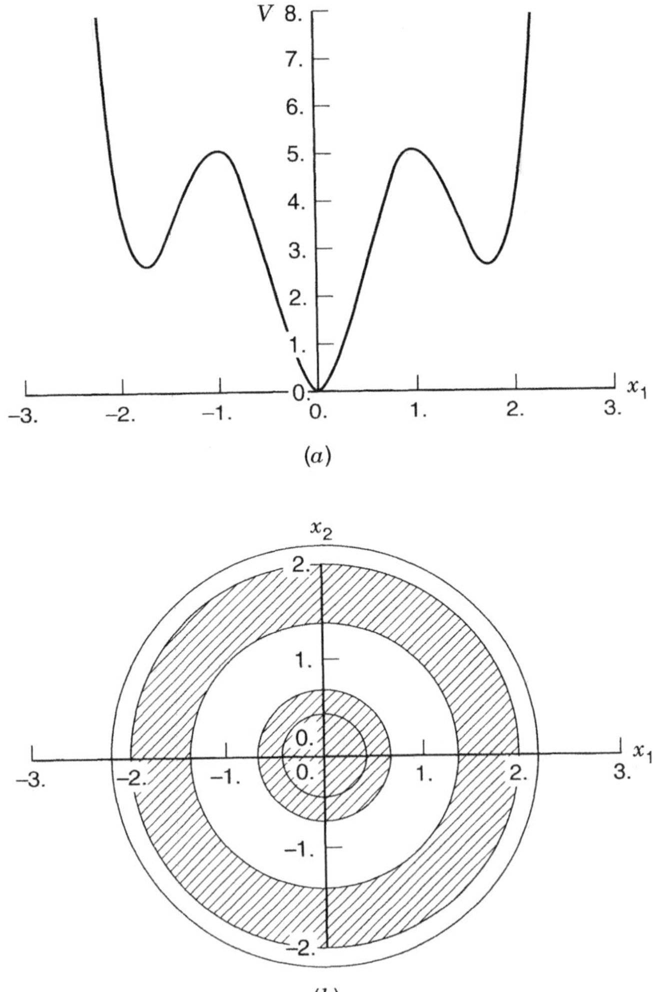

Figure 4.3-2 Nested regions $V(\mathbf{x}) = 6r^4 - 22r^3 + 21r^2 \leq c$. (*a*) Radial cross section. (*b*) Contours.

Lyapunov Sufficiency Conditions

Positive definite functions that satisfy certain additional conditions provide a means for testing stability in nonlinear dynamical systems. The procedure employs the following results:

Lyapunov's Stability Theorem: *The equilibrium solution at* $\mathbf{x} = \mathbf{0}$ *for (4.3-1) is stable if a* C^1 *function* $V(\mathbf{x})$ *can be found satisfying the following conditions in some open region* X *containing the origin as an interior point:*

(i) $V(\mathbf{0}) = 0$,
(ii) $V(\mathbf{x}) > 0$ *for all nonzero* $\mathbf{x} \in X$,
(iii) $\dot{V}(\mathbf{x}) \leq 0$ *for all* $\mathbf{x} \in X$ *where*

$$\dot{V}(\mathbf{x}) \triangleq \frac{\partial V(\mathbf{x})}{\partial \mathbf{x}} \mathbf{f}(\mathbf{x}) \tag{4.3-5}$$

is called the **Lyapunov derivative** [*at a point* $\mathbf{x}$ *on a trajectory* $\mathbf{x}(t)$]. *The origin for (4.3-1) is asymptotically stable if a function can be found such that* (i)–(iii) *hold, and*

(iv) $\dot{V}(\mathbf{x}) \leq 0$ *is not identically zero on any trajectory* $\mathbf{x}(t)$ *in* X, *except at the equilibrium* $\mathbf{x} = \mathbf{0}$.
Alternatively, and more restrictively:
(v) $\dot{V}(\mathbf{x}) < 0$ *for all nonzero* $\mathbf{x} \in X$.
If, in addition, the "growth" condition,
(vi) $V(\mathbf{x}) \to \infty$ *as* $\|\mathbf{x}\| \to \infty$,
is satisfied, regardless of direction, then the origin is globally asymptotically stable.

A C^1 function $V(\mathbf{x})$ satisfying conditions (i)–(iii) above is called a **Lyapunov function** on the open region X for the system (4.3-1). Thus the problem of establishing stability of the origin for the system (4.3-1) is to find a positive definite function $V(\mathbf{x})$ with at least a negative semidefinite Lyapunov derivative $\dot{V}(\mathbf{x})$ on some open region X containing the origin. Often X is taken to be a region of the form $X = \{\mathbf{x} | V(\mathbf{x}) \leq c\}$. For global asymptotic stability X needs to be the entire state space.

Proof of the Lyapunov's stability theorem can be found in LaSalle and Lefschetz [1961]. However, we have already outlined the basic ideas of the proof, in Example 4.3-1 and in our discussion of positive definite functions and bounded and nested $V(\mathbf{x}) \leq c$ regions. In essence, conditions (i)–(iii) ensure that, at least for sufficiently small $c > 0$, the regions $V(\mathbf{x}) \leq c$ are bounded positively invariant sets. Hence the origin is stable. The additional conditions (iv) or (v) imply that trajectories starting in $V(\mathbf{x}) \leq c$ penetrate surfaces of constant $V(\mathbf{x})$, from higher to lower values of $V(\mathbf{x})$. Therefore, at least for small c values, the trajectories approach the origin, where $V(\mathbf{x})$ has a global minimum, and the origin is thus asymptotically stable. Finally, the growth condition (vi) ensures that all of the regions $V(\mathbf{x}) \leq c$ are nested and bounded, especially for very large c values. Therefore none of the trajectories can approach infinity while decreasing $V[\mathbf{x}(t)]$, so the asymptotic stability is global.

Example 4.3-4: Energy as a Lyapunov Function In many physical systems the total energy of the system is a good candidate for a Lyapunov function. Consider the nonlinear system introduced in Example 4.2-5, with state equations

$$\begin{aligned} \dot{x}_1 &= x_2 \\ \dot{x}_2 &= -\omega^2 x_1 - 2\zeta\omega x_2^n, \end{aligned}$$

where $n \geq 1$ and ω and ζ are positive. The energy per unit mass

$$V(\mathbf{x}) = \tfrac{1}{2}\, x_2^2 + \tfrac{1}{2}\, \omega^2 x_1^2$$

is a positive definite quadratic form, so the stability conditions (i), (ii), and the growth condition (vi) are satisfied. Along system trajectories we have

$$\dot{V}(\mathbf{x}) = \omega^2 x_1 \dot{x}_1 + x_2 \dot{x}_2 = -2\zeta\omega x_2^{n+1}.$$

Thus for n odd (iii) is satisfied and the origin is stable. Furthermore, since $\dot{x}_2 \neq 0$ on $x_2 = 0$ except at the origin, condition (iv) is also satisfied. Thus for the case where n is an odd integer, the origin is globally asymptotically stable. Figure 4.3-3 illustrates the geometry of the Lyapunov function and the system trajectories for $n = 3$, with $\omega = 2$ and $\zeta = 1$. ■

Lyapunov functions for a given system are not unique. Furthermore, while the conditions in Lyapunov's stability theorem are sufficient conditions, they are not constructive conditions; they do not tell us how to find a Lyapunov function for a particular system. It turns out that conditions (i)–(iii) are also necessary conditions for stability [Kalman and Bertram, 1960]. While it may be reassuring to know that, if the origin is stable, then a Lyapunov function does indeed exist, this fact is of no direct help in the search for a Lyapunov function.

It should also be noted that failure to find a Lyapunov function for a given system does not necessarily imply that the system is unstable, only that we may need to try some other candidate $V(\cdot)$ function. There are, however, various **instability theorems** available to help detect an unstable equilibrium. For example, if we find a positive definite function $V(\mathbf{x})$ with $\dot{V}(\mathbf{x}) > 0$ for all nonzero $\mathbf{x}$ in a neighborhood of the origin, then the origin is clearly unstable.

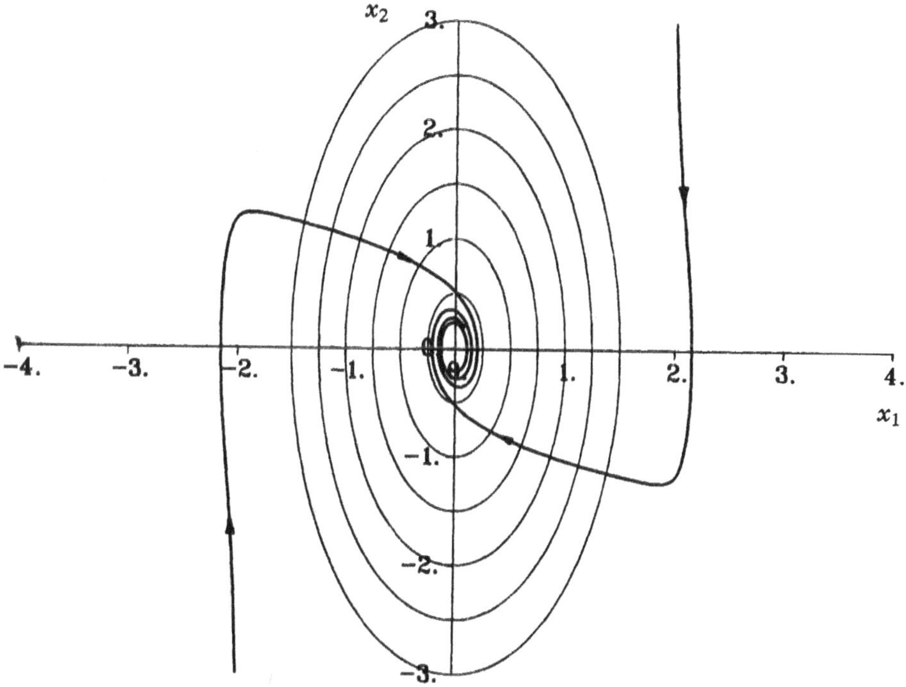

Figure 4.3-3 Lyapunov contours and trajectories for Example 4.3-4.

For less restrictive instability conditions see LaSalle and Lefschetz [1961, pp. 38–40].

The growth condition (vi) in Lyapunov's stability theorem provides a test for global asymptotic stability. It is usually applied with $V(\mathbf{x})$ functions that also satisfy $\partial V(\mathbf{x})/\partial \mathbf{x} \neq \mathbf{0}^T$ for $\mathbf{x} \neq \mathbf{0}$, so that the $V(\mathbf{x}) \leq c$ regions are topologically equivalent to nested spherical regions. For example, the $V(\mathbf{x})$ function in Example 4.3-3 could not be used to establish global asymptotic stability, because of the valley outside of the ridge in Figure 4.3-2*a*. Even though the growth condition (vi) is satisfied it would be impossible to satisfy conditions (iv) or (v), which imply that $V[\mathbf{x}(t)]$ always decreases. The $V(\mathbf{x})$ function in Example 4.3-3 can only establish local asymptotic stability.

The geometry of the growth condition (vi) is demonstrated in the following example.

Example 4.3-5: Unbounded Lyapunov Contours To illustrate the importance of the growth condition (vi) consider a two-dimensional state space and the function

$$V(\mathbf{x}) = x_2^2 + \frac{x_1^2}{1 + x_1^2}. \tag{4.3-6}$$

We use this function, analogous to one used in Barbishin and Krasovskiĭ[1953], Letov [1961], and Barbishin [1970] to demonstrate the effect of not having bounded contours for a Lyapunov function. The $V(\mathbf{x})$ = constant contours of this function are plotted in Figure 4.3-4.

The function given by (4.3-6) is positive definite, but it only has bounded contours for $V(\mathbf{x}) < 1$. This can be seen by noting that

$$V(\mathbf{x}) \rightarrow 1 + x_2^2 \qquad \text{as } |x_1| \rightarrow \infty.$$

Thus the $V(\mathbf{x}) = c$ contours do not cross the x_1-axis for $c \geq 1$. Along the x_1-axis

$$V(x_1, 0) \rightarrow 1 \qquad \text{as } |x_1| \rightarrow \infty.$$

Thus the growth condition (vi) is not satisfied on the x_1-axis. Also note that the gradient

$$\frac{\partial V(\mathbf{x})}{\partial \mathbf{x}} = \begin{bmatrix} \dfrac{2x_1}{(1 + x_1^2)^2} & 2x_2 \end{bmatrix}$$

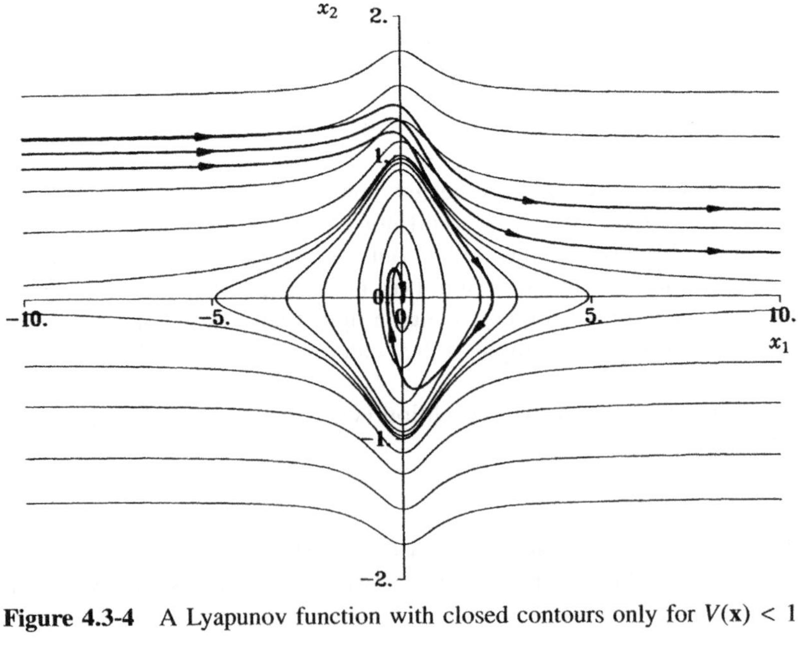

Figure 4.3-4 A Lyapunov function with closed contours only for $V(\mathbf{x}) < 1$.

is zero not only at $(x_1, x_2) = (0, 0)$, as it must be since $V(\mathbf{x})$ is positive definite and therefore has a minimum at the origin, but the gradient also approaches zero as $\|\mathbf{x}\| \to \infty$ along the x_1-axis. This is because $V(\mathbf{x})$ approaches a constant value along the x_1-axis.

Now, consider the dynamical system

$$\dot{x}_1 = x_2$$

$$\dot{x}_2 = -\frac{x_1}{(1 + x_1^2)^2} - \frac{x_2}{(1 + x_1^2 + x_2^2)^2}.$$

The origin is the only equilibrium point, and near the origin the linearized equations of motion are the same as the damped linear harmonic oscillator in Example 4.3-1. Thus the nonlinear system is at least locally asymptotically stable.

Using the $V(\cdot)$ function in (4.3-6) we have

$$\dot{V}(\mathbf{x}) = \frac{-2x_2^2}{(1 + x_1^2 + x_2^2)^2} \leq 0 \qquad \text{for all } \mathbf{x}.$$

Furthermore, since the x_1-axis is not a trajectory for the system, $V[\mathbf{x}(t)]$ is always decreasing along any trajectory $\mathbf{x}(t)$, except at the origin.

Since $V(\mathbf{x})$ is positive definite and $V[\mathbf{x}(t)]$ is always decreasing, we might be tempted to conclude (falsely) that the origin is globally asymptotically stable. However, because the growth condition (vi) is not satisfied, we cannot make such a conclusion. The difficulty is that, without the growth condition, some of the $V(\mathbf{x}) \leq c$ regions may be unbounded. As a consequence, even if $V[\mathbf{x}(t)]$ always decreases, some of the trajectories may approach infinity instead of approaching the origin. This is exactly what happens in this example, as illustrated in Figure 4.3-4. Since $\dot{V}(\mathbf{x}) \to 0$ as $\|\mathbf{x}\| \to \infty$, trajectories that are sufficiently far away from the origin asymptotically approach the $V(\mathbf{x})$ contours, which in turn approach infinity as $|x_1| \to \infty$ for $V(\mathbf{x}) \geq 1$. ■

4.4 CONSTRUCTING LYAPUNOV FUNCTIONS

Lyapunov's direct method is a powerful tool for stability analysis, since the approach does not require that we integrate the equations of motion and the method can be applied to higher dimensional nonlinear systems. The main problem in using the method lies in finding a Lyapunov function for a given system. In this section we present some procedures for constructing Lyapunov functions for various types of systems.

Matrix Lyapunov Equation for Linear Systems

To start with a familiar, but very broad class of systems, consider an n_x-dimensional constant-coefficient linear system

$$\dot{\mathbf{x}} = \mathbf{A}\mathbf{x}. \tag{4.4-1}$$

For such systems, as we will see, a necessary and sufficient condition for the origin to be stable is that there be a Lyapunov function expressible as a positive definite quadratic form

$$V(\mathbf{x}) = \mathbf{x}^{\mathrm{T}}\mathbf{P}\mathbf{x}, \tag{4.4-2}$$

where $\mathbf{P}$ is a constant symmetric positive definite $n_x \times n_x$ matrix.

At a point $\mathbf{x}$ on a trajectory of (4.4-1) we have

$$\begin{aligned} \dot{V}(\mathbf{x}) &= \mathbf{x}^{\mathrm{T}}\mathbf{P}\dot{\mathbf{x}} + \dot{\mathbf{x}}^{\mathrm{T}}\mathbf{P}\mathbf{x} \\ &= \mathbf{x}^{\mathrm{T}}[\mathbf{P}\mathbf{A} + \mathbf{A}^{\mathrm{T}}\mathbf{P}]\mathbf{x} \\ &= -\mathbf{x}^{\mathrm{T}}\mathbf{Q}\mathbf{x}, \end{aligned}$$

where

$$\mathbf{P}\mathbf{A} + \mathbf{A}^{\mathrm{T}}\mathbf{P} = -\mathbf{Q}. \tag{4.4-3}$$

Equation (4.4-3) is called the **matrix Lyapunov equation**. For a given symmetric matrix $\mathbf{Q}$, (4.4-3) is a linear equation involving the $n_x(n_x + 1)/2$ unknown elements of the symmetric matrix $\mathbf{P} = [p_{ij}]$, $p_{ij} = p_{ji}$. While this equation is not too difficult to solve by hand for two-dimensional systems, solutions for higher-dimensional systems would be very tedious. The MATLAB command lyap($\mathbf{A}^{\mathrm{T}}, \mathbf{Q}$) is very useful in this regard. We also present a numerical procedure for solving the Lyapunov equation later in this section.

If the equilibrium $\mathbf{x}(t) \equiv \mathbf{0}$ for the linear system (4.4-1) is stable, then it is globally stable. Therefore, in order for the quadratic form $V(\mathbf{x})$ to be a Lyapunov function, we need $\mathbf{P}$ to be positive definite, so that

$$V(\mathbf{x}) > 0 \qquad \text{for all } \mathbf{x} \neq \mathbf{0},$$

and we need $\mathbf{Q}$ to be at least positive semidefinite, to get

$$\dot{V}(\mathbf{x}) = -\mathbf{x}^{\mathrm{T}}\mathbf{Q}\mathbf{x} \leq 0,$$

with an additional requirement that $\dot{V}(\mathbf{x})$ is not identically zero along a nonequilibrium trajectory $\mathbf{x}(t)$. This last requirement is satisfied automatically if $\mathbf{Q}$ is positive definite, since we then have $\dot{V}(\mathbf{x}) < 0$ for $\mathbf{x} \neq \mathbf{0}$.

To construct a Lyapunov function for the linear system (4.4-1), we can choose $\mathbf{Q}$ as any convenient $n_x \times n_x$ symmetric matrix that is nonzero and at least positive semidefinite, $\mathbf{Q} \geq 0$. In order to obtain $\dot{V}(\mathbf{x}) < 0$ for $\mathbf{x} \neq \mathbf{0}$, we choose $\mathbf{Q}$ positive definite. Frequently it is convenient to choose $\mathbf{Q}$ to be the n_x-dimensional identity matrix.

For a given $\mathbf{Q} \geq 0$ we seek a symmetric positive definite matrix $\mathbf{P} > 0$ that satisfies the matrix Lyapunov equation (4.4-3). If, and only if, such a matrix $\mathbf{P}$ can be found, then the corresponding quadratic form $V(\mathbf{x})$ in (4.4-2) will be a Lyapunov function, establishing the stability of the origin for (4.4-1).

Concerning the existence and nature of solutions to the matrix Lyapunov equation, we have [Kalman and Bertram, 1960]:

Bellman's Lemma: *The Lyapunov matrix equation (4.4-3) has a unique solution* $\mathbf{P}$ *(not necessarily positive definite) for any given* $\mathbf{Q}$ *whenever the eigenvalues* $\{\mu_j\}$ *of* $\mathbf{A}$ *satisfy*

$$\mu_j \neq -\mu_k \qquad \text{for all } j, k = 1, \ldots, n_x. \tag{4.4-4}$$

This condition would hold, for example, if all of the eigenvalues were real and negative (or positive). For a proof of this result, see (4.4-15) later in this section. Note that condition (4.4-4) is not satisfied, for example, if there are a conjugate pair of purely imaginary eigenvalues, or if zero is a repeated eigenvalue.

Lyapunov's Lemma: *The origin for (4.4-1) is asymptotically stable,* $Re(\mu_j) < 0$ *for all* j*, if and only if for any symmetric positive definite* $\mathbf{Q}$ *there exists a symmetric positive definite* $\mathbf{P}$ *satisfying (4.4-3).*

Note that Lyapunov's Lemma requires $\mathbf{Q} > 0$ for asymptotic stability. Another useful result can also be given: if a nonzero symmetric positive definite solution $\mathbf{P}$ exists for the Lyapunov matrix equation (4.4-3) with a symmetric matrix $\mathbf{Q} \geq 0$, then origin for (4.4-1) is stable. This result follows from the fact that $\dot{V}$ is then guaranteed to be at least negative semidefinite. These results imply that for a linear system (4.4-1), quadratic forms are the only candidate Lyapunov functions that we need to consider.

Example 4.4-1: Damped Harmonic Oscillator Consider the linear system

$$\ddot{y} + 2\zeta\omega_n\dot{y} + \omega_n^2 y = 0.$$

In matrix form, with $x_1 = y$ and $x_2 = \dot{y}$, we have

$$\begin{bmatrix} \dot{x}_1 \\ \dot{x}_2 \end{bmatrix} = \begin{bmatrix} 0 & 1 \\ -\omega_n^2 & -2\zeta\omega_n \end{bmatrix} \begin{bmatrix} x_1 \\ x_2 \end{bmatrix}.$$

To investigate quadratic forms as candidate Lyapunov functions we set

$$\mathbf{P} = \begin{bmatrix} p_{11} & p_{12} \\ p_{12} & p_{22} \end{bmatrix} > 0 \qquad \mathbf{Q} = \begin{bmatrix} q_{11} & 0 \\ 0 & q_{22} \end{bmatrix} \geq 0.$$

We consider two such Lyapunov functions, to illustrate the difference between positive definite and positive semidefinite choices for $\mathbf{Q}$. For the first Lyapunov function we choose $\mathbf{Q} \geq 0\,(q_{11} \geq 0, q_{22} \geq 0$, not both zero) and for the second we choose $\mathbf{Q} > 0$ (q_{11}, q_{22} both positive). In either case, for the solution $\mathbf{P}$ of (4.4-3) to be positive definite we require

$$0 < p_{11}$$
$$0 < p_{11}p_{22} - p_{12}^2.$$

For the left-hand side of (4.4-3) we have

$$\mathbf{PA} + \mathbf{A}^{\mathrm{T}}\mathbf{P} = \begin{bmatrix} p_{11} & p_{12} \\ p_{12} & p_{22} \end{bmatrix} \begin{bmatrix} 0 & 1 \\ -\omega_n^2 & -2\zeta\omega_n \end{bmatrix} + \begin{bmatrix} 0 & -\omega_n^2 \\ 1 & -2\zeta\omega_n \end{bmatrix} \begin{bmatrix} p_{11} & p_{12} \\ p_{12} & p_{22} \end{bmatrix}$$
$$= \begin{bmatrix} -2p_{12}\omega_n^2 & p_{11} - 2p_{12}\zeta\omega_n - p_{22}\omega_n^2 \\ p_{11} - 2p_{12}\zeta\omega_n - p_{22}\omega_n^2 & 2p_{12} - 4p_{22}\zeta\omega_n \end{bmatrix}.$$

For a specified matrix $\mathbf{Q}$, (4.4-3) yields a system of linear equations for the $n_x(n_x + 1)/2 = 3$ unknowns (p_{11}, p_{12}, p_{22}):

$$\begin{aligned} & -2\omega_n^2 p_{12} & & = -q_{11} \\ p_{11} \quad & -2\zeta\omega_n p_{12} & -\omega_n^2 p_{22} & = 0 \\ & 2p_{12} & -4\zeta\omega_n p_{22} & = -q_{22}. \end{aligned}$$

Now, for example, if we choose

$$\mathbf{Q} = \mathbf{Q}_1 = \begin{bmatrix} 0 & 0 \\ 0 & 2\zeta\omega_n \end{bmatrix},$$

then

$$\mathbf{P} = \mathbf{P}_1 = \frac{1}{2}\begin{bmatrix} \omega_n^2 & 0 \\ 0 & 1 \end{bmatrix}$$

and

$$V_1(\mathbf{x}) = \mathbf{x}^{\mathrm{T}}\mathbf{P}_1\mathbf{x} = \tfrac{1}{2}\omega_n^2 x_1^2 + \tfrac{1}{2}x_2^2$$

is the energy (per unit mass), with

$$\dot{V}_1(\mathbf{x}) = -\mathbf{x}^{\mathrm{T}}\mathbf{Q}_1\mathbf{x} = -2\zeta\omega_n x_2^2.$$

For $\omega_n \neq 0$ we have $\mathbf{P}_1 > 0$, so that $V_1(\mathbf{x}) > 0$ for all $\mathbf{x} \neq \mathbf{0}$, and for $\zeta\omega_n > 0$ we have $\mathbf{Q}_1 \geq 0$ and thus $\dot{V}_1(\mathbf{x}) \leq 0$ for all $\mathbf{x} \neq \mathbf{0}$. Hence the equilibrium at $\mathbf{x} = \mathbf{0}$ is (at least) Lyapunov stable. Furthermore the equations of motion imply that $x_2(t) \not\equiv 0$, except at the origin, so $\dot{V}_1[\mathbf{x}(t)] \not\equiv 0$. Thus, the origin is, in fact, globally asymptotically stable if $\omega_n \neq 0$ and $\zeta\omega_n > 0$.

In general the choice of a Lyapunov function is not unique. As an alternative for our example, which completely avoids the possibility of $\dot{V}(\mathbf{x}) = 0$, we could select any positive definite $\mathbf{Q} > 0$, so that $\dot{V}(\mathbf{x}) < 0$ for $\mathbf{x} \neq \mathbf{0}$. For instance, we could choose

$$\mathbf{Q} = \mathbf{Q}_2 = \mathbf{I},$$

yielding

$$\mathbf{P} = \mathbf{P}_2 = \begin{bmatrix} \dfrac{1 + 4\zeta^2 + \omega_n^2}{4\zeta\omega_n} & \dfrac{1}{2\omega_n^2} \\ \dfrac{1}{2\omega_n^2} & \dfrac{1 + \omega_n^2}{4\zeta\omega_n^3} \end{bmatrix}.$$

From the principal minor determinants of $\mathbf{P}_2$, and Sylvester's theorem (4.3-3), the matrix $\mathbf{P}_2$ is positive definite if and only if $\zeta\omega_n > 0$ and $\omega_n \neq 0$. If so, then

$$V_2(\mathbf{x}) = \mathbf{x}^{\mathrm{T}}\mathbf{P}_2\mathbf{x} > 0$$

and

$$\dot{V}_2(\mathbf{x}) = -\mathbf{x}^{\mathrm{T}}\mathbf{Q}_2\mathbf{x} < 0$$

for all $\mathbf{x} \neq \mathbf{0}$ and again we see that the origin is globally asymptotically stable if $\zeta\omega_n > 0$ and $\omega_n \neq 0$.

Figure 4.4-1 illustrates $V(\mathbf{x})$ = constant contours for the two Lyapunov functions used in this example, for the case where $\zeta = \omega_n = 1/\sqrt{2}$. The first Lyapunov function is appealing because it represents the energy of the system. However, because of the $\dot{x}_1 = x_2$ equation of motion, trajectories of the system are tangent to the $V_1(\mathbf{x})$ contour on the x_1-axis, yielding $\dot{V}_1(\mathbf{x}) = 0$. Whenever this situation occurs we must be concerned with whether or not the trajectory actually penetrates the contour. The second Lyapunov function avoids this problem, since it yields $\dot{V}_2(\mathbf{x}) < 0$ everywhere away from the origin.

We can verify the asymptotic stability conditions, by applying **Routh's procedure** for linear systems [Grantham and Vincent, 1993, p. 119] to the characteristic equation

$$0 = |\mu\mathbf{I} - \mathbf{A}| = \mu^2 + 2\zeta\omega_n\mu + \omega_n^2,$$

which, for a stable system, allows no sign changes in the first column of the Routh array

$$\begin{matrix} 1 & \omega_n^2 \\ 2\zeta\omega_n & 0 \\ \omega_n^2. & \end{matrix}$$

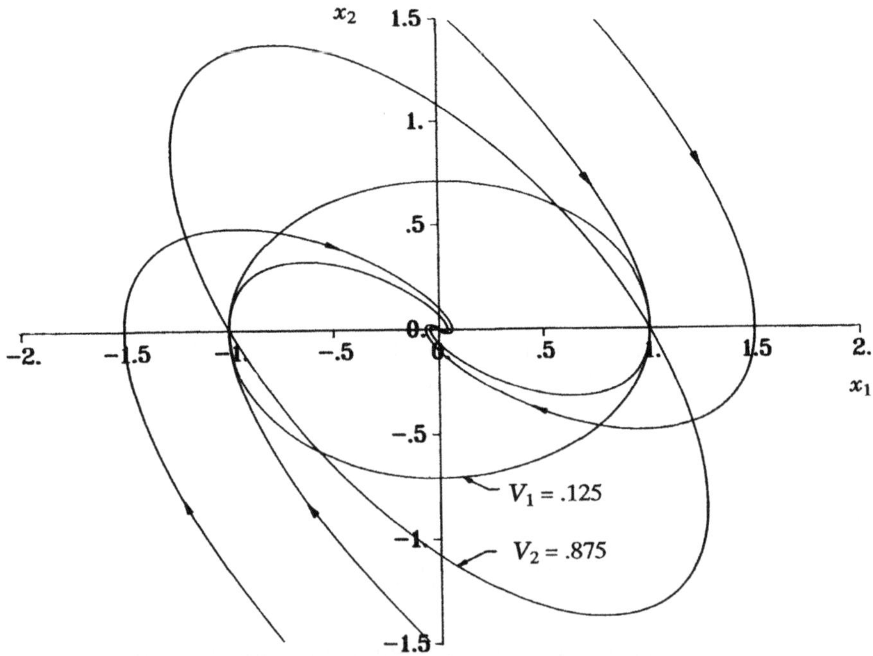

Figure 4.4-1 Lyapunov contours for Example 4.4-1.

Since local asymptotic stability for linear systems implies global asymptotic stability, we again arrive at the result that the origin is globally asymptotically stable if $\zeta\omega_n > 0$ and $\omega_n \neq 0$. ■

Numerical Solution of the Matrix Lyapunov Equation

The matrix Lyapunov equation (4.4-3) occurs often in modern control theory because of its central role in Lyapunov stability theory for linear systems. In this section we present a numerical procedure for solving (4.4-3).

Newton's Identities The algorithm begins by determining the coefficients $\{p_j\}$ in the polynomial characteristic equation

$$\begin{aligned} 0 = |\mu \mathbf{I} - \mathbf{A}| &= \mu^{n_x} + p_{n_x-1}\mu^{n_x-1} + \cdots + p_1\mu + p_0 \\ &= (\mu - \mu_1)(\mu - \mu_2) \cdots (\mu - \mu_{n_x}). \end{aligned} \tag{4.4-5}$$

The coefficients can be computed easily using **Newton's identities,** one form of which follows from expanding (4.4-5) and equating coefficients of powers of μ:

$$\begin{aligned} p_{n_x-1} &= -\sum_j \mu_j \\ p_{n_x-2} &= \sum_{i \neq j} \mu_i\mu_j \\ p_{n_x-3} &= -\sum_{i \neq j \neq k} \mu_i\mu_j\mu_k \\ &\vdots \\ p_0 &= (-1)^{n_x}\mu_1\mu_2 \cdots \mu_{n_x}. \end{aligned} \tag{4.4-6}$$

For our purposes an alternate form is more convenient. Let

$$s_k = \mathrm{Tr}[\mathbf{A}^k], \tag{4.4-7}$$

where the **trace** of a square $n \times n$ matrix $\mathbf{M} = [m_{ij}]$, denoted by $\mathrm{Tr}[\mathbf{M}] = m_{11} + \cdots + m_{nn}$, is the sum of the diagonal elements of $\mathbf{M}$. In terms of the trace of successive powers of $\mathbf{A}$, **Newton's identities** may be written equivalently as

$$
\begin{aligned}
0 &= s_1 + p_{n_x - 1} \\
0 &= s_2 + p_{n_x - 1}s_1 + 2p_{n_x - 2} \\
&\vdots \\
0 &= s_{n_x - 1} + p_{n_x - 1}s_{n_x - 2} + \cdots + p_2 s_1 + (n_x - 1)p_1 \\
0 &= s_{n_x} + p_{n_x - 1}s_{n_x - 1} + \cdots + p_1 s_1 + n_x p_0.
\end{aligned}
\tag{4.4-8}
$$

Equations (4.4-8) can be solved successively for $p_{n_x - 1}, p_{n_x - 2}, \ldots, p_0$.

Jameson's Algorithm Now consider the matrix Lyapunov equation

$$
\mathbf{PA} + \mathbf{A}^{\mathrm{T}}\mathbf{P} = \mathbf{C} \triangleq -\mathbf{Q}, \tag{4.4-9}
$$

where $\mathbf{Q}$ is a specified $n_x \times n_x$ matrix. If $\mathbf{Q}$ is symmetric, then the solution $\mathbf{P}$ is also symmetric.

Jameson's algorithm [Rothschild and Jameson, 1970] for constructing a solution $\mathbf{P}$ consists of forming the sequence of matrices

$$
\begin{array}{lcl}
\mathbf{C}_0 & \triangleq & \mathbf{0} \\
\mathbf{C}_1 & \triangleq & \mathbf{C} = \mathbf{PA} + \mathbf{A}^{\mathrm{T}}\mathbf{P} \\
\mathbf{C}_2 & \triangleq & \mathbf{C}_1\mathbf{A} - \mathbf{A}^{\mathrm{T}}\mathbf{C}_1 + \mathbf{A}^{\mathrm{T}}\mathbf{C}_0\mathbf{A} = \mathbf{PA}^2 - (\mathbf{A}^{\mathrm{T}})^2\mathbf{P} \\
\vdots & \vdots & \qquad\qquad\qquad\qquad\qquad \vdots \\
\mathbf{C}_{n_x} & \triangleq & \mathbf{C}_{n_x - 1}\mathbf{A} - \mathbf{A}^{\mathrm{T}}\mathbf{C}_{n_x - 1} + \mathbf{A}^{\mathrm{T}}\mathbf{C}_{n_x - 2}\mathbf{A} = \mathbf{PA}^{n_x} - (-1)^{n_x}(\mathbf{A}^{\mathrm{T}})^{n_x}\mathbf{P}.
\end{array}
$$

This sequence of matrices can be generated using the recurrence relation

$$
\mathbf{C}_k \triangleq \mathbf{CA}^{k-1} - \mathbf{A}^{\mathrm{T}}\mathbf{C}_{k-1}, \qquad k = 2, \ldots, n_x, \tag{4.4-10}
$$

where $\mathbf{C}_0 \triangleq \mathbf{0}$ and $\mathbf{C}_1 \triangleq \mathbf{C} = -\mathbf{Q}$. In terms of these matrices we have

$$
\begin{aligned}
\mathbf{C}_{n_x} - p_{n_x - 1}\mathbf{C}_{n_x - 1} + \cdots + (-1)^{n_x - 2}p_2\mathbf{C}_2 + (-1)^{n_x - 1}p_1\mathbf{C}_1 \\
= \mathbf{P}\{\mathbf{A}^{n_x} - p_{n_x - 1}\mathbf{A}^{n_x - 1} + \cdots + (-1)^{n_x - 2}p_2\mathbf{A}^2 + (-1)^{n_x - 1}p_1\mathbf{A}\} \\
+ (-1)^{n_x}\{-(\mathbf{A}^{\mathrm{T}})^{n_x} - p_{n_x - 1}(\mathbf{A}^{\mathrm{T}})^{n_x - 1} - \cdots - p_2(\mathbf{A}^{\mathrm{T}})^2 - p_1\mathbf{A}^{\mathrm{T}}\}\mathbf{P}.
\end{aligned}
\tag{4.4-11}
$$

From the Cayley–Hamilton theorem both $\mathbf{A}$ and $\mathbf{A}^{\mathrm{T}}$ satisfy the characteristic equation (4.4-5). Therefore the sum in the last set of braces on the right-hand side of (4.4-11) is equal to $p_0\mathbf{I}$. Substituting this result into (4.4-11) and solving for $\mathbf{P}$ yields

$$\mathbf{P} = \mathbf{W}\mathbf{J}^{-1}, \tag{4.4-12}$$

where

$$\mathbf{W} = \mathbf{C}_{n_x} - p_{n_x-1}\mathbf{C}_{n_x-1} + \cdots + (-1)^{n_x-2}p_2\mathbf{C}_2 + (-1)^{n_x-1}p_1\mathbf{C}_1 \tag{4.4-13}$$

$$\mathbf{J} = \mathbf{A}^{n_x} - p_{n_x-1}\mathbf{A}^{n_x-1} + \cdots + (-1)^{n_x-1}p_1\mathbf{A} + (-1)^{n_x}p_0\mathbf{I}. \tag{4.4-14}$$

This algorithm also provides a proof of Bellman's lemma on the uniqueness of solutions to the matrix Lyapunov equation. Using Newton's identities in the form (4.4-6) it follows from (4.4-14) that

$$\mathbf{J} = (\mathbf{A} + \mu_1\mathbf{I})(\mathbf{A} + \mu_2\mathbf{I}) \cdots (\mathbf{A} + \mu_{n_x}\mathbf{I}). \tag{4.4-15}$$

Since the determinant of a product of square matrices is the product of the individual determinants, the inverse of $\mathbf{J}$ exists and (4.4-9) has a unique solution, if and only if $-\mu_j$ is not an eigenvalue of $\mathbf{A}$ for some j, that is, if and only if (4.4-4) holds.

Example 4.4-2: An Application of Jameson's Algorithm Consider the two-dimensional system $\dot{\mathbf{x}} = \mathbf{A}\mathbf{x}$, with

$$\mathbf{A} = \begin{bmatrix} 0 & 1 \\ -1 & -1 \end{bmatrix} \qquad \mathbf{A}^2 = \begin{bmatrix} -1 & -1 \\ 1 & 0 \end{bmatrix}.$$

To establish that the origin is globally asymptotically stable we seek a Lyapunov function

$$V(\mathbf{x}) = \mathbf{x}^{\mathrm{T}}\mathbf{P}\mathbf{x},$$

where $\mathbf{P} > 0$ is the solution to

$$\mathbf{P}\mathbf{A} + \mathbf{A}^{\mathrm{T}}\mathbf{P} = -\mathbf{Q}.$$

For simplicity we choose $\mathbf{Q} = \mathbf{I}$.

The coefficients of the characteristic equation

$$0 = \mu^2 + p_1\mu + p_0$$

can be determined easily by direct evaluation of $|\mu\mathbf{I} - \mathbf{A}|$ for low-order systems. However, to illustrate the procedure for higher-order systems, we use Newton's identities

$$p_1 = -s_1$$
$$p_0 = -\tfrac{1}{2}\{s_2 + p_1 s_1\},$$

where

$$\begin{aligned} s_1 &= \mathrm{Tr}[\mathbf{A}] &= -1 \\ s_2 &= \mathrm{Tr}[\mathbf{A}^2] &= -1. \end{aligned}$$

Thus

$$p_0 = p_1 = 1.$$

From (4.4-10) we compute the matrices

$$\mathbf{C}_0 = \mathbf{0}$$
$$\mathbf{C}_1 = \mathbf{C} = -\mathbf{I}$$
$$\mathbf{C}_2 = \mathbf{C}\mathbf{A}^1 - \mathbf{A}^{\mathrm{T}}\mathbf{C}_1 = -\mathbf{A} + \mathbf{A}^{\mathrm{T}} = \begin{bmatrix} 0 & -2 \\ 2 & 0 \end{bmatrix}.$$

From these results and (4.4-12)–(4.4-14) we have

$$\mathbf{W} = \mathbf{C}_2 - p_1\mathbf{C}_1 = \begin{bmatrix} 1 & -2 \\ 2 & 1 \end{bmatrix}$$
$$\mathbf{J} = \mathbf{A}^2 - p_1\mathbf{A} + p_0\mathbf{I} = \begin{bmatrix} 0 & -2 \\ 2 & 2 \end{bmatrix}$$
$$\mathbf{P} = \mathbf{W}\mathbf{J}^{-1} = \tfrac{1}{2}\begin{bmatrix} 3 & 1 \\ 1 & 2 \end{bmatrix} > 0.$$

Then

$$V(\mathbf{x}) = \mathbf{x}^{\mathrm{T}}\mathbf{P}\mathbf{x} = \tfrac{3}{2}x_1^2 + x_1x_2 + x_2^2 > 0$$

and

$$\dot{V}(\mathbf{x}) = -\mathbf{x}^{\mathrm{T}}\mathbf{Q}\mathbf{x} = -(x_1^2 + x_2^2) < 0$$

for all $\mathbf{x} \neq \mathbf{0}$, so the origin is globally asymptotically stable. ■

Lur'e's Integral Lyapunov Function

For general nonlinear systems there is no simple procedure for determining a Lyapunov function. However, certain techniques do exist that are applicable to various types of nonlinear systems.

As we have seen, in Example 4.3-4 and in the first part of Example 4.4-1, the total energy in a physical system is often a good candidate for a Lyapunov function, since damping dissipates the energy in the system. Energy concepts, particularly the notion of potential energy (in which a force is generated opposite to the gradient of the potential energy function), can also be applied to other types of nonlinear systems. One such application yields a Lyapunov function involving an integral term.

Consider the class of systems governed by **Liénard's equation**

$$\ddot{y} + h(y)\dot{y} + g(y) = 0, \tag{4.4-16}$$

where $h(\cdot)$ and $g(\cdot)$ are specified continuous functions of y only. In addition to continuity, we make the following assumptions:

$$h(y) > 0 \qquad \text{for all } y \neq 0, \tag{4.4-17}$$

$$yg(y) > 0 \qquad \text{for all } y \neq 0, \tag{4.4-18}$$

$$G(y) \triangleq \int_0^y g(\xi)d\xi \to \infty \qquad \text{as } |y| \to \infty. \tag{4.4-19}$$

Notice that the **sector condition** (4.4-18) requires $g(y) > 0$ for $y > 0$ and $g(y) < 0$ for $y < 0$; that is, the graph of $g(y)$ versus y lies entirely in the first and third quadrants. Since $g(y)$ is continuous, we have $g(y) = 0$ at $y = 0$ (and only there). From these results we see that (4.4-16) has a unique equilibrium solution, given by $y(t) \equiv 0$.

Thinking in terms of position $y(t)$, velocity $\dot{y}(t)$, and acceleration $\ddot{y}(t)$, the second and third terms in (4.4-16) correspond to nonlinear damping and spring forces, respectively. Condition (4.4-17) requires a positive damping coefficient. Condition (4.4-18) allows any continuous nonlinear function $g(y)$ whose graph $g(y)$ versus y lies within the first and third quadrants. For example, $g(y)$ could represent the force required to stretch a (hard, soft, or linear) spring. The function $G(y)$ corresponds to potential energy, producing a spring force $-\partial G(y)/\partial y = -g(y)$. Condition (4.4-19) will lead to a Lyapunov growth condition, allowing us to establish global asymptotic stability.

In terms of **phase plane** state variables $(x_1, x_2) = (y, \dot{y})$, the equations of motion for **Liénard's system** can be written as

$$\begin{aligned}\dot{x}_1 &= x_2\\ \dot{x}_2 &= -g(x_1) - h(x_1)x_2.\end{aligned} \tag{4.4-20}$$

Because of the sector condition (4.4-18) this system has a unique equilibrium at $\mathbf{x} = \mathbf{0}$. As we show, the origin is globally asymptotically stable for any continuous nonlinear functions $g(x_1)$ and $h(x_1)$ satisfying (4.4-17)–(4.4-19).

As a Lyapunov function we use the kinetic plus potential energy

$$V(\mathbf{x}) = \tfrac{1}{2}x_2^2 + G(x_1) = \tfrac{1}{2}x_2^2 + \int_0^{x_1} g(\xi)d\xi. \tag{4.4-21}$$

This function, known as **Lur'e's integral** Lyapunov function, satisfies

$$V(\mathbf{0}) = 0,$$

since $G(0) = 0$. Furthermore, the integral term $G(x_1)$ satisfies $G(x_1) > 0$ for all $x_1 \neq 0$. To verify this fact, note that, if $x_1 > 0$, then $d\xi > 0$ during the integration from 0 to x_1, while $x_1 g(x_1) > 0$ for all $x_1 > 0$ implies $g(\xi) > 0$ for $0 < \xi \leq x_1$. Similarly, if $x_1 < 0$, then $d\xi < 0$ and $g(\xi) < 0$ during the integration. Thus we have

$$V(\mathbf{x}) > 0 \qquad \text{for all } \mathbf{x} \neq \mathbf{0}.$$

From (4.4-19) and (4.4-21) we also have

$$V(\mathbf{x}) \longrightarrow \infty \qquad \text{as} \|\mathbf{x}\| \longrightarrow \infty.$$

Finally, in view of (4.4-17), along trajectories of (4.4-20) we have

$$\begin{aligned}\dot{V}(\mathbf{x}) &= g(x_1)\dot{x}_1 + x_2\dot{x}_2\\ &= -h(x_1)x_2^2\\ &\leq 0 \qquad \text{for all } \mathbf{x} \neq \mathbf{0}.\end{aligned}$$

Since $h(x_1) > 0$ for $x_1 \neq 0$ and neither $x_1(t) \equiv 0$ nor $x_2(t) \equiv 0$ is a solution to the equations of motion (4.4-20), except at $\mathbf{x}(t) \equiv \mathbf{0}$, we have $\dot{V}[\mathbf{x}(t)] \leq 0$ and $\dot{V}[\mathbf{x}(t)] \not\equiv 0$.

Thus for any state-space system governed by Liénard's equations (4.4-20), the origin is globally asymptotically stable, if (and only if) the continuous functions $h(y)$ and $g(y)$ satisfy conditions (4.4-17)–(4.4-19).

Example 4.4-3: Hardening Cubic Spring Consider a spring-mass-damper system

$$\ddot{y} + \dot{y} + y^3 = 0$$

with a hardening cubic spring. In the $\mathbf{x} = (y, \dot{y})$ phase plane the state equations are

$$\begin{aligned} \dot{x}_1 &= x_2 \\ \dot{x}_2 &= -x_1^3 - x_2. \end{aligned} \tag{4.4-22}$$

This is a Liénard system (4.4-20) with

$$g(x_1) = x_1^3 \tag{4.4-23}$$

$$h(x_1) = 1. \tag{4.4-24}$$

The functions $g(\cdot)$ and $h(\cdot)$ satisfy conditions (4.4-17)–(4.4-19), namely,

$$h(x_1) = 1 > 0, \tag{4.4-25}$$

$$x_1 g(x_1) = x_1^4 > 0 \qquad \text{for } x_1 \neq 0 \tag{4.4-26}$$

$$G(x_1) = \int_0^{x_1} \xi^3 d\xi = \tfrac{1}{4} x_1^4 \to \infty \qquad \text{as } |x_1| \to \infty. \tag{4.4-27}$$

Thus the origin is globally asymptotically stable. ■

Krasovskii's Method

Other dissipation functions besides energy can be used as candidate Lyapunov functions. For example, consider a system

$$\dot{\mathbf{x}} = \mathbf{f}(\mathbf{x}) \tag{4.4-28}$$

having a single equilibrium at $\mathbf{x} = \mathbf{0}$, that is,

$$\mathbf{f}(\mathbf{0}) = \mathbf{0} \tag{4.4-29}$$

and

$$\mathbf{f}(\mathbf{x}) \neq \mathbf{0} \qquad \text{for } \mathbf{x} \neq \mathbf{0}. \tag{4.4-30}$$

Suppose we could show that the "speed" $\|\dot{\mathbf{x}}\|$ decreases toward zero along all trajectories $\mathbf{x}(t)$. Then we could conclude that every trajectory approaches an equilibrium state, namely, in view of (4.4-29) and (4.4-30), the unique equilibrium at $\mathbf{x} = \mathbf{0}$.

For systems satisfying (4.4-28)–(4.4-30) **Krasovskii's method** [Krasovskii, 1963] chooses a candidate Lyapunov function of the form

$$V(\mathbf{x}) = \mathbf{f}^{\mathrm{T}}\mathbf{P}\mathbf{f} > 0 \qquad \text{for } \mathbf{x} \neq \mathbf{0}, \tag{4.4-31}$$

where **P** is a constant symmetric positive definite matrix to be chosen. For this $V(\cdot)$ function we have

$$\dot{V}(\mathbf{x}) = -\mathbf{f}^{\mathrm{T}}\mathbf{R}\mathbf{f} < 0 \qquad \text{for } \mathbf{x} \neq \mathbf{0}, \tag{4.4-32}$$

provided the symmetric matrix

$$\mathbf{R} = -\left\{\mathbf{P}\left[\frac{\partial \mathbf{f}(\mathbf{x})}{\partial \mathbf{x}}\right] + \left[\frac{\partial \mathbf{f}(\mathbf{x})}{\partial \mathbf{x}}\right]^{\mathrm{T}}\mathbf{P}\right\} \tag{4.4-33}$$

is positive definite. If so, the origin is globally asymptotically stable. Note that (4.4-33) is similar to the matrix Lyapunov equation for linear systems, except that $\mathbf{R} = \mathbf{R}(\mathbf{x})$ and (4.4-33) applies to nonlinear systems.

Example 4.4-4: Krasovskii's Method The nonlinear system

$$\begin{aligned} \dot{x}_1 &= -2x_1 - x_1^3 + x_2 &&= f_1(\cdot) \\ \dot{x}_2 &= x_1 - x_2 &&= f_2(\cdot) \end{aligned} \tag{4.4-34}$$

has a unique equilibrium at $\mathbf{x} = \mathbf{0}$. Setting

$$\mathbf{P} = \begin{bmatrix} p_{11} & p_{12} \\ p_{12} & p_{22} \end{bmatrix} \tag{4.4-35}$$

and using

$$\frac{\partial \mathbf{f}(\mathbf{x})}{\partial \mathbf{x}} = \begin{bmatrix} -2 - 3x_1^2 & 1 \\ 1 & -1 \end{bmatrix},$$

we compute **R** from (4.4-33) as

$$\mathbf{R} = \begin{bmatrix} 2p_{11}(2 + 3x_1^2) - 2p_{12} & -p_{11} - p_{22} + 3p_{12}(1 + x_1^2) \\ -p_{11} - p_{22} + 3p_{12}(1 + x_1^2) & 2(p_{22} - p_{12}) \end{bmatrix}. \tag{4.4-36}$$

For simplicity we take $\mathbf{P} = \mathbf{I} > 0$, yielding

$$\mathbf{R} = \begin{bmatrix} 2(2+3x_1^2) & -2 \\ -2 & 2 \end{bmatrix}. \tag{4.4-37}$$

Applying Sylvester's theorem, which is a sufficient condition for positive definiteness [but no longer necessary when the elements of **R** are functions of **x** (see Exercise 4-7)], establishes that $\mathbf{R} > 0$ for all **x**. Thus, since $\mathbf{P} = \mathbf{I}$,

$$V(\mathbf{x}) = \mathbf{f}^{\mathrm{T}}\mathbf{f} = (-2x_1 - x_1^3 + x_2)^2 + (x_1 - x_2)^2 > 0$$

and

$$\dot{V}(\mathbf{x}) = -\mathbf{f}^{\mathrm{T}}\mathbf{R}\mathbf{f} < 0$$

for all $\mathbf{x} \neq \mathbf{0}$. In addition, $V(\mathbf{0}) = 0$ and $V(\mathbf{x}) \rightarrow \infty$ as $\|\mathbf{x}\| \rightarrow \infty$. Therefore the equilibrium at $\mathbf{x} = \mathbf{0}$ is globally asymptotically stable. ■

Variable Gradient Method

So far the techniques that we have examined for finding Lyapunov functions have focused on choosing the form of a positive definite function $V(\mathbf{x})$ and then forcing $\dot{V}(\mathbf{x})$ to be negative definite (or at least negative semidefinite).

For the system

$$\dot{\mathbf{x}} = \mathbf{f}(\mathbf{x}), \tag{4.4-38}$$

with $\mathbf{f}(\mathbf{0}) = \mathbf{0}$, the **variable gradient method** [Schultz, 1965] begins by assuming a form for the (row) gradient vector

$$\frac{\partial V}{\partial \mathbf{x}} = \mathbf{x}^{\mathrm{T}}\boldsymbol{\Psi}(\mathbf{x}), \tag{4.4-39}$$

where

$$\boldsymbol{\Psi}(\mathbf{x}) = \begin{bmatrix} \psi_{11}(\mathbf{x}) & \cdots & \psi_{1n_x}(\mathbf{x}) \\ \vdots & \ddots & \vdots \\ \psi_{n_x1}(\mathbf{x}) & \cdots & \psi_{n_xn_x}(\mathbf{x}) \end{bmatrix}. \tag{4.4-40}$$

This form for $\partial V(\mathbf{x})/\partial \mathbf{x}$ ensures that the gradient is zero at the origin, as it must be if $V(\mathbf{x})$ is positive definite with a minimum value at the origin.

The matrix $\boldsymbol{\Psi}(\mathbf{x})$, which need not be symmetric, is chosen to try to satisfy

$$\dot{V}(\mathbf{x}) = \frac{\partial V}{\partial \mathbf{x}} \dot{\mathbf{x}} = \mathbf{x}^T \boldsymbol{\Psi} \mathbf{f} < 0 \tag{4.4-41}$$

and

$$V(\mathbf{x}) > 0$$

for $\mathbf{x} \neq \mathbf{0}$, with $V(\mathbf{x})$ being determined by integrating the gradient vector:

$$V(\mathbf{x}) = \int_{\mathbf{0}}^{\mathbf{x}} \frac{\partial V(\boldsymbol{\xi})}{\partial \mathbf{x}} \, d\boldsymbol{\xi}. \tag{4.4-42}$$

In order for the path integral (4.4-42) to define a function $V(\mathbf{x})$ whose value depends only on the point $\mathbf{x}$, the value of the integral must be **independent of the path** of integration. A necessary and sufficient condition for this is that $V(\mathbf{x})$ is C^2 (twice continuously differentiable), which implies

$$\frac{\partial V(\mathbf{x})}{\partial x_i \, \partial x_j} = \frac{\partial V(\mathbf{x})}{\partial x_j \, \partial x_i}$$

for all i, j. Thus the $\boldsymbol{\Psi}(\mathbf{x})$ matrix must satisfy the **curl conditions**

$$\frac{\partial}{\partial x_i} \left[\frac{\partial V(\mathbf{x})}{\partial x_j} \right] = \frac{\partial}{\partial x_j} \left[\frac{\partial V(\mathbf{x})}{\partial x_i} \right], \tag{4.4-43}$$

where

$$\frac{\partial V(\mathbf{x})}{\partial x_k} = \mathbf{x}^T \boldsymbol{\Psi}_k(\mathbf{x})$$

with $\boldsymbol{\Psi}_k(\mathbf{x})$ denoting the kth column of $\boldsymbol{\Psi}(\mathbf{x})$. In choosing $\boldsymbol{\Psi}(\mathbf{x})$ to satisfy these conditions it is sufficient to consider $\boldsymbol{\Psi}(\mathbf{x})$ matrices having diagonal elements that are at most functions of the corresponding $\mathbf{x}$ component:

$$\psi_{ii}(\mathbf{x}) = \psi_{ii}(x_i). \tag{4.4-44}$$

Furthermore, to scale the resulting $V(\mathbf{x})$ function, we can choose, for example,

$$\psi_{n_x n_x} = 2. \tag{4.4-45}$$

A convenient path for evaluating the integral in (4.4-42) is to move first along the x_1-axis, from the origin to x_1, and then sequentially move parallel to each of the remaining coordinate axes. This path sequence yields

$$V(\mathbf{x}) = \int_0^{x_1} \frac{\partial V(\xi_1, 0, \ldots, 0)}{\partial x_1} d\xi_1 + \int_0^{x_2} \frac{\partial V(x_1, \xi_2, 0, \ldots, 0)}{\partial x_2} d\xi_2$$
$$+ \cdots + \int_0^{x_{n_x}} \frac{\partial V(x_1, \ldots, x_{n_x - 1}, \xi_{n_x})}{\partial x_{n_x}} d\xi_{n_x}. \tag{4.4-46}$$

Example 4.4-5: Hardening Cubic Spring The nonlinear spring-mass-damper system

$$\dot{x}_1 = x_2$$
$$\dot{x}_2 = -x_1^3 - x_2 \tag{4.4-47}$$

has a unique equilibrium at the origin. Choosing

$$\frac{\partial V}{\partial \mathbf{x}} = \mathbf{x}^{\mathrm{T}} \boldsymbol{\Psi}(\mathbf{x}) = [\psi_{11} x_1 + \psi_{21} x_2 \quad \psi_{12} x_1 + 2x_2] \tag{4.4-48}$$

and evaluating

$$\dot{V} = \mathbf{x}^{\mathrm{T}} \boldsymbol{\Psi} \dot{\mathbf{x}}$$

yields

$$\dot{V} = (\psi_{11} x_1 + \psi_{21} x_2) x_2 - (\psi_{12} x_1 + 2x_2)(x_1^3 + x_2)$$
$$= -\psi_{12} x_1^4 - (2 - \psi_{21}) x_2^2 + (\psi_{11} - \psi_{12} - 2x_1^2) x_1 x_2. \tag{4.4-49}$$

The curl conditions (4.4-43), along with (4.4-44), require that

$$\psi_{12} + x_1 \frac{\partial \psi_{12}}{\partial x_1} = \psi_{21} + x_2 \frac{\partial \psi_{21}}{\partial x_2}. \tag{4.4-50}$$

This condition can be satisfied, for example, by choosing

$$\psi_{12} = \psi_{21} = k = \text{constant}. \tag{4.4-51}$$

For simplicity in making $\dot{V} < 0$, we can eliminate the cross-product term in (4.4-49) by choosing

$$\psi_{11} = \psi_{12} + 2x_1^2. \tag{4.4-52}$$

Then

$$\dot{V} = -kx_1^4 - (2-k)x_2^2 \tag{4.4-53}$$

will be negative definite provided

$$0 < k < 2. \tag{4.4-54}$$

Using (4.4-51) and (4.4-52) in (4.4-48) we have

$$\frac{\partial V}{\partial \mathbf{x}} = [k(x_1 + x_2) + 2x_1^3 \quad kx_1 + 2x_2]$$

and we compute $V(\mathbf{x})$, from (4.4-46), as

$$\begin{aligned} V(\mathbf{x}) &= \int_0^{x_1} (k[\xi_1 + 0] + 2\xi_1^3)\, d\xi_1 + \int_0^{x_2} (kx_1 + 2\xi_2)\, d\xi_2 \\ &= \frac{k}{2} x_1^2 + \frac{1}{2} x_1^4 + kx_1x_2 + x_2^2. \end{aligned}$$

This result can be written as

$$V(\mathbf{x}) = \mathbf{x}^T \mathbf{P} \mathbf{x},$$

where

$$\mathbf{P}(\mathbf{x}) = \frac{1}{2} \begin{bmatrix} k + x_1^2 & k \\ k & 2 \end{bmatrix}.$$

From Sylvester's inequalities,

$$\begin{aligned} &0 < k + x_1^2 \\ &0 < 2(k + x_1^2) - k^2 = k(2-k) + 2x_1^2, \end{aligned}$$

we see that $\mathbf{P}(\mathbf{x})$ is positive definite for all $\mathbf{x} \neq \mathbf{0}$ provided

$$0 < k < 2.$$

Comparing this result, for $V(\mathbf{x}) > 0$, with the condition (4.4-54) for $\dot{V}(\mathbf{x}) < 0$ (which are identical only by coincidence), we see that for any such k, $V(\mathbf{x})$ is a Lyapunov function establishing global asymptotic stability for the equilibrium at $\mathbf{x}(t) \equiv \mathbf{0}$. ■

4.5 ESTIMATING ASYMPTOTIC STABILITY REGIONS

In most of the stability examples considered so far, the equilibrium point of interest has been globally asymptotically stable. This will not be the case in general. A typical situation concerns the design of a feedback controller for a nonlinear dynamical system. In designing a controller, suppose that our objective is to transfer every possible initial state to some specified equilibrium. The extent to which this goal is accomplished is determined by the set of initial states that actually do get transferred to the target state.

Consider a dynamical system of the usual form

$$\dot{\mathbf{x}} = \mathbf{f}(\mathbf{x}) \tag{4.5-1}$$

and assume that $\mathbf{x}(t) \equiv \hat{\mathbf{x}}$ is a locally asymptotically stable equilibrium solution. If the system has other equilibrium solutions besides $\mathbf{x}(t) \equiv \hat{\mathbf{x}}$, then clearly $\hat{\mathbf{x}}$ will not be globally asymptotically stable. In any event the problem is to determine (or estimate) the **domain of attraction** $\mathcal{D}$ to $\hat{\mathbf{x}}$, defined as the set of all initial states $\mathbf{x}(0)$ such that $\mathbf{x}(t)$ approaches $\hat{\mathbf{x}}$ as $t \to \infty$.

Lyapunov Estimates

A Lyapunov function can be used to estimate the domain of attraction. The basis for this approach lies in a result due to LaSalle and Lefschetz [1961, p. 59].

Attractor Theorem: *Let $V(\mathbf{x})$ be a continuously differentiable function and let $\mathcal{D}_c$ denote the region where $V(\mathbf{x}) < c$. Assume that within $\mathcal{D}_c$*

$$V(\mathbf{x}) > 0 \qquad \text{for } \mathbf{x} \neq \hat{\mathbf{x}} \tag{4.5-2}$$

$$\dot{V}(\mathbf{x}) \leq 0. \tag{4.5-3}$$

In addition, assume that $\mathcal{D}_c$ is bounded. Let $\mathcal{Z}$ denote the set of points within $\mathcal{D}_c$ at which $\dot{V}(\mathbf{x}) = 0$, and let I denote the invariant sets contained in $\mathcal{Z}$. Then every solution $\mathbf{x}(t)$ in $\mathcal{D}_c$ tends to I as $t \to \infty$.

Corollary: *If (4.5-3) is replaced by*

$$\dot{V}(\mathbf{x}) < 0 \qquad \text{for all } \mathbf{x} \neq \hat{\mathbf{x}} \text{ in } \mathcal{D}_c, \tag{4.5-4}$$

then the equilibrium at $\hat{\mathbf{x}}$ is asymptotically stable and every solution $\mathbf{x}(t)$ in $\mathcal{D}_c$ approaches $\hat{\mathbf{x}}$ as $t \to \infty$.

The proof of the attractor theorem relies on the fact that, if a trajectory $\mathbf{x}(t)$ is bounded for $t \geq 0$, then its **positive limit set** γ^+ (defined as the set of points that $\mathbf{x}(t)$ approaches as $t \to \infty$) is a nonempty, compact (closed and bounded), invariant set. A trajectory's positive limit set might be a point, such as an asymptotically stable equilibrium point, or a limit cycle, or a surface, such as a "strange attractor."

To outline a proof of the attractor theorem, note that (4.5-2) and (4.5-3) imply that $V[\mathbf{x}(t)]$ is nonnegative and nonincreasing. Thus every solution $\mathbf{x}(t)$ starting in $\mathcal{D}_c$ remains in it. Because $\mathcal{D}_c$ is bounded, so is $\mathbf{x}(t)$ for $t \geq 0$. As a consequence $\mathbf{x}(t)$ must approach its positive limit set γ^+ as $t \to \infty$. Furthermore, since $\dot{V}[\mathbf{x}(t)] \leq 0$ and $V[\mathbf{x}(t)] \geq 0$ for $t \geq 0$, $V[\mathbf{x}(t)]$ must approach a constant value as $t \to \infty$. In particular $\dot{V}[\mathbf{x}(t)] \to 0$ and thus $\mathbf{x}(t) \to \mathcal{Z}$ as $t \to \infty$. Therefore γ^+ is in $\mathcal{Z}$ and, since γ^+ is an invariant set, we conclude that $\mathbf{x}(t) \to I$ as $t \to \infty$, which establishes the attractor theorem. The corollary follows from the fact that, if $\dot{V}[\mathbf{x}(t)]$ is negative on $\mathcal{D}_c - \{\hat{\mathbf{x}}\}$, then $\mathcal{Z}$ is the point $\hat{\mathbf{x}}$ and so is I, so that $\mathbf{x}(t) \to \hat{\mathbf{x}}$ as $t \to \infty$.

Using the attractor theorem, or its corollary, we can develop conservative estimates of the domain of attraction for an equilibrium point, by finding a Lyapunov function $V(\mathbf{x})$ and determining any (particularly the largest) $V(\mathbf{x}) < c$ region $\mathcal{D}_c$ for which $\dot{V}(\mathbf{x}) \leq 0$ (and not identically zero) within the region. Such an estimate will be conservative in the sense that the domain of attraction $\mathcal{D}$ will contain the estimated region $\mathcal{D}_c$ as a subset.

Example 4.5-1: An Application of the Attractor Theorem In Example 4.3-5 we examined the equilibrium at the origin for the system

$$\begin{aligned} \dot{x}_1 &= x_2 \\ \dot{x}_2 &= -\frac{x_1}{(1+x_1^2)^2} - \frac{x_2}{(1+x_1^2+x_2^2)^2} \end{aligned} \tag{4.5-5}$$

using the Lyapunov function

$$V(\mathbf{x}) = x_2^2 + \frac{x_1^2}{1+x_1^2}. \tag{4.5-6}$$

The $V(\mathbf{x}) = c$ contours of this positive definite function have been plotted previously in Figure 4.3-4.

For this $V(\cdot)$ function we have

$$\dot{V}(\mathbf{x}) = \frac{-2x_2^2}{(1+x_1^2+x_2^2)^2} \leq 0 \qquad \text{for all } \mathbf{x}, \tag{4.5-7}$$

but the $V(\mathbf{x}) < c$ regions are bounded only for $0 \leq c < 1$. Since $\dot{V}(\mathbf{x})$ is not negative definite in the regions $V(\mathbf{x}) < c < 1$, we cannot use the corollary to the attractor theorem. However, we can use the attractor theorem itself to show that the origin is asymptotically stable and to develop an estimate for its domain of attraction.

For $0 < c < 1$, conditions (4.5-7) implies that trajectories starting in the bounded region $V(\mathbf{x}) < c$ remain in the region. Thus the origin is stable. From the equations of motion, the region $\mathcal{Z}$, where $\dot{V}(\mathbf{x}) = 0$ and $V(\mathbf{x}) < c < 1$, is that portion of the x_1-axis defined by

$$\mathcal{Z} = \left\{ \mathbf{x} \middle| x_1^2 < \frac{c}{1-c},\ x_2 = 0 \right\}, \tag{4.5-8}$$

and the only invariant set in $\mathcal{Z}$ is the origin. Thus $I = \{\mathbf{0}\}$ and it follows from the attractor theorem that every trajectory starting in $V(\mathbf{x}) < c < 1$ approaches the origin. Therefore, the origin is asymptotically stable and its domain of attraction includes at least the region defined by $V(\mathbf{x}) < 1$. From Figure 4.3-4 it is clear, however, that the region $V(\mathbf{x}) < 1$ is not all of the domain of attraction. ■

In the previous example we could not use the corollary to the attractor theorem because $\dot{V}(\mathbf{x})$ was not negative definite. Instead, we had to employ the attractor theorem itself. In the following example we can use the more direct results available through the corollary.

Example 4.5-2: A Prey-Predatory System Consider the prey (n_1) and predator (n_2) population dynamics model given by

$$\begin{aligned} \frac{dn_1}{dT} &= n_1[a(n_1^* - n_1) - bn_2] \\ \frac{dn_2}{dT} &= n_2[cn_1 - dn_2], \end{aligned} \tag{4.5-9}$$

where T denotes time and n_1^*, a, b, c, and d are positive constants.

This system is a similar to a Lotka–Volterra competition model that models limited resources, such as food, within each species. The term containing a in (4.5-9) is a logistics term modeling an environment that can only sustain a maximum equilibrium prey population n_1^* (called the "carrying capacity"), in the absence of predators. The term containing b in (4.5-9) represents the rate at which the predator species captures prey. The right-hand side of the second equation in (4.5-9) is also a logistics-type term, except that the predator carrying capacity depends on the prey population level. This system was used by Grantham [1981*a*,*b*] for studying the possibility of an extinction in renewable resource systems such as a krill-whale fishery, in which large schools of tiny shrimp-like krill constitute the major source of food for whales off of the coast of Antarctica.

In analyzing a dynamical system that contains several parameters, the number of parameters that need to be specified can be minimized by using **nondimensional variables**. Defining nondimensional state and time variables

$$x_1 \triangleq \frac{n_1}{\overline{n}_1}, \qquad x_2 \triangleq \frac{n_2}{\overline{n}_2}, \qquad t \triangleq \frac{T}{\overline{T}},$$

we can choose any convenient values for $\overline{n}_1$, $\overline{n}_2$, and $\overline{T}$ to eliminate three of the five parameters in (4.5-9). Using these definitions we have

$$\dot{x}_1 = x_1\left(a\overline{n}_1\overline{T}\left[\frac{n_1^*}{\overline{n}_1} - x_1\right] - b\overline{T}\,\overline{n}_2 x_2\right)$$

$$\dot{x}_2 = x_2(c\overline{T}\,\overline{n}_1 x_1 - d\overline{T}\,\overline{n}_2 x_2),$$

where $(\dot{\cdot})$ denotes $d(\quad)/dt$. Choosing

$$\overline{n}_1 = n_1^*$$

$$\overline{n}_2 = \frac{c}{d}\,n_1^*$$

$$\overline{T} = \frac{1}{an_1^*}$$

and defining the parameters

$$\alpha = \frac{bc}{ad}, \qquad \beta = \frac{c}{a},$$

we have

$$\begin{aligned} \dot{x}_1 &= x_1(1 - x_1 - \alpha x_2) \\ \dot{x}_2 &= \beta x_2(x_1 - x_2). \end{aligned} \tag{4.5-10}$$

The system (4.5-10) has three equilibrium points: one at $\mathbf{x} = \mathbf{0}$, which corresponds to extinction of both species; one at $\mathbf{x} = [1 \quad 0]^T$, which corresponds to extinct predators with the prey at their equilibrium carrying capacity; and an equilibrium $\hat{\mathbf{x}}$ at which neither population is extinct, given by

$$\hat{x}_1 = \hat{x}_2 = \frac{1}{1+\alpha}.$$

The existence of multiple equilibrium points guarantees that none of them are globally asymptotically stable. However, for population analysis, we are concerned only with nonnegative populations and in particular with the domain of attraction $\mathcal{D}$ for the equilibrium point $\hat{\mathbf{x}}$ in the positive quadrant

$$R^+ \triangleq \{(x_1, x_2) | x_1 > 0, x_2 > 0\}.$$

The stability question is whether or not the equilibrium at $\hat{\mathbf{x}}$ is asymptotically stable, and if so, whether its domain of attraction $\mathcal{D}$ is all of the positive quadrant R^+.

In Example 4.1-2 we showed that for the underlying Lotka–Volterra system (4.1-15), the "energy" function $V(\mathbf{x})$ in (4.1-17) is constant along trajectories of the Lotka–Volterra system. Based on our previous discussions of energy dissipation and the use of energy as a Lyapunov function in physical systems, consider the function [Goh, 1980]

$$V(\mathbf{x}) = p\left[\frac{x_1 - \hat{x}_1}{\hat{x}_1} - \ln\left(\frac{x_1}{\hat{x}_1}\right)\right] + q\left[\frac{x_2 - \hat{x}_2}{\hat{x}_2} - \ln\left(\frac{x_2}{\hat{x}_2}\right)\right], \qquad (4.5\text{-}11)$$

where p and q are positive parameters to be chosen. This $V(\mathbf{x})$ function has the same form as in (4.1-17), except that a constant term has been added so that $V(\hat{\mathbf{x}}) = 0$. The function in (4.5-11) also satisfies the condition $V(\mathbf{x}) > 0$ for all $\mathbf{x} \in R^+$ with $\mathbf{x} \neq \hat{\mathbf{x}}$. Thus $V(\mathbf{x})$ has a local minimum at $\hat{\mathbf{x}}$. Furthermore, $V(\mathbf{x}) \to \infty$ as $\mathbf{x} \to \partial R^+$ and the gradient

$$\frac{\partial V(\mathbf{x})}{\partial \mathbf{x}} = \left[p\,\frac{x_1 - \hat{x}_1}{x_1\hat{x}_1} \quad q\,\frac{x_2 - \hat{x}_2}{x_2\hat{x}_2}\right] \qquad (4.5\text{-}12)$$

is nonzero on R^+, except at $\hat{\mathbf{x}}$. Thus $V(\mathbf{x})$ has a global minimum on R^+ at $\hat{\mathbf{x}}$ and the contours of constant $V(\mathbf{x})$ in the positive quadrant are all concentric closed curves surrounding $\hat{\mathbf{x}}$. In particular, the regions $V(\mathbf{x}) < c < \infty$ in the positive quadrant are all bounded.

In light of these properties and the attractor theorem corollary, we can now establish that the equilibrium at $\hat{\mathbf{x}}$ is asymptotically stable throughout R^+ by showing that $\dot{V}(\mathbf{x}) < 0$ for all $\mathbf{x} \in R^+$ other than at $\hat{\mathbf{x}}$.

From the equations of motion (4.5-10) and the gradient (4.5-12) we have

$$\dot{V}(\mathbf{x}) = p\,\frac{x_1 - \hat{x}_1}{\hat{x}_1}\,[1 - x_1 - \alpha x_2] + q\beta\,\frac{x_2 - \hat{x}_2}{\hat{x}_2}\,[x_1 - x_2].$$

From the definition of the equilibrium point $\hat{\mathbf{x}}$, the bracketed terms are equal to zero when evaluated at $\mathbf{x} = \hat{\mathbf{x}}$. We can subtract such zero-valued terms from $\dot{V}(\mathbf{x})$ to get an equivalent but more convenient representation, given by

$$\dot{V}(\mathbf{x}) = p\,\frac{x_1 - \hat{x}_1}{\hat{x}_1}\,[-(x_1 - \hat{x}_1) - \alpha(x_2 - \hat{x}_2)]$$
$$+\, q\beta\,\frac{x_2 - \hat{x}_2}{\hat{x}_2}\,[(x_1 - \hat{x}_1) - (x_2 - \hat{x}_2)].$$

For simplicity in achieving $\dot{V}(\mathbf{x}) < 0$, we eliminate the cross-product terms by choosing

$$p = \hat{x}_1, \qquad q = \frac{\alpha}{\beta}\,\hat{x}_2,$$

which yields

$$V(\mathbf{x}) = \left[x_1 - \hat{x}_1 - \hat{x}_1 \ln\left(\frac{x_1}{\hat{x}_1}\right)\right] + \frac{\alpha}{\beta}\left[x_2 - \hat{x}_2 - \hat{x}_2 \ln\left(\frac{x_2}{\hat{x}_2}\right)\right] \tag{4.5-13}$$

and

$$\dot{V}(\mathbf{x}) = -(x_1 - \hat{x}_1)^2 - \alpha(x_2 - \hat{x}_2)^2 < 0 \tag{4.5-14}$$

for all $\mathbf{x} \in R^+$ with $\mathbf{x} \neq \hat{\mathbf{x}}$.

Thus we see that the equilibrium at $\hat{\mathbf{x}}$ is asymptotically stable throughout the positive quadrant. Furthermore, by inspection of the equations of motion, the x_1- and x_2-axes are invariant sets, which no trajectory can cross. Thus the positive quadrant R^+ is the entire domain of attraction for $\hat{\mathbf{x}}$. Trajectories and contours for the Lyapunov function $V(\mathbf{x})$ are shown in Figure 4.5-1 for the case where $\alpha = 1$ and $\beta = 1/2$. ■

An Optimization Procedure for Lyapunov Estimates

In Example 4.5-2 we were able to find a Lyapunov function that yielded the entire domain of attraction. More often, a Lyapunov estimate will only be a subset of the domain of attraction. Let us consider the problem of determining the largest estimate for the domain of attraction based on a given Lyapunov function. For simplicity, we consider the case where the equilibrium point is at the origin.

We begin by finding a Lyapunov function. Assume that a control law $\mathbf{u}(\mathbf{x})$ has been specified so that the controlled system is of the form (4.1-11). Any of the construction methods that we have previously discussed can be used. In general they will yield different Lyapunov functions and, therefore, different estimates for the domain of attraction. If (4.1-11) is a linear system, then a $V(\mathbf{x})$ function can be obtained directly by solving Lyapunov's equation. If (4.1-11)

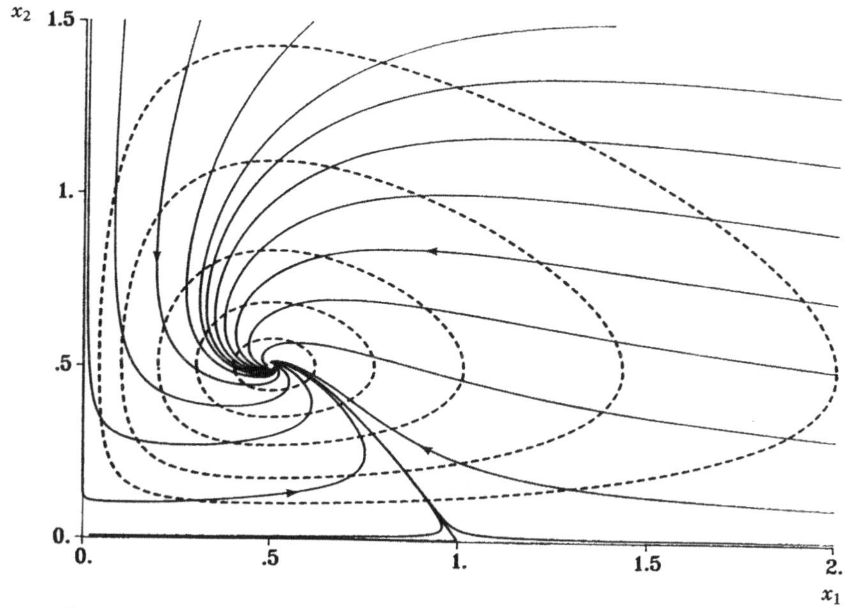

Figure 4.5-1 Trajectories and Lyapunov contours for Example 4.5-2.

is nonlinear, then one method for obtaining a Lyapunov function would be to linearize the system about the equilibrium point and then obtain a $V(\mathbf{x})$ function as in the linear case.

The problem of finding the largest $V(\mathbf{x}) < c$ estimate for the domain of attraction can be addressed by considering a related optimization problem. The region $V(\mathbf{x}) < c$ that we seek is the largest such region for which $\dot{V}(\mathbf{x})$ is negative definite (or negative semidefinite but not identically zero). As a consequence, the $V(\mathbf{x}) = c$ and $\dot{V}(\mathbf{x}) = 0$ surfaces will satisfy a local optimality condition: they must be tangent to each other at some point $\mathbf{x}^*$. Note that, if linearization has been used to obtain $V(\mathbf{x})$, the derivative $\dot{V}(\mathbf{x})$ must be evaluated using the original nonlinear equations of motion, including any bounds on the control.

The optimization procedure is the following. For an equilibrium at the origin, choose a positive definite Lyapunov function $V(\mathbf{x})$ for which $\dot{V}(\mathbf{x})$ is negative semidefinite (preferably negative definite) at least in some neighborhood of the origin (excluding the origin). Then solve the following optimization problem:

$$\min V(\mathbf{x}) \qquad \text{subject to } \dot{V}(\mathbf{x}) \geq 0. \tag{4.5-15}$$

If this problem has a solution $\mathbf{x}^* \neq \mathbf{0}$, and if the region $V(\mathbf{x}) < V^* \triangleq V(\mathbf{x}^*)$ is bounded, with $\dot{V}(\mathbf{x}) \leq 0$ in the region and not identically zero except at the origin, then the region $V(\mathbf{x}) < V^*$ is the largest estimate for the domain of attraction, based on the specified Lyapunov function $V(\mathbf{x})$.

Candidate solutions for the optimization problem (4.5-15) can be found by using the first-order necessary conditions for optimality from Chapter 3. Assuming that $\partial V(\mathbf{x}^*)/\partial \mathbf{x} \neq 0$, the necessary conditions reduce to

$$\dot{V}(\mathbf{x}^*) = 0 \tag{4.5-16}$$

$$\frac{\partial V(\mathbf{x}^*)}{\partial \mathbf{x}} = \gamma \frac{\partial \dot{V}(\mathbf{x}^*)}{\partial \mathbf{x}} \tag{4.5-17}$$

where $\gamma > 0$. These results imply that the surfaces $V(\mathbf{x}) = V^*$ and $\dot{V}(\mathbf{x}) = 0$ are tangent at $\mathbf{x}^*$.

In estimating the domain of attraction using (4.5-15), it might seem odd that we find the smallest value of $V(\mathbf{x})$ subject to $\dot{V}(\mathbf{x}) \geq 0$ in order to determine the largest $V(\mathbf{x}) < c$ region for which $\dot{V}(\mathbf{x}) \leq 0$. The reason for this is that we want to search *outside* of the $\dot{V}(\mathbf{x}) < 0$ region. The resulting point $\mathbf{x}^*$ will be on the boundary, that is, $\dot{V}(\mathbf{x}^*) = 0$, and every point inside the $V(\mathbf{x}^*)$ contour will be inside the $\dot{V}(\mathbf{x}) \leq 0$ region. The rationale becomes clear from the geometry illustrated in the following example.

Example 4.5-3: Retro-Time Van der Pol System For the backward-time Van der Pol system

$$\begin{aligned} \dot{x}_1 &= -x_2 \\ \dot{x}_2 &= x_1 - \beta(1 - x_1^2)x_2, \end{aligned} \tag{4.5-18}$$

the origin is asymptotically stable for $\beta > 0$ and its domain of attraction is bounded by a limit cycle. However, a closed-form solution for the limit cycle is unknown, so we will use the optimization procedure to estimate the domain of attraction. To use this procedure, consider the linearized system

$$\dot{\mathbf{x}} = \mathbf{A}\mathbf{x},$$

where

$$\mathbf{A} = \frac{\partial \mathbf{f}(\mathbf{0})}{\partial \mathbf{x}} = \begin{bmatrix} 0 & -1 \\ 1 & -\beta \end{bmatrix}.$$

We construct a quadratic Lyapunov function

$$V(\mathbf{x}) = \mathbf{x}^T \mathbf{P} \mathbf{x}$$

by solving Lyapunov's equation

$$\mathbf{PA} + \mathbf{A}^{\mathrm{T}}\mathbf{P} = -\mathbf{Q},$$

where $\mathbf{Q}$ is any convenient symmetric positive definite matrix. For $\mathbf{Q} = \mathbf{I}$ we get

$$V(\mathbf{x}) = \frac{1}{2\beta}\{(\beta^2 + 2)x_1^2 - 2\beta x_1 x_2 + 2x_2^2\}, \tag{4.5-19}$$

which is positive definite for any $\beta > 0$.

For this $V(\mathbf{x})$ function, with the nonlinear system (4.5-18), we have

$$\dot{V}(\mathbf{x}) = -x_1^2 - \beta x_1^3 x_2 - (1 - 2x_1^2)x_2^2,$$

and the necessary conditions for the optimization problem (4.5-15) are

$$0 = \frac{\partial V}{\partial x_1} - \gamma\frac{\partial \dot{V}}{\partial x_1} = \frac{\beta^2 + 2}{\beta}x_1 - x_2 - \gamma(-2x_1 - 3\beta x_1^2 x_2 + 4x_1 x_2^2)$$

$$0 = \frac{\partial V}{\partial x_2} - \gamma\frac{\partial \dot{V}}{\partial x_2} = -x_1 + \frac{2}{\beta}x_2 - \gamma\,[-\beta x_1^3 - 2(1 - 2x_1^2)x_2]$$

$$0 = \dot{V}(\mathbf{x}) \qquad = -x_1^2 - \beta x_1^3 x_2 - (1 - 2x_1^2)x_2^2.$$

For example, with $\beta = 1$, solving these equations numerically for x_1, x_2, and γ yields four tangent points:

$$x_1 = \pm 0.8559 \qquad V = 2.304$$
$$x_2 = \mp 0.7505 \qquad \gamma = 1.778$$

and

$$x_1 = \pm 1.0822 \qquad V = 2.420$$
$$x_2 = \pm 1.5187 \qquad \gamma = 0.696,$$

with the first pair corresponding to two symmetric values for the optimizing solution $\mathbf{x}^*$. Thus, as illustrated in Figure 4.5-2, the region $V(\mathbf{x}) < 2.304$ is guaranteed to be a subset of the domain of attraction for the case where $\beta = 1$. ■

Zubov's Analytic Continuation Method

In Zubov [1961] two results are developed that can be used to find a Lyapunov function $V(\mathbf{x})$ for a system

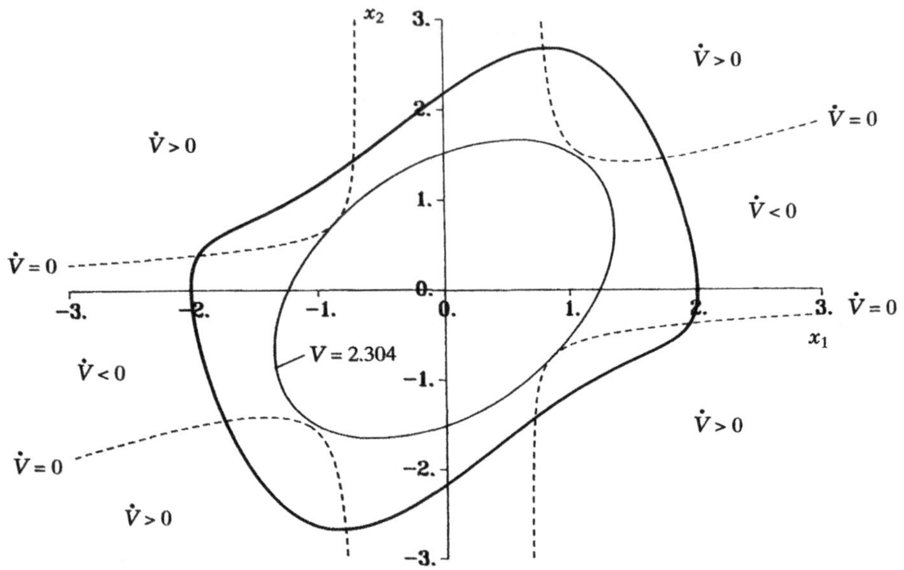

Figure 4.5-2 Domain of attraction estimate for Example 4.5-3.

$$\dot{\mathbf{x}} = \mathbf{f}(\mathbf{x}), \qquad \mathbf{f}(\mathbf{0}) = \mathbf{0}, \tag{4.5-20}$$

which will yield the entire domain of attraction to the origin. The first result is a partial differential equation for $V(\mathbf{x})$, under an assumption that $\mathbf{f}(\mathbf{x})$ is continuous and (4.5-20) has unique solutions that are continuous functions of the initial state. The second result is a recursive construction procedure for $V(\mathbf{x})$ for the case where $\mathbf{f}(\mathbf{x})$ is **analytic**, that is, where $\mathbf{f}(\mathbf{x})$ is continuous and has continuous partial derivatives of all orders.

Zubov's Partial Differential Equation

Zubov's Theorem: *For a region $\mathcal{D}$, with $\mathbf{x} = \mathbf{0}$ interior to $\mathcal{D}$, to be the domain of attraction to the origin, it is necessary and sufficient that there exist two functions $V(\mathbf{x})$ and $\psi(\mathbf{x})$ such that:*

(i) *$\psi(\mathbf{x})$ is continuous for all $\mathbf{x}$ and $V(\mathbf{x})$ is continuous on $\mathcal{D}$;*

(ii) *$V(\mathbf{0}) = \psi(\mathbf{0}) = 0$;*

(iii) *$\psi(\mathbf{x}) > 0$ for all nonzero $\mathbf{x}$;*

(iv) *$0 < V(\mathbf{x}) < 1$ for all nonzero $\mathbf{x} \in \mathcal{D}$;*

(v) *If $\mathbf{x}' \in \partial\mathcal{D}$ (boundary of $\mathcal{D}$), then*

$$\lim_{\mathbf{x} \to \mathbf{x}'} V(\mathbf{x}) = 1.$$

If $\|\mathbf{x}\| \to \infty$, $\mathbf{x} \in \mathcal{D}$, *then*

$$\lim_{\|\mathbf{x}\| \to \infty} V(\mathbf{x}) = 1;$$

(vi) $V(\mathbf{x})$ *satisfies the partial differential equation*

$$\dot{V}(\mathbf{x}) \triangleq \frac{\partial V}{\partial \mathbf{x}} \mathbf{f}(\mathbf{x}) = -\psi(\mathbf{x})[1 - V]\sqrt{1 + \mathbf{f}^T\mathbf{f}}. \tag{4.5-21}$$

The basic idea behind a proof of this result is that $\dot{V} < 0$ in the region $0 < V(\mathbf{x}) < 1$, by virtue of (4.5-21), implies that all trajectories in this region penetrate the $V(\mathbf{x})$ contours and approach the origin. In addition, $\dot{V}(\mathbf{x}) = 0$ on the surface $V(\mathbf{x}) = 1$ implies that this surface is the boundary of the domain of attraction. For the details of the proof see Zubov [1961, pp. 63–67].

Unless otherwise specified, we assume that the system (4.5-20) does not have a **finite escape time** (see Section 1.4). For this case (4.5-21) can be replaced with

$$\frac{\partial V}{\partial \mathbf{x}} \mathbf{f}(\mathbf{x}) = -\psi(\mathbf{x})[1 - V]. \tag{4.5-22}$$

To determine the domain of attraction for the origin, we choose a positive definite continuous function $\psi(\mathbf{x})$ and solve the partial differential equation (4.5-22) for $V(\mathbf{x})$, subject to the boundary condition $V(\mathbf{0}) = 0$. The domain of attraction is the region defined by $0 \le V(\mathbf{x}) < 1$.

The function $\psi(\mathbf{x})$ can be chosen as needed, to facilitate solving (4.5-22). However, it is sufficient to choose $\psi(\mathbf{x})$ as a quadratic form. Furthermore, in view of the attractor theorem, the requirement that $\psi(\mathbf{x})$ be positive definite can be relaxed to $\psi(\mathbf{x}) \ge 0$, provided $\psi[\mathbf{x}(t)]$ is not identically zero along a trajectory in the region $0 < V(\mathbf{x}) < 1$.

In seeking an analytic solution to (4.5-22) it may be helpful to note that, by defining

$$W(\mathbf{x}) \triangleq -\ln[1 - V(\mathbf{x})], \tag{4.5-23}$$

(4.5-22) can be transformed to

$$\frac{\partial W}{\partial \mathbf{x}} \mathbf{f}(\mathbf{x}) = -\psi(\mathbf{x}), \tag{4.5-24}$$

with the boundary condition $W(\mathbf{0}) = 0$. Then the domain of attraction for the origin is the region defined by $W(\mathbf{x}) \geq 0$.

Example 4.5-4: Zubov's Partial Differential Equation The following system was considered by Zubov:

$$\begin{aligned} \dot{x}_1 &= -x_1 + 2x_1^2 x_2 \\ \dot{x}_2 &= -x_2. \end{aligned} \tag{4.5-25}$$

Choosing

$$\psi(\mathbf{x}) = x_1^2 + x_2^2, \tag{4.5-26}$$

the partial differential equation (4.5-22) becomes

$$\frac{\partial V}{\partial x_1}(2x_1^2 x_2 - x_1) - \frac{\partial V}{\partial x_2} x_2 = -(x_1^2 + x_2^2)[1 - V]. \tag{4.5-27}$$

At this point it is not at all clear how one would solve this nonlinear partial differential equation. For example, separation of variables does not work, even with the transformation to (4.5-24). Nevertheless it is easy to verify that the solution for $V(\mathbf{0}) = 0$ is given by

$$V(\mathbf{x}) = 1 - \exp\left[-\frac{1}{2}x_2^2 - \frac{x_1^2}{2(1 - x_1 x_2)}\right]. \tag{4.5-28}$$

From the condition $V(\mathbf{x}) = 1$, we see that the boundary of the domain of attraction is given by the curve

$$x_1 x_2 = 1,$$

as illustrated in Figure 4.5-3. ■

Zubov's Recursive Procedure It is often not possible to find a closed-form solution to Zubov's partial differential equation. However, for a very large class of systems, there is a recursive procedure for constructing the solution.

Consider the case where the right-hand sides of the differential equations (4.5-20) possess continuous partial derivatives of all orders. Then $\mathbf{f}(\mathbf{x})$ can be expanded in a Taylor series about $\mathbf{x} = \mathbf{0}$, yielding

$$\dot{\mathbf{x}} = \mathbf{A}\mathbf{x} + \mathbf{g}(\mathbf{x}), \tag{4.5-29}$$

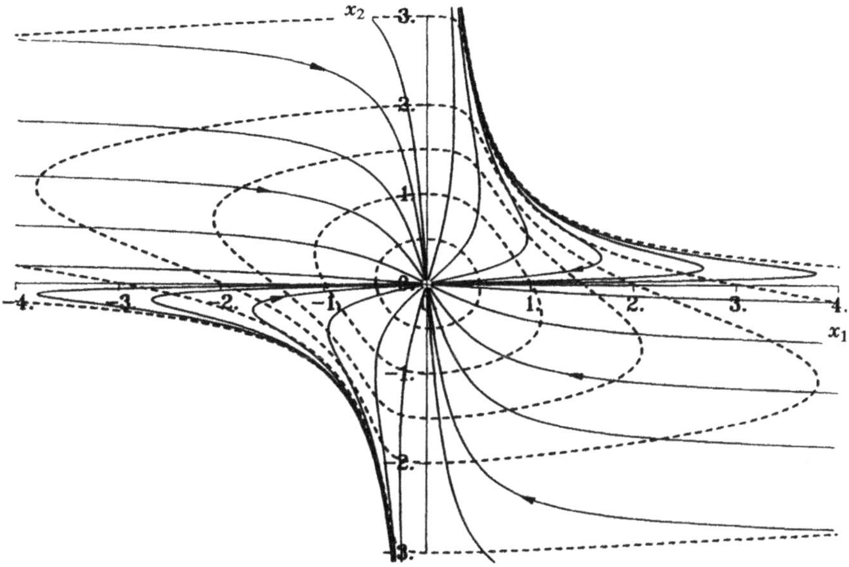

Figure 4.5-3 Domain of attraction for Example 4.5-4.

where

$$\mathbf{A} = \frac{\partial \mathbf{f}(\mathbf{0})}{\partial \mathbf{x}} \tag{4.5-30}$$

is the matrix for the linearized equations of motion and $\mathbf{g}(\mathbf{x})$ is a power series

$$\mathbf{g}(\mathbf{x}) = \mathbf{g}_2(\mathbf{x}) + \mathbf{g}_3(\mathbf{x}) + \cdots, \tag{4.5-31}$$

where $\mathbf{g}_m(\mathbf{x})$ consists of all terms that are homogeneous of degree $m \geq 2$, that is,

$$\mathbf{g}_m(\mathbf{x}) = \sum_{\Sigma n_j = m} [\mathbf{c}_{n_1 \ldots n_{n_x}}] x_1^{n_1} x_2^{n_2} \cdots x_{n_x}^{n_{n_x}}. \tag{4.5-32}$$

For example, for $n_x = 2$ and $m = 3$, $\mathbf{g}_3(\mathbf{x})$ would contain vectors multiplied by cubic terms like x_2^3, $x_1 x_2^2$, $x_1^2 x_2$, and so on.

Choosing $\psi(\mathbf{x})$ as a quadratic form, the solution to Zubov's partial differential equation

$$\frac{\partial V}{\partial \mathbf{x}}\,[\mathbf{Ax} + \mathbf{g}(\mathbf{x})] = -\psi(\mathbf{x})[1 - V], \tag{4.5-33}$$

with boundary condition $V(\mathbf{0}) = 0$, can then be written as an infinite series

$$V(\mathbf{x}) = V_2(\mathbf{x}) + V_3(\mathbf{x}) + \cdots, \tag{4.5-34}$$

where each function $V_m(\mathbf{x})$ in the series is a homogeneous form of degree m:

$$V_m(\mathbf{x}) = \sum_{\Sigma n_j = m} b_{n_1 \ldots n_{n_x}}\, x_1^{n_1} x_2^{n_2} \cdots x_{n_x}^{n_{n_x}}. \tag{4.5-35}$$

The subscript notation for the coefficients in $V_m(\mathbf{x})$ can be remembered by noting that the sum of the subscripts is equal to the degree m of $V_m(\mathbf{x})$ and that the first subscript is the power of x_1, the second is the power of x_2, and so on. This notation will facilitate the development of numerical procedures for a general n_x-dimensional system.

The coefficients for each $V_m(\mathbf{x})$ can be determined by substituting (4.5-31) and (4.5-34) into (4.5-33), which gives

$$\left[\frac{\partial V_2}{\partial \mathbf{x}} + \frac{\partial V_3}{\partial \mathbf{x}} + \cdots\right] [\mathbf{Ax}+\mathbf{g}_2(\mathbf{x})+\mathbf{g}_3(\mathbf{x})+\cdots] = -\psi(\mathbf{x})[1-V_2-V_3-\cdots]. \tag{4.5-36}$$

For a quadratic $\psi(\mathbf{x})$, expanding both sides of (4.5-36) and equating terms of the same degree produces the recursive relations

$$\begin{aligned}
\frac{\partial V_2}{\partial \mathbf{x}}\,\mathbf{Ax} &= -\psi \\
\frac{\partial V_3}{\partial \mathbf{x}}\,\mathbf{Ax} &= -\frac{\partial V_2}{\partial \mathbf{x}}\,\mathbf{g}_2 \\
\frac{\partial V_4}{\partial \mathbf{x}}\,\mathbf{Ax} &= \psi V_2 - \frac{\partial V_2}{\partial \mathbf{x}}\,\mathbf{g}_3 - \frac{\partial V_3}{\partial \mathbf{x}}\,\mathbf{g}_2 \qquad (4.5\text{-}37) \\
&\;\;\vdots \\
\frac{\partial V_m}{\partial \mathbf{x}}\,\mathbf{Ax} &= \psi V_{m-2} - \sum_{j=2}^{m-1} \frac{\partial V_j}{\partial \mathbf{x}}\,\mathbf{g}_{m+1-j}.
\end{aligned}$$

Note that the left- and right-hand sides of the first of equations (4.5-37) are quadratic in $\mathbf{x}$, the second equation in (4.5-37) is cubic, and so on. Substituting $V_2(\mathbf{x})$ from (4.5-35) into the first of equations (4.5-37), and equating the coefficients of corresponding terms on the left- and right-hand sides produces a system of three linear equations for the coefficients in $V_2(\mathbf{x})$. After determining $V_2(\mathbf{x})$, the second of equations (4.5-37) can then be used to determine the coefficients in $V_3(\mathbf{x})$. This process is repeated to construct each successive function in the series expansion of $V(\mathbf{x})$. At a general stage, finding the coefficients for $V_m(\mathbf{x})$ reduces to solving a system of $m + 1$ linear equations, a process that can easily be carried out on a computer.

If the series representation (4.5-34) for $V(\mathbf{x})$ can be recognized and synthesized to a closed-form representation for $V(\mathbf{x})$, then the domain of attraction corresponds to the region defined by $0 \le V(\mathbf{x}) < 1$. Otherwise, determination of the domain of attraction must rely on (a truncation of) the series representation itself.

Example 4.5-5: Logistics Equation As a one-dimensional example, consider the backward-time logistics equation

$$\dot{x} = -x(1 - x).$$

For the equilibrium at $x = 0$, let

$$\begin{aligned} \psi(x) &= x^2 \\ V(x) &= V_2(x) + V_3(x) + \cdots, \end{aligned}$$

where

$$V_m(x) = b_m x^m.$$

Then Zubov's partial differential equation becomes

$$\left(\frac{\partial V_2}{\partial \mathbf{x}} + \frac{\partial V_3}{\partial \mathbf{x}} + \cdots\right)(-x + x^2) = -x^2(1 - V_2 - V_3 - \cdots),$$

and the recursive relations (4.5-37) produce

$$\begin{aligned} -2b_2x^2 &= -x^2 \\ -3b_3x^3 &= -2b_2x^3 \\ -4b_4x^4 &= (b_2 - 3b_3)x^4 \\ &\vdots \\ -mb_mx^m &= (b_{m-2} - [m-1]b_{m-1})x^m. \end{aligned}$$

Thus we have

$$b_2 = \frac{1}{2} = \frac{1}{2!}$$

$$b_3 = \frac{2}{3}\, b_2 = \frac{2}{3!}$$

$$b_4 = \frac{1}{4}\,(3b_3 - b_2) = \frac{3}{4!} \tag{4.5-38}$$

$$\vdots$$

$$b_m = \frac{1}{m}\,[(m-1)b_{m-1} - b_{m-2}] = \frac{m-1}{m!},$$

which yields

$$V(x) = \frac{1}{2!}\,x^2 + \frac{2}{3!}\,x^3 + \frac{3}{4!}\,x^4 + \cdots. \tag{4.5-39}$$

Since

$$e^x = 1 + x + \frac{1}{2!}\,x^2 + \frac{1}{3!}\,x^3 + \cdots, \tag{4.5-40}$$

we suspect that the closed-form representation for $V(x)$ will involve e^x. Using the form of the last of equations (4.5-38) for each of the coefficients in (4.5-39), we have

$$\begin{aligned} V(x) &= \frac{2-1}{2!}\,x^2 + \frac{3-1}{3!}\,x^3 + \frac{4-1}{4!}\,x^4 + \cdots \\ &= x\left(x + \frac{1}{2!}\,x^2 + \frac{1}{3!}\,x^3 + \cdots\right) - \left(\frac{1}{2!}\,x^2 + \frac{1}{3!}\,x^3 + \cdots\right) \\ &= x(e^x - 1) - (e^x - 1 - x). \end{aligned}$$

Thus we have

$$V(x) = 1 - (1 - x)e^x. \tag{4.5-41}$$

Noting that $V(x)$ is positive definite and that $V(x) \to 1$ as $x \to 1$ or as $x \to -\infty$, we see that the domain of attraction is the region $-\infty < x < 1$, corresponding to $0 \le V(x) < 1$. ■

In the previous example we were able to synthesize a closed-form representation for $V(\mathbf{x})$. In lieu of trying to determine the function $V(\mathbf{x})$ that corresponds to the infinite series (4.5-34), we can stop the recursive process after any step and approximate $V(\mathbf{x})$ by

$$V(\mathbf{x}) \approx \overline{V}_m(\mathbf{x}) \triangleq V_2(\mathbf{x}) + V_3(\mathbf{x}) + \cdots + V_m(\mathbf{x}). \tag{4.5-42}$$

Because we have truncated the infinite series, the domain of attraction does not correspond to the region $\overline{V}(\mathbf{x}) < 1$. For the truncated representation, the optimization procedure for estimating the domain of attraction can be used to find the largest $\overline{V}_m(\mathbf{x}) < c$ region for which $[\partial\overline{V}_m/\partial\mathbf{x}]\mathbf{f} \leq 0$ (and not identically zero). We refer to the resulting largest estimate region $\overline{V}_m(\mathbf{x}) < \overline{V}_m^*$ as the mth degree **Zubov approximation** to the domain of attraction.

Example 4.5-6: Soft Cubic Spring The system

$$\begin{aligned} \dot{x}_1 &= x_2 \\ \dot{x}_2 &= -(1 - x_1^2)x_1 - x_2 \end{aligned}$$

is analogous to a spring-mass-damper system with a "softening" nonlinear cubic spring. The system has three equilibrium points: $\hat{\mathbf{x}} = \mathbf{0}$ and $\hat{\mathbf{x}} = [\pm 1 \quad 0]^T$. The linearized equations of motion indicate that the origin is at least locally asymptotically stable. We use Zubov's recursive procedure to estimate the domain of attraction for the origin.

Choosing the positive semidefinite function

$$\psi(\mathbf{x}) = x_1^2$$

and using

$$V(\mathbf{x}) = V_2(\mathbf{x}) + V_3(\mathbf{x}) + \cdots$$

in (4.5-22) yields

$$\left(\frac{\partial V_2}{\partial x_1} + \frac{\partial V_3}{\partial x_1} + \cdots\right) x_2 + \left(\frac{\partial V_2}{\partial x_2} + \frac{\partial V_3}{\partial x_2} + \cdots\right)(-x_1 - x_2 + x_1^3)$$
$$= -\psi[1 - V_2 - V_3 - \cdots].$$

Expanding this and equating terms of the same degree, the recursive relations (4.5-37) become

$$\frac{\partial V_2}{\partial x_1} x_2 + \frac{\partial V_2}{\partial x_2}(-x_1 - x_2) = -\psi$$

$$\frac{\partial V_3}{\partial x_1} x_2 + \frac{\partial V_3}{\partial x_2}(-x_1 - x_2) = 0 \tag{4.5-43}$$

$$\vdots$$

$$\frac{\partial V_m}{\partial x_1} x_2 + \frac{\partial V_m}{\partial x_2}(-x_1 - x_2) = \psi V_{m-2} - x_1^3 \frac{\partial V_{m-2}}{\partial x_2}, \qquad m \geq 4.$$

For the quadratic function

$$V_2(\mathbf{x}) = b_{20}x_1^2 + b_{11}x_1x_2 + b_{02}x_2^2,$$

equating coefficients in the first of equations (4.5-43),

$$(2b_{20}x_1 + b_{11}x_2)x_2 + (b_{11}x_1 + 2b_{02}x_2)(-x_1 - x_2) = -x_1^2,$$

yields

$$\begin{bmatrix} 0 & -1 & 0 \\ 2 & -1 & -2 \\ 0 & 1 & -2 \end{bmatrix} \begin{bmatrix} b_{20} \\ b_{11} \\ b_{02} \end{bmatrix} = \begin{bmatrix} -1 \\ 0 \\ 0 \end{bmatrix} \Rightarrow \begin{bmatrix} b_{20} \\ b_{11} \\ b_{02} \end{bmatrix} = \begin{bmatrix} 1 \\ 1 \\ 0.5 \end{bmatrix}.$$

Thus

$$V_2(\mathbf{x}) = x_1^2 + x_1x_2 + \tfrac{1}{2}x_2^2.$$

For the cubic function

$$V_3(\mathbf{x}) = b_{30}x_1^3 + b_{21}x_1^2x_2 + b_{12}x_1x_2^2 + b_{03}x_2^3$$

we get

$$\begin{bmatrix} 0 & -1 & 0 & 0 \\ 3 & -1 & -2 & 0 \\ 0 & 2 & -2 & -3 \\ 0 & 0 & 1 & -3 \end{bmatrix} \begin{bmatrix} b_{30} \\ b_{21} \\ b_{12} \\ b_{03} \end{bmatrix} = \begin{bmatrix} 0 \\ 0 \\ 0 \\ 0 \end{bmatrix} \Rightarrow \begin{bmatrix} b_{30} \\ b_{21} \\ b_{12} \\ b_{03} \end{bmatrix} = \begin{bmatrix} 0 \\ 0 \\ 0 \\ 0 \end{bmatrix},$$

so that

$$V_3(\mathbf{x}) \equiv 0.$$

Repeating the process we find that, for this example system, $V_m(\mathbf{x}) \equiv 0$ for m odd. For m even we get a "tridiagonal" system of equations for the coefficients of $V_m(\mathbf{x})$ in terms of the coefficients of $V_{m-2}(\mathbf{x})$, given by

$$\begin{bmatrix} 0 & -1 & & & & & \\ m & -1 & -2 & & & \mathbf{0} & \\ & m-1 & -2 & -3 & & & \\ & & & \ddots & & & \\ & \mathbf{0} & & & 2 & -(m-1) & -m \\ & & & & & 1 & -m \end{bmatrix} \begin{bmatrix} b_{m,0} \\ b_{m-1,1} \\ b_{m-2,2} \\ \vdots \\ b_{1,m-1} \\ b_{0,m} \end{bmatrix} = \begin{bmatrix} b_{m-2,0} - b_{m-3,1} \\ b_{m-3,1} - 2b_{m-4,2} \\ b_{m-4,2} - 3b_{m-5,3} \\ \vdots \\ b_{1,m-3} - (m-2)b_{0,m-2} \\ b_{0,m-2} \\ 0 \\ 0 \end{bmatrix}.$$

It should be emphasized that this formula applies only to this particular example and to our choice for $\psi(\mathbf{x})$. However, for any case, the result at each stage will be a linear system of equations.

Table 4.5-1 shows the results through $m = 14$. For any corresponding value of m, the associated estimate for the domain of attraction is determined by solving the optimization problem (4.5-15) with $V(\mathbf{x})$ approximated by $\overline{V}_m(\mathbf{x})$ from (4.5-42). The resulting stability estimates are shown in Figure 4.5-4.

The dashed lines in Figure 4.5-4 indicate the actual boundary of the domain of attraction. This boundary was generated by integrating the equations of motion backwards from near the two equilibrium points at $\hat{\mathbf{x}} = [\pm 1 \quad 0]^T$, both of which are saddle points. At each of these points two trajectories approach equilibrium along two corresponding branches of the saddle. Forward in time these trajectories are unstable, but backward in time they become stable. This allows us to plot the stability boundary, at least in two dimensions, by starting the integration very close to the saddle points. Note that this retro-time procedure is practical only in two or three dimensions, and that it does not yield a formula, $V(\mathbf{x}) < 1$, for the domain of attraction to the origin. ■

From Example 4.5-6 we note several facts about Zubov's recursive procedure. First of all, the process is well suited for digital computation, particularly if the nonlinear terms $\mathbf{g}(\mathbf{x})$ in the equations of motion (4.5-29) can be specified easily, instead of having to use a series expansion. The coefficients for each of the $V_m(\mathbf{x})$ functions in (4.5-42) can be computed from the recursive relations

Table 4.5-1 Zubov Recursion Results for Example 4.5-6

V_m	V_2	V_4	V_6	V_8	V_{10}	V_{12}	V_{14}
x_1^2	±1	±1.1637	±0.6485	±1.5665	±0.7647	±1.7874	±0.8650
x_2^2	0	∓1.4142	±0.6183	∓2.2492	±0.4372	∓2.7376	±0.2591
$\overline{V}_m^*$	1	0.3542	0.9851	0.5866	0.9992	0.6835	0.9998
Type	Tangent	Corner	Tangent	Corner	Tangent	Corner	Tangent
$b_{m,0}$	1	$-.7143E-1$	$.8201E-1$	$-.1024E-1$	$.6376E-4$	$-.3009E-3$	$-.5554E-4$
$b_{m-1,1}$	1	0	$.7143E-1$	$-.1058E-1$	$-.3438E-3$	$-.4076E-3$	$-1.066E-3$
$b_{m-2,2}$	0.5	$-.1429$	$.6746E-1$	$-.3916E-2$	$.1866E-2$	$.4361E-3$	$.3044E-3$
$b_{m-3,3}$		$-.7143E-1$	$.5026E-1$	$.5697E-2$	$.4727E-2$	$.2320E-2$	$.1509E-2$
$b_{m-4,4}$		$-.1786E-1$	$.2976E-1$	$.7049E-2$	$.5812E-2$	$.3981E-2$	$.3182E-2$
$b_{m-5,5}$			$.9921E-2$	$.4026E-2$	$.4584E-2$	$.4412E-2$	$.4390E-2$
$b_{m-6,6}$			$.1653E-2$	$.1345E-2$	$.2667E-2$	$.3534E-2$	$.4444E-2$
$b_{m-7,7}$				$.3362E-3$	$.1132E-2$	$.2134E-2$	$.3465E-2$
$b_{m-8,8}$				$.4203E-4$	$.3424E-3$	$.9838E-3$	$.2128E-2$
$b_{m-9,9}$					$.6848E-4$	$.3416E-3$	$.1036E-2$
$b_{m-10,10}$					$.6848E-5$	$.8604E-4$	$.3966E-3$
$b_{m-11,11}$						$.1434E-4$	$.1168E-3$
$b_{m-12,12}$						$.1195E-5$	$.2519E-4$
$b_{m-13,13}$							$.3599E-5$
$b_{m-14,14}$							$.2570E-6$

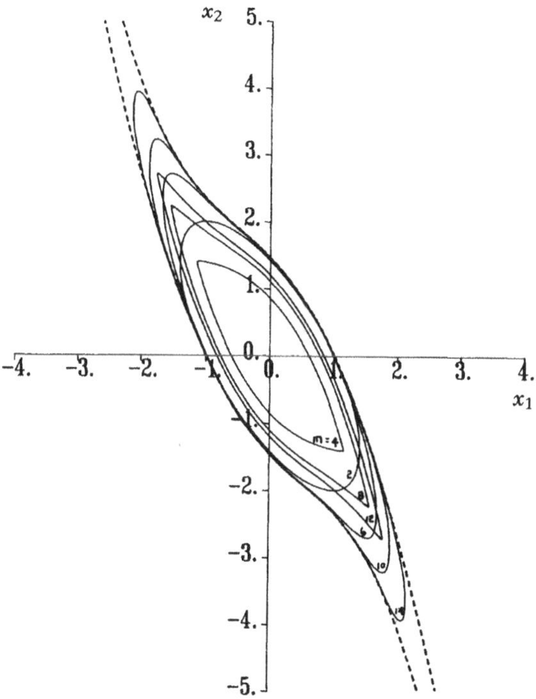

Figure 4.5-4 Zubov stability estimates for Example 4.5-6.

(4.5-37) by solving a corresponding system of $m + 1$ linear equations, for example, by using Gaussian elimination. The optimization procedure for finding the largest $\overline{V}_m(\mathbf{x}) < c$ estimate can, of course, be done numerically, for example by using Newton's method to solve the nonlinear equations (4.5-16) and (4.5-17). However, this process would be difficult or impossible to completely automate. In general, the surface in state space defined by $d\overline{V}_m(\mathbf{x})/dt = 0$ is not connected, but consists of several such surfaces. Furthermore, there are often several points where the surface $\overline{V}_m(\mathbf{x})$ = constant and $d\overline{V}_m(\mathbf{x})/dt = 0$ are tangent, as well as corner points where the gradient $\partial\overline{V}_m(\mathbf{x})/\partial\mathbf{x}$ is equal to zero. Thus the optimization problem often has multiple local solutions. Care must be taken to ensure that the global optimum has been found for the problem, that is, that the smallest $\overline{V}_m(\mathbf{x}) < c$ region has been found, subject to $d\overline{V}_m(\mathbf{x})/dt \geq 0$. For a further discussion of the numerical aspects of Zubov's recursive procedure, see Rodden [1964].

As Example 4.5-6 illustrates, the greatest shortcoming of Zubov's recursive procedure is that the convergence of the approximation is not monotonic. In general, having determined the mth degree approximation, there is no assurance that the $(m + 1)$th degree approximation will give a better estimate of the domain of attraction. It may actually give a worse estimate. Indeed, many iterations may be required in order to achieve a better estimate than the intial quadratic approximation. We should also note that the convergence properties of the process depend on the choice of the function $\psi(\mathbf{x})$. Also, the recursive procedure applied to the alternate representation (4.5-24) of Zubov's partial differential equation has different convergence properties than the procedure that we have developed for (4.5-22).

EXERCISES

4-1. Determine the solution $r(t)$ of the system

$$\dot{r} = -r(1 - r)$$

with $r(0) = r_0$, for:

(a) $0 < r_0 < 1$. Show that $r(t) \to 0$ as $t \to \infty$.

(b) $r_0 > 1$. Show that $r(t) \to \infty$ as t increases.

4-2. For the system

$$\dot{X}_1 = X_2$$

$$\dot{X}_2 = -\frac{X_1}{(1 + X_1^2)^2} - \frac{X_2}{(1 + X_1^2 + X_2^2)^2}$$

write the linearized state perturbation differential equations for small perturbations $\mathbf{x}(t)$ from the equilibrium at $\mathbf{X}(t) = 0$.

4-3. For a scalar-valued function $\phi(\mathbf{X})$ of a vector argument $\mathbf{X}$, consider the gradient system

$$\dot{\mathbf{X}} = -\left(\frac{\partial\phi(\mathbf{X})}{\partial\mathbf{X}}\right)^{\mathrm{T}}.$$

By linearizing the gradient system about an equilibrium solution $\hat{\mathbf{X}}$, with $\mathbf{X} = \hat{\mathbf{X}} + \mathbf{x}$, show that the local stability of $\hat{\mathbf{X}}$ depends on the eigenvalues of

$$\left.\frac{\partial^2\phi}{\partial\mathbf{X}^2}\right|_{\mathbf{X}=\hat{\mathbf{X}}}.$$

State the stability and asymptotic stability conditions for $\hat{\mathbf{X}}$ in terms of these eigenvalues. Discuss whether or not a perturbed solution $\mathbf{x}(t)$ could exhibit oscillatory (or damped oscillatory) motion.

4-4. For the following system determine the equilibrium points and discuss the stability of each equilibrium using Lyapunov's (indirect) linearization method:

$$\dot{x}_1 = x_1(1 - x_1) - \tfrac{1}{2}x_1x_2$$
$$\dot{x}_2 = x_2(1 - x_2) + \tfrac{1}{2}x_1x_2.$$

Classify each equilibrium point (node, saddle, etc.) and use the results to sketch the flow of trajectories in the (x_1, x_2) plane.

4-5. Use Lyapunov's (indirect) linearization method and center manifold theory to investigate the stability of the origin for the system shown below. Sketch the trajectories on the center manifold. From these results make a three-dimensional sketch of the trajectories near the origin.

$$\dot{X}_1 = -X_2 - X_3$$
$$\dot{X}_2 = X_1X_2$$
$$\dot{X}_3 = X_3(X_1 - 1).$$

4-6. For $r^2 = x_1^2 + x_2^2$ show that the function

$$V(\mathbf{x}) = 6r^4 - 22r^3 + 21r^2$$

is positive definite everywhere.

4-7. For $\mathbf{x} = [x_1 \quad x_2]^\mathrm{T}$ consider the function $V(\mathbf{x})$ defined by

$$V(\mathbf{x}) = \mathbf{x}^\mathrm{T}\mathbf{P}\mathbf{x}, \qquad \text{with } \mathbf{P} = \begin{bmatrix} 1 - x_2^2 & x_1x_2 \\ x_1x_2 & 1 \end{bmatrix}.$$

Show that $V(\mathbf{x}) > 0$ for all $\mathbf{x} \neq \mathbf{0}$, despite the fact that Sylvester's sufficient conditions (4.3-3) are not satisfied for all nonzero $\mathbf{x}$. Are the eigenvalues of $\mathbf{P}$ real and positive for all $\mathbf{x}$?

4-8. Write the quadratic form

$$V(\mathbf{x}) = \mathbf{x}^\mathrm{T}\mathbf{P}\mathbf{x}, \qquad \mathbf{P} = \begin{bmatrix} 1 & 2 & 3 \\ 4 & 5 & 6 \\ 7 & 8 & 9 \end{bmatrix}$$

in terms of an equivalent symmetric matrix $\mathbf{P}_s$. Is $V(\mathbf{x})$ positive definite?

4-9. Show that Newton's identities (4.4-6) can be written equivalently as in (4.4-8).

4-10. For the system

$$\dot{\mathbf{x}} = \mathbf{A}\mathbf{x},$$

solve the matrix Lyapunov equation

$$\mathbf{PA} + \mathbf{A}^{\mathrm{T}}\mathbf{P} = -\mathbf{Q}$$

for **P** for the following cases:

(a)

$$\mathbf{A} = \begin{bmatrix} 1 & 2 \\ 4 & 1 \end{bmatrix} \qquad \mathbf{Q} = \begin{bmatrix} 1 & 0 \\ 0 & 1 \end{bmatrix}.$$

(b)

$$\mathbf{A} = \begin{bmatrix} -1 & -2 \\ 2 & -1 \end{bmatrix} \qquad \mathbf{Q} = \begin{bmatrix} 1 & 0 \\ 0 & 1 \end{bmatrix}.$$

(c)

$$\mathbf{A} = \begin{bmatrix} 0 & 1 & 0 \\ 2 & 0 & 2 \\ -1 & -1 & -1 \end{bmatrix} \qquad \mathbf{Q} = \begin{bmatrix} 1 & 0 & 0 \\ 0 & 1 & 0 \\ 0 & 0 & 1 \end{bmatrix}.$$

For each case answer the following questions:

(i) Is the solution unique?

(ii) Is there a positive definite solution? If so, find it.

(iii) Using Lyapunov's second method, show whether or not the origin is asymptotically stable.

(iv) Repeat (iii) using eigenvalue analysis. Are the results in agreement?

(v) Assume the linear system resulted from linearization of a nonlinear system near an equilibrium at the origin. What conclusions can be drawn about the stability of the origin for the nonlinear system?

4-11. Consider the system

$$\dot{x}_1 = x_2$$
$$\dot{x}_2 = -x_1^3 - x_1^2 x_2.$$

(a) Use Lyapunov's first method to investigate the stability of the origin. Is the origin stable for the linearized system? From these results, what can be concluded about the stability of the origin for the nonlinear system?

(b) Use Lyapunov's second method to investigate the stability of the origin for the nonlinear system. If the origin is asymptotically stable, what is its domain of attraction?

4-12. Use the variable gradient method to construct a Lyapunov function for the system

$$\dot{x}_1 = -2x_1 - x_1^3 + x_2$$
$$\dot{x}_2 = x_1 - x_2.$$

Use the Lyapunov function to show that the origin is globally asymptotically stable.

4-13. Consider the system

$$\dot{x} = y$$
$$\dot{y} = y - \tfrac{1}{3}y^3 + x - x^3.$$

(a) Determine the location, stability, and geometry (node, focus, etc.) for each equilibrium point and sketch the (x, y) trajectory flow for the overall system.

(b) **Proposition:** The trajectories $[x(t), y(t)]$ are bounded. Assuming the proposition is true, investigate the possible existence of a limit cycle for the system.

(c) Prove that the proposition in (b) is true by using the variable gradient method to construct a suitable function $V(x, y)$. List all conditions that $V(\cdot)$ should have to satisfy in order to prove the proposition. Verify that your $V(\cdot)$ function satisfies them.

4-14. An alternate approach to estimating a domain of attraction (not necessarily the largest) is to

$$\max \dot{V} \qquad \text{subject to } V(\mathbf{x}) = V_m,$$

where V_m is a Lyapunov level curve defining the estimated domain as given by

$$V(\mathbf{x}) \le V_m.$$

If $\max \dot{V} < 0$, then the estimate is valid. Suppose that a nonlinear system has been linearized in the form

$$\dot{\mathbf{x}} = \mathbf{A}\mathbf{x} + \mathbf{B}\mathbf{u}$$

and that LQR design methods have been used to determine feedback gains $\mathbf{K}$ such that $\mathbf{u} = -\mathbf{K}\mathbf{x}$ results in the stable system

$$\dot{\mathbf{x}} = \hat{\mathbf{A}}\mathbf{x}.$$

Solving the Lyapunov equation (4.4-3) results in the Lyapunov function

$$V = \mathbf{x}^{\mathrm{T}}\mathbf{P}\mathbf{x}.$$

The necessary conditions for maximizing $\dot{V}$ on the surface $V(\mathbf{x}) = V_m$ are given by

$$\frac{\partial L}{\partial \mathbf{x}} = 0$$

$$\mathbf{x}^{\mathrm{T}}\mathbf{P}\mathbf{x} = V_m$$

where $L = -\mathbf{x}^{\mathrm{T}}\mathbf{Q}\mathbf{x} + \lambda(\mathbf{x}^{\mathrm{T}}\mathbf{P}\mathbf{x} - V_m)$ and λ is a Lagrange multiplier. Show that these necessary conditions reduce to the eigenvalue problem

$$\hat{\mathbf{P}}\mathbf{x} = \mu\mathbf{x}$$

where $\hat{\mathbf{P}} = \mathbf{Q}^{-1}\mathbf{P}$ and $\mu = V_m/\mathbf{x}^{\mathrm{T}}\mathbf{Q}\mathbf{x}$. The maximum solution is where the eigenvector corresponding to the largest eigenvalue intersects the V_m level curve. Explain why this is the "obvious" answer. Could this result be useful in solving the nonlinear problem?

4-15. Use the optimization procedure of Section 4.5 to estimate the domain of attraction for Example 2.3-4. For this problem the controllable set is the entire state space. One approach for a more comprehensive control design would be to use the closed-loop state feedback controller only inside the domain of attraction estimate and to use an open-loop control $u(t)$ outside the region such that the system will be driven to the domain of attraction estimate. In Chapter 7 we discuss an optimal control method for doing this. Another possibility is to use a $u(t)$ that will generate chaotic control. Demonstrate numerically that there exists an A and an ω such that the control $u = A \sin \omega t$ will drive this system in a chaotic fashion. Furthermore, demonstrate that this chaotic controller will always drive the system into the domain of attraction estimate.

4-16. Consider the system (a retro-time Rayleigh system) given by

$$\dot{x}_1 = -x_2$$

$$\dot{x}_2 = x_1 - x_2 + x_2^3.$$

(a) Construct a Lyapunov function $V(x_1, x_2)$ and use it to show that the origin is at least locally asymptotically stable.

(b) For the Lyapunov function in (a), determine the largest estimate, of the form $V(x_1, x_2) < c$, for the domain of attraction.

4-17. For the system in Exercise 4-16, use Zubov's recursive procedure to determine the domain of attraction. Carry out the calculations only far enough to determine $\overline{V}_3(\mathbf{x})$, that is, through degree three for the Lyapunov function.

5

LYAPUNOV CONTROL SYSTEM DESIGN

5.1 UNCONSTRAINED LINEAR SYSTEMS

In this chapter some of the Lyapunov stability concepts from Chapter 4 are applied to the problem of designing a feedback controller $\mathbf{u}(\mathbf{x})$ for systems of the form

$$\dot{\mathbf{x}} = \mathbf{f}(\mathbf{x}, \mathbf{u}), \qquad \mathbf{u} \in \mathcal{U}, \tag{5.1-1}$$

so that the origin becomes an asymptotically stable equilibrium point for

$$\dot{\mathbf{x}} = \mathbf{f}[\mathbf{x}, \mathbf{u}(\mathbf{x})]. \tag{5.1-2}$$

If we can construct a Lyapunov function such that $V(\mathbf{0}) = 0$ and for $\mathbf{x} \neq \mathbf{0}$, $V(\mathbf{x}) > 0$ and

$$\dot{V}(\mathbf{x}) = \frac{\partial V(\mathbf{x})}{\partial \mathbf{x}} \mathbf{f}[\mathbf{x}, \mathbf{u}(\mathbf{x})]$$

is negative definite (or negative semidefinite, but not identically zero on a trajectory) for all points in some open neighborhood containing the origin, then the origin will be asymptotically stable.

We first investigate Lyapunov-based stabilizing controllers for linear systems

$$\dot{\mathbf{x}} = \mathbf{A}\mathbf{x} + \mathbf{B}\mathbf{u}, \tag{5.1-3}$$

where $\mathbf{x}$ is an n_x-dimensional state vector and $\mathbf{u}$ is an unconstrained n_u-dimensional control vector. Under **state vector feedback control**

$$\mathbf{u} = -\mathbf{K}\mathbf{x} \tag{5.1-4}$$

the resulting closed-loop system will be of the form

$$\dot{\mathbf{x}} = \hat{\mathbf{A}}\mathbf{x}, \tag{5.1-5}$$

where

$$\hat{\mathbf{A}} = \mathbf{A} - \mathbf{B}\mathbf{K}. \tag{5.1-6}$$

To achieve asymptotic stability, the **feedback gain matrix K** is chosen so that Lyapunov's equation

$$\mathbf{P}\hat{\mathbf{A}} + \hat{\mathbf{A}}^{\mathrm{T}}\mathbf{P} = -\mathbf{Q} \tag{5.1-7}$$

has a symmetric positive definite solution $\mathbf{P}$ corresponding to any conveniently chosen symmetric positive definite matrix $\mathbf{Q}$. From Lyapunov's lemma in Section 4.4, the closed-loop system (5.1-5) will be asymptotically stable if, and only if, such a solution $\mathbf{P}$ exists. For such a solution the positive definite quadratic form

$$V(\mathbf{x}) = \mathbf{x}^{\mathrm{T}}\mathbf{P}\mathbf{x} \tag{5.1-8}$$

will be a Lyapunov function for the system (5.1-5), since

$$\frac{dV[\mathbf{x}(t)]}{dt} = -\mathbf{x}^{\mathrm{T}}\mathbf{Q}\mathbf{x} < 0 \qquad \text{for } \mathbf{x} \neq \mathbf{0}. \tag{5.1-9}$$

Example 5.1-1: State Vector Feedback As in Example 2.2-1, consider the linearized inverted pendulum system given by

$$\dot{x}_1 = x_2$$
$$\dot{x}_2 = x_1 + u.$$

For $u = 0$ the origin is unstable. However, the system satisfies the Kalman controllability criterion (2.2-2), so we know that a linear state feedback controller can be designed to stabilize the system. In fact, the closed-loop eigenvalues can be arbitrarily placed.

Using linear state feedback for the control,

$$u = -k_1 x_1 - k_2 x_2,$$

yields the closed-loop system

$$\begin{bmatrix} \dot{x}_1 \\ \dot{x}_2 \end{bmatrix} = \begin{bmatrix} 0 & 1 \\ 1 - k_1 & -k_2 \end{bmatrix} \begin{bmatrix} x_1 \\ x_2. \end{bmatrix}.$$

Let $\mathbf{Q} = \mathbf{I}$, so that (5.1-7) becomes

$$\begin{bmatrix} p_{11} & p_{12} \\ p_{12} & p_{22} \end{bmatrix} \begin{bmatrix} 0 & 1 \\ 1 - k_1 & -k_2 \end{bmatrix} + \begin{bmatrix} 0 & 1 - k_1 \\ 1 & -k_2 \end{bmatrix} \begin{bmatrix} p_{11} & p_{12} \\ p_{12} & p_{22} \end{bmatrix} = \begin{bmatrix} -1 & 0 \\ 0 & -1 \end{bmatrix}.$$

For asymptotic stability, we need to choose k_1 and k_2 such that a positive definite solution for $\mathbf{P}$ is obtained. This Lyapunov equation yields a system of three linear equations for the three elements of $\mathbf{P}$:

$$\begin{aligned} -1 &= 2(1 - k_1)p_{12} \\ 0 &= p_{11} - k_2 p_{12} + (1 - k_1)p_{22} \\ -1 &= 2p_{12} - 2k_2 p_{22}, \end{aligned}$$

from which we obtain

$$p_{12} = \frac{1}{2(k_1 - 1)}$$

$$p_{22} = \frac{k_1}{2k_2(k_1 - 1)}$$

$$p_{11} = \frac{k_2}{2(k_1 - 1)} + \frac{k_1}{2k_2}.$$

Thus we have

$$\mathbf{P} = \begin{bmatrix} \dfrac{k_1^2 + k_2^2 - k_1}{2k_2(k_1 - 1)} & \dfrac{1}{2(k_1 - 1)} \\ \dfrac{1}{2(k_1 - 1)} & \dfrac{k_1}{2k_2(k_1 - 1)} \end{bmatrix}.$$

For $\mathbf{P}$ to be positive definite we require

$$0 < p_{11} = \frac{k_1^2 + k_2^2 - k_1}{2k_2(k_1 - 1)}$$

$$0 < |\mathbf{P}| = \frac{k_1^2 + k_2^2}{4k_2^2(k_1 - 1)}.$$

Both conditions are satisfied if, and only if,

$$k_1 > 1$$
$$k_2 > 0.$$

For any such k_1 and k_2, the origin will be globally asymptotically stable under the closed-loop state variable feedback controller.

Comparing this result with an eigenvalue analysis, we note that the closed-loop characteristic equation

$$0 = |\mu\mathbf{I} - \hat{\mathbf{A}}| = \begin{vmatrix} \mu & -1 \\ k_1 - 1 & \mu + k_2 \end{vmatrix} = \mu^2 + k_2\mu + k_1 - 1$$

also yields eigenvalues with $Re(\mu) < 0$ if, and only if, $k_1 > 1$ and $k_2 > 0$. ■

Another method for choosing the feedback gain matrix $\mathbf{K}$ would be to use the LQR design method introduced in Section 2.2. Under LQR design the feedback gains are given by

$$\mathbf{K} = \mathbf{R}^{-1}\mathbf{B}^T\overline{\mathbf{P}}$$

where $\overline{\mathbf{P}}$ is the solution to the Riccati equation

$$\overline{\mathbf{P}}\mathbf{A} + \mathbf{A}^T\overline{\mathbf{P}} - \overline{\mathbf{P}}\mathbf{B}\mathbf{R}^{-1}\mathbf{B}^T\overline{\mathbf{P}} = -\overline{\mathbf{Q}}$$

for specified positive-definite matrices $\mathbf{R}$ and $\overline{\mathbf{Q}}$. As we have shown previously, $\overline{\mathbf{P}}$ will satisfy a Lyapunov equation [see (2.2-6)]. Thus $\overline{\mathbf{P}}$ could be used to define a Lyapunov function $V(\mathbf{x}) = \mathbf{x}^T\overline{\mathbf{P}}\mathbf{x}$. In general $\overline{\mathbf{P}} \neq \mathbf{P}$ and the gain matrices $\mathbf{K}$ differ for the two methods.

5.2 LINEAR SYSTEMS WITH BOUNDED CONTROL

One approach to designing controllers for linear systems with bounded controls is to design a linear state feedback controller using, say, LQR design, and then let the control value "saturate" at the upper or lower bound whenever the bound is exceeded. While saturating state variable feedback often yields an effective controller, the domain of attraction may, in many cases, be somewhat less than the entire controllable set. The following two examples illustrate this situation. In the first example we show, using a Lyapunov function, that the domain of attraction under saturating state variable is the entire state space. In the second example saturating state variable feedback provides a domain of attraction that is a subset of the controllable set.

Example 5.2-1: Design of a Disk Drive Controller The read/write head of a computer disk drive moves in and out radially to read or write narrow "track" positions on a platter of spinning disks. The head, of mass m, moves under the action of an applied force u, with bounds $|u| \leq u_{\max}$, and a damping force that we model as a linear viscous damping force $-b\dot{r}$. Our objective is to design a controller to position the head as quickly and as accurately as possible in response to an input command to move to the track at radius $\hat{r}$.

The system is modeled by the equation of motion

$$m\ddot{r} = u - b\dot{r}.$$

In terms of the **error** $e(t) = \hat{r} - r(t)$ between the commanded and the current positions, the equation of motion becomes

$$m\ddot{e} + b\dot{e} + u = 0, \tag{5.2-1}$$

where m and b are positive constants and $u(\cdot)$ has the constraint

$$|u| \leq u_{\max}.$$

The control system design problem is to choose the function $u(\cdot)$ so that $e(t) \to 0$ as rapidly as possible. We could design a system using an open-loop control $u(t)$ and "stepper motors," with the discrete motor positions preset to correspond to the tracks on the disks, but any misalignment, due to wear, disk warpage, or other reasons, could cause the head to be unable to read or write its data properly. Current high performance disk drive controllers employ continuous rather than discrete positioning, using a device such as a "voice-coil" actuator, coupled with a feedback controller. We assume that the computer can accurately determine the head position relative to the center of a track so that the error signal $e(t)$ is available for feedback.

A primary performance measure for disk drives is the average "seek" time, that is, the average time required to move the head from one track to another.

Typical values for personal computers are on the order of 10–50 msec. To be competitive, the disk drive manufacturer would like to have a controller that minimizes the time required to move the head from one track to any other specified track. As we show in Chapter 7, because the bounded control u appears linearly in the equation of motion (5.2-1), a minimum-time controller is a discontinuous switching type controller, in which the control force u, at certain times, switches instantaneously between its maximum and minimum values. Such instantaneous switching cannot be achieved exactly in a physical system. Moreover, switching control tends to produce at least a small limit cycle about the desired track position, due to small time delays in the system.

To avoid limit cycles, consider using a **saturating proportional controller** with a high gain. The controller is given by

$$u(e) = \begin{cases} u_{\max} & \text{if } Ke \geq u_{\max} \\ Ke & \text{if } |Ke| \leq u_{\max} \\ -u_{\max} & \text{if } Ke \leq -u_{\max}. \end{cases} \tag{5.2-2}$$

This is a nonlinear controller that will approximate on-off relay action if the gain K is chosen sufficiently large. Having selected the nonlinear controller, we must now show that it can actually transfer the head from any initial track to any target track. First examine the linearized behavior near a target track. When the force is linear, $u = Ke$, (5.2-1) becomes a linear system

$$m\ddot{e} + b\dot{e} + Ke = 0 \tag{5.2-3}$$

with eigenvalues

$$u_{1,2} = \frac{1}{2m}\left(-b \pm \sqrt{b^2 - 4Km}\right). \tag{5.2-4}$$

For $K > 0$ the linearized system is asymptotically stable, since we assume that m and b are positive. The **return time**, or system **time constant**, is given by

$$T = \frac{1}{\min\{|Re(\mu_i)|\}} = \frac{2m}{b - \sqrt{b^2 - 4Km}} \tag{5.2-5}$$

for $0 < K \leq b^2/(4m)$. We can decrease the return time by increasing K, up to $K = b^2/(4m)$, but the minimum return time is fixed by the mass m and the damping constant b.

For the nonlinear saturating controller given by (5.5-2), the equation of motion (5.2-1) is a special case of Liénard's equation (4.4-16), with $h(e) = b/m$ and $g(e) = u(e)/m$, and conditions (4.4-17) and (4.4-18) are satisfied with $b > 0$ and $K > 0$. For condition (4.4-19) we have

$$G(e) = \frac{1}{m}\int_0^e u(\xi)d\xi = \begin{cases} \dfrac{K}{2m}e^2 & \text{if } |e| \le \dfrac{u_{\max}}{K} \\ \dfrac{u_{\max}}{m}\left(|e| - \dfrac{u_{\max}}{2K}\right) & \text{if } |e| \ge \dfrac{u_{\max}}{K}. \end{cases} \tag{5.2-6}$$

Thus $G(e) \to \infty$ as $|e| \to \infty$.

For a Lyapunov function we use the energy per unit mass, given by the positive definite Lur'e integral function

$$V(e, \dot{e}) = \tfrac{1}{2}\dot{e}^2 + G(e), \tag{5.2-7}$$

yielding

$$\dot{V} = -b\dot{e}^2. \tag{5.2-8}$$

Since $\dot{V} \le 0$ and not identically zero, we conclude that the equilibrium at $(e, \dot{e}) = (0, 0)$ is globally asymptotically stable. ■

Example 5.2-2: Saturating Minimum Integral-Square Error Feedback Consider again the problem of stabilizing the origin for the linearized inverted pendulum system

$$\dot{x}_1 = x_2$$
$$\dot{x}_2 = x_1 + u,$$

only now with the control constraint

$$|u| \le 1.$$

As illustrated in Example 2.2-4, the controllable set to the origin for this system is the region $|x_1 + x_2| < 1$. Our basic objective is to design a controller that will have a domain of attraction equal to the controllable set $|x_1 + x_2| < 1$ to the origin.

In Example 3.3-2 it has been shown that, in the absence of any control constraints, the linear state feedback controller

$$u = r - \mathbf{Kx},$$

with the feedback gain matrix

$$\mathbf{K} = [2 \quad 1],$$

minimizes the integral-square error

$$J = \int_0^\infty (\hat{x}_1 - x_1)^2 dt$$

in response to a step command input $r(t) = \hat{x}_1$, starting from $\mathbf{x}(0) = \mathbf{0}$. Using the same gains to stabilize the origin, starting from some nonzero initial state, the corresponding control is

$$u = -2x_1 - x_2.$$

To account for the control bounds, we use **saturating state feedback** control as given by (1.4-7). Figure 5.2-1 is a plot of the resulting trajectories for this controller. Those portions of the trajectories where the control is not saturated, in the region $|2x_1 + x_2| < 1$, are drawn with a thicker line. Only on these portions of the trajectories is the motion of the closed-loop system equivalent to a damped oscillator with a minimum integral-square error damping ration $\zeta = \frac{1}{2}$. Note that this particular controller fails to transfer all initial states in the controllable set $|x_1 + x_2| < 1$ to the origin. However, this failure is not an inherent characteristic of saturating state feedback controllers. With a suitable choice for the feedback gain matrix $\mathbf{K}$ (so that the control is saturated at the proper value everywhere along the controllability boundaries), a saturating linear state feedback control can be constructed that will transfer the entire controllable set to the origin for this example. ■

5.3 FUNCTION OPTIMIZING FEEDBACK CONTROLS

In this section we study two Lyapunov optimizing approaches for designing feedback controllers. The combination of function minimization and Lyapunov stability techniques represents a powerful approach to closed-loop nonlinear control design [Gutman and Leitmann, 1976; Grantham, 1981*c*, 1982, 1986; Grantham and Chingcuanco, 1984; Anderson and Grantham, 1989]. The two related methods that we present, **steepest descent control** and **quickest descent control**, also provide a basis for and simpler alternatives to the optimal control design methods that we study later in Chapter 7.

Lyapunov Optimizing Feedback Controls

The fact that the saturating feedback controller in Example 5.2-2 has a domain of attraction less than the controllable set can be traced to the fact that the controller was designed with no concern for the constraints on the control. The constraints were handled after the fact by simply letting the linear feedback

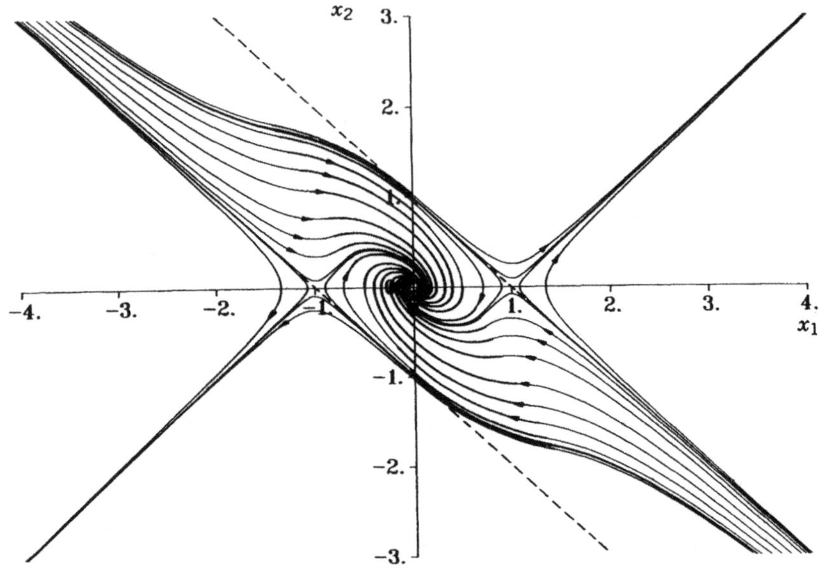

Figure 5.2-1 Saturating minimum integral-square error control.

stabilization controller saturate at the control bounds. We need an approach that allows us to include the constraints as an integral part of the control design. The concept of a gradient system provides a basis for such an approach.

Gradient System In Section 4.1 we have presented a preview of Lyapunov stability methods by introducing the notion of linear feedback control systems design to generate a linear gradient system. For nonlinear systems we can try to do the same thing. Suppose we could develop a controller $\mathbf{u}(\mathbf{x})$, satisfying control constraints, that yields a gradient system

$$\dot{\mathbf{x}} = \mathbf{f}[\mathbf{x}, \mathbf{u}(\mathbf{x})] = -\nabla\phi(\mathbf{x}),$$

where $\phi(\mathbf{x})$ is a chosen differentiable scalar-valued function, satisfying

(i)

$$\phi(\mathbf{0}) = 0, \tag{5.3-1a}$$

(ii)

$$\phi(\mathbf{x}) > 0 \qquad \text{for } \mathbf{x} \neq \mathbf{0}, \tag{5.3-1b}$$

(iii)

$$\nabla\phi(\mathbf{x}) \neq 0 \qquad \text{for } \mathbf{x} \neq \mathbf{0}, \tag{5.3-1c}$$

with

$$\nabla\phi(\mathbf{x}) \triangleq \left[\frac{\partial\phi(\mathbf{x})}{\partial\mathbf{x}}\right]^{\mathrm{T}}.$$

By construction, $\phi(\mathbf{x})$ has a global minimum at $\mathbf{x} = \mathbf{0}$ and its gradient vector is nonzero away from the origin. Thus the origin is a unique equilibrium point. Furthermore, away from the origin, we have

$$\dot{\phi}[\mathbf{x}(t)] = -\|\nabla\phi[\mathbf{x}(t)]\|^2 < 0.$$

Therefore $\phi[\mathbf{x}(t)]$ is decreasing and we conclude that $\mathbf{x}(t) \to 0$ as $t \to \infty$.

In a certain sense the "velocity potential function" $\phi(\mathbf{x})$ would be the ultimate Lyapunov function; it would not only allow us to establish asymptotic stability of the origin without having to integrate the equations of motion, but it would also allow us to determine the system trajectories without integrating the equations of motion. This follows, since the gradient vector $\nabla\phi(\mathbf{x})$ is normal to the surface $\phi(\mathbf{x})$ = constant and the velocity vector $\dot{\mathbf{x}} = -\nabla\phi(\mathbf{x})$ is opposite to the gradient vector. Thus trajectories $\mathbf{x}(t)$ coincide with the curves that are orthogonal to the surface of constant $\phi(\mathbf{x})$.

The main geometric feature of interest with a gradient system is that the trajectories $\mathbf{x}(t)$ penetrate the Lyapunov surfaces $\phi(\mathbf{x})$ = constant, so the $\phi(\mathbf{x}(t)]$ decreases, and they do so in a "steepest descent" fashion, perpendicular to the Lyapunov surfaces and opposite to the gradient vector $\nabla\phi(\mathbf{x})$. In general, it may not be possible to construct a gradient system, because the structure of the equations of motion and the control constraints may be such that we are unable to achieve $\dot{\mathbf{x}} = -\nabla\phi(\mathbf{x})$ at every state $\mathbf{x}$. However, the idea of trying to penetrate Lyapunov-type surfaces as much as possible can still be employed.

Descent Function $W(\mathbf{x})$ The basic approach for **Lyapunov optimizing feedback controls** is to choose a candidate Lyapunov-type function $W(\mathbf{x})$, such as distance to the target, and then choose $\mathbf{u}(\mathbf{x})$ so that $W[\mathbf{x}(t)]$ decreases as much as possible along trajectories of (5.1-2). There is more freedom in this process than is normally available in standard Lyapunov stability analysis. This is because at every state point $\mathbf{x}$, the control system (5.1-1) allows for a set of possible velocity vectors $\dot{\mathbf{x}} \in \mathbf{f}(\mathbf{x}, \mathcal{U})$, rather than a single velocity vector $\dot{\mathbf{x}} = \mathbf{f}(\mathbf{x}) \leftarrow \mathbf{f}[\mathbf{x}, \mathbf{u}(\mathbf{x})]$, for making $W[\mathbf{x}(t)]$ decrease.

We need to be more precise in our meaning of a "candidate" Lyapunov-type function. The conditions (5.3-1) on $\phi(\mathbf{x})$ would suffice, but weaker sets of conditions can be used depending on the circumstances. We use the following general definition:

A function $W(\mathbf{x})$ is a **descent function** for the target set $\mathcal{X}$ if the following conditions hold at points that are controllable to the target:

(i) $W(\mathbf{x})$ is continuous and continuously differentiable outside the target set and on the boundary of the target set.

(ii) The regions $W(\mathbf{x}) \leq c$ are nested, that is, for $c_1 < c_2$,

$$\{\mathbf{x}|W(\mathbf{x}) \leq c_1\} \subset \{\mathbf{x}|W(\mathbf{x}) \leq c_2\}.$$

(iii) The target is contained in one of the $W(\mathbf{x}) \leq c$ regions.

In addition, if the target set $\mathcal{X}$ is bounded, then we also require:

(iv) The regions $W(\mathbf{x}) \leq c$ are bounded.

Condition (iv) is imposed to exclude the possibility, as in Example 4.3-5, that trajectories might decrease the value of $W[\mathbf{x}(t)]$ but with $\mathbf{x}(t)$ approaching infinity instead of approaching the target set $\mathcal{X}$.

A descent function can be thought of as some measure of "distance" to the target. If we can make its value decrease along a trajectory, then the state must approach the target. As an example, if the target is the origin, then as discussed in Section 4.3, any continuously differentiable positive-definite function $W(\mathbf{x})$ for which $W(\mathbf{x}) \to \infty$ as $\|\mathbf{x}\| \to \infty$ could be used as a descent function since its $W(\mathbf{x}) \leq c$ regions are bounded, nested, and contain the origin. As another example, if the target set is the region $x_1 \leq 0$, we might choose $W(\mathbf{x}) = x_1$. For this target, $W(\mathbf{x})$ is a perfectly valid descent function, even though it is not positive definite.

Once a specific descent function $W(\mathbf{x})$ has been selected, the feedback control $\mathbf{u}(\mathbf{x})$ is chosen so that $W[\mathbf{x}(t)]$ decreases as much as possible at each state $\mathbf{x}$, hopefully penetrating $W(\mathbf{x}) =$ constant surfaces. If the resulting controller does make $W[\mathbf{x}(t)]$ decrease everywhere outside the target, and the target is the origin, then $W(\mathbf{x})$ may be a Lyapunov function for the feedback control system. If this is the case we also have a built-in proof of the effectiveness of the controller, since this is a sufficient condition for asymptotic stability. However we do not require this. There are many applications where a Lyapunov optimizing control is effective even though $W[\mathbf{x}(t)]$ does not decrease everywhere or does not provide asymptotic stability. The controller we are seeking will try to penetrate the $W(\mathbf{x}) =$ constant surfaces as much as possible, even though $W(\mathbf{x})$ itself may not turn out to be a Lyapunov function for the feedback system.

There are two principal ways that we can try to make $W[\mathbf{x}(t)]$ decrease as much as possible. We can either try to make trajectories penetrate surfaces of constant $W(\mathbf{x})$ as quickly as possible (minimize $dW[\mathbf{x}(t)]/dt$) or as steeply as possible (i.e., opposite to the gradient vector $\partial W(\mathbf{x})/\partial \mathbf{x}$). These two approaches generally lead to two different feedback controllers, but each is computed by solving a corresponding function minimization problem at each state $\mathbf{x}$.

Steepest Descent Control

The objective under steepest descent control [Grantham, 1981*c*, 1982, 1986; Grantham and Chingcuanco, 1984; Anderson and Grantham, 1989] is to try to produce a gradient system. Given the control system (5.1-1) and the control constraint $\mathbf{u} \in \mathcal{U}$, this may or may not be possible. In any event, at each point of a system trajectory $\mathbf{x}$, if a feedback control $\mathbf{u}(\mathbf{x})$ exists such that the velocity vector $\dot{\mathbf{x}}$ can be made to point in the direction opposite to the gradient $\nabla W \triangleq [\partial W(\mathbf{x})/\partial \mathbf{x}]^T$, then that is the control that will be used. Otherwise, $\mathbf{u}(\mathbf{x})$ is chosen so that $\dot{\mathbf{x}}$ is as close to the negative gradient direction as possible. Thus choosing a descent function $W(\mathbf{x})$ for steepest descent control is equivalent to specifying a **preferred direction of motion** at each point in state space. Under steepest descent control, the design objective is to produce a "flow" of trajectories in state space, perpendicular to the surfaces $W(\mathbf{x}) = $ constant. This flow objective can be visualized easily by the designer, even though the desired flow may not be achieved everywhere because of limitations imposed by the equations of motion or by the control constraints.

For a specified descent function $W(\mathbf{x})$, a feedback control law $\mathbf{u}^*(\mathbf{x})$ is a Lyapunov optimizing **steepest descent control** at $\mathbf{x}$ if

$$\mathbf{f}[\mathbf{x}, \mathbf{u}^*(\mathbf{x})] \neq \mathbf{0} \tag{5.3-2}$$

and

$$W'[\mathbf{x}, \mathbf{u}^*(\mathbf{x})] < W'(\mathbf{x}, \mathbf{u}) \qquad \text{for all } \mathbf{u} \subset \mathcal{U}, \tag{5.3-3}$$

where

$$W'(\mathbf{x}, \mathbf{u}) \triangleq \frac{\partial W(\mathbf{x})}{\partial \mathbf{x}} \frac{\mathbf{f}(\mathbf{x}, \mathbf{u})}{\|\mathbf{f}(\mathbf{x}, \mathbf{u})\|} \tag{5.3-4}$$

is the dot product of the gradient with the unit velocity vector.

Since our objective is to transfer all possible initial states to the target, condition (5.3-2) is imposed to ensure that the control law does not create an equilibrium point at a state that can be transferred to the target.

Condition (5.3-3) requires that the control penetrate the $W(\mathbf{x})$ contours as steeply as possible. To see this note that, from the definition of a "dot product" $\mathbf{a} \cdot \mathbf{b} \triangleq \|\mathbf{a}\| \|\mathbf{b}\| \cos\theta$, a geometric interpretation of $W'(\mathbf{x}, \mathbf{u})$ is given by

$$W'(\mathbf{x}, \mathbf{u}) = \|\nabla W\| \cos\theta, \tag{5.3-5}$$

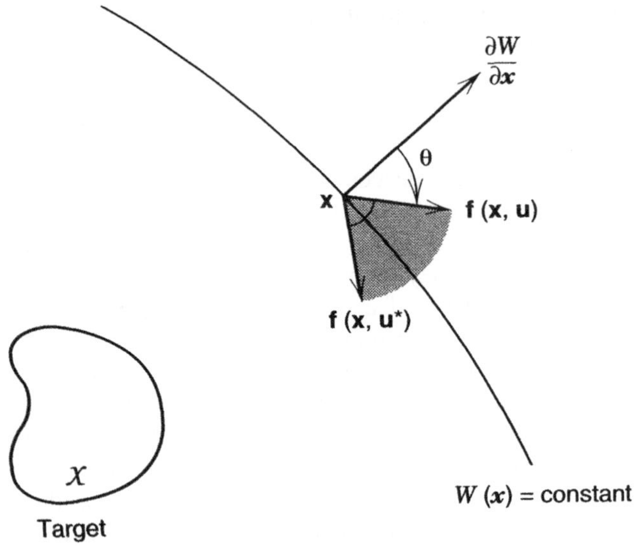

Figure 5.3-1 Geometry for steepest descent control.

where θ is the angle from the gradient vector ∇W to the velocity vector $\mathbf{f}(\mathbf{x}, \mathbf{u})$, as illustrated in Figure 5.3-1. Also shown in this figure is the $\boldsymbol{u}$-vectogram at the point $\mathbf{x}$ (shown shaded) and the velocity vector $\mathbf{f}(\mathbf{x}, \mathbf{u}^*)$ corresponding to steepest descent control. The steepest descent nature of the resulting controller can be seen by noting that $W'(\mathbf{x}, \mathbf{u})$ is the rate of change of $W[\mathbf{x}(t)]$ with respect to **arc length** s along a trajectory; that is,

$$W'(\mathbf{x}, \mathbf{u}) \triangleq \frac{dW[\mathbf{x}(t)]}{ds} = \frac{\partial W(\mathbf{x})}{\partial \mathbf{x}} \frac{d\mathbf{x}}{dt} \frac{dt}{ds} = \frac{\partial W(\mathbf{x})}{\partial \mathbf{x}} \frac{\mathbf{f}(\mathbf{x}, \mathbf{u})}{\|\mathbf{f}(\mathbf{x}, \mathbf{u})\|},$$

since the differential arc length ds along a trajectory $\mathbf{x}(t)$ is given by

$$ds \triangleq \sqrt{dx_1^2 + \cdots + dx_{n_x}^2} = \sqrt{\dot{x}_1^2 + \cdots + \dot{x}_{n_x}^2}\,dt = \|\mathbf{f}(\mathbf{x}, \mathbf{u})\|dt. \tag{5.3-6}$$

Since we are explicitly avoiding equilibrium points outside the target set, the differential change in position ds is nonzero and can be used to compute a corresponding distance rate of change dW/ds.

Assuming for convenience that $\partial W(\mathbf{x})/\partial \mathbf{x} \neq 0$, a steepest descent control $\mathbf{u}(\mathbf{x})$ that minimizes $W'(\mathbf{x}, \mathbf{u})$ also minimizes

$$G(\mathbf{x}, \mathbf{u}) = \frac{1}{\|\nabla W\|} \frac{\partial W(\mathbf{x})}{\partial \mathbf{x}} \frac{\mathbf{f}(\mathbf{x}, \mathbf{u})}{\|\mathbf{f}(\mathbf{x}, \mathbf{u})\|} = \cos\theta. \tag{5.3-7}$$

For states at which a Lyapunov optimizing steepest descent control $\mathbf{u}^*(\mathbf{x})$ is interior to the control constraint set $\mathcal{U}$, the necessary conditions corresponding to (3.3-5)–(3.3-9) reduce to

$$\mathbf{0}^{\mathrm{T}} = \frac{\partial G[\mathbf{x}, \mathbf{u}^*(\mathbf{x})]}{\partial \mathbf{u}} = \frac{1}{\|\nabla W\|} \frac{\partial W(\mathbf{x})}{\partial \mathbf{x}} \left[\frac{1}{\|\mathbf{f}\|} \frac{\partial \mathbf{f}}{\partial \mathbf{u}} - \frac{1}{\|\mathbf{f}\|^2} \mathbf{f} \frac{\partial \|\mathbf{f}\|}{\partial \mathbf{u}} \right].$$

Since $\|\mathbf{f}\|^2 = \mathbf{f}^{\mathrm{T}}\mathbf{f}$ implies that

$$\frac{\partial \|\mathbf{f}\|}{\partial \mathbf{u}} = \frac{\mathbf{f}^{\mathrm{T}}}{\|\mathbf{f}\|} \frac{\partial \mathbf{f}}{\partial \mathbf{u}},$$

a more convenient computational result is given by

$$\mathbf{0}^{\mathrm{T}} = \frac{\partial G[\mathbf{x}, \mathbf{u}^*(\mathbf{x})]}{\partial \mathbf{u}} = \frac{1}{\|\mathbf{f}\|^3 \|\nabla W\|} \frac{\partial W(\mathbf{x})}{\partial \mathbf{x}} [\|\mathbf{f}\|^2 \mathbf{I} - \mathbf{f}\mathbf{f}^{\mathrm{T}}] \frac{\partial \mathbf{f}}{\partial \mathbf{u}}. \tag{5.3-8}$$

The steepest descent interpretation becomes clear if we write (5.3-8) in the equivalent form

$$\mathbf{0}^{\mathrm{T}} = \frac{\partial G[\mathbf{x}, \mathbf{u}^*(\mathbf{x})]}{\partial \mathbf{u}} = \frac{1}{\|\mathbf{f}\|} \left[\frac{\partial W(\mathbf{x})/\partial \mathbf{x}}{\|\nabla W\|} - G \frac{\mathbf{f}^{\mathrm{T}}}{\|\mathbf{f}\|} \right] \frac{\partial \mathbf{f}}{\partial \mathbf{u}}. \tag{5.3-9}$$

If the velocity vector $\dot{\mathbf{x}}$ is opposite to the gradient vector ∇W, that is, if

$$\mathbf{f}^{\mathrm{T}}[\mathbf{x}, \mathbf{u}^*(\mathbf{x})] = -\alpha(\mathbf{x}) \left[\frac{\partial W(\mathbf{x})}{\partial \mathbf{x}} \right] \qquad \alpha(\mathbf{x}) > 0,$$

then $G[\mathbf{x}, \mathbf{u}^*(\mathbf{x})] = -1 = \cos\theta$, which implies $\theta = \pi$, and the row vector in brackets in (5.3-9) is the zero vector.

Example 5.3-1: Steepest Descent Control for Zermelo's Problem In Zermelo's problem, introduced in Example 2.2-6, a swimmer tries to reach a circular island in a river, as illustrated in Figure 2.2-6. In terms of the swimmer's heading angle u from the positive x_1-axis, the equations of motion can be written as

$$\dot{x}_1 = -2 + \cos u$$
$$\dot{x}_2 = \sin u,$$

and the target set is chosen as

$$\mathcal{X} = \{(x_1, x_2) | x_1^2 + x_2^2 \leq 1\}.$$

For simplicity we take our descent function as the square of the distance from the origin,

$$W(\mathbf{x}) = x_1^2 + x_2^2,$$

which yields

$$W'(\mathbf{x}, u) = \frac{\partial W(\mathbf{x})}{\partial \mathbf{x}} \frac{\mathbf{f}(\mathbf{x}, u)}{\|\mathbf{f}(\mathbf{x}, u)\|} = \frac{2x_1(-2 + \cos u) + 2x_2 \sin u}{\sqrt{(-2 + \cos u)^2 + \sin^2 u}}.$$

Since there are no constraints on the heading angle u, the necessary condition for minimizing $W'(\mathbf{x}, u)$ is given by

$$0 = \frac{\partial W'[\mathbf{x}, u^*(\mathbf{x})]}{\partial u} = \frac{1}{\|\mathbf{f}\|^3} \frac{\partial W(\mathbf{x})}{\partial \mathbf{x}} [\|\mathbf{f}\|^2 \mathbf{I} - \mathbf{f}\mathbf{f}^{\mathrm{T}}] \frac{\partial \mathbf{f}}{\partial u}$$

which, for two-dimensional systems, reduces to

$$0 = \frac{1}{\|\mathbf{f}\|^3} \left\{ \frac{\partial W}{\partial x_1} f_2 - \frac{\partial W}{\partial x_2} f_1 \right\} \left\{ f_2 \frac{\partial f_1}{\partial u} - f_1 \frac{\partial f_2}{\partial u} \right\}. \tag{5.3-10}$$

Note that the expression in the first pair of braces in (5.3-10) can be written as the "cross product"

$$\boldsymbol{\nabla} W \times \mathbf{f} \triangleq \|\boldsymbol{\nabla} W\| \|\mathbf{f}\| \sin \theta,$$

where θ is the angle from the descent function gradient vector $\boldsymbol{\nabla} W \triangleq [\partial W / \partial \mathbf{x}]^{\mathrm{T}}$ to the velocity vector $\mathbf{f}$. If this cross product is zero at a minimizing point, then $\theta = \pi$ and the velocity vector is opposite to the gradient vector. However, (5.3-10) could also be satisfied by having the quantity in the second set of braces be zero, without the velocity vector being opposite to the gradient vector. Both of these situations occur in Zermelo's problem.

For Zermelo's system, condition (5.3-10) yields

$$0 = \{x_1 \sin u - x_2(-2 + \cos u)\}\{2 \cos u - 1\}, \tag{5.3-11}$$

and we have two cases. Using polar coordinates, we let

$$x_1 = r\cos\psi$$
$$x_2 = r\sin\psi.$$

If the first set of braces in (5.3-11), representing $\nabla W \times \mathbf{f}$, is zero, then we have

$$0 = \sin(u - \psi) + 2\sin\psi. \tag{5.3-12}$$

Figure 5.3-2 shows a plot of this function of u for various values of ψ, both of which are plotted in degrees. We see that (5.3-12) can be satisfied only for locations corresponding to $|\psi| \le \pi/6$, and that there is a unique solution $u = \mp\pi/3$ at $\psi = \pm\pi/6$. For $|\psi| < \pi/6$ condition (5.3-12) has two solutions. Both solutions produce a unique unit velocity vector $\mathbf{f}/\|\mathbf{f}\|$, opposite to the gradient vector ∇W, but they are associated with two distinct actual velocity vectors $\mathbf{f}$. One has a slow speed, corresponding to the near side of the circle of possible velocity vectors at $\mathbf{x}$, shown in Figure 2.2-6, and the other has a faster speed, corresponding to the far side of the circle of velocity vectors. Choosing the faster speed we can write a steepest descent heading angle as

$$u^* = \pi - \psi + \sin^{-1}(2\sin\psi) \qquad |\psi| \le \pi/6, \tag{5.3-13}$$

where $\sin^{-1}(\cdot)$ is the principle arcsine, with values in the range $-\pi/2$ to $\pi/2$.

For the other case in (5.3-10), yielding $2\cos u - 1 = 0$ in (5.3-11), we have

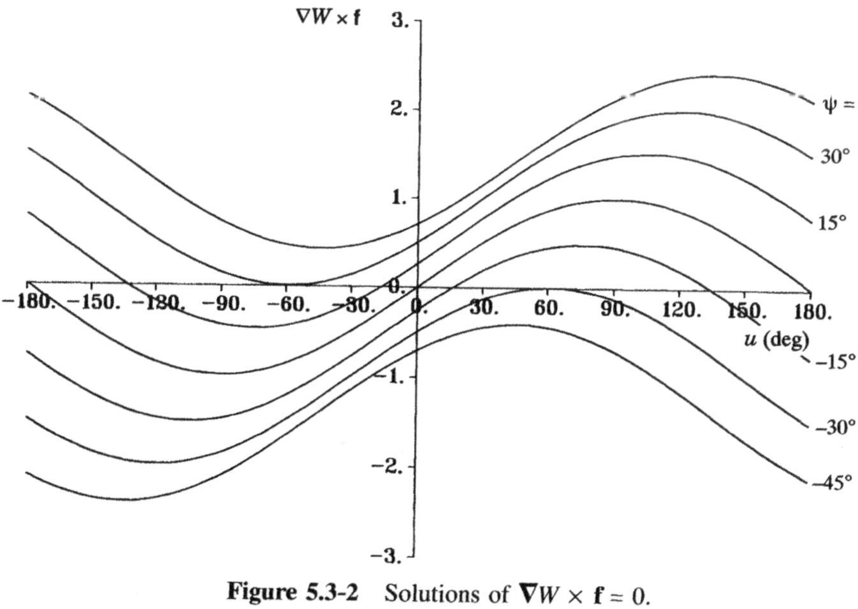

Figure 5.3-2 Solutions of $\nabla W \times \mathbf{f} = 0$.

$$u^* = \pm\frac{\pi}{6},$$

which corresponds to the cases $\psi = \mp\pi/6$ in (5.3-13). Combining both cases produces the steepest descent control:

$$u^* = \begin{cases} -\dfrac{\pi}{3} & \text{if } \dfrac{\pi}{6} \le \psi \le \pi \\ \pi - \psi + \sin^{-1}(2\sin\psi) & \text{if } |\psi| \le \dfrac{\pi}{6} \\ \dfrac{\pi}{3} & \text{if } -\pi < \psi \le \dfrac{\pi}{6}. \end{cases} \tag{5.3-14}$$

Figure 5.3-3 shows trajectories that reach the target for this controller. A comparison with the controllable set presented in Figure 2.2-6 shows that the **domain of effectiveness** for this controller (that is, the set of initial states that get transferred to the target) is the entire controllable set. This desirable result is a consequence of the choice for the $W(\cdot)$ function and, in general, is not guaranteed [Grantham, 1981*c*, 1982]. Also note that, while trajectories throughout the controllable set are "attracted" to the target, they do not, for this example, stay in the target. Instead, they pass on through, as the swimmer is swept downstream. ■

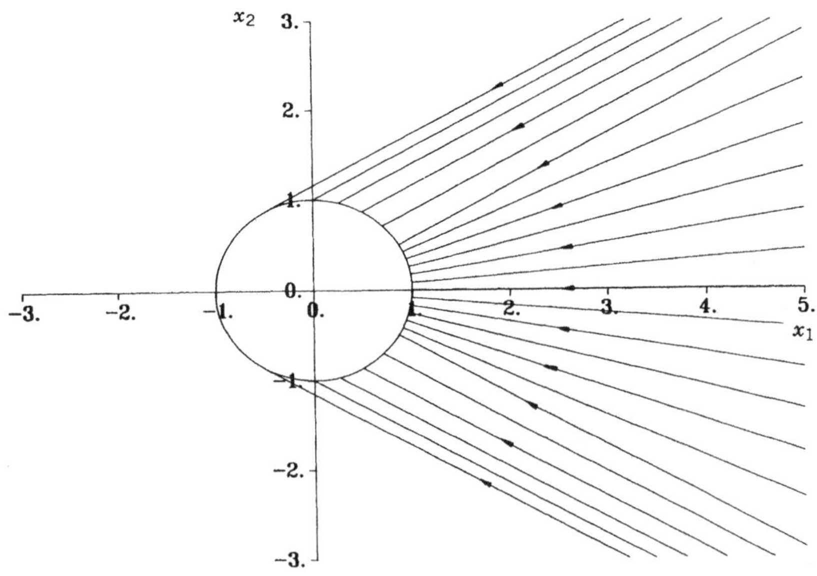

Figure 5.3-3 Steepest descent trajectories for Zermelo's problem.

Quickest Descent Control

Another way to try to make a descent function $W[\mathbf{x}(t)]$ decrease along trajectories of (5.1-1) is to choose the feedback control $\mathbf{u}(\mathbf{x})$ to minimize $dW[\mathbf{x}(t)]/dt$ at each state $\mathbf{x}$. A feedback control law $\mathbf{u}(\mathbf{x})$ is a Lyapunov optimizing **quickest descent** control for a specified descent function $W(\mathbf{x})$ if

$$\dot{W}[\mathbf{x}, \mathbf{u}(\mathbf{x})] \leq \dot{W}(\mathbf{x}, \mathbf{u}) \qquad \text{for all } u \in \mathcal{U}, \tag{5.3-15}$$

where

$$\dot{W}(\mathbf{x}, \mathbf{u}) \triangleq \frac{\partial W(\mathbf{x})}{\partial \mathbf{x}} \mathbf{f}(\mathbf{x}, \mathbf{u}). \tag{5.3-16}$$

Notice that the difference between minimizing $\dot{W}(\mathbf{x}, \mathbf{u})$ for quickest descent control and $W'(\mathbf{x}, \mathbf{u})$ for steepest descent control is that in computing $W'(\mathbf{x}, \mathbf{u})$ the velocity vector $\mathbf{f}(\mathbf{x}, \mathbf{u})$ has been normalized to a unit vector by dividing by $\|\mathbf{f}(\mathbf{x}, \mathbf{u})\|$. For quickest descent control, the quantity $\dot{W}(\mathbf{x}, \mathbf{u})$ is the time rate of change of $W[\mathbf{x}(t)]$ along a trajectory $\mathbf{x}(t)$. For steepest descent control, $W'(\mathbf{x}, \mathbf{u})$ is the rate of change of $W[\mathbf{x}(t)]$ with respect to arc length s along a trajectory. Steepest descent control focuses on the direction of motion, while quickest descent control uses information about both direction and speed. Since quickest descent control uses more information, it has the potential to produce a better feedback control law. Indeed, as we see later in Chapter 7, the function $\dot{W}(\mathbf{x}, \mathbf{u})$ for quickest descent control has exactly the same structure as the optimization function associated with Pontryagin's minimum principal for minimum-time optimal control problems.

Despite its potential, if $W(\mathbf{x})$ is not carefully chosen, quickest descent control may result in an undesirable trade-off of direction of motion in exchange for an increase in speed.

Example 5.3-2: Quickest Descent Control for Zermelo's Problem Consider the same system as in Example 5.3-1,

$$\dot{x}_1 = -2 + \cos u$$
$$\dot{x}_2 = \sin u,$$

with the same target, a unit circle centered at the origin, and the same descent function

$$W(\mathbf{x}) = x_1^2 + x_2^2.$$

For quickest descent control we minimize

$$\dot{W}(\mathbf{x}, u) = \frac{\partial W}{\partial x_1} f_1 + \frac{\partial W}{\partial x_2} f_2 = 2x_1(-2 + \cos u) + 2x_2 \sin u.$$

From the necessary condition

$$0 = \frac{\partial \dot{W}}{\partial u} = -2x_1 \sin u + 2x_2 \cos u,$$

we get the following feedback control law:

$$u^*(\mathbf{x}) = \pi + \tan^{-1}\left(\frac{x_2}{x_1}\right), \tag{5.3-17}$$

where $\tan^{-1}(\cdot)$ is the Fortran ATAN2$(\cdot, \cdot)$ two-argument arctangent function.

Figure 5.3-4 shows the quickest descent trajectories that reach the target for Zermelo's problem. Notice that the domain of effectiveness for the quickest descent controller is not all of the controllable set. The reason for this, as illustrated in the lower portion of the figure, is that quickest descent minimization of this descent function results in a trade-off between the direction of motion and speed. As a result, trajectories starting in the controllable set but near the controllability boundary ultimately leave the controllable set. ■

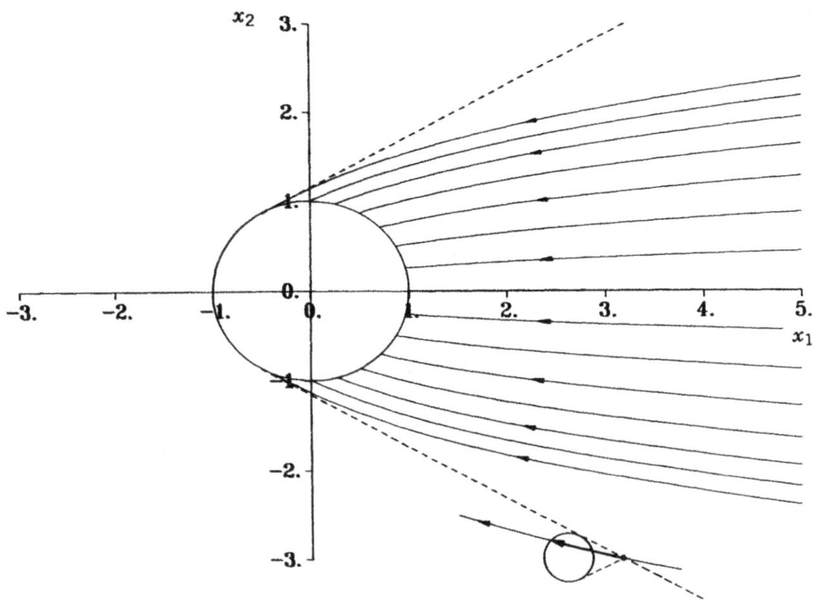

Figure 5.3-4 Quickest descent trajectories for Zermelo's problem.

There is a strong connection between quickest descent controls and optimal controls. As we see in Chapter 7, the "optimal return" function, which is the "minimum cost-to-go" to the target for the optimal controller, is also a descent function. In particular, if the descent function is taken as minimum time-to-go, then minimizing $\dot{W}$ will result in minimum time trajectories to the target. It is for this reason that the region of effectiveness under quickest descent control can always be made equal to the controllable set by using a suitable choice for the descent function. However, without solving the optimal control problem, it is not obvious how to do this. Later we present some necessary conditions and sufficient conditions to aid in selecting a descent function.

The selection of the descent function for quickest descent control is not as intuitive as the choice of a descent function for steepest descent control. This was the principle reason for introducing steepest descent control. The steepest descent approach allows the designer to focus on controlling a desired direction of motion.

A Comparison of Three Optimizing Feedback Controllers

In this section we compare the saturating minimum integral-square error controller, previously worked out for the linearized inverted pendulum in Example 5.2-2, with a steepest descent controller and a quickest descent controller for the same problem using the same descent function. The controllers obtained in the following two examples illustrate some possible difficulties associated with Lyapunov optimizing feedback controls. In particular, a quickest descent controller is obtained that produces "chatter," with corresponding "sliding mode" trajectories, and a steepest descent controller is obtained that produces unwanted near-equilibrium points.

Example 5.3-3: Steepest Descent Control of the Linearized Inverted Pendulum The equations of motion from Example 5.2-2 are given in matrix form by

$$\begin{bmatrix} \dot{x}_1 \\ \dot{x}_2 \end{bmatrix} = \begin{bmatrix} 0 & 1 \\ 1 & 0 \end{bmatrix} \begin{bmatrix} x_1 \\ x_2 \end{bmatrix} + \begin{bmatrix} 0 \\ 1 \end{bmatrix} u, \tag{5.3-18}$$

with the control constraint $|u| \leq 1$. From the results in Example 2.2-4, the controllable set to the origin is the region $|x_1 + x_2| < 1$. Thus in designing a feedback controller we are not concerned with states outside of this region, but the domain of attraction or domain of effectiveness for the controller should be all of the controllable set.

Since the system (5.3-18) is linear except for the control bounds, we use a positive definite quadratic form for the descent function,

$$W(\mathbf{x}) = \mathbf{x}^{\mathrm{T}}\mathbf{P}\mathbf{x}, \tag{5.3-19}$$

where $\mathbf{P} > 0$ is symmetric and positive definite. Contours of constant $W(\mathbf{x})$ are ellipses surrounding the origin. In picking a $\mathbf{P}$ matrix for steepest descent control, we note that for $u = 0$ the origin is an unstable saddle-point equilibrium, with eigenvalues $\mu = \pm 1$ and corresponding eigenvectors $\boldsymbol{\xi} = [1 \quad \mu]^{\mathrm{T}}$. For the stable eigenvalue $\mu = -1$, the associated eigenvector $\boldsymbol{\xi} = [1 \quad -1]^{\mathrm{T}}$ is in a direction parallel to the controllability boundaries illustrated in Figure 5.2-1. The system trajectories that approach the origin also lie on this eigenvector. For this reason, we choose a descent function that will produce trajectories in this preferred direction of motion. In particular, we choose a real, symmetric, positive definite $\mathbf{P}$ matrix with $\boldsymbol{\xi} = [1 \quad -1]^{\mathrm{T}}$ as one of its eigenvectors, associated with the smallest eigenvalue of $\mathbf{P}$, in which case $\boldsymbol{\xi}$ lies on the semimajor axis of the ellipses. Since the eigenvalues of $\mathbf{P}$ are real and positive, the eigenvectors will be orthogonal for distinct eigenvalues and will be aligned with the axes of the ellipse. The desired $\mathbf{P}$ matrix must satisfy the eigenvector equations

$$\begin{bmatrix} p_{11} & p_{12} \\ p_{12} & p_{22} \end{bmatrix} \begin{bmatrix} 1 & 1 \\ -1 & 1 \end{bmatrix} = \begin{bmatrix} 1 & 1 \\ -1 & 1 \end{bmatrix} \begin{bmatrix} a & 0 \\ 0 & b \end{bmatrix}$$

for eigenvalues $0 < a < b$. Thus we see that there is a family of such $\mathbf{P}$ matrices, given by

$$\mathbf{P} = \alpha \begin{bmatrix} \beta + 1 & \beta - 1 \\ \beta - 1 & \beta + 1 \end{bmatrix} \qquad \alpha > 0, \beta > 1. \tag{5.3-20}$$

For illustrative purposes we choose $\alpha = \frac{1}{2}$ and $\beta = 4$, yielding the positive definite quadratic descent function

$$W(\mathbf{x}) = \tfrac{1}{2}[x_1 \quad x_2] \begin{bmatrix} 5 & 3 \\ 3 & 5 \end{bmatrix} \begin{bmatrix} x_1 \\ x_2 \end{bmatrix} = \tfrac{5}{2}x_1^2 + 3x_1x_2 + \tfrac{5}{2}x_2^2. \tag{5.3-21}$$

A Lyapunov optimizing **steepest descent control** can be computed by minimizing $W'(\mathbf{x}, u) = dW/ds$, which from (5.3-4) reduces to

$$W'(\mathbf{x}, u) = \frac{(5x_1 + 3x_2)x_2 + (3x_1 + 5x_2)(x_1 + u)}{\sqrt{x_2^2 + (x_1 + u)^2}}. \tag{5.3-22}$$

For the case where the steepest descent control is interior to the control constraint $|u| \leq 1$, from the right-hand side of (5.3-8) multiplied by $\|\boldsymbol{\nabla} W\|$, we have

$$0 = \frac{\partial W'[\mathbf{x}, u^*(\mathbf{x})]}{\partial u} = \frac{1}{\|\mathbf{f}\|^3} \frac{\partial W}{\partial \mathbf{x}} [\|\mathbf{f}\|^2 \mathbf{I} - \mathbf{f}\mathbf{f}^{\mathrm{T}}] \frac{\partial \mathbf{f}}{\partial u}, \tag{5.3-23}$$

which, using (5.3-10) for the two-dimensional linearized inverted pendulum system, reduces to

$$\begin{aligned} 0 = \frac{\partial W'}{\partial u} &= \frac{f_1}{\|\mathbf{f}\|^3} \left\{ \frac{\partial W}{\partial x_2} f_1 - \frac{\partial W}{\partial x_1} f_2 \right\} \\ &= \frac{x_2\{(3x_1 + 5x_2)x_2 - (5x_1 + 3x_2)(x_1 + u)\}}{[x_2^2 + (x_1 + u)^2]^{3/2}}. \end{aligned} \tag{5.3-24}$$

In addition, for an interior minimum of $W'(\mathbf{x}, u)$ for the linearized inverted pendulum, we have the second-order necessary condition

$$0 \le \frac{\partial^2 W'(\mathbf{x}, u)}{\partial u^2} = \frac{-f_1(\partial W/\partial x_1)}{\|\mathbf{f}\|^3} = \frac{-x_2(3x_1 + 5x_2)}{\|\mathbf{f}\|^3}. \tag{5.3-25}$$

Denoting the solution to (5.3-24) by

$$u^\circ(\mathbf{x}) = -x_1 + \frac{\partial W/\partial x_2}{\partial W/\partial x_1} f_1 = -x_1 + \frac{3x_1 + 5x_2}{5x_1 + 3x_2} x_2, \tag{5.3-26}$$

one possible steepest descent control can be written as

$$u^*(\mathbf{x}) = \begin{cases} -1 & \text{if } x_2 \ge 0 \text{ and } 3x_1 + 5x_2 > 0 \\ \mathrm{sat}[u^\circ(\mathbf{x})] & \text{if } x_2 \ne 0 \text{ and } x_2(3x_1 + 5x_2) \le 0 \\ 1 & \text{if } x_2 \le 0 \text{ and } 3x_1 + 5x_2 < 0. \end{cases} \tag{5.3-27}$$

The right-hand conditions in (5.3-27) were not obtained using the saturation boundary for $u^*(\mathbf{x})$, but rather, they were chosen to satisfy the sign requirements of (5.3-25). In addition, these conditions will maximize the speed at the x_1-axis, where all unit velocity vectors are the same except for sign. Figure 5.3-5 shows the resulting trajectories for this controller, with the thicker portions of the trajectories indicating where the control is an interior minimum, with $u^*(\mathbf{x}) = u^\circ(\mathbf{x})$ and $|u^*(\mathbf{x})| < 1$. The entire controllable set is transferred to the origin with the trajectories approaching the origin along the semimajor axis of the $W(\mathbf{x})$ = constant ellipses, which as noted earlier, is the chosen preferred direction of approach.

Interior control, with $u^*(\mathbf{x}) = u^\circ(\mathbf{x})$ and $|u^*(\mathbf{x})| < 1$, is obtained when the term in braces in (5.3-24) is equal to zero,

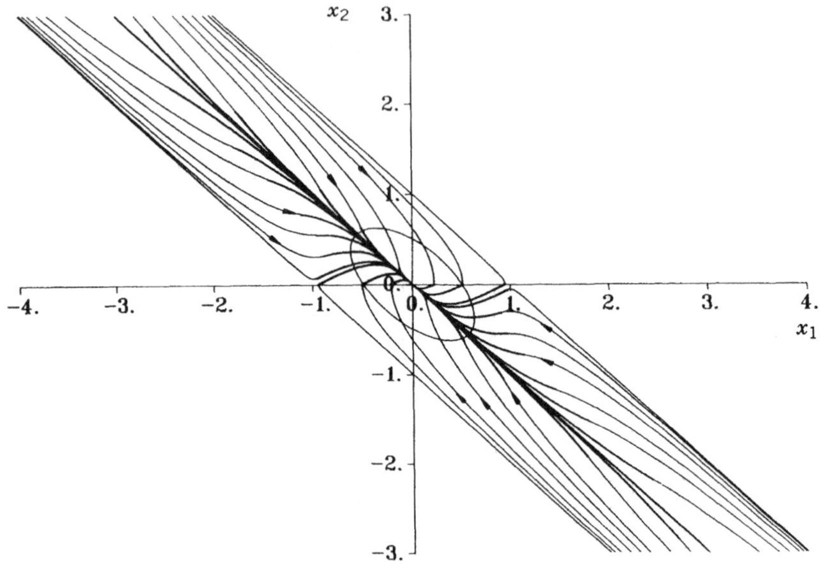

Figure 5.3-5 Steepest descent trajectories for Example 5.3-3.

$$0 = \frac{\partial W}{\partial x_2} f_1 - \frac{\partial W}{\partial x_1} f_2. \tag{5.3-28}$$

In this case $f_2 \to 0$ if $f_1 \to 0$, yielding equilibrium points (see Anderson and Grantham [1989]). This occurs on the x_1-axis in the range $|x_1| \leq 1$. However, on the x_1-axis in this range, all nonzero velocity vectors yield the same unit velocity vector and we have chosen a particular steepest descent control from these multiple solutions, with $u^*(\mathbf{x}) = -\text{sgn}(x_1)$ on the x_1-axis. Thus our steepest descent control does not produce equilibrium points, other than the origin. However, the control $u^*(\mathbf{x})$ is discontinuous across the x_1-axis for $|x_1| < 1$, and immediately after a trajectory crosses this portion of the x_1-axis the trajectory behaves as if it were arbitrarily close to an equilibrium point. Thus the speed, while nonzero, becomes very small just after crossing the x_1-axis in the region $|x_1| \leq 1$. This difficulty, caused by the fact that steepest descent control uses only the direction of motion instead of both direction and speed, could be overcome by choosing a different descent function, in which the region where the term in braces in (5.3-24) is zero is well separated from the region where $f_1 = 0$.

As demonstrated in Anderson and Grantham [1989], this same approach can also be applied to a low-torque multiple-revolution nonlinear inverted pendulum. The steepest descent controller for this case also yields a domain of attraction identical to the controllable set, which is all of state space. ■

Example 5.3-4: Quickest Descent Control of the Linearized Inverted Pendulum Consider again the system of the previous example. Using the same descent function $W(\mathbf{x})$, **quickest descent control** is obtained by choosing $u^*(\mathbf{x})$ to minimize

$$\dot{W}(\mathbf{x}, u) = \frac{\partial W}{\partial x_1} f_1 + \frac{\partial W}{\partial x_2} f_2$$

$$= (5x_1 + 3x_2)x_2 + (3x_1 + 5x_2)(x_1 + u) \qquad (5.3\text{-}29)$$

subject to the constraint $|u| \leq 1$.

Using the necessary conditions for a minimum, we first form the L function (3.3-9)

$$L = (5x_1 + 3x_2)x_2 + (3x_1 + 5x_2)x_1 + u\sigma(\mathbf{x}) - \gamma_1(1 - u) - \gamma_2(u + 1),$$

where

$$\sigma(\mathbf{x}) \triangleq \frac{\partial \dot{W}}{\partial u} = 3x_1 + 5x_2 \qquad (5.3\text{-}30)$$

is a **switching function** that is nonzero everywhere except on the **switching surface** $\sigma(\mathbf{x}) = 0$. From (3.3-5) we obtain

$$\frac{\partial L}{\partial u} = \sigma(\mathbf{x}) + \gamma_1 - \gamma_2 = 0. \qquad (5.3\text{-}31)$$

If $\sigma(\mathbf{x}) = 0$, if follows that (5.3-31) is satisfied by $\gamma_1 = \gamma_2 = 0$ for any value of u satisfying $|u| \leq 1$. In other words, u^* is not unique. If $\sigma(\mathbf{x}) \neq 0$, it then follows from (5.3-31) and the necessary conditions (3.3-6)–(3.3-8) that $u^* = -1$ if $\sigma(\mathbf{x}) > \mathbf{0}$ and $u^* = +1$ if $\sigma(\mathbf{x}) < \mathbf{0}$. In a physical system or in a numerical simulation, the condition $\sigma(\mathbf{x}) = 0$ would rarely be obtained precisely. Thus for all practical purposes, the quickest descent control is given by

$$u^*(\mathbf{x}) = -\text{sgn}[\sigma(\mathbf{x})] \qquad \text{for } \sigma(\mathbf{x}) = 3x_1 + 5x_2 \neq 0. \qquad (5.3\text{-}32)$$

While the signum function sgn(·) is not defined when its argument is zero, some mathematical software programs such as BASIC and MATLAB will assign the value zero.

For completeness we consider the case where $\sigma(\mathbf{x}) = 0$ occurs. When a trajectory $\mathbf{x}(t)$ crosses the switching surface $\sigma(\mathbf{x}) = 0$, the control switches sign, from ± 1 to ∓ 1. After the switch there are two possibilities: either (1) the new velocity vector points ahead beyond the switching surface, in which case the trajectory penetrates through the surface, or (2) the new velocity vector points back toward the switching surface, in which case the control "chatters" between ± 1 and the trajectory "slides" along the switching surface. In the first case (5.3-32) can be used and the control has a simple switch in sign. In the second case the chattering control produces an **effective velocity vector**, in the sense of Filippov (see Section 1.1), given by the convex combination

$$\dot{\mathbf{x}} = \alpha \mathbf{f}(\mathbf{x}, 1) + (1 - \alpha)\mathbf{f}(\mathbf{x}, -1) = \begin{bmatrix} x_2 \\ x_1 + 2\alpha - 1 \end{bmatrix}, \tag{5.3-33}$$

where $\alpha \in (0, 1)$ is determined so that $\dot{\mathbf{x}}$ is tangent to the switching surface. Thus

$$0 = \frac{\partial \sigma}{\partial \mathbf{x}} \dot{\mathbf{x}} = [3 \quad 5] \begin{bmatrix} x_2 \\ x_1 + 2\alpha - 1 \end{bmatrix},$$

which implies

$$\alpha = 0.5 - 0.5x_1 - 0.3x_2.$$

Since the set of velocity vectors $\mathbf{f}(\mathbf{x}, u)$, $|u| \le 1$, is in fact convex for a linear system, the effective Filippov velocity corresponds to an actual velocity vector $\dot{\mathbf{x}} = \mathbf{f}(\mathbf{x}, u)$, with

$$u = 2\alpha - 1 = -x_1 - 0.6x_2,$$

which reduces to

$$u = -0.64x_1 \tag{5.3-34}$$

on the switching surface $\sigma(\mathbf{x}) = 3x_1 + 5x_2 = 0$. Because of the control constraint $|u| \le 1$, the switching surface is a chattering trajectory, with $\sigma[\mathbf{x}(t)] \equiv 0$, only in the range $|x_1| \le 1.5625$. For any trajectory that reaches the switching surface outside of this range, the control simply switches sign and the trajectory continues on through the switching surface.

Because the effective chattering velocity vector corresponds to an actual velocity vector $\dot{\mathbf{x}} = \mathbf{f}(\mathbf{x}, u)$ for some u, the corresponding control value can be calculated in an easier manner. If $\sigma[\mathbf{x}(t)] \equiv 0$ for a nonzero time interval, then all time derivatives of $\sigma[\mathbf{x}(t)]$ must also be zero. Thus

$$0 = \sigma = 3x_1 + 5x_2$$
$$0 = \dot{\sigma} = 3\dot{x}_1 + 5\dot{x}_2 = 3x_2 + 5x_1 + 5u,$$

which implies

$$u = -x_1 - 0.6x_2 = -0.64x_1$$

and agrees with (5.3-34). Note that a chattering trajectory along the switching surface will asymptotically approach the origin.

The quickest descent control may now be stated more precisely as

$$u^*(\mathbf{x}) = \begin{cases} -\text{sgn}[\sigma(\mathbf{x})] & \text{if } \sigma(\mathbf{x}) \neq 0 \\ -\text{sat}[0.64x_1] & \text{if } \sigma(\mathbf{x}) = 0, \end{cases} \tag{5.3-35}$$

where

$$\sigma(\mathbf{x}) = 3x_1 + 5x_2.$$

As we show in Chapters 7 and 8, optimal controllers include switching control whenever the control variable appears linearly in the problem formulation. Switching (or "bang-bang," "on-off") controllers, in which the control switches between two extreme values as a trajectory crosses a switching surface in state space, have been studied extensively in Flügge-Lotz, [1968] and more recently as a special case of **variable structure** control systems [Itkin, 1976; Dorling and Zinober, 1986], in which a feedback control has different structures in different regions of state space. In particular, chattering motion along a switching surface could be viewed as desirable, at least mathematically, since the chatter occurs on a surface of lower dimension than the overall state space, thereby reducing the dimension of the control design problem.

Figure 5.3-6 is a plot of the trajectories produced by this controller, with the thin trajectories corresponding to $u = -1$ and the thick trajectories corresponding to $u = +1$. We see that quickest descent control also produces a preferred direction of approach to the target, along a chattering switching surface corresponding to $\partial W(\mathbf{x})/\partial x_2 = 0$. However, this quickest descent control has a domain of attraction that is less than the controllable set. Trajectories starting at some points in the controllable set leave the set, never to return. This difficulty can be eliminated by changing the **P** matrix in the descent function to yield a switching surface $\sigma(\mathbf{x}) = x_1 + x_2$ that is parallel to the controllability boundaries $x_1 + x_2 = \pm 1$. ■

Comparing the Three Control Designs Figure 5.3-7 shows the output response $\theta(t) = x_1(t)$ of the linearized inverted pendulum, starting from $\mathbf{x}(0) = [0.9 \quad 0]^T$, for the three function-minimizing controllers that we have investigated: (1) integral-square error with saturation, (2) steepest descent, and (3) quickest descent. The relatively slow response time for steepest descent control can be improved significantly by letting $\beta \rightarrow \infty$ in the **P** matrix of (5.3-

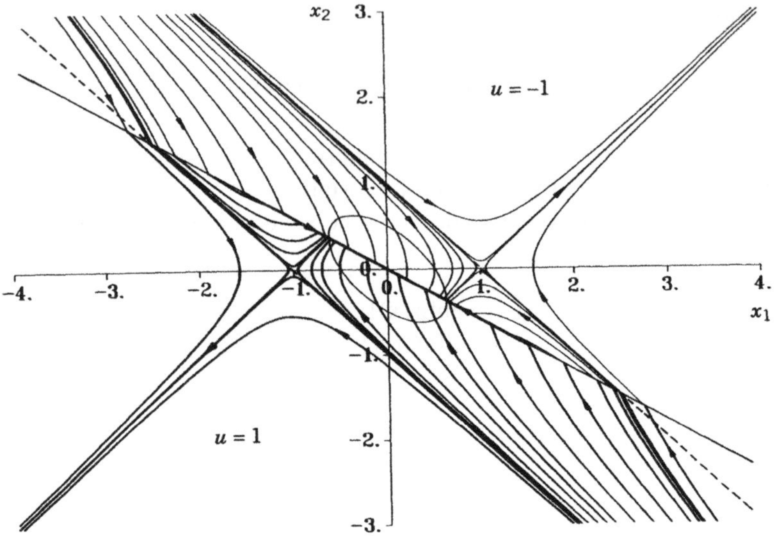

Figure 5.3-6 Quickest descent trajectories for Example 5.3-4.

20). This will cause the descent function contours to become very long, skinny ellipses. For large values of β, the switching surface for quickest descent control will approach the semimajor axis of the ellipses and in the limit both steepest descent and quickest descent control will be effective throughout the controllable set and will have the same preferred direction of approach to the origin.

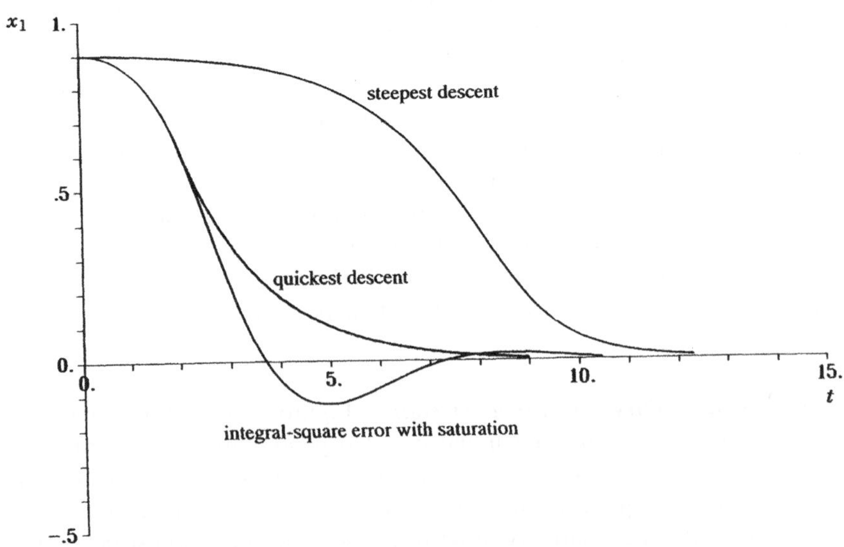

Figure 5.3-7 Step responses for three function-minimizing controllers.

However, steepest descent trajectories will approach in a gradient-like fashion, while quickest descent trajectories will continue to approach the origin along a chattering switching surface. Figure 5.3-8 shows, as functions of time, the steepest descent and minimum square error controls used to produce the output responses plotted in Figure 5.3-7. The quickest descent control is not plotted because after about 2 seconds of control at $u = -1$, the control starts to chatter and remains chattering, between $u = 1$ and $u = -1$, all the way to the equilibrium solution. A chattering controller is not desirable in a physical system. However, chattering can be eliminated by modifying (5.3-35) so that

$$u^*(\mathbf{x}) = \begin{cases} -\text{sgn}[\sigma(\mathbf{x})] & \text{if } |\sigma(\mathbf{x})| \geq \varepsilon \\ -\text{sat}[0.64x_1] & \text{if } |\sigma(\mathbf{x})| < \varepsilon, \end{cases} \tag{5.3-36}$$

where ε is a small positive number, producing a domain of attraction for u^* nearly identical to the domain of attraction for the quickest descent controller with chatter.

Quickest descent control has a structure similar to the minimum-time optimal controllers introduced in Chapter 7. Quickest descent feedback controls, and indeed many chattering variable structure controllers, tend to be **robust**, in that they are able to accomplish their function despite system uncertainties. This is explored in more detail in Section 5.4.

The principal advantage of Lyapunov optimizing controllers is that they directly produce feedback control laws, without having to solve the nonlinear two-point boundary-value problems that are associated with the optimal con-

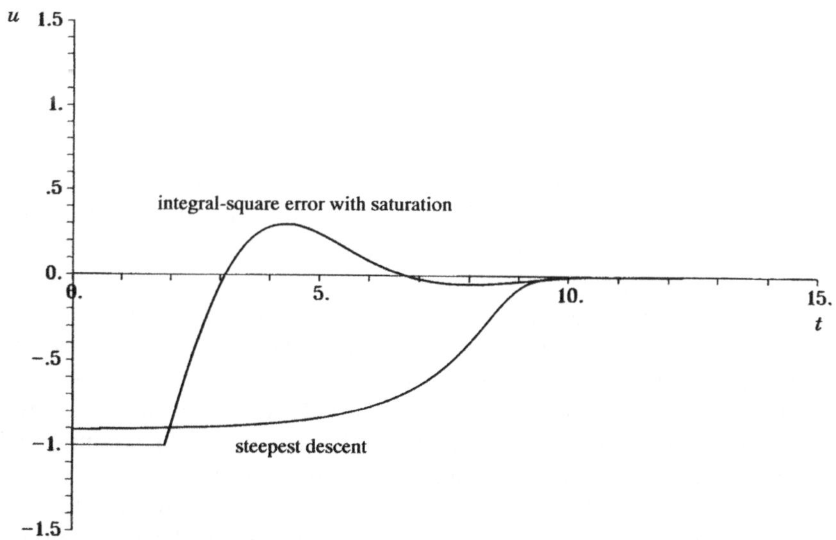

Figure 5.3-8 Steepest descent and minimum integral-square error controls.

trollers of Chapters 7 and 8. On the other hand, well-posed optimal control problems result in controllers that are always effective throughout the controllable set.

Lyapunov Optimizing Control Using Trajectory Following

Since the above example problems could be solved analytically, a closed-loop controller could be obtained by direct application of the necessary conditions for a minimum. However, if an analytical solution cannot be found, then some numerical technique must be used in order to solve the minimization problem at every point of a trajectory. Depending on the application, this might be difficult to do in a continuous fashion in order to produce the desired closed-loop control. One way to avoid continuously solving a numerical optimization problem is to use the trajectory following method discussed in Chapter 3. Using this method produces what could be called Lyapunov suboptimizing control. We outline the procedure for the case of interior controls. Constraints on the controls can be handled as in Section 3.5.

The idea is to solve the numerical optimization problem only once, at the initial point of the trajectory. For all other points $\mathbf{x}$ along a trajectory, the (interior) controls are determined from the differential equations

$$\dot{\mathbf{u}} = -\left[\frac{\partial W'}{\partial \mathbf{u}}\right]^{\mathrm{T}}$$

for steepest descent control and

$$\dot{\mathbf{u}} = -\left[\frac{\partial \dot{W}}{\partial \mathbf{u}}\right]^{\mathrm{T}}$$

for quickest descent control. The initial conditions for the differential equations are determined from the initial numerical optimization problem, and the gradients are calculated numerically so that only W' or $\dot{W}$ need be specified. How well these solutions agree with the actual solutions for the control, as the system moves along the trajectory, depends on how well these equations track the optimal solutions. Gains can be introduced into these equations to adjust the rates. Generally, they will do a good job of tracking the Lyapunov optimizing result if the control does not change too rapidly. However, even in cases where part of the Lyapunov optimizing control is bang-bang, as in Example 5.3-4, good results can be obtained. In fact (see Exercise 5-9), if trajectory following is used, instead of obtaining a continuously switching control, a finite time interval switching control is obtained (which depends on the integration step size), resulting in a zigzag trajectory along the switching curve.

Minimum Cost-Descent Control

The feedback controllers that we have considered so far require only that the state of a system

$$\dot{\mathbf{x}} = \mathbf{f}(\mathbf{x}, \mathbf{u}), \qquad \mathbf{u} \in \mathcal{U} \tag{5.3-37}$$

be transferred to a target set $\mathcal{X}$. An **optimal controller** must do so while minimizing a cumulative "cost" functional

$$J[\mathbf{u}(\cdot)] = \int_0^{t_f} f_0(\mathbf{x}, \mathbf{u}) dt, \tag{5.3-38}$$

where t_f is the time at which the state reaches the target, $\mathbf{x}(t_f) \in \mathcal{X}$. This is similar to calculus of variations problems introduced in Chapter 3. However, the tools we need to properly solve optimal control problems are developed in Chapter 7. For now, we simply take the opportunity to introduce this type of problem and to investigate Lyapunov optimizing feedback controls that consider a "cost" associated with transferring the state to a target.

At time t let

$$x_0(t) = \int_0^{t} f_0(\mathbf{x}, \mathbf{u}) dt \tag{5.3-39}$$

denote the accumulated cost along a trajectory of (5.3-37). This cost accumulates at a rate

$$\dot{x}_0 = f_0(\mathbf{x}, \mathbf{u}) \tag{5.3-40}$$

and has the initial condition $x_0(0) = 0$. Let $W(\mathbf{x})$ be a descent function for the target $\mathcal{X}$ and let

$$W_0 = x_0 + W(\mathbf{x})$$

be an **augmented descent function**. It follows that

$$\dot{W}_0 = \frac{\partial W}{\partial x_0} \dot{x}_0 + \frac{\partial W_0}{\partial \mathbf{x}} \dot{\mathbf{x}} = f_0(\mathbf{x}, \mathbf{u}) + \frac{\partial W_0}{\partial \mathbf{x}} \mathbf{f}(\mathbf{x}, \mathbf{u}).$$

A control $\mathbf{u}(\mathbf{x}) \in \mathcal{U}$ is a **minimum cost-descent control** if, at each state $\mathbf{x}$ in the controllable set $\mathcal{C}$ to the target $\mathcal{X}$, the control minimizes $\dot{W}_0$. In keeping with the notation used in Chapter 7, define

$$H(\mathbf{x}, \mathbf{u}, \nabla W) \triangleq f_0(\mathbf{x}, \mathbf{u}) + \frac{\partial W(\mathbf{x})}{\partial \mathbf{x}} \mathbf{f}(\mathbf{x}, \mathbf{u}), \tag{5.3-41}$$

where $\nabla W \triangleq [\partial W(\mathbf{x})/\partial \mathbf{x}]^{\mathrm{T}}$. The function $H(\mathbf{x}, \mathbf{u}, \nabla W)$ represents a trade-off between the current penetration rate dW/dt for the descent function and the current cost accumulation rate $\dot{x}_0$, with quickest descent control ($f_0 = 0$) being a special case. Minimum cost-descent control is obtained by minimizing $H(\mathbf{x}, \mathbf{u}, \nabla W)$ subject to the control constraint $\mathbf{u} \in \mathcal{U}$ at every point $\mathbf{x}$ of the resulting trajectory. As a guide in selecting the cost rate function $f_0(\cdot)$ and the descent function $W(\cdot)$ for constructing $H(\cdot)$, we have the following two results [Grantham, 1981c]:

Controllability Necessary Condition: *Suppose the interior $\mathring{C}$ of the controllable set C to the target X is of the form*

$$\mathring{C} = \{\mathbf{x} \in R^{n_x} \,|\, \psi(\mathbf{x}) < 0\}, \tag{5.3-42}$$

where the scalar-valued function $\psi(\cdot)$ is C^1 in a neighborhood of the controllability boundary $\partial C = \{\mathbf{x} \in R^{n_x} \,|\, \psi(\mathbf{x}) = 0\}$. Then in order for an admissible control $\mathbf{u}(\mathbf{x}) \in \mathcal{U}$ to have a **domain of effectiveness** *equal to the controllable set (i.e., transfer every initial state in C to the target X), it is necessary that*

$$\frac{\partial \psi(\mathbf{x})}{\partial \mathbf{x}} \mathbf{f}[\mathbf{x}, \mathbf{u}(\mathbf{x})] = 0 \qquad \text{for all } \mathbf{x} \in \partial C. \tag{5.3-43}$$

This result follows from some basic properties of controllable sets that we present in Chapter 6. The crux of a proof is that:

(a) The quantity in (5.3-43) cannot be negative on ∂C; otherwise trajectories could come from outside the controllable set to inside it.

(b) The quantity in (5.3-43) also cannot be positive on ∂C; otherwise $\mathbf{u}(\mathbf{x})$ would cause trajectories to leave the controllable set. Once that happens, they cannot return to the controllable set, which contains the target.

Minimum Cost-Descent Sufficiency Conditions: *Let $\mathbf{u}(\mathbf{x})$ be a minimum cost-descent control for $H(\cdot)$ in (5.3-41), with descent function $W(\mathbf{x})$ and with $\mathbf{u}(\mathbf{x})$ satisfying the controllability necessary condition (5.3-43). Let C be the controllable set to X. If, for all $\mathbf{x} \in C - X$,*

(i)

$$0 \geq \dot{W}[\mathbf{x}, \mathbf{u}(\mathbf{x})] \triangleq \frac{\partial W(\mathbf{x})}{\partial \mathbf{x}} \mathbf{f}[\mathbf{x}, \mathbf{u}(\mathbf{x})] \not\equiv 0,$$

and for all $\mathbf{x} \in \mathring{C} - X$ *in a neighborhood of* ∂C,

(ii)

$$0 > \frac{\partial \psi(\mathbf{x})}{\partial \mathbf{x}} \mathbf{f}[\mathbf{x}, \mathbf{u}(\mathbf{x})],$$

then the trajectory $\mathbf{x}(t) \to X$.

Condition (ii) and continuity of solutions to (5.3-37) with respect to initial conditions imply that all trajectories starting in the controllable set C remain in C. Condition (i) implies that $W[\mathbf{x}(t)]$ decreases along trajectories $\mathbf{x}(t)$. Since $W(\mathbf{x})$ is a descent function for the target X, we conclude that $\mathbf{x}(t) \to X$ for every trajectory starting in the controllable set C.

Example 5.3-5: Minimum Effort Control Consider the linearized inverted pendulum system

$$\begin{aligned} \dot{x}_1 &= x_2 \\ \dot{x}_2 &= x_1 + u \\ |u| &\le 1. \end{aligned} \tag{5.3-44}$$

For $u = 0$ the origin is unstable, and for $|u| \le 1$ the controllable set to the origin does not contain its boundary points and is given by (see Example 2.2-4)

$$C = \{\mathbf{x} \in R^2 | \psi(\mathbf{x}) < 0\}, \tag{5.3-45}$$

where

$$\psi(\mathbf{x}) = |x_1 + x_2| - 1. \tag{5.3-46}$$

We wish to transfer the system to the origin, starting from any point in the controllable set C, with a penalty on the amount of control effort u. In particular, we choose $u(\mathbf{x})$ to minimize a cost-descent function of the form

$$\begin{aligned} H &= \frac{1}{2} u^2 + \frac{\partial W}{\partial x_1} \dot{x}_1 + \frac{\partial W}{\partial x_2} \dot{x}_2 \\ &= \frac{1}{2} u^2 + \frac{\partial W}{\partial x_1} x_2 + \frac{\partial W}{\partial x_2} (x_1 + u), \end{aligned} \tag{5.3-47}$$

where $W(\mathbf{x})$ is a positive definite function to be chosen. Using the necessary conditions for a minimum, we form

$$L = H - \gamma_1(1-u) - \gamma_2(u+1)$$

to obtain

$$\frac{\partial L}{\partial u} = u + \frac{\partial W}{\partial x_2} + \gamma_1 - \gamma_2 = 0.$$

We conclude that the necessary conditions for a minimum are satisfied by

$$\begin{aligned}
&u + \frac{\partial W}{\partial x_2} = 0, &&\gamma_1 = \gamma_2 = 0 \quad \Rightarrow u = -\frac{\partial W}{\partial x_2} &&\text{provided that } \left|\frac{\partial W}{\partial x_2}\right| \le 1\\
&u + \frac{\partial W}{\partial x_2} > 0, &&\gamma_1 = 0, \gamma_2 > 0 \Rightarrow u = -1 &&\text{provided that } \frac{\partial W}{\partial x_2} > 1\\
&u + \frac{\partial W}{\partial x_2} < 0, &&\gamma_1 > 0, \gamma_2 = 0 \Rightarrow u = +1 &&\text{provided that } \frac{\partial W}{\partial x_2} < -1.
\end{aligned}$$

More simply, the minimum cost-descent control is given by

$$u(\mathbf{x}) = -\text{sat}\left(\frac{\partial W}{\partial x_2}\right). \tag{5.3-48}$$

In order to keep trajectories from leaving the controllable set, the necessary condition (5.3-43) on $x_1 + x_2 = 1$ requires

$$0 = \frac{\partial \psi}{\partial \mathbf{x}}\dot{\mathbf{x}} = [1 \quad 1]\begin{bmatrix} x_2 \\ x_1 + u \end{bmatrix} = 1 + u \Rightarrow u = -1,$$

and on $-(x_1 + x_2) = 1$ it requires

$$0 = \frac{\partial \psi}{\partial \mathbf{x}}\dot{\mathbf{x}} = [-1 \quad -1]\begin{bmatrix} x_2 \\ x_1 + u \end{bmatrix} = 1 - u \Rightarrow u = 1.$$

Thus in (5.3-48) we need $\partial W/\partial x_2$ to saturate on the boundary of the controllable set. For simplicity, we take

$$\frac{\partial W}{\partial x_2} = k(x_1 + x_2), \qquad k \ge 1.$$

Integrating this equation we have

$$W(\mathbf{x}) = \alpha(x_1) + kx_1x_2 + \frac{k}{2} x_2^2.$$

We can make $W(\mathbf{x})$ be a positive definite quadratic descent function by choosing $\alpha(x_1) = ax_1^2$, yielding

$$W(\mathbf{x}) = ax_1^2 + kx_1x_2 + \frac{k}{2} x_2^2 = \frac{1}{2}[x_1 \quad x_2]\begin{bmatrix} 2a & k \\ k & k \end{bmatrix}\begin{bmatrix} x_1 \\ x_2 \end{bmatrix}. \tag{5.3-49}$$

For $W(\mathbf{x})$ to be positive definite, with $k \geq 1$, we need

$$1 \leq k < 2a.$$

The resulting control

$$u(\mathbf{x}) = -\text{sat}[k(x_1 + x_2)]$$

yields

$$\begin{aligned} \dot{W}(\mathbf{x}, u) &= \frac{\partial W}{\partial x_1} x_2 + \frac{\partial W}{\partial x_2} (x_1 + u) \\ &= (2ax_1 + kx_2)x_2 + k(x_1 + x_2)(x_1 + u) \\ &= (2a - k)x_1x_2 + k(x_1 + x_2)(x_1 + x_2 + u). \end{aligned}$$

In the region $|x_1 + x_2| \leq 1/k$ we have $u = -k(x_1 + x_2)$ and

$$\begin{aligned} \dot{W}(\mathbf{x}, u) &= (2a - k)x_1x_2 - k(k - 1)(x_1 + x_2)^2 \\ &= [2a - k - 2k(k - 1)]x_1x_2 - k(k - 1)(x_1^2 + x_2^2) \\ &= [2a - k(2k - 1)]x_1x_2 - k(k - 1)(x_1^2 + x_2^2). \end{aligned}$$

We can null the cross product term, make $\dot{W}(\mathbf{x}, u) < 0$ away from the origin in the region $|x_1 + x_2| \leq 1/k$, and make $W(\mathbf{x})$ positive definite by taking

$$2a = k(2k - 1), \qquad k > 1. \tag{5.3-50}$$

For example, $k = 1 + \epsilon$, $\epsilon > 0$, would do. In the limit as $\epsilon \to 0$ the region $|x_1 + x_2| < 1/(1 + \epsilon)$ becomes the entire controllable set. Thus we conclude that the minimum cost-descent control

$$u(\mathbf{x}) = -\text{sat}[k(x_1 + x_2)], \qquad k > 1 \tag{5.3-51}$$

makes the origin asymptotically stable in (5.3-44), with a domain of attraction that is the entire controllable set. ■

In general, minimizing H at each point $\mathbf{x}$ along a trajectory $\mathbf{x}(t)$ will *not* solve the corresponding optimal control problem of minimizing the integral in (5.3-38). This is because minimizing the current rate at which cost accumulates will generally not be optimal in the long term; it may be better to take a cost "penalty" now in order to move the state along a different trajectory whose total cost is less. In this sense, the optimal control analysis that we use in Chapter 7 has the ability to "look ahead."

As we show in Chapter 7, for a particular choice of $W(\mathbf{x})$ in (5.3-41), a minimum cost-descent control $\mathbf{u}(\mathbf{x})$ that is effective throughout the controllable set will be identical to optimal control for some cost function. Unfortunately, the "optimal return" (minimum cost-to-go) function cannot be known until after the optimal control problem has been solved. Nevertheless, we now have an alternate interpretation of the "descent" function $W(\mathbf{x})$. Instead of viewing it as a "distance" to the target, we can view it as a desired minimum cost-to-go function.

Example 5.3-6: Steel Production Consider the problem [Isaacs, 1965, pp. 116–124] of a nation embarking on a program of steel production. During any time interval, the current supply S of steel is to be allocated to one of three uses: (1) as an ingredient in the manufacture of new steel, (2) in the manufacture of more steel mills M, or (3) for the stockpiling of steel, which we wish to maximize after a given interval T. Let u_1 be the fraction of steel allocated to building more mills, and u_2 the fraction devoted to making more steel. Thus

$$u_1 \geq 0, \qquad u_2 \geq 0, \qquad u_1 + u_2 \leq 1. \tag{5.3-52}$$

We take the rate of mill building as

$$\dot{M} = cu_1 S, \tag{5.3-53}$$

where $c > 0$ is a constant parameter. In a unit of time let $a > 1$ be the quantity of new steel that can be made from a unit of old steel. The net rate at which steel is stockpiled is

$$\dot{S} = au_2 S - u_2 S - u_1 S = [(a-1)u_2 - u_1]S, \tag{5.3-54}$$

where au_2S is the rate at which new steel is produced, u_2S the rate at which old steel is used to manufacture new steel, and u_1S the rate at which steel is used to manufacture new mills. The number of existing mills constrains the rate of new steel production. Thus for some $b > 0$, we have the additional constraint $au_2S \leq bM$, or

$$u_2 \leq \frac{bM}{aS}. \tag{5.3-55}$$

Starting from an initial supply of steel $S(0)$ and mills $M(0)$, the optimal control problem over a fixed time interval $0 \leq t \leq t_f = T$ is to maximize

$$S(t_f) = \int_0^{t_f} \dot{S}\, dt \tag{5.3-56}$$

subject to the differential equations (5.3-53) and (5.3-54) and the control constraints (5.3-52) and (5.3-55).

In its current form, the fixed-time minimum cost-descent control problem (5.3-52)–(5.3-56) has no specified state-space target. Thus we have no need for the descent function $W(\mathbf{x})$ to drive the state to a target. However, there is a "target" inherent in the performance integral (5.3-56): we would like to drive $S(t)$ as high as possible and therefore take the cost-descent function H as the cost rate, $H = -\dot{S}$. The corresponding controls are determined by choosing $\mathbf{u}(\mathbf{x})$ to minimize

$$H = -[(a-1)u_2 - u_1]S, \tag{5.3-57}$$

subject to (5.3-52)–(5.3-55). This is equivalent to maximizing the current rate at which steel is stockpiled. We leave the solution to this problem to Exercises 5-10 and 7-16. This allows us to compare short-term maximization of the stockpiling rate $\dot{S}$ with the long-term option of using some of the newly produced steel to make new mills and thus possibly increase our long-term ability to produce and stockpile steel. ■

5.4 FUNCTION MIN-MAX FEEDBACK CONTROL

Consider the problem of designing a feedback controller $\mathbf{u}(\mathbf{x})$ to transfer $\mathbf{x}(t)$ to the origin for a system of the form

$$\dot{\mathbf{x}} = \mathbf{f}(\mathbf{x}, \mathbf{u}, \mathbf{v}) = \mathbf{g}(\mathbf{x}) + \mathbf{B}[\mathbf{u} + \mathbf{v}], \tag{5.4-1}$$

where each element of the unconstrained control input vector $\mathbf{u} \in R^{n_u}$ appears linearly in (5.4-1) and is subjected to a **disturbance** input represented by the corresponding element in the **uncertainty** vector $\mathbf{v}(t)$. The elements of $\mathbf{v}(t)$ might correspond to noise or to unmodeled features of the control actuators, such as backlash, hysteresis, coulomb friction, and so on. The control $\mathbf{u}$ has no constraints imposed on it and we assume that, for $\mathbf{v}(t) \equiv \mathbf{0}$, the state $\mathbf{x}$ is globally controllable to the origin. In other words, the system does not contain any controllability boundaries.

We treat the disturbance or uncertainty vector $\mathbf{v}(t)$ as a completely unknown function, except that each component of $\mathbf{v}(t)$ is bounded:

$$\mathbf{v}(t) \in \mathcal{V} = \{\mathbf{v} \in R^{n_u} | v_{i_{\min}} \leq v_i(t) \leq v_{i_{\max}}\} \tag{5.4-2}$$

with these upper and lower bounds known. We also assume that the system

$$\dot{\mathbf{x}} = \mathbf{g}(\mathbf{x}) \tag{5.4-3}$$

has a globally asymptotically stable equilibrium at $\mathbf{x} = \mathbf{0}$. Thus [Kalman and Bertram, 1960; Krasovskiĭ, 1961] there is a Lyapunov function $V(\mathbf{x})$ for (5.4-3) such that $V(\mathbf{x}) \to \infty$ as $\|\mathbf{x}\| \to \infty$ and

$$0 \geq \frac{\partial V(\mathbf{x})}{\partial \mathbf{x}} \mathbf{g}(\mathbf{x}) \not\equiv 0 \qquad \text{for all } \mathbf{x} \neq \mathbf{0}. \tag{5.4-4}$$

We assume that the Lyapunov function is known. The asymptotic stability assumption is not as restrictive as it might seem. If (5.4-3) is unstable but (5.4-1) is controllable for $\mathbf{v} = \mathbf{0}$, with the controllable set to the origin being all of state space, then a portion of the unbounded control $\mathbf{u}(\mathbf{x})$ can be used to stabilize (5.4-1), yielding a new system of the form (5.4-1) with the origin being globally asymptotically stable for (5.4-3).

We seek a feedback controller $\mathbf{u}(\mathbf{x})$ that makes the origin for (5.4-1) globally asymptotically stable against all $\mathbf{v}(t)$ satisfying (5.4-2); that is, we seek **guaranteed global asymptotic stability**.

A Min-Max Controller for Uncertain Systems

The controller that we present is adapted from Gutman and Leitmann [1976], Leitmann [1981b], and Corless and Leitmann [1981]. For each control component u_i, $i = 1, \ldots, n_u$, we artificially impose bounds

$$|u_i| \leq \rho_i \triangleq \max\{|v_{i_{\min}}|, |v_{i_{\max}}|\}, \tag{5.4-5}$$

that is, $\mathbf{u} \in \mathcal{U} = \{\mathbf{u} \in R^{n_u} | |u_i| \leq \rho_i\}$, reflecting the fact that, because of the stability of (5.4-3), we need no more u_i effort than the worst-case bound on the corresponding v_i in order to stabilize (5.4-1).

Given the Lyapunov function $V(\mathbf{x})$ we compute the quickest descent control $\mathbf{u}(\mathbf{x})$ against a worst-case "quickest ascent" uncertainty $\mathbf{v}(\mathbf{x})$. Specifically, for

$$\dot{V}(\mathbf{x}, \mathbf{u}, \mathbf{v}) = \frac{\partial V(\mathbf{x})}{\partial \mathbf{x}} \mathbf{f}(\mathbf{x}, \mathbf{u}, \mathbf{v}) = \frac{\partial V(\mathbf{x})}{\partial \mathbf{x}} \{\mathbf{g}(\mathbf{x}) + \mathbf{B}[\mathbf{u} + \mathbf{v}]\} \tag{5.4-6}$$

we calculate a **min-max descent control** $\mathbf{u}(\mathbf{x})$ as the solution to

$$\min_{\mathbf{u}\in\mathcal{U}} \max_{\mathbf{v}\in\mathcal{V}} \dot{V}(\mathbf{x},\mathbf{u},\mathbf{v}).$$

The resulting control can be written as

$$u_i(\mathbf{x}) = \begin{cases} -\rho_i \operatorname{sgn}[\sigma_i(\mathbf{x})] & \text{if } \sigma_i(\mathbf{x}) \neq 0 \\ \in \{u_i \mid |u_i| \leq \rho_i\} & \text{if } \sigma_i(\mathbf{x}) = 0, \end{cases} \tag{5.4-7}$$

where

$$\sigma_i(\mathbf{x}) = \frac{\partial V}{\partial \mathbf{x}} \mathbf{b}_i, \tag{5.4-8}$$

and $\mathbf{b}_i$ is the ith column of $\mathbf{B} = [\mathbf{b}_1 \cdots \mathbf{b}_{n_u}]$. Notice that each control component $u_i(\mathbf{x})$ is a switching-type control, with **switching surface** $\sigma_i(\mathbf{x}) = 0$. Thus we have a generalized dynamical system, with solutions $\mathbf{x}(t)$ defined in the sense of Filippov as in Chapter 1, and the trajectories $\mathbf{x}(t)$ may exhibit chatter, with $\sigma_i[\mathbf{x}(t)] \equiv 0$.

To verify that the controller yields global asymptotic stability of the origin for (5.4-1), for any disturbance satisfying (5.4-2), we note that

$$\begin{aligned}
\frac{dV}{dt} &= \frac{\partial V}{\partial \mathbf{x}} \mathbf{g} + \sum_{i=1}^{n_u} \frac{\partial V}{\partial \mathbf{x}} \mathbf{b}_i(u_i + v_i) \\
&= \frac{\partial V}{\partial \mathbf{x}} \mathbf{g} + \sum_{i=1}^{n_u} \sigma_i(-\rho_i \operatorname{sgn}(\sigma_i) + v_i) \\
&= \frac{\partial V}{\partial \mathbf{x}} \mathbf{g} + \sum_{i=1}^{n_u} (-|\sigma_i|\rho_i + \sigma_i v_i) \\
&\leq \frac{\partial V}{\partial \mathbf{x}} \mathbf{g} + \sum_{i=1}^{n_u} |\sigma_i|(-\rho_i + |v_i|) \\
&\leq \frac{\partial V}{\partial \mathbf{x}} \mathbf{g} + \sum_{i=1}^{n_u} |\sigma_i|(-\rho_i + \rho_i) \\
&\leq 0 \text{ and } \not\equiv 0 \qquad \text{for } \mathbf{x} \neq \mathbf{0}.
\end{aligned}$$

From Lyapunov's stability theorem we conclude that the origin is globally asymptotically stable. However, since the switching control produces differential equations with discontinuous right-hand sides, we are taking a certain degree of liberty in applying Lyapunov's sufficiency theorem. The justification is that we have explicitly restricted the class of admissible controls to those that are piecewise C^1 (except across certain switching surfaces) and produce solutions (in the sense of Filippov) having the usual existence, uniqueness, and continuity properties. For such trajectories Lyapunov's stability theorem does apply. In particular, the above result for dV/dt (which now is generally only piecewise continuous) holds below, above, and on the switching surface $\sigma(\mathbf{x}) = 0$. Thus $V[\mathbf{x}(t)] \to 0$ and we conclude that the origin is globally asymptotically stable.

Example 5.4-1: Linearized Pendulum with Disturbance Input Consider the problem of stabilizing the origin for the linearized inverted pendulum system

$$\begin{bmatrix} \dot{x}_1 \\ \dot{x}_2 \end{bmatrix} = \begin{bmatrix} 0 & 1 \\ 1 & 0 \end{bmatrix} \begin{bmatrix} x_1 \\ x_2 \end{bmatrix} + \begin{bmatrix} 0 \\ 1 \end{bmatrix} (u + v), \tag{5.4-9}$$

where the control input u is corrupted by a bounded disturbance input v, with $|v| \leq \rho = 1$.

For $u = v = 0$ the system with no inputs is unstable. Thus the first step is to use part of the unbounded control $u(\mathbf{x})$ to stabilize the system for the case where $v(t) \equiv 0$. As shown in Example 2.2-1, the system with $v(t) \equiv 0$ can be stabilized by choosing

$$u = \overline{u}(\mathbf{x}) \triangleq -k_1 x_1 - k_2 x_2.$$

For example, choosing

$$\overline{u}(\mathbf{x}) = -2x_1 - x_2$$

yields a positive definite $\mathbf{P}$ matrix in Example 5.1-1 and

$$V(\mathbf{x}) = \mathbf{x}^{\mathrm{T}} \mathbf{P} \mathbf{x} = \tfrac{3}{2} x_1^2 + x_1 x_2 + x_2^2 \tag{5.4-10}$$

is a Lyapunov function for (5.4-9) with $u = \overline{u}(\mathbf{x})$ and $v = 0$, since

$$\dot{V}(\mathbf{x}) = -(x_1^2 + x_2^2). \tag{5.4-11}$$

To guarantee global asymptotic stability in (5.4-9) against any disturbance $v(t)$ satisfying $|v(t)| \leq \rho = 1$, we compute the min-max descent control for $\dot{V}(\mathbf{x}, \mathbf{u}, \mathbf{v})$ in (5.4-6). In particular, we define the switching function

$$\sigma(\mathbf{x}) = \frac{\partial V(\mathbf{x})}{\partial \mathbf{x}} \mathbf{b} = [3x_1 + x_2 \quad x_1 + 2x_2] \begin{bmatrix} 0 \\ 1 \end{bmatrix} = x_1 + 2x_2$$

and use the combined linear state feedback plus switching control

$$u(\mathbf{x}) = -[2 \quad 1]\mathbf{x} + \hat{u}(\mathbf{x}), \tag{5.4-12}$$

where

$$\hat{u}(\mathbf{x}) = \begin{cases} -\rho \ \mathrm{sgn}(\sigma) & \text{if } \sigma(\mathbf{x}) \neq 0 \\ \in \{u| \ |u| \leq \rho\} & \text{if } \sigma(\mathbf{x}) = 0. \end{cases} \tag{5.4-13}$$

Applying this control to (5.4-9) yields

$$\begin{bmatrix} \dot{x}_1 \\ \dot{x}_2 \end{bmatrix} = \begin{bmatrix} 0 & 1 \\ -1 & -1 \end{bmatrix} \begin{bmatrix} x_1 \\ x_2 \end{bmatrix} + \begin{bmatrix} 0 \\ 1 \end{bmatrix} (\hat{u} + v). \tag{5.4-14}$$

With $V(\mathbf{x})$ given by (5.4-10), global asymptotic stability of the origin is established, since for any $v(t)$ satisfying $|v(t)| \leq \rho$, $dV/dt < 0$ for $\mathbf{x} \neq \mathbf{0}$.

Figure 5.4-1*a* is a plot of the state-space trajectories and a typical $V(\mathbf{x})$ contour for the case of no disturbance input, but with $\rho = 1$ to place bounds on u. From (5.4-13) we have $\hat{u}(\mathbf{x}) = 1$ below the switching surface $x_1 + 2x_2 = 0$, $\hat{u}(\mathbf{x}) = -1$ above the switching surface, and we choose $\hat{u}(\mathbf{x}) = 0$ on the switching surface, for the following reason. Well away from the origin, trajectories that reach the switching surface penetrate the surface and $\hat{u}(\mathbf{x})$ simply switches from one boundary value to the other. Nearer the origin, however, trajectories approach the switching surface from both sides and $\hat{u}(\mathbf{x})$ chatters, yielding trajectories that slide along the switching surface as they approach the origin, with a Filippov velocity $\dot{\mathbf{x}}$ corresponding, in this case, to $\hat{u} = 0$. Note that, despite this chattering phenomenon, Figure 5.4-1*a* demonstrates that the trajectories always penetrate the $V(\mathbf{x})$ contours, as we have previously demonstrated analytically.

While we cannot plot trajectories for the general case of an uncertain input, since $v(t)$ is unknown, Figure 5.4-1*b* shows trajectories for the case of $v(t) = \sin 10t$. Chatter, of a wavy sort, also occurs in this case, but we are still assured that all of the trajectories in Figure 5.4-1*b* penetrate the $V(\mathbf{x})$ contours and approach the origin. ■

Instantaneous Two-Player Games

If we add a "cost" aspect to our min-max controllers, we obtain min-max cost-descent controllers analogous to those for the **differential games** (two-player min-max optimal control problems) that we consider in Chapter 9. Min-max

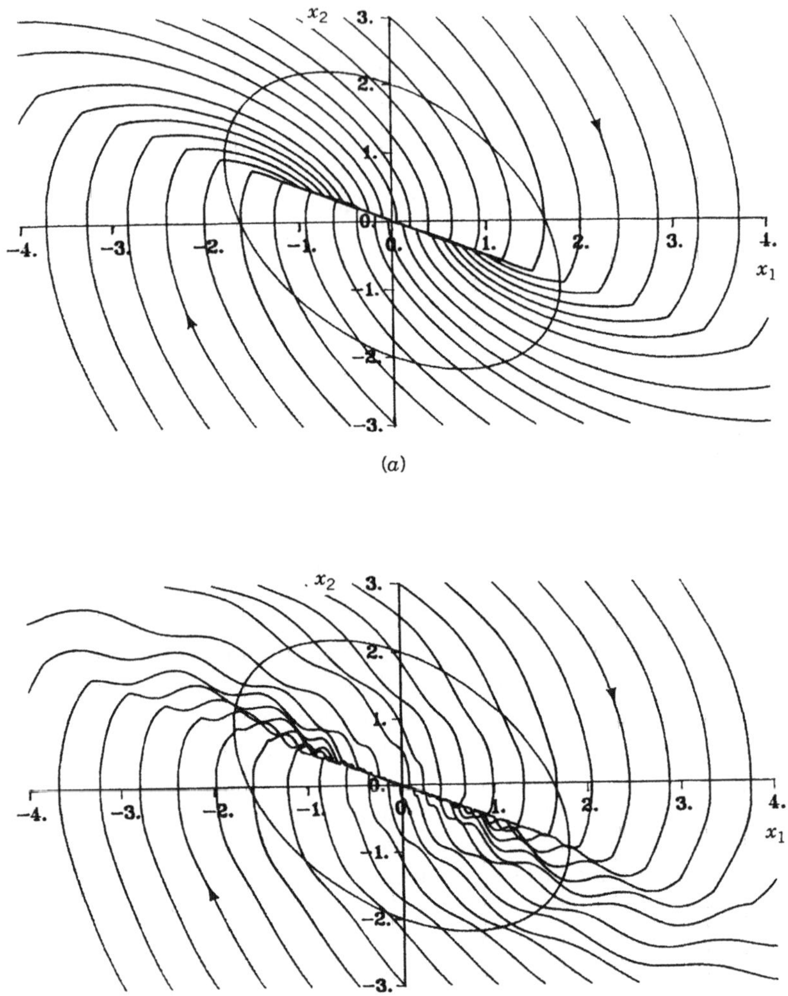

Figure 5.4-1 Trajectories for Example 5.4-1. (*a*) $v(t) = 0$. (*b*) $v(t) = \sin 10t$.

cost-descent controls $\mathbf{u}(\mathbf{x}) \in \mathcal{U}$ and "worst-case" uncertainties $\mathbf{v}(\mathbf{x}) \in \mathcal{V}$ are defined by

$$\min_{\mathbf{u} \in \mathcal{U}} \max_{\mathbf{v} \in \mathcal{V}} \left\{ H(\mathbf{x}, \mathbf{u}, \mathbf{v}, \nabla W) \triangleq f_0(\mathbf{x}, \mathbf{u}, \mathbf{v}) + \frac{\partial W(\mathbf{x})}{\partial \mathbf{x}} \mathbf{f}(\mathbf{x}, \mathbf{u}, \mathbf{v}) \right\}, \qquad (5.4\text{-}15)$$

where

$$J = \int_0^{t_f} f_0(\mathbf{x}, \mathbf{u}, \mathbf{v})dt \tag{5.4-16}$$

is the cost to the **u**-player and "payoff" to the **v**-player, t_f is the time at which $\mathbf{x}(t)$ reaches a specified target set $\mathcal{X}$, and $W(\mathbf{x})$ is a descent function for the target.

Example 5.4-2: War of Attrition and Attack Consider a differential game involving two nations in a protracted war [Isaacs, 1965, pp. 330–335]. Country i has a supply of weapons x_i, manufactured at a rate m_i. Each country must decide how to allocate its weapons between "attrition" (depleting the enemy's rate of weapon supply) or "attack" (entering weapons into the battlefield). Let u be the fraction of weapons Country 1 allocates to attack, and v be the fraction allocated to attack by Country 2. Let the dynamics be given by

$$\begin{aligned} \dot{x}_1 &= m_1 - c_2(1-v)x_1x_2 \\ \dot{x}_2 &= m_2 - c_1(1-u)x_1x_2, \end{aligned} \tag{5.4-17}$$

where c_i is the effectiveness of Country i's weapons. Country 1 seeks u to minimize, and Country 2 seeks v to maximize, the payoff to Country 2, which is the excess "attack" weapons Country 2 puts into battle:

$$J = \int_0^{t_f} (vx_2 - ux_1)dt, \tag{5.4-18}$$

subject to the constraints

$$\begin{aligned} 0 \le u \le 1 \\ 0 \le v \le 1. \end{aligned} \tag{5.4-19}$$

In Isaacs' original formulation this game was considered with a fixed time duration T. Thus there is no state-space target. Instead, we give each country a multiple-objective target of maximizing its supply of weapons, minimizing the enemy's, and maximizing the total of its weapons allocated to attack. In particular, assuming the war goes on indefinitely, the equations of motion could yield an equilibrium $(\hat{x}_1, \hat{x}_2)$. Country 1 would like to maximize $\hat{x}_1$, minimize $\hat{x}_2$, and minimize J, while Country 2 would like to maximize $\hat{x}_2$, minimize $\hat{x}_1$, and maximize J.

Picking as a min-max descent function

$$W(\mathbf{x}) = x_2 - x_1, \tag{5.4-20}$$

$u(\mathbf{x})$ and $v(\mathbf{x})$ are chosen to min-max

$$\begin{aligned} H &= f_0(\mathbf{x}, u, v) + \frac{\partial W(\mathbf{x})}{\partial x_1}\dot{x}_1 + \frac{\partial W(\mathbf{x})}{\partial x_2}\dot{x}_2 \\ &= vx_2(1 - c_2x_1) - ux_1(1 - c_1x_2) \\ &\quad + m_2 - m_1 + (c_2 - c_1)x_1x_2. \end{aligned} \tag{5.4-21}$$

We leave the results as Exercise 5-15. However, the function min-max results are not likely to be the same as the functional min-max results for the corresponding differential game. ■

EXERCISES

5-1. A Lorenz system, with a control input u, has the dynamics

$$\begin{aligned} \dot{x} &= s(y - x) \\ \dot{y} &= (u + r - z)x - y \\ \dot{z} &= xy - bz, \end{aligned}$$

with $s = 10$, $b = 8/3$, $r = 28$. For $u = 0$ the system is chaotic. Linearize the system about the origin and design a feedback controller $u(x, y, z)$ so that the origin for the linear system is asymptotically stable. Using a Lyapunov function based on the linear system, estimate the domain of attraction for the controller when applied to the nonlinear system.

5-2. Consider the system

$$\begin{aligned} \dot{x}_1 &= x_2 \\ \dot{x}_2 &= -x_1 - x_2 + x_1x_2. \end{aligned}$$

(a) Use Lyapunov's first method to investigate the stability of the origin. Is the origin locally stable?

(b) Find a Lyapunov function for the linearized system.

(c) Use the Lyapunov function of part **(b)** to obtain conditions defining a region about the origin from which trajectories asymptotically approach the origin for the nonlinear system.

5-3. Consider the control system

$$\begin{aligned} \dot{x}_1 &= x_2 \\ \dot{x}_2 &= x_1 + u, \end{aligned}$$

where $|u| \leq 1$. Use the W function

$$W = [x_1 \quad x_2]\begin{bmatrix} 1 & 0.5 \\ 0.5 & 1 \end{bmatrix}\begin{bmatrix} x_1 \\ x_2 \end{bmatrix} = x_1^2 + x_1x_2 + x_2^2.$$

(a) Show that lines of constant W are ellipses about the origin. Draw a sketch.

(b) Design a quickest descent controller for the system.

(c) Show that $\dot{W} < 0$ for all points along the switching curve (except the origin).

(d) For the quickest descent controller, is the domain of attraction to the origin the entire state space? Why?

5-4. An object with dynamics

$$\dot{x} = \cos u$$
$$\dot{y} = \sin u$$

moves on a surface

$$z = g(x, y) = 6r^4 - 22r^3 + 21r^2,$$

where $r = \sqrt{x^2 + y^2}$. Design a Lyapunov steepest descent control for u to steer the system to the global minimum point for z. Note that $g(x, y)$ cannot be used as a descent function, as is evident from Figure 4.3-2*a*.

5-5. Consider the linear system

$$\begin{bmatrix} \dot{x}_1 \\ \dot{x}_2 \end{bmatrix} = \begin{bmatrix} a_{11} & a_{12} \\ a_{21} & a_{22} \end{bmatrix}\begin{bmatrix} x_1 \\ x_2 \end{bmatrix} + \begin{bmatrix} b_1 \\ b_2 \end{bmatrix} u,$$

but with

$$u_{\min} \leq u \leq u_{\max}.$$

Determine a closed-loop feedback control law to drive the state to the origin for this system using the quickest descent method, with descent function

$$W(\mathbf{x}) = \tfrac{1}{2}\,\mathbf{x}^T\mathbf{P}\mathbf{x}, \qquad \mathbf{P} = \begin{bmatrix} p_{11} & p_{12} \\ p_{12} & p_{22} \end{bmatrix},$$

where $\mathbf{P}$ is symmetric and positive definite.

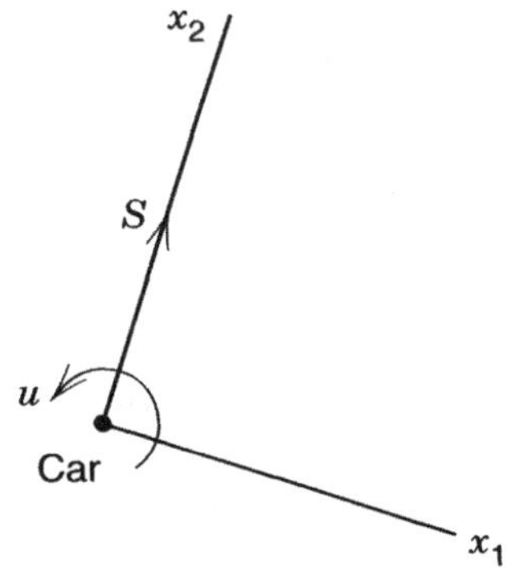

Figure E5-6 Rotating coordinate system for Exercise 5-6.

5-6. Consider a planar "suicidal pedestrian" pursuit problem, illustrated in Figure E5-6, in which a car with a fixed speed S and bounded turning rate u tries to run down a stationary pedestrian. Let (x_1, x_2) denote the position of the pedestrian relative to the car, in a rotating coordinate system centered at the car, with the x_2-axis aligned with the car's velocity vector. As the car turns the coordinate system rotates with angular velocity u, positive in the counterclockwise direction. That is, positive u corresponds to a left turn. The position of the pedestrian relative to the car is given by

$$\dot{x}_1 = ux_2$$
$$\dot{x}_2 = -S - ux_1.$$

The car's turning rate is bounded by

$$|u| \leq u_{\max}.$$

The car's objective is to drive the state to the origin. In the following, use $S = u_{\max} = 1$.

(a) Design a Lyapunov function minimizing controller $u(\mathbf{x})$ for the car.

(b) Plot trajectories throughout the state space and from these determine the region of effectiveness for the controller in (a), that is, the set of states that get transferred to the origin.

(c) Redesign the controller in (a), by using different "descent" functions in different regions of state space, so that the resulting region of effectiveness for the new controller is all of state space.

5-7. A system is moving under the simple dynamics given by

$$\dot{x}_1 = \cos u$$
$$\dot{x}_2 = \sin u.$$

(a) Using the quickest descent method, obtain a control law that will drive this system from any point in state space to the target defined by

$$(x_1 - 1)^2 + 2x_2^2 - 1 = 0.$$

You will want to use the W function

$$W = (x_1 - 1)^2 + 2x_2^2.$$

(b) What value of the control is used at the point $(x_1, x_2) = (6, 5)$?

(c) Determine the equation of the path from this point and draw a sketch of the trajectory from $(x_1, x_2) = (6, 5)$ to the target.

(d) Set up the differential equation for u, if a trajectory following algorithm is used to determine u after the initial minimization at $(x_1, x_2) = (6, 5)$. What is $\dot{u}$ initially? Will it change?

5-8. Solve Example 5.3-2 using the trajectory following method.

5.9. Solve Example 5.3-4 using the constrained trajectory following method. Then use (5.3-36) to eliminate any tendency for your controller to chatter and compare results.

5.10. For the steel production problem in Example 5.3-6 and the $H(\cdot)$ function in (5.3-57), determine the minimum cost-descent control $\mathbf{u}(\mathbf{x})$, where $\mathbf{x} = [S \quad M]^{\mathrm{T}}$. Use parameter values $a = 2$, $b = 3$, $c = 1$, $t_f = 10$. Plot the resulting trajectories from throughout the S, $M > 0$ quadrant. On the same graph plot representative curves of constant $W[\mathbf{x}, \mathbf{u}(\mathbf{x})]$ and curves of constant $S(t_f)$, that is, curves of $\mathbf{x}(0)$ having the same final $S(t_f)$ value.

5-11. Consider the following one-dimensional control system

$$\dot{x} = 3x + u + v,$$

where u is a control input and v is an uncertain input. Lyapunov's (second) stability method is to be used to answer all questions. A useful Lyapunov function for this problem is given by

$$V(x) = x^2.$$

(a) With $v = 0$, $u = -kx$, and u unbounded, determine the values of k for which the domain of attraction to the origin is unbounded.

(b) With $v = 0$, $u = -4x$, and $|u| \leq 8$, estimate the domain of attraction.

(c) With $|v| \leq 2$ and u unbounded, design a controller that will guarantee global asymptotic stability of the origin.

5-12. The relationship between the input voltage E and the shaft rotation angle θ for a low inductance armature-controlled DC motor is given by

$$\frac{RJ}{K_T}\ddot{\theta} + K_B\dot{\theta} = E,$$

where R = armature resistance, J = rotor moment of inertia, and K_T and K_B are motor constants. For a certain motor

$$\frac{RJ}{K_T} = 0.05 \text{ volt-sec}^2$$

$$K_B = 0.1 \text{ volt-sec.}$$

It is possible to measure both θ and $\dot{\theta}$ (using a shaft encoder and a tachometer).

(a) Determine $\mathbf{A}$ and $\mathbf{b}$ in an equivalent state-space representation

$$\dot{\mathbf{x}} = \mathbf{A}\mathbf{x} + \mathbf{b}u,$$

with $x_1 = \theta$, $x_2 = \dot{\theta}$, and $u = E$.

(b) Determine the rank of the Kalman controllability matrix and comment on system controllability.

(c) Using a controller of the form $u = -k_1x_1 - k_2x_2$, choose k_1 and k_2 so that the controlled system is equivalent to a second-order system $\ddot{\theta} + 2\zeta\omega_n\dot{\theta} + \omega_n^2\theta = 0$, with $\zeta = 0.5$ and $\omega_n = 5$ rad/sec.

(d) It was discovered that an uncertain input $v(t)$, with $|v| \leq \rho$, also enters through E. In order to maintain asymptotic stability of the origin, an additional control $\hat{u}$ will now be added. The total input is thus given by

$$u = v - k_1x_1 - k_2x_2 + \hat{u}.$$

Using the same k_1 and k_2 as determined in (c), design a min-max controller $\hat{u}(\mathbf{x})$ so that asymptotic stability of the origin is assured.

5-13. As in Corless and Leitmann [1981] consider a forced nonlinear pendulum of length l with an uncertain input $\ddot{x}_p(t)$, $|\ddot{x}_p| \leq \rho l$, corresponding to a horizontal acceleration of its pivot point. The equations of motion are

$$\dot{x}_1 = x_2$$

$$\dot{x}_2 = -\omega^2 \sin x_1 + u - \frac{\ddot{x}_p(t)}{l} \cos x_1,$$

where $\mathbf{x} = [\theta \quad \dot{\theta}]^{\mathrm{T}}$, $\omega > 0$, and θ is the angle from the downward vertical to the pendulum. This system, with $u = \ddot{x}_p = 0$, has multiple equilibrium points $(\theta, \dot{\theta}) = (\pm k\pi, 0)$, $k = 0, 1, \ldots$. Treating the $-(\ddot{x}_p/l)\cos x_1$ term as a bounded uncertain input $v(t)$ with $|v| \leq \rho$, design a feedback control $u(\mathbf{x})$ so that the origin is guaranteed globally asymptotically stable despite the uncertain input term. Verify your result by finding a Lyapunov function and applying Lyapunov's stability theorem.

5.14. A particular system is given by

$$\dot{x}_1 = x_1 - 2x_2$$

$$\dot{x}_2 = x_1 + x_2 + u + v$$

$$|u| \leq 1 \qquad |v| \leq 0.5.$$

Design a controller $u(\mathbf{x})$ that will provide asymptotic stability for the origin in some neighborhood of the origin. By a sketch, give an estimate of the resulting domain of attraction.

5.15. For the war of attrition and attack in Example 5.4-2 and the $H(\cdot)$ function in (5.4-21), determine the min-max cost-descent controls $u(\mathbf{x})$ and $v(\mathbf{x})$. Use the parameter values $m_1 = 2$, $m_2 = 1$, $c_1 = 1$, $c_2 = 2$, $t_f = 10$. Plot the resulting trajectories from throughout the $x_1, x_2 > 0$ quadrant.

6

CONTROLLABILITY OF NONLINEAR SYSTEMS

6.1 CONTROLLABLE SETS AND REACHABLE SETS

In this chapter we examine the problem of determining the set of initial states that can be transferred to the target. The results that we develop provide a basis for evaluating and designing effective controllers. At an even more fundamental level, the controllability results that we develop provide some basic tools needed for our subsequent development of Pontryagin's minimum principle for optimal control problems in Chapter 7.

Controllable Set

Given the general nonlinear control system

$$\dot{\mathbf{x}} = \mathbf{f}(\mathbf{x}, \mathbf{u}), \tag{6.1-1}$$

with a specified target set $\mathcal{X}$, control constraint $\mathbf{u} \in \mathcal{U}$, and the definition of admissible controls presented in Section 2.1, the **controllable set** $\mathcal{C}$ to the target $\mathcal{X}$ is defined as the set of all initial states that can be transferred to the target in finite time using some admissible control law. Note that the definition of the controllable set depends only on the target set $\mathcal{X}$ and the capabilities of the system (6.1-1), subject to the control constraint set $\mathcal{U}$. While a control choice is not specified, there must exist at least one admissible control law $\mathbf{u}(\mathbf{x})$ such that for each initial state $\mathbf{x}(0)$ in $\mathcal{C}$ there is a time $0 \leq t_f < \infty$ at which the corresponding solution $\mathbf{x}(t)$ to (6.1-1) satisfies $\mathbf{x}(t_f) \in \mathcal{X}$. In other words, the **domain of effectiveness** for this control law is equal to the controllable set $\mathcal{C}$.

The controllable set is a useful design tool for comparing the effectiveness of different control algorithms. As we have previously noted, given control laws might not be effective throughout the controllable set, and comparing the domains of effectiveness (or regions of attraction) for these control laws to the controllable set provides a meaningful performance measure. Furthermore, knowledge of the controllable set provides useful information on what cannot be done. If initial conditions of interest lie outside the controllable set, then there is no point in designing a controller for that system. For example, a skyrocket in a fireworks display does not have sufficient thrust and fuel capability to achieve Earth orbit. Alternately one can use the controllable set as a design tool for modifying the control system capabilities so that initial conditions of interest do lie inside the controllable set. (Replace the skyrocket with an Atlas missile?)

Reachable Set

In many applications, the question is not "Which states can be transferred to the target?" but instead, "Which states can be reached, starting from some initial set?" For the control system (6.1-1), an initial condition set $\mathcal{X}$, the control constraint $\mathbf{u} \in \mathcal{U}$, and the definition of admissible controls, the **reachable set** $\mathcal{R}$ from $\mathcal{X}$ is defined as the set of all states $\mathbf{x}(t)$, $t \geq 0$, that can be achieved, starting from an initial state in $\mathcal{X}$, by using some admissible control law. For each point $\mathbf{x}_f \in \mathcal{R}$ there exists at least one associated admissible control law $\mathbf{u}(\mathbf{x})$ and a time $0 \leq t_f < \infty$ such that the corresponding solution $\mathbf{x}(t)$ to (6.1-1), from some initial state $\mathbf{x}(0) \in \mathcal{X}$, satisfies $\mathbf{x}(t_f) = \mathbf{x}_f$. For points outside $\mathcal{R}$, no such control exists. Like the controllable set, the reachable set is not defined in terms of a particular control choice. Instead, it reflects the capabilities associated with all possible control choices. The reachable set is also a useful design tool in clarifying the performance of different control designs or determining the capabilities of a given control system.

Forward and Backward Integration

One important connection between controllable sets and reachable sets is the following: The controllable set $\mathcal{C}$ to $\mathcal{X}$ for (6.1-1) is the *same set* as the reachable set $\mathcal{R}$ from $\mathcal{X}$ for the **retro-time** system

$$\frac{d\mathbf{x}}{d\tau} = -\mathbf{f}(\mathbf{x}, \mathbf{u}), \qquad \mathbf{u} \in \mathcal{U}, \tag{6.1-2}$$

where $\tau \triangleq t_f - t$ is the **time-to-go** to the target for (6.1-1).

This result was proven in Snow [1967] for open-loop controls $\mathbf{u}(t)$. For such a control, the forward-time solution $\mathbf{x}(t)$ to (6.1-1), with $\mathbf{x}(t_f) = \mathbf{x}_f \in \mathcal{X}$, and the retro-time solution $\hat{\mathbf{x}}(\tau)$ to (6.1-2), generated by $\hat{\mathbf{u}}(\tau) \triangleq \mathbf{u}(t_f - \tau)$ and satisfying $\hat{\mathbf{x}}(0) = \mathbf{x}_f$, are related by

$$\mathbf{x}(t) = \hat{\mathbf{x}}(t_f - t) = \hat{\mathbf{x}}(\tau). \tag{6.1-3}$$

Thus moving backward in time along $\mathbf{x}(t)$ is the same as increasing the time-to-go τ along $\hat{\mathbf{x}}(\tau)$. Hence points that can be controlled to $\mathcal{X}$ for (6.1-1) are the same as the points that can be reached from $\mathcal{X}$ for (6.1-2).

The same result is valid for closed-loop controls $\mathbf{u}(\mathbf{x})$. However, the proof is a bit more complicated. Since admissible controls $\mathbf{u}(\mathbf{x})$ may be discontinuous, we generally have to consider solutions in the sense of Filippov (Section 1.2). In particular, a trajectory $\mathbf{x}(t)$ may "chatter" or "slide" along a switching surface, with a velocity $\dot{\mathbf{x}}$ that may not correspond to any $\mathbf{u} \in \mathcal{U}$. Furthermore, while admissible controls produce unique trajectories forward in time, the solutions backward in time may not be unique.

Let $\mathbf{u}(\mathbf{x})$ be an admissible control that generates a (unique) solution $\mathbf{x}(t)$, $0 \le t \le t_f$, to (6.1-1), transferring the state from $\mathbf{x}(0) \in \mathcal{C}$ to $\mathbf{x}(t_f) = \mathbf{x}_f \in \mathcal{X}$. For the retro-time system (6.1-2), define a corresponding closed-loop control $\hat{\mathbf{u}}(\mathbf{x}) \triangleq \mathbf{u}(\mathbf{x})$ along the trajectory $\mathbf{x}(t)$, $0 \le t \le t_f$, except that along a sliding-mode trajectory segment, $\mathbf{u}(\mathbf{x})$ is replaced in the definition of $\hat{\mathbf{u}}(\mathbf{x})$ by a control corresponding to the equivalent Filippov velocity. Then the (unique) retro-time solution $\hat{\mathbf{x}}(\tau)$ to (6.1-2), generated by $\hat{\mathbf{u}}(\hat{\mathbf{x}})$ and satisfying $\hat{\mathbf{x}}(0) = \mathbf{x}_f$, is related to the forward-time solution $\mathbf{x}(t)$ to (6.1-1) by (6.1-3). Thus, again, moving backward in time along $\mathbf{x}(t)$ is the same as increasing the time-to-go τ along $\hat{\mathbf{x}}(\tau)$. In particular, every point $\mathbf{x}(0) \in \mathcal{C}$ is reachable from $\mathcal{X}$ for the retro-time system (6.1-2). Reversing the argument shows that every point in $\mathcal{R}$ for the retro-time system is controllable to $\mathcal{X}$ for the forward-time system (6.1-1). Hence the points that can be controlled to $\mathcal{X}$ for (6.1-1) are the same as the points that can be reached from $\mathcal{X}$ for (6.1-2).

There are two things worth noting: (1) The closed-loop control $\mathbf{u}(\cdot)$ may be the same function for every forward-time trajectory, but the control $\hat{\mathbf{u}}(\cdot)$ may have to be a different function for each retro-time trajectory. (2) Along a sliding-mode forward-time trajectory segment, using a Filippov-equivalent control in the definition of $\hat{\mathbf{u}}(\mathbf{x})$ might yield $\hat{\mathbf{u}}(\mathbf{x}) \notin \mathcal{U}$. This technical difficulty in the proof can be overcome by defining an admissible retro-time control $\mathbf{u}'(\mathbf{x})$, in a neighborhood of the sliding-mode segment, that yields the same Filippov retro-velocity as $\hat{\mathbf{u}}(\mathbf{x})$.

Example 6.1-1: Equivalence of the Controllable and Retro-Reachable Sets As in Example 1.2-2, consider the system

$$\dot{x}_1 = 1$$
$$\dot{x}_2 = u, \qquad u = \pm 1,$$

with the origin as the target. It is evident, from the definition and by direct construction, that the controllable set to the origin is

$$C = \{(x_1, x_2) | x_1 \leq 0, |x_2| \leq -x_1\}$$

and that there are various admissible controls that would be effective throughout C. We examine two such controls in demonstrating that C is the same as the reachable set $\mathcal{R}$ from the origin for the retro-time system

$$\frac{dx_1}{d\tau} = -1$$

$$\frac{dx_2}{d\tau} = -u, \qquad u = \pm 1.$$

Case 1: Consider the switching control

$$u(\mathbf{x}) = \begin{cases} 1 & \text{if } x_1 + x_2 < 0 \\ -1 & \text{if } x_1 + x_2 \geq 0, \end{cases}$$

with trajectories shown in Figure 6.1-1*a*. If we were to take $\hat{u}(\mathbf{x}) \equiv u(\mathbf{x})$ and use this control in the retro-time system, we would get only a single trajectory $\hat{\mathbf{x}}(\tau)$ from the origin, corresponding to the upper boundary of C in Figure 6.1-1*a*. To fill in the reachable set for the retro-time system, we need a different controller for $\hat{u}(\mathbf{x})$, corresponding to each forward-time trajectory $\mathbf{x}(t)$. Let $\mathbf{x}_s = [x_{1_s} \quad x_{2_s}]^{\mathrm{T}}$ denote the point at which $\mathbf{x}(t)$ switches control. Then, for example, defining

$$\hat{u}(\mathbf{x}) = \begin{cases} 1 & \text{if } x_1 < x_{1_s} \\ -1 & \text{if } x_1 \geq x_{1_s} \end{cases}$$

and using $\hat{u}(\hat{\mathbf{x}})$ will produce a retro-time trajectory $\hat{\mathbf{x}}(\tau) = \mathbf{x}(t)$, with the direction of motion reversed from that shown in the figure. Repeating this process for each forward-time trajectory $\mathbf{x}(t)$ yields $C = \mathcal{R}$.

Case 2: Consider the "chattering" control

$$u(\mathbf{x}) = \begin{cases} 1 & \text{if } x_2 < 0 \\ -1 & \text{if } x_2 \geq 0. \end{cases}$$

The forward-time trajectories for this controller are illustrated in Figure 6.1-1*b*. As in Case 1, if we were to use $\hat{u}(\mathbf{x}) \equiv u(\mathbf{x})$, we would get a single retro-time trajectory $\hat{\mathbf{x}}(\tau)$ from the origin, namely, the upper portion of the boundary of the controllable set C in Figure 6.1-1*b*. We need a different $\hat{u}(\mathbf{x})$ for each forward-time trajectory $\mathbf{x}(t)$. However, in this case, constructing the retro-time feedback controls is somewhat more complicated than in Case 1.

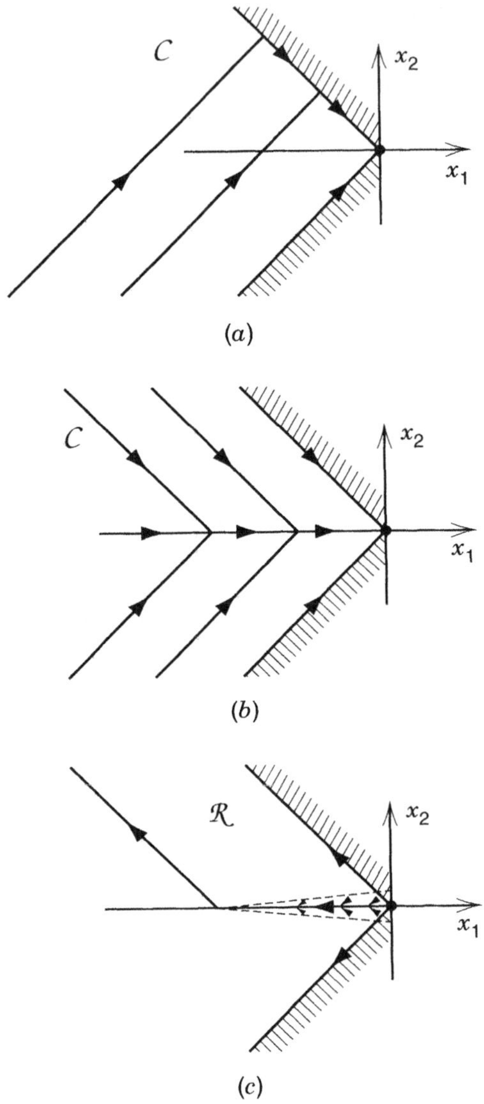

Figure 6.1-1 Forward- and retro-time trajectories for Example 6.1-1. (*a*) Switching control. (*b*) Chattering control. (*c*) Retro-time trajectory.

Note that forward-time motion, on the switching surface $x_2 = 0$, is the same as if an effective control $u = u_{\text{eff}} = 0$ were used. Letting $\mathbf{x}_s = [x_{1_s} \quad x_{2_s}]^T$ denote the point where a forward-time trajectory reaches the switching surface, we could use

$$\hat{u}(\mathbf{x}) = \begin{cases} 0 & \text{if } x_1 > x_{1_s} \text{ and } x_2 = 0 \\ \pm 1 & \text{if } x_1 = x_{1_s} \text{ and } x_2 = 0 \\ u(\mathbf{x}) & \text{elsewhere.} \end{cases}$$

However, on the switching surface this control would not satisfy the control constraints $\hat{u} = \pm 1$. To get an admissible feedback control for the retro-time system, we can use a small wedge-shaped ε-neighborhood (see Figure 6.1-1*c*) about the chattering segment so that the retro-time trajectory $\hat{\mathbf{x}}(\tau)$ from the origin has the same chattering behavior with the same effective control $u_{\text{eff}} = 0$. This can be accomplished, for example, by using the control

$$\hat{u}(\mathbf{x}) = \begin{cases} \text{sgn}(x_2) & \text{if } x_1 > x_{1_s} \text{ and } 0 < |x_2| < \varepsilon(x_1 - x_{1_s}) \\ \pm 1 & \text{if } x_1 = x_{1_s} \text{ and } x_2 = 0 \\ u(\mathbf{x}) & \text{elsewhere.} \end{cases}$$

Taking the limit as $\varepsilon \to 0$ and repeating this process for each forward-time trajectory, we again find that the reachable set $\mathcal{R}$ from the origin for the retro-time system is the same as the controllable set $\mathcal{C}$ to the origin for the forward-time system. ■

The Intersection of Controllable and Reachable Sets

For a given system (6.1-1) and a given set $\mathcal{X}$, the controllable set $\mathcal{C}$ to $\mathcal{X}$ and the reachable set $\mathcal{R}$ from $\mathcal{X}$ are generally two different regions in state space, although they do have (at least) $\mathcal{X}$ in common. The intersection of the controllable and reachable sets has some interesting properties. If the controllable set and reachable set have points in common other than just the target/initial set, then the intersection not only contains an invariant set, but one with the property that each point in the set is controllable to and reachable from every other point in the set. This result is applicable to the **patrol problem**, of finding the largest region in state space that the system can reach and still be able to return to the starting point.

Let $\mathbf{x}(t) = \boldsymbol{\xi}(t, \mathbf{y})$, $\boldsymbol{\xi}(0, \mathbf{y}) = \mathbf{y}$, $t \geq 0$, denote a solution to (6.1-1) from $\mathbf{x}(0) = \mathbf{y}$, generated by an admissible control $\mathbf{u}(\mathbf{x})$. Let $\mathcal{R}(\mathbf{y})$ be the reachable set from a point $\mathbf{y}$ and let $\mathcal{C}(\mathbf{y})$ be the controllable set to $\mathbf{y}$. We are interested in properties of the set $\mathcal{R}(\mathbf{y}) \cap \mathcal{C}(\mathbf{y})$, particularly if this set contains more that the point $\mathbf{y}$ itself.

We begin by extending the notion of an invariant set, to handle control systems of the form (6.1-1). A set $\mathcal{I}$ in state space is **positively invariant** if, and only if, for each $\mathbf{y} \in \mathcal{I}$ there exists an admissible control $\mathbf{u}(\mathbf{x})$ such that $\boldsymbol{\xi}(t, \mathbf{y}) \in \mathcal{I}$ for all $t \geq 0$. Note that this definition does not mean that trajectories cannot leave $\mathcal{I}$, just that there exists at least one admissible control $\mathbf{u}(\mathbf{x})$ such that every trajectory starting in $\mathcal{I}$ remains in $\mathcal{I}$.

A related concept is that of a maneuverable set. A nonempty, connected set $\mathcal{M}$ in state space is said to be **maneuverable** if, and only if, for each pair of points $\mathbf{y}$, $\mathbf{x}_f \in \mathcal{M}$ there exist an admissible control $\mathbf{u}(\mathbf{x})$ and a time $t_f \geq 0$ such that $\boldsymbol{\xi}(t_f, \mathbf{y}) = \mathbf{x}_f$ and $\boldsymbol{\xi}(t, \mathbf{y}) \in \mathcal{M}$ for $0 \leq t \leq t_f$. Thus every point in a maneuverable set $\mathcal{M}$ can be transferred to any other point in $\mathcal{M}$, along a trajectory that does not leave $\mathcal{M}$. It follows that a maneuverable set is positively invariant, but a positively invariant set need not be maneuverable.

Let $\mathcal{M} \triangleq \mathcal{R}(\mathbf{y}) \cap \mathcal{C}(\mathbf{y})$; we then have the following results for open-loop controls $\mathbf{u}(t)$ [Gayek and Vincent, 1986]:

(a) $\mathcal{M}$ is a connected set and is independent of its defining point $\mathbf{y}$; every point in $\mathcal{M}$ has the same controllable and reachable sets.

(b) If $\mathcal{M} - \{\mathbf{y}\} \neq \varnothing$, then $\mathcal{M}$ is maneuverable.

(c) If a given nonempty, connected, compact set X in state space is maneuverable, then $\mathcal{M} = \mathcal{R} \cap \mathcal{C}$ is the largest maneuverable set containing X, where $\mathcal{R}$ is the reachable set from X and $\mathcal{C}$ is the controllable set to X.

It turns out that the arguments carry over to closed-loop controls $\mathbf{u}(\mathbf{x})$ as well. The crux of the proofs is the fact that for $\mathbf{x} \in \mathcal{R}(\mathbf{y})$ there is a trajectory from $\mathbf{y}$ to $\mathbf{x}$. If $\mathbf{x}$ is also in $\mathcal{C}(\mathbf{y})$, then there is also a trajectory from $\mathbf{x}$ back to $\mathbf{y}$. Instead of presenting formal proofs here, we illustrate the results with an example.

Example 6.1-2: Patrol Problem The problem of patrol from a target is to determine the largest region in state space that can be reached by the system and still allow it to return to the starting point. This region must be maneuverable and can be determined by finding the maneuverable set. From (c) above, all we need to do is find the intersection of the reachable and controllable sets from the starting point.

Consider the system

$$\begin{bmatrix} \dot{x}_1 \\ \dot{x}_2 \end{bmatrix} = \begin{bmatrix} 1 & 0 \\ 0 & -1 \end{bmatrix} \begin{bmatrix} x_1 \\ x_2 \end{bmatrix} + \begin{bmatrix} u_1 \\ u_2 \end{bmatrix},$$

with $|u_1| \leq 1$ and $|u_2| \leq 2$. Let the starting point be the origin. Because the state equations are decoupled, we can easily determine the reachable set from the origin,

$$\mathcal{R} = \{(x_1, x_2) \mid |x_2| < 2\},$$

and the controllable set to the origin,

$$\mathcal{C} = \{(x_1, x_2) \mid |x_1| < 1\}.$$

Then the maneuverable set is the open rectangular region

$$\mathcal{M} = \mathcal{R} \cap \mathcal{C} = \{(x_1, x_2) \mid |x_1| < 1 \text{ and } |x_2| < 2\}.$$

Let $\mathbf{y}$ be any point in $\mathcal{M}$. It can easily be verified, by construction, that $\mathcal{R}(\mathbf{y}) = \mathcal{R}$ and $\mathcal{C}(\mathbf{y}) = \mathcal{C}$. Under various possible controls a trajectory $\mathbf{x}(t)$ may leave $\mathcal{M}$, but there must always exist control laws $\mathbf{u}(\mathbf{x})$ that not only can keep $\mathbf{x}(t)$ in $\mathcal{M}$, but can also transfer the state between any pair of points in $\mathcal{M}$. ■

Controllability Boundaries

For the examples that we have considered so far we have been able to determine controllable or reachable sets by direct examination of possible admissible controls. This technique is clearly not feasible for all problems. We now present a general method that will allow us to construct the controllable set to a target or the reachable set from an initial conditions "target." As noted previously, the reachable set for a control system of the form (6.1-1) can be obtained by finding the controllable set for the retro-time system (6.1-2). Hence we focus on just one: the controllable set for the forward-time system (6.1-1).

When the controllable set is not all of state space, it will have a boundary. We are particularly concerned with the problem of determining the **boundary of the controllable set,** designated by $\partial\mathcal{C}$. As previous examples illustrate, we can use the boundaries to define the set.

Our approach to determining $\partial\mathcal{C}$ is to develop necessary conditions governing controls $\mathbf{u}(\mathbf{x})$ that will yield trajectories $\mathbf{x}(t)$ that lie in $\partial\mathcal{C}$, and then use these trajectories to plot the boundary of the controllable set. For most practical problems such boundary trajectories do, in fact, exist. Later we give a boundary trajectory theorem that gives conditions that guarantee existence. While this method of determining $\mathcal{C}$ by running trajectories along $\partial\mathcal{C}$ is practical only for two- or three-dimensional problems, the geometric approach that we develop provides a powerful basis for our later development of optimal control theory.

We make very few assumptions concerning the geometry of the controllable set $\mathcal{C}$. In general, $\mathcal{C}$ may be either an open set or a closed set, so that $\partial\mathcal{C}$ may or may not be in $\mathcal{C}$. For simplicity we assume later that $\partial\mathcal{C}$ is a smooth $(n_x - 1)$-dimensional surface in state space, so that it has a well-defined tangent plane. This assumption can be relaxed to allow for possible "corners" in $\partial\mathcal{C}$ [Grantham, 1973; Grantham and Vincent, 1975], but we focus on the basic geometry and avoid the mathematical complications that would be required to handle all aspects of the most general case.

The necessary conditions we develop are based on the fact that, while a trajectory $\mathbf{x}(t)$ may leave the controllable set $\mathcal{C}$, it can never return to $\mathcal{C}$ (or even to $\partial\mathcal{C}$) once it departs. In particular, the following **properties of controllable sets** apply:

(i) A solution $\mathbf{x}(t)$ to (6.1-1) starting outside of C can never reach C;

(ii) If $\mathbf{x}(t)$ starts from ∂C, it can never reach the interior of $C(\mathring{C})$;

(iii) If $\mathbf{x}(t)$ starts in the exterior of C, it can never reach ∂C;

(iv) If $\mathbf{x}(0) \in \partial C$ and $\mathbf{x}(t_f) \in \partial C$, then $\mathbf{x}(t) \in \partial C$ for all $0 \leq t \leq t_f$.

(v) If ∂C contains **corners** (points where a tangent plane to ∂C is not unique), then trajectories in ∂C cannot cross such corners.

Condition (i) follows from the definition of the controllable set C. If a trajectory could start outside C (where there is no path to the target) and reach some point in C (where there is a path to the target), then there would exist a path to the target starting from outside C, which is a contradiction.

Conditions (ii), (iii), and (v) follow from Condition (i) and the fact that solutions $\mathbf{x}(t)$ to (6.1-1) are continuous functions of their initial conditions. Consider Condition (ii). As illustrated in Figure 6.1-2, if a trajectory $\mathbf{x}(t)$ were to exist from ∂C to the interior of C, then the same control $\mathbf{u}(\mathbf{x})$, applied from a point near $\mathbf{x}(0) \in \partial C$ but outside C, would provide a trajectory that starts outside C and reaches the interior of C, which violates Condition (i). A similar argument establishes Condition (iii), which in turn establishes Condition (iv). The same argument also establishes Condition (v).

Condition (iv) provides a basis for constructing ∂C using a feedback control $\mathbf{u}(\mathbf{x})$ that generates trajectories $\mathbf{x}(t)$ in ∂C. The fact that ∂C is composed of trajectories can be demonstrated most easily for the case where the controllable set is closed, so that $\partial C \subseteq C$. Let $\mathbf{u}(\mathbf{x})$ be an admissible control that is effective throughout the controllable set. In other words, every trajectory $\mathbf{x}(t)$ starting in the controllable set reaches the target under this control law. Consider a point $\mathbf{x}(0) \in \partial C$ in the controllable set C. Since this point is controllable to the target, using the control $\mathbf{u}(\mathbf{x})$, there exists a time $t_f \geq 0$ and a corresponding trajectory $\mathbf{x}(t)$ such that $\mathbf{x}(t)$ remains in C for $t \in [0, t_f]$ and reaches the target at $\mathbf{x}(t_f) \in X$. From Condition (ii) $\mathbf{x}(t_f)$ cannot be in the interior of C. Thus we conclude that

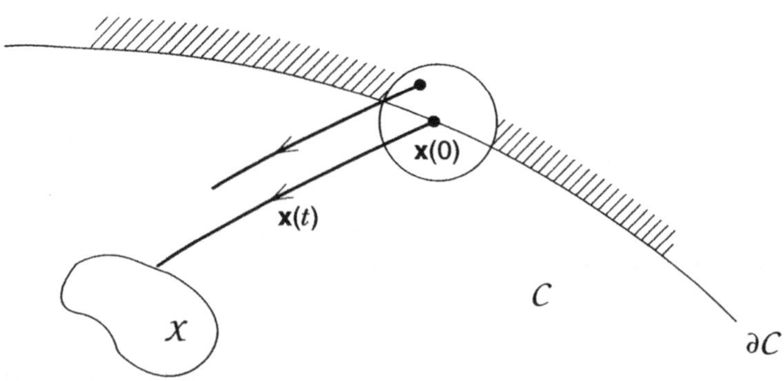

Figure 6.1-2 Geometry of the controllable set (showing an impossible trajectory).

$\mathbf{x}(t_f) \in \partial C$. Therefore, from Condition (iv), the trajectory $\mathbf{x}(t)$ must lie in ∂C for the entire time interval $0 \leq t \leq t_f$. A more general existence result for trajectories in ∂C is presented in Section 6.3

As a consequence of Conditions (i)–(v) we see that ∂C, the boundary of the controllable set, forms a separating surface in state space that is a **semipermeable membrane**. It is a surface composed of trajectories generated by some control law $\mathbf{u}(\mathbf{x})$ such that trajectories generated by other controls may cross this surface from inside the controllable set to outside, but they cannot penetrate ∂C from outside of the controllable set to inside.

6.2 CONTROLLABILITY MINIMUM PRINCIPLE

Using the properties of controllable sets, we show that a control $\mathbf{u}(\mathbf{x})$ that will drive the system along the boundary of the controllable set must satisfy a controllability minimum principle. Such a control law, at every point $\mathbf{x} \in \partial C$, chooses a $\mathbf{u} \in \mathcal{U}$ that comes as close as possible to penetrating the controllability boundary from the outside to the inside. Since penetration of ∂C from outside C to inside C is not possible, the resulting trajectories travel along ∂C. To establish the controllability minimum principle, we must first develop some supporting geometric conditions for trajectories that lie in the controllability boundary.

Tangent Planes to a Trajectory Surface

In this section, we are interested in some properties of trajectories that form an $(n_x - 1)$-dimensional surface in state space, which is called a Σ surface or just Σ for short. It will turn out that the boundary of the controllable set must be a Σ surface. However, not all Σ surfaces need be controllability boundaries. Other similar surfaces will play central roles in the development of optimal controls, introduced in Chapter 7, and in differential games, introduced in Chapter 9.

We need to make some assumptions about Σ that involve **tangent vectors**. Recall from Section 3.2 that a tangent vector to a surface (or other set) represents an extension of the familiar concept of a tangent vector to a line. Let $\Sigma \subseteq R^{n_x}$ be an arcwise connected set (such as a surface, a curve, etc.), and let $\mathbf{x} \in \Sigma$ be a point in the set. We say that a vector $\mathbf{e} \in R^{n_x}$ is **tangent** to Σ at $\mathbf{x}$ if, and only if, a continuous vector-valued function $\boldsymbol{\eta}(\alpha)$ exists such that for all scalar values α in some interval $(0, \beta)$, with $\beta > 0$, the following conditions hold:

(i)

$$\mathbf{x} + \alpha\boldsymbol{\eta}(\alpha) \in \Sigma$$

(ii)

$$\boldsymbol{\eta}(\alpha) \rightarrow \mathbf{e} \qquad \text{as } \alpha \rightarrow 0.$$

For a locally smooth $(n_x - 1)$-dimensional surface Σ, the set of all vectors tangent to Σ at $\mathbf{x}$ defines an $(n_x - 1)$-dimensional plane called the **tangent plane** $\mathcal{T}$ to Σ at a point $\mathbf{x} \in \Sigma$.

Given a control law $\mathbf{u}(\mathbf{x})$, consider a number of trajectories $\mathbf{x}(t)$ generated by the differential equations

$$\dot{\mathbf{x}} = \mathbf{f}[\mathbf{x}, \mathbf{u}(\mathbf{x})] \tag{6.2-1}$$

by varying initial conditions along some curve $\mathcal{S}$ in state space. The resultant trajectories define a Σ surface. For simplicity, we assume that at every point along a trajectory $\mathbf{x}(t)$ in Σ, the surface Σ is smooth and thus has a unique $(n_x - 1)$-dimensional tangent plane. A more general proof of the minimum principle may be obtained without regard to tangent planes [Pontryagin et al., 1962; Grantham and Vincent, 1975]. However, assuming Σ is smooth avoids a lot of complications. In what follows, we determine differential equations that define how the tangent plane and a corresponding vector $\boldsymbol{\lambda}(t)$ normal to Σ change as we move along a trajectory $\mathbf{x}(t)$.

Let $\mathbf{u}(\mathbf{x})$ be the control that generates $\mathbf{x}(t)$. Let $\mathbf{x}(t)$, $0 \leq t \leq t_f$, denote a particular trajectory in a surface Σ, with initial condition $\mathbf{x}(0)$. For now we assume that the closed-loop control $\mathbf{u}(\mathbf{x})$ is continuous and continuously differentiable along $\mathbf{x}(t)$, $0 \leq t \leq t_f$. In the next section we show how this assumption can be relaxed to include the class of controls $\mathbf{u}(\mathbf{x})$ that are continuous and continuously differentiable along $\mathbf{x}(t)$, except across certain switching surfaces in state space.

Consider now neighboring trajectories $\hat{\mathbf{x}}(t)$ that lie in Σ, obtained from solutions to (6.2-1) using the same control $\mathbf{u}(\mathbf{x})$ but with initial conditions that lie in a neighborhood of $\mathbf{x}(0)$, along a smooth but otherwise arbitrary curve $\mathcal{S}$ in Σ that passes through $\mathbf{x}(0)$. Figure 6.2-1 illustrates a Σ surface, which contains the curve $\mathcal{S}$, and also shows the reference trajectory $\mathbf{x}(t)$ and the neighboring trajectories $\hat{\mathbf{x}}(t)$. The initial condition for any of the neighboring trajectories can be written as

$$\hat{\mathbf{x}}(0) = \mathbf{x}(0) + \alpha\delta\mathbf{x}(0, \alpha), \tag{6.2-2}$$

where $\delta\mathbf{x}(0, \alpha)$ is a vector pointing from $\mathbf{x}(0)$ to $\hat{\mathbf{x}}(0)$. Various initial conditions on $\mathcal{S}$ are obtained by varying α. From the definition of a tangent vector it follows that as $\alpha \rightarrow 0$, $\delta\mathbf{x}(0, \alpha) \rightarrow \mathbf{e}(0)$, where $\mathbf{e}(0)$ is a tangent vector to $\mathcal{S}$ and thus to Σ at $\mathbf{x}(0)$, since $\mathcal{S}$ lies in Σ.

We can now obtain a set of conditions that describe the transformation of a tangent plane to Σ at $\mathbf{x}(0)$ to a tangent plane to Σ at any other point along the reference trajectory $\mathbf{x}(t)$. Let the solution for a neighboring trajectory in Σ be

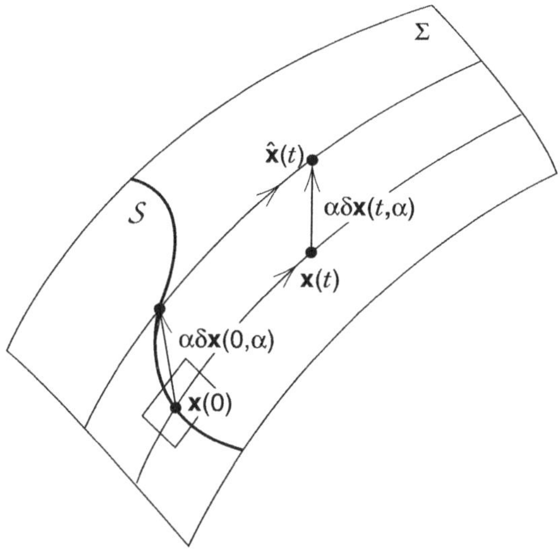

Figure 6.2-1 Transformation of tangent planes in a trajectory surface.

given by

$$\hat{\mathbf{x}}(t) = \mathbf{x}(t) + \alpha\,\delta\mathbf{x}(t, \alpha). \tag{6.2-3}$$

Using the notation $\delta\dot{\mathbf{x}} \triangleq \partial(\delta\mathbf{x})/\partial t$, the solution $\hat{\mathbf{x}}(t)$ must satisfy the differential equation

$$\dot{\mathbf{x}} + \alpha\,\delta\dot{\mathbf{x}} = \mathbf{f}[\mathbf{x} + \alpha\,\delta\mathbf{x}, \mathbf{u}(\mathbf{x} + \alpha\,\delta\mathbf{x})] \tag{6.2-4}$$

at each point where $\mathbf{u}(\mathbf{x})$ is continuous, with the initial condition given by (6.2-2). We use Taylor's theorem to expand the right-hand side of (6.2-4) and assume that $\mathbf{u}(\mathbf{x})$ is C^1 (continuous and continuously differentiable) along $\mathbf{x}(t)$; this yields

$$\dot{\mathbf{x}} + \alpha\,\delta\dot{\mathbf{x}} = \mathbf{f}(\mathbf{x}, \mathbf{u}) + \alpha\left[\frac{\partial \mathbf{f}}{\partial \mathbf{x}} + \frac{\partial \mathbf{f}}{\partial \mathbf{u}}\frac{\partial \mathbf{u}}{\partial \mathbf{x}}\right]\delta\mathbf{x} + \mathbf{O}(\alpha^2),$$

where the partial derivatives are evaluated along the reference trajectory and $\mathbf{O}(\alpha^2)$ denotes terms of order α^2 or higher. Using (6.2-1) to cancel $\dot{\mathbf{x}}$ and $\mathbf{f}(\mathbf{x}, \mathbf{u})$ and dividing by α yields

$$\delta\dot{\mathbf{x}} = \left[\frac{\partial \mathbf{f}}{\partial \mathbf{x}} + \frac{\partial \mathbf{f}}{\partial \mathbf{u}}\frac{\partial \mathbf{u}}{\partial \mathbf{x}}\right]\delta\mathbf{x} + \mathbf{O}(\alpha), \tag{6.2-5}$$

where the bracketed terms are evaluated along the reference trajectory $\mathbf{x}(t)$.

Now, both $\mathbf{x}(t)$ and the neighboring solution $\hat{\mathbf{x}}(t) = \mathbf{x}(t)+\alpha\,\delta\mathbf{x}(t,\alpha)$ are in the Σ surface. Thus in the limit as $\alpha \to 0$, $\delta\mathbf{x}(t,\alpha) \to \mathbf{e}(t)$ is a tangent vector to Σ at $\mathbf{x}(t)$. Taking the limit as $\alpha \to 0$ in (6.2-5) we see that the tangent vector $\mathbf{e}(t)$ satisfies the **state-perturbation equations**

$$\dot{\mathbf{e}} = \left[\frac{\partial \mathbf{f}}{\partial \mathbf{x}} + \frac{\partial \mathbf{f}}{\partial \mathbf{u}}\frac{\partial \mathbf{u}}{\partial \mathbf{x}}\right]\mathbf{e}, \tag{6.2-6}$$

with the initial condition

$$\mathbf{e}(0) = \lim_{\alpha \to 0} \delta\mathbf{x}(0,\alpha).$$

The vector $\mathbf{e}(0)$ is tangent to both $\mathcal{S}$ and Σ at $\mathbf{x}(0)$. Since the initial conditions curve $\mathcal{S}$ can be any smooth curve in Σ, we conclude that every tangent vector $\mathbf{e}(0)$ to Σ at $\mathbf{x}(0)$ is transformed by (6.2-6) to a vector $\mathbf{e}(t)$ tangent to Σ at $\mathbf{x}(t)$. Furthermore, since the solution to the linear time-varying differential equations (6.2-6) is of the form

$$\mathbf{e}(t) = \mathbf{\Phi}(t,\tau)\mathbf{e}(\tau) \qquad \text{for all } t \text{ and } \tau, \tag{6.2-7}$$

where $\mathbf{\Phi}(t,\tau)$ is the nonsingular **state transition matrix** for (6.2-6), it follows that (6.2-6) transforms the $(n_x - 1)$-dimensional tangent plane to Σ at $\mathbf{x}(0)$ to an $(n_x - 1)$-dimensional plane at $\mathbf{x}(t)$, namely, the tangent plane to Σ at $\mathbf{x}(t)$.

Consider now an n_x-dimensional column vector $\boldsymbol{\lambda}(t)$ that satisfies the **adjoint equations**

$$\dot{\boldsymbol{\lambda}}^{\mathrm{T}} = -\boldsymbol{\lambda}^{\mathrm{T}}\left[\frac{\partial \mathbf{f}}{\partial \mathbf{x}} + \frac{\partial \mathbf{f}}{\partial \mathbf{u}}\frac{\partial \mathbf{u}}{\partial \mathbf{x}}\right]. \tag{6.2-8}$$

It follows from (6.2-6) and (6.2-8) that

$$\frac{d}{dt}(\boldsymbol{\lambda}^{\mathrm{T}}\mathbf{e}) \equiv 0,$$

so that

$$\boldsymbol{\lambda}^{\mathrm{T}}(t)\mathbf{e}(t) = \text{constant}. \tag{6.2-9}$$

If the initial condition for $\boldsymbol{\lambda}(t)$ is chosen such that $\boldsymbol{\lambda}(0)$ is perpendicular to Σ at $\mathbf{x}(0)$ resulting in a zero constant for (6.2-9), then since this equation applies to every $\mathbf{e}(t)$ in the tangent plane to Σ at $\mathbf{x}(t)$, it follows that $\boldsymbol{\lambda}(t)$ must be perpendicular to Σ at each point along the reference trajectory $\mathbf{x}(t)$. The evolution of a vector $\boldsymbol{\lambda}(t)$ perpendicular to Σ provides a convenient means for describing a vector that is normal to both Σ and the tangent plane to Σ along a trajectory $\mathbf{x}(t)$ in the surface Σ.

We are particularly interested in an outward-directed normal vector. In this regard: let $\mathcal{A}$ be an arcwise connected set in state space, with a boundary $\partial\mathcal{A}$ that has a unique tangent plane at each point in the boundary. A nonzero vector $\mathbf{n} \in R^{n_x}$ is an **outward normal** to $\mathcal{A}$ at a boundary point $\mathbf{x} \in \partial\mathcal{A}$ if the tangent plane to $\partial\mathcal{A}$ consists of all vectors $\mathbf{e}$ satisfying

$$0 = \mathbf{n}^{\mathrm{T}}\mathbf{e}$$

and, in addition, any vector $\mathbf{e}$ satisfying

$$0 < \mathbf{n}^{\mathrm{T}}\mathbf{e}$$

is tangent to the exterior of $\mathcal{A}$.

Jump Conditions Crossing a Switching Surface

In this section the assumption that the closed-loop control $\mathbf{u}(\mathbf{x})$ is continuous and continuously differentiable along $\mathbf{x}(t)$, $0 \le t \le t_f$, is relaxed. We begin by noting that solutions $\boldsymbol{\lambda}(t)$ defined by the adjoint differential equations (6.2-8) are **absolutely continuous** (see Section 1.2), even if the terms in brackets in (6.2-8) are only piecewise continuous along $\mathbf{x}(t)$, that is, even if $\partial\mathbf{u}[\mathbf{x}(t)]/\partial\mathbf{x}$ is only piecewise continuous. However, as we have seen in Section 1.2, discontinuous closed-loop controls $\mathbf{u}(\mathbf{x})$ could possibly violate conditions on the existence, uniqueness, and continuity with respect to initial conditions for solutions of differential equations.

The basic difficulty with discontinuous closed-loop controls $\mathbf{u}(\mathbf{x})$ lies with the representation of neighboring trajectories. For discontinuous feedback controls $\mathbf{u}(\mathbf{x})$ the state-perturbation equations (6.2-6), while still applicable between discontinuities, do not by themselves completely describe the evolution of neighboring trajectories. If a reference trajectory $\mathbf{x}(t)$ crosses a surface of discontinuity in $\mathbf{u}(\mathbf{x})$, then a certain "jump condition" must be imposed on the state-perturbation vectors [Blaquière et al., 1969; Grantham and Lee, 1993]. This condition will not affect the results that we develop for controllability in this chapter or for optimal control problems in Chapter 7, but it will play a role for certain types of differential games that we consider in Chapter 9.

Let $\sigma(\mathbf{x})$ be a continuous and continuously differentiable scalar-valued function, with $\sigma(\mathbf{x}) = 0$ defining an $(n_x - 1)$-dimensional **switching surface** across which $\mathbf{u}(\mathbf{x})$ is discontinuous. As illustrated in Figure 6.2-2, let $\mathbf{x}(t)$, $0 \le t \le t_f$, be a reference trajectory for

$$\dot{\mathbf{x}} = \mathbf{f}(\mathbf{x}) \leftarrow \mathbf{f}[\mathbf{x}, \mathbf{u}(\mathbf{x})] \tag{6.2-10}$$

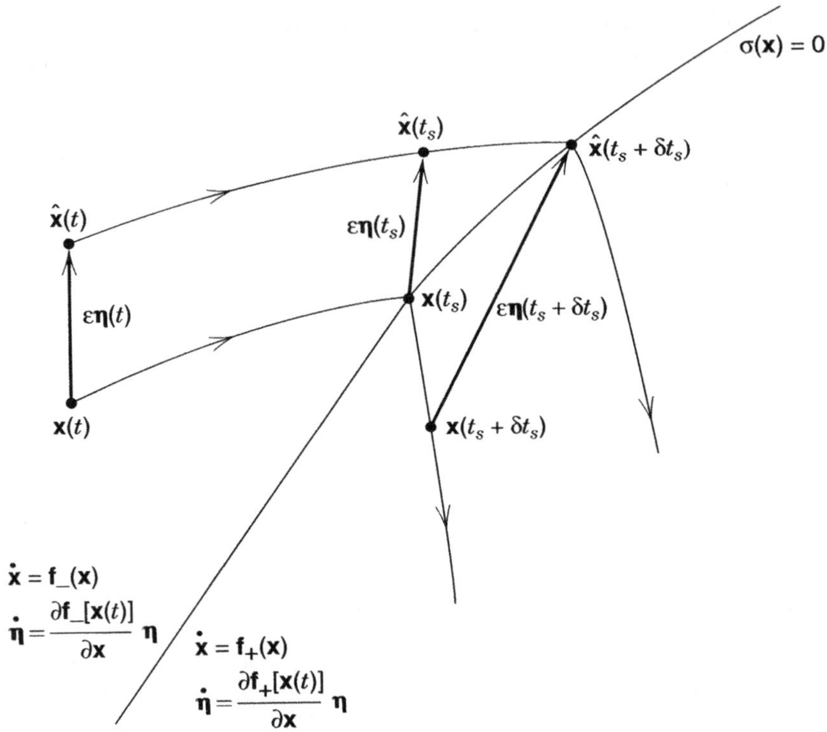

Figure 6.2-2 Geometry of the state-perturbation jump condition.

that crosses $\sigma(\mathbf{x}) = 0$ nontangentially at time t_s. Let $\hat{\mathbf{x}}(t)$ be a neighboring trajectory generated by $\mathbf{u}(\mathbf{x})$, but starting from an initial state near $\mathbf{x}(0)$ given by

$$\hat{\mathbf{x}}(0) = \mathbf{x}(0) + \varepsilon\boldsymbol{\eta}(0),$$

where ε is vanishingly small and $\boldsymbol{\eta}(0)$ is arbitrary. In particular, $\boldsymbol{\eta}(0)$ is not necessarily required to be tangent to the surface Σ discussed earlier in this section.

Off of the switching surface, the perturbed trajectory is given by

$$\hat{\mathbf{x}}(t) = \mathbf{x}(t) + \varepsilon\boldsymbol{\eta}(t) + \mathbf{O}(\varepsilon^2), \tag{6.2-11}$$

where $\boldsymbol{\eta}(t)$ satisfies the **state-perturbation equations**

$$\dot{\boldsymbol{\eta}} = \frac{\partial \mathbf{f}[\mathbf{x}(t)]}{\partial \mathbf{x}}\,\boldsymbol{\eta}. \tag{6.2-12}$$

Under closed-loop control $\mathbf{u}(\mathbf{x})$ the reference and perturbed trajectories will generally switch the control $\mathbf{u}(\mathbf{x})$ at slightly different times, depending on when they reach the switching surface. Let the subscripts "−" and "+" denote conditions just before and after the reference trajectory crosses $\sigma(\mathbf{x}) = 0$, respectively. As illustrated in Figure 6.2-2, the point where the perturbed trajectory reaches $\sigma(\mathbf{x}) = 0$, at time $t_s+\delta t_s$, can be calculated in two ways. Using first-order Taylor series expansions and neglecting higher-order terms in ε and δt_s yields

$$\begin{aligned}\hat{\mathbf{x}}(t_s + \delta t_s) &= \hat{\mathbf{x}}(t_s) + \mathbf{f}_-[\hat{\mathbf{x}}(t_s)]\ \delta t_s \\ &= \mathbf{x}(t_s) + \varepsilon\boldsymbol{\eta}(t_s) + \mathbf{f}_-[\mathbf{x}(t_s)]\ \delta t_s, \\ \hat{\mathbf{x}}(t_s + \delta t_s) &= \mathbf{x}(t_s + \delta t_s) + \varepsilon\boldsymbol{\eta}(t_s + \delta t_s) \\ &= \mathbf{x}(t_s) + \mathbf{f}_+[\mathbf{x}(t_s)]\delta t_s + \varepsilon\boldsymbol{\eta}(t_s + \delta t_s).\end{aligned}$$

Using the first expression and expanding $\sigma[\hat{\mathbf{x}}(t_s + \delta t_s)] = 0$ yields

$$\frac{\delta t_s}{\varepsilon} = \frac{-(\partial\sigma/\partial\mathbf{x})\boldsymbol{\eta}_-}{(\partial\sigma/\partial\mathbf{x})\mathbf{f}_-},$$

and equating the two expressions for $\hat{\mathbf{x}}(t_s + \delta t_s)$ yields the **state-perturbation jump condition**

$$\boldsymbol{\eta}_+ = \boldsymbol{\eta}_- + \{\mathbf{f}_- - \mathbf{f}_+\}\ \frac{\delta t_s}{\varepsilon}, \tag{6.2-13}$$

where the quantities $\mathbf{f}_+(\mathbf{x})$, $\mathbf{f}_-(\mathbf{x})$, and $\partial\sigma(\mathbf{x})/\partial\mathbf{x}$ are evaluated at $\mathbf{x} = \mathbf{x}(t_s)$.

Thus we see that in order to represent neighboring trajectories using (6.2-11) and the state-perturbation equations (6.2-12), the jump condition (6.2-13) must be applied to the perturbation vector $\boldsymbol{\eta}(t)$ whenever the reference trajectory $\mathbf{x}(t)$ crosses a surface of discontinuity for the control $\mathbf{u}(\mathbf{x})$. Ignoring this jump condition may lead to significant errors, for example in the calculation of Lyapunov exponents in chaotic systems [Grantham and Lee, 1993].

To investigate the effect of the jump condition on the relationship between adjoint vectors and state-perturbation vectors, let $\boldsymbol{\lambda}(t)$ be a corresponding solution to the **adjoint equations**

$$\dot{\boldsymbol{\lambda}}^{\mathrm{T}} = -\boldsymbol{\lambda}^{\mathrm{T}}\ \frac{\partial\mathbf{f}[\mathbf{x}(t)]}{\partial\mathbf{x}}. \tag{6.2-14}$$

Because of the jump discontinuity in perturbation vectors $\boldsymbol{\eta}(t)$, the condition $\boldsymbol{\lambda}^{\mathrm{T}}(t)\boldsymbol{\eta}(t) \equiv$ constant, from (6.2-9), does not generally hold across a switching surface, although it does hold off of the switching surface. That is, the "constant" in (6.2-9) also generally undergoes a jump discontinuity across a switching surface. The adjoint vector $\boldsymbol{\lambda}(t)$ is continuous at a switching time t_s, so after crossing a switching surface we have

$$\boldsymbol{\lambda}^{\mathrm{T}}(t)\boldsymbol{\eta}(t) \equiv \boldsymbol{\lambda}^{\mathrm{T}}(t_s)\boldsymbol{\eta}_+ = \boldsymbol{\lambda}^{\mathrm{T}}(t_s)\left\{\boldsymbol{\eta}_- + \{\mathbf{f}_- - \mathbf{f}_+\}\frac{\delta t_s}{\varepsilon}\right\} \tag{6.2-15}$$

for $t > t_s$, or at least until a subsequent switch occurs in $\mathbf{u}(\mathbf{x})$.

If $\boldsymbol{\lambda}(t)$, $0 \le t \le t_f$, is perpendicular to a smooth trajectory surface Σ along $\mathbf{x}(t)$, and if $\boldsymbol{\eta}(t) = \mathbf{e}(t)$ is tangent to Σ, then the condition

$$\boldsymbol{\lambda}^{\mathrm{T}}(t)\mathbf{e}(t) \equiv 0$$

from (6.2-9) still holds, even at a switching surface $\sigma(\mathbf{x}) = 0$. This follows from the fact that at the switching surface we have $\boldsymbol{\lambda}^{\mathrm{T}}\{\mathbf{f}_- - \mathbf{f}_+\} = 0$ in the jump condition (6.2-15), since $\mathbf{x}(t)$ lies in Σ, and thus the velocity vectors $\mathbf{f}_-$ and $\mathbf{f}_+$ are tangent to the trajectory surface Σ, even though they may change direction discontinuously tangent to the surface. Thus for the Σ surface defined in this chapter, the state perturbation jump condition will have no effect. The results we obtained assuming that the closed-loop control $\mathbf{u}(\mathbf{x})$ is continuous and continuously differentiable along $\mathbf{x}(t)$, $0 \le t \le t_f$, remain valid, even if $\mathbf{u}(\mathbf{x})$ is discontinuous as the trajectory crosses a switching surface.

In a more general setting, in which $\mathbf{f}_+$ is not tangent to Σ, $\mathbf{x}(t)$ leaves the surface Σ after crossing the switching surface, $\sigma(\mathbf{x}) = 0$. Then $\boldsymbol{\lambda}^{\mathrm{T}}(t_s)\mathbf{f}_+ \neq 0$ implies $\boldsymbol{\lambda}^{\mathrm{T}}(t)\boldsymbol{\eta}(t) \equiv \boldsymbol{\lambda}^{\mathrm{T}}(t_s)\boldsymbol{\eta}_+ \neq 0$ for $t > t_s$.

Controllability Path Necessary Conditions

We have shown that ∂C must be a semipermeable surface, consisting of trajectories that lie in the surface and that cannot penetrate the surface from the outside of C to the inside of C, nor can trajectories in ∂C cross any "corners" in ∂C, where the tangent plane to ∂C is nonunique. We summarize our results so far as follows:

Controllability Path Conditions: *Let* $\mathbf{u}(\mathbf{x})$ *be an admissible control law that generates a trajectory* $\mathbf{x}(t)$ *satisfying*

$$\dot{\mathbf{x}} = \mathbf{f}[\mathbf{x}, \mathbf{u}(\mathbf{x})], \qquad \mathbf{u}(\mathbf{x}) \in \mathcal{U}, \tag{6.2-16}$$

with $\mathbf{x}(t)$ *lying in the boundary* ∂C *of the controllable set* C *to* X *for some time interval* $0 \le t \le t_f$. *Then there exists a nonzero continuous* n_x*-dimensional vector* $\boldsymbol{\lambda}(t)$ *such that, for all* $t \in [0, t_f]$,

$$0 = \mathcal{H}[\mathbf{x}(t), \mathbf{u}[\mathbf{x}(t)], \boldsymbol{\lambda}(t)] \le \mathcal{H}[\mathbf{x}(t), \mathbf{u}, \boldsymbol{\lambda}(t)] \qquad \text{for all } \mathbf{u} \in \mathcal{U}, \tag{6.2-17}$$

with $\boldsymbol{\lambda}(t)$ *satisfying the adjoint equations*

$$\dot{\boldsymbol{\lambda}}^{\mathrm{T}} = -\frac{\partial \mathcal{H}}{\partial \mathbf{x}} - \frac{\partial \mathcal{H}}{\partial \mathbf{u}} \frac{\partial \mathbf{u}}{\partial \mathbf{x}}, \tag{6.2-18}$$

where the function $\mathcal{H}(\cdot)$ is defined by

$$\mathcal{H}(\mathbf{x}, \mathbf{u}, \boldsymbol{\lambda}) \triangleq \boldsymbol{\lambda}^{\mathrm{T}} \mathbf{f}(\mathbf{x}, \mathbf{u}). \tag{6.2-19}$$

Note that (6.2-8) and (6.2-18) are equivalent. Figure 6.2-3 illustrates the geometry of the controllability path conditions.

To establish the controllability path conditions, assume that the boundary of the controllable set has a unique tangent plane at each point $\mathbf{x} \in \partial C$ and let $\boldsymbol{\lambda}(0)$ denote an outward normal to ∂C at $\mathbf{x}(0) \in \partial C$. As we have shown previously, the control $\mathbf{u}(\mathbf{x})$ and the state-perturbation equations (6.2-6) transform the tangent plane to ∂C at $\mathbf{x}(0)$ to a plane tangent to ∂C at each point $\mathbf{x}(t)$ along a trajectory in ∂C. Thus it follows from (6.2-9) that a vector $\boldsymbol{\lambda}(t)$ satisfying the adjoint equations (6.2-18) is an outward normal to ∂C at each point along $\mathbf{x}(t)$. In terms of this outward normal, the minimization condition in (6.2-17) follows from the fact that, since no trajectory can go from ∂C to the interior of C, at $\mathbf{x} = \mathbf{x}(t)$ the velocity vectors $\dot{\mathbf{x}} = \mathbf{f}(\mathbf{x}, \mathbf{u})$, $\mathbf{u} \in \mathcal{U}$, cannot point to the interior of C; that is, they cannot produce a negative value for the dot product $\mathcal{H}(\mathbf{x}, \mathbf{u}, \boldsymbol{\lambda}) = \boldsymbol{\lambda}^{\mathrm{T}} \mathbf{f}(\mathbf{x}, \mathbf{u})$. Furthermore, a control $\mathbf{u}(\mathbf{x})$ that generates a trajectory $\mathbf{x}(t)$ in ∂C yields a velocity vector that is tangent to ∂C. Thus $\mathcal{H}[\mathbf{x}(t), \mathbf{u}[\mathbf{x}(t)], \boldsymbol{\lambda}(t)] \equiv 0$ along $\mathbf{x}(t)$. The controllability path conditions can also be obtained under weaker assumptions [Grantham, 1973].

In some cases controllability boundaries may have corners, where the tangent plane to ∂C is not unique. This may happen, for example, at an equilibrium point in ∂C. However, the controllability path conditions still apply, since trajectories in ∂C cannot cross corners and still remain in ∂C. If they could, continuity of

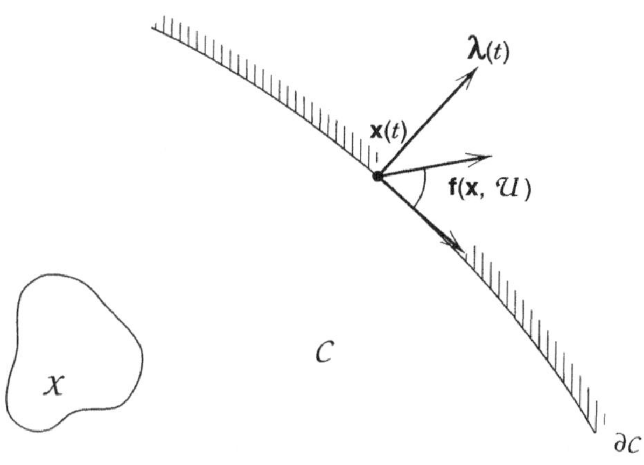

Figure 6.2-3 Geometry of the controllability path conditions.

solutions to (6.2-16) with respect to initial conditions would imply that the same control $\mathbf{u}(\mathbf{x})$ that generates $\mathbf{x}(t) \in \partial\mathcal{C}$ would generate a neighboring trajectory $\hat{\mathbf{x}}(t)$ that penetrates from outside the controllable set to inside the controllable set as $\mathbf{x}(t)$ turns, crossing through the corner. This inability to cross corners in $\partial\mathcal{C}$ is also reflected in the fact that the adjoint vector $\boldsymbol{\lambda}(t)$ is a continuous outward normal to $\partial\mathcal{C}$ along $\mathbf{x}(t)$.

The control $\mathbf{u}(\mathbf{x})$ that minimizes $\mathcal{H}$ may be discontinuous at certain points. However, as we have previously demonstrated, $\dot{\mathbf{x}} = \mathbf{f}[\mathbf{x}, \mathbf{u}(\mathbf{x})]$ is tangent to $\partial\mathcal{C}$ before and after a control discontinuity. Since $\boldsymbol{\lambda}(t)$ is continuous, it follows that $\mathcal{H}_+ = \mathcal{H}_- = 0$ across a switching surface for any trajectory in $\partial\mathcal{C}$.

Example 6.2-1: Kalman's Controllability Condition for Linear Systems In this example we demonstrate that the controllability path conditions can be used to obtain Kalman's controllability criterion for linear systems, previously introduced in Section 2.2. Consider a linear system

$$\dot{\mathbf{x}} = \mathbf{A}\mathbf{x} + \mathbf{B}\mathbf{u}, \tag{6.2-20}$$

with no constraints on the control $\mathbf{u}$. The definition of **complete controllability** for linear systems is that for every initial state and any desired final state, there exists some control law that will accomplish the transfer in finite time. Thus a linear system is either completely controllable or it has a controllability boundary separating some initial states from the target state.

Suppose a controllability boundary $\partial\mathcal{C}$ exists for this linear system and let $\mathbf{u}(\mathbf{x})$ be a feedback control that generates a trajectory $\mathbf{x}(t)$ in $\partial\mathcal{C}$. The $\mathcal{H}(\cdot)$ function for this problem is given by

$$\mathcal{H}(\mathbf{x}, \mathbf{u}, \boldsymbol{\lambda}) = \boldsymbol{\lambda}^{\mathrm{T}}[\mathbf{A}\mathbf{x} + \mathbf{B}\mathbf{u}]. \tag{6.2-21}$$

The necessary conditions for minimizing $\mathcal{H}[\mathbf{x}(t), \mathbf{u}, \boldsymbol{\lambda}(t)]$ are that

$$\boldsymbol{\sigma}(t) \triangleq \left[\frac{\partial\mathcal{H}}{\partial\mathbf{u}}\right]^{\mathrm{T}} = \mathbf{B}^{\mathrm{T}}\boldsymbol{\lambda} \equiv \mathbf{0}, \tag{6.2-22}$$

so that the adjoint equations (6.2-18) reduce to

$$\dot{\boldsymbol{\lambda}}^{\mathrm{T}} = -\frac{\partial\mathcal{H}}{\partial\mathbf{x}} = -\boldsymbol{\lambda}^{\mathrm{T}}\mathbf{A}. \tag{6.2-23}$$

Then from (6.2-22) and the adjoint equations (6-2.23), using the notation $\boldsymbol{\sigma}^{(n)} \triangleq d^n(\boldsymbol{\sigma})/dt^n$, we have

$$
\begin{aligned}
\mathbf{0} &= \boldsymbol{\sigma}(t) = \mathbf{B}^{\mathrm{T}}\boldsymbol{\lambda} \\
\mathbf{0} &= \dot{\boldsymbol{\sigma}}(t) = -\mathbf{B}^{\mathrm{T}}\mathbf{A}^{\mathrm{T}}\boldsymbol{\lambda} = -[\mathbf{AB}]^{\mathrm{T}}\boldsymbol{\lambda} \\
&\;\;\vdots \\
\mathbf{0} &= \boldsymbol{\sigma}^{(n_x-1)}(t) = (-1)^{n_x-1}[\mathbf{A}^{n_x-1}\mathbf{B}]^{\mathrm{T}}\boldsymbol{\lambda},
\end{aligned}
\tag{6.2-24}
$$

which can be written equivalently as

$$
\mathbf{0}^{\mathrm{T}} = \boldsymbol{\lambda}^{\mathrm{T}}\mathcal{P}, \tag{6.2-25}
$$

where $\mathcal{P}$ is the $n_x \times n_x n_u$ **controllability matrix**

$$
\mathcal{P} \triangleq [\mathbf{B}, \mathbf{AB}, \ \ldots, \ \mathbf{A}^{n_x}\mathbf{B}]. \tag{6.2-26}
$$

In order for a controllability boundary to exist, (6.2-25) must have a nonzero solution $\boldsymbol{\lambda}$, which requires rank$[\mathcal{P}] < n_x$. That is, the rows of $\mathcal{P}$ would have to be linearly dependent. Conversely, if rank$[\mathcal{P}] = n_x$, then (6.2-25) has only the solution $\boldsymbol{\lambda} = \mathbf{0}$, which does not satisfy the controllability path condition that $\boldsymbol{\lambda}$ be nonzero. Thus if **Kalman's controllability condition** for linear systems

$$
\mathrm{rank}[\mathcal{P}] = n_x \tag{6.2-27}
$$

is satisfied, then a controllability boundary does not exist. The controllability path conditions and Kalman's controllability criterion yield the same result: the linear system (6.2-20) is "completely controllable" if (and only if) rank$[\mathcal{P}] = n_x$. ■

Control Constraints

Generally, control constraints could be state-dependent of the form

$$
\mathbf{u} \in \mathcal{U}(\mathbf{x}) = \{\mathbf{u} | \mathbf{h}(\mathbf{x}, \mathbf{u}) \geq 0\}, \tag{6.2-28}
$$

where the elements of $\mathbf{h} = [h_1 \ \cdots \ h_{n_h}]^{\mathrm{T}}$ depend explicitly on the state $\mathbf{x}$. For applications involving such constraints, the controllability path conditions may be difficult to implement. The problem lies with the second term in the adjoint equations (6.2-18). If $\partial \mathcal{H}/\partial \mathbf{u} \neq \mathbf{0}$, then $\partial \mathbf{u}(\mathbf{x})/\partial \mathbf{x}$ must be evaluated in order to solve the adjoint equations. This is a severe limitation, since we do not have a priori knowledge of the function $\mathbf{u}(\cdot)$. Fortunately this difficulty can be overcome for a very large class of problems. As we show below, if the control constraints are state-independent of the form

$$
\mathbf{u} \in \mathcal{U} = \{\mathbf{u} | \mathbf{h}(\mathbf{u}) \geq \mathbf{0}\}, \tag{6.2-29}
$$

where $\partial \mathbf{h}/\partial \mathbf{x} = \mathbf{0}^{\mathrm{T}}$, the second term in (6.2-18) is identically zero.

From Section 3.3, with $G(\mathbf{u}) \triangleq \mathcal{H}(\mathbf{x}, \mathbf{u}, \boldsymbol{\lambda})$, the necessary conditions for minimizing $\mathcal{H}(\mathbf{x}, \mathbf{u}, \boldsymbol{\lambda})$ with respect to $\mathbf{u}$ at $\mathbf{u} = \mathbf{u}(\mathbf{x})$ are given by

$$\mathbf{0}^{\mathrm{T}} = \frac{\partial \mathcal{H}}{\partial \mathbf{u}} - \boldsymbol{\gamma}^{\mathrm{T}} \frac{\partial \mathbf{h}}{\partial \mathbf{u}}, \tag{6.2-30}$$

where $\boldsymbol{\gamma} = [\gamma_1 \quad \cdots \quad \gamma_{n_h}]^{\mathrm{T}}$ is a vector of Lagrange multipliers, with

$$\begin{aligned} \gamma_i &\geq 0 \qquad \text{if } h_i[\mathbf{x}, \mathbf{u}(\mathbf{x})] = 0 \\ \gamma_i &= 0 \qquad \text{if } h_i[\mathbf{x}, \mathbf{u}(\mathbf{x})] > 0. \end{aligned} \tag{6.2-31}$$

For this $\boldsymbol{\gamma}$, consider the function $\psi(\mathbf{x}, \mathbf{u}) \triangleq \boldsymbol{\gamma}^{\mathrm{T}}\mathbf{h}(\mathbf{x}, \mathbf{u})$. From (6.2-31) we have $\boldsymbol{\gamma}^{\mathrm{T}}\mathbf{h}[\mathbf{x}, \mathbf{u}(\mathbf{x})] = 0$. Now, suppose $\mathbf{x}$ is interior to any state-constraint region (see Section 7.7), so that we can make an arbitrary small change in $\mathbf{x}$, to $\mathbf{x} + \Delta\mathbf{x}$, without violating the constraints (6.2-28). Thus, $\mathbf{h}[\mathbf{x} + \Delta\mathbf{x}, \mathbf{u}(\mathbf{x} + \Delta\mathbf{x})] \geq \mathbf{0}$ for arbitrary small $\Delta\mathbf{x}$. Note that this move may or may not change the set of active constraints. In either case, since $\boldsymbol{\gamma} \geq \mathbf{0}$, we have $\psi[\mathbf{x}+\Delta\mathbf{x}, \mathbf{u}(\mathbf{x}+\Delta\mathbf{x})] \geq 0$. Thus $\psi(\mathbf{x}, \mathbf{u})$ has a stationary minimum at $\mathbf{u} = \mathbf{u}(\mathbf{x})$ [Blaquière and Leitmann, 1969, pp. 49–52], which yields

$$\boldsymbol{\gamma}^{\mathrm{T}} \left[\frac{\partial \mathbf{h}}{\partial \mathbf{x}} + \frac{\partial \mathbf{h}}{\partial \mathbf{u}} \frac{\partial \mathbf{u}}{\partial \mathbf{x}} \right] = \mathbf{0}^{\mathrm{T}}. \tag{6.2-32}$$

Using (6.2-30) and (6.2-32), the second term in the adjoint equations (6.2-18) is given by

$$\frac{\partial \mathcal{H}}{\partial \mathbf{u}} \frac{\partial \mathbf{u}}{\partial \mathbf{x}} = \boldsymbol{\gamma}^{\mathrm{T}} \frac{\partial \mathbf{h}}{\partial \mathbf{u}} \frac{\partial \mathbf{u}}{\partial \mathbf{x}} = \boldsymbol{\gamma}^{\mathrm{T}} \frac{\partial \mathbf{h}}{\partial \mathbf{x}}. \tag{6.2-33}$$

Thus for the state-independent control constraints (6.2-29), we have

$$\frac{\partial \mathcal{H}}{\partial \mathbf{u}} \frac{\partial \mathbf{u}}{\partial \mathbf{x}} = \mathbf{0}^{\mathrm{T}}$$

and the **adjoint equations** reduce to

$$\dot{\boldsymbol{\lambda}}^{\mathrm{T}} = -\frac{\partial \mathcal{H}}{\partial \mathbf{x}}. \tag{6.2-34}$$

For those situations where the control constraints do depend explicitly on the state, see Leitmann [1981a, pp. 167–174]. We also consider this situation again in Section 7.7 when we discuss optimal control problems with state constraints.

Target Conditions

According to the controllability path conditions, the control $\mathbf{u}$ is calculated from (6.2-17) as a function $\mathbf{u}[\mathbf{x}(t), \boldsymbol{\lambda}(t)]$ of the current state and adjoint vectors, obtained by simultaneously integrating the state and adjoint differential equations. Therefore, boundary conditions are needed for both $\mathbf{x}(t)$ and $\boldsymbol{\lambda}(t)$. In particular, we need to find a point $\mathbf{x}(0)$ in the boundary of the controllable set and a corresponding outward normal $\boldsymbol{\lambda}(0)$ to the controllability boundary at the point $\mathbf{x}(0)$. If the boundary of the controllable set intersects the target set, then there exists a relationship between $\boldsymbol{\lambda}$ and the target at the point of intersection that determines $\boldsymbol{\lambda}$ at that point

Suppose that a trajectory $\mathbf{x}(t)$ in $\partial\mathcal{C}$ reaches the target set $\mathcal{X}$, at a point $\mathbf{x}(t_f) = \mathbf{x}_f$ where $\partial\mathcal{C}$ intersects the boundary $\partial\mathcal{X}$ of the target. Then at $\mathbf{x}_f$ the following condition on $\boldsymbol{\lambda}_f \triangleq \boldsymbol{\lambda}(t_f)$ must hold:

$$\boldsymbol{\lambda}_f^{\mathrm{T}}\mathbf{e} \leq 0 \qquad \text{for all } \mathbf{e} \text{ tangent to } \mathcal{X}. \tag{6.2-35}$$

The geometry of condition (6.2-35) is illustrated in Figure 6.2-4. To establish this result, choose a time interval $[0, t_f]$ so that we do not have to consider switches in $\mathbf{u}[\mathbf{x}(t)]$ and suppose there were a vector $\mathbf{e}_f$ tangent to $\mathcal{X}$ at $\mathbf{x}_f$, but with $\boldsymbol{\lambda}_f^{\mathrm{T}}\mathbf{e}_f > 0$. The same control $\mathbf{u}(\mathbf{x})$ that generates $\mathbf{x}(t)$ would create a neighboring trajectory of the form

$$\hat{\mathbf{x}}(t) = \mathbf{x}(t) + \alpha\mathbf{e}(t) + \mathbf{O}(\alpha^2), \tag{6.2-36}$$

where $\mathbf{e}(t)$ is the solution to the state-perturbation equations (6.2-6) with $\mathbf{e}(t_f) = \mathbf{e}_f$ and α is an arbitrarily small positive number. Since $\boldsymbol{\lambda}(t)$ satisfies the adjoint equations, given by (6.2-8), and we have assumed that $\boldsymbol{\lambda}^{\mathrm{T}}(t_f)\mathbf{e}(t_f) > 0$, it follows from condition (6.2-9) that $\boldsymbol{\lambda}^{\mathrm{T}}(t)\mathbf{e}(t) = \text{constant}$. Hence $\boldsymbol{\lambda}^{\mathrm{T}}(0)\mathbf{e}(0)$

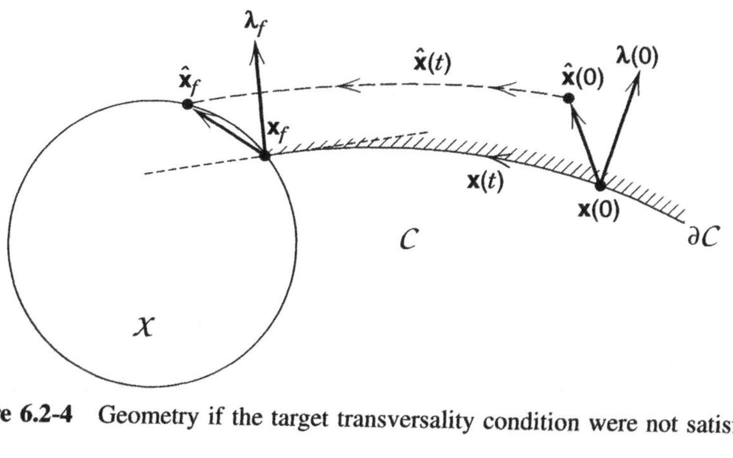

Figure 6.2-4 Geometry if the target transversality condition were not satisfied.

$= \boldsymbol{\lambda}^{\mathrm{T}}(t_f)\mathbf{e}(t_f) > 0$. Since $\boldsymbol{\lambda}(t)$ is an outward normal to the boundary of the controllable set, we see that $\mathbf{e}(0)$ would have to be tangent to the exterior of the controllable set. Thus the neighboring trajectory $\hat{\mathbf{x}}(t)$ would start from a point $\mathbf{x}(0)$ outside the controllable set and reach a point $\mathbf{x}(t_f)$ in the target set, which would contradict the definition of the controllable set.

In order to implement (6.2-35), assume that the **target** is given by a set of n_g equality conditions of the form

$$\mathcal{X} = \{\mathbf{x} \in R^{n_x} \,|\, \mathbf{g}(\mathbf{x}) = \mathbf{0}\}, \tag{6.2-37}$$

where $\mathbf{g} = [g_1 \ \cdots \ g_{n_g}]^{\mathrm{T}}$ is a continuous n_g-dimensional vector function of the state $\mathbf{x}$, with continuous partial derivatives with respect to $\mathbf{x}$. Also assume that the target gradient vectors $\partial g_i/\partial \mathbf{x}$, $i = 1, \ldots, n_g$, are linearly independent on $\mathcal{X}$. For target sets of the form (6.2-37) we have $\partial\mathcal{X} = \mathcal{X}$ and every point of $\mathcal{X}$ is a **regular** point (see Section 3.2). That is, the tangent set to $\mathcal{X}$ at a point $\mathbf{x} \in \mathcal{X}$ is given by

$$\mathcal{T} = \left\{\mathbf{e} \in R^{n_x} \,\middle|\, \frac{\partial \mathbf{g}(\mathbf{x})}{\partial \mathbf{x}}\,\mathbf{e} = \mathbf{0}\right\}. \tag{6.2-38}$$

In terms of the tangent set $\mathcal{T}$, the condition (6.2-35) may be restated as: $\boldsymbol{\lambda}_f^{\mathrm{T}}\mathbf{e} \le 0$ for all $\mathbf{e} \in \mathcal{T}$. Thus the system of inequalities

$$\boldsymbol{\lambda}^{\mathrm{T}}\mathbf{e} > 0 \tag{6.2-39}$$

$$\frac{\partial \mathbf{g}(\mathbf{x}_f)}{\partial \mathbf{x}}\,\mathbf{e} = \mathbf{0} \tag{6.2-40}$$

must *not* have a solution $\mathbf{e}$. From the **theorem of the alternative** [Vincent and Grantham, 1981, p. 22], it follows that the row vector $\boldsymbol{\lambda}^{\mathrm{T}}$ must be a linear combination of the rows of $\partial \mathbf{g}(\mathbf{x}_f)/\partial \mathbf{x}$.

In summary, if a trajectory in the boundary of the controllable set reaches the target set, then there exists a nonzero vector $\boldsymbol{\rho} = [\rho_1 \ \cdots \ \rho_{n_g}]^{\mathrm{T}}$ such that the following **target transversality conditions** hold:

$$\mathbf{0} = \mathbf{g}(\mathbf{x}_f) \tag{6.2-41}$$

$$\boldsymbol{\lambda}_f^{\mathrm{T}} = \boldsymbol{\rho}^{\mathrm{T}}\,\frac{\partial \mathbf{g}(\mathbf{x}_f)}{\partial \mathbf{x}} = \sum_{j=1}^{n_g} \rho_j\,\frac{\partial g_j(\mathbf{x}_f)}{\partial \mathbf{x}}. \tag{6.2-42}$$

The name comes from the fact that, if the boundary of the target has an $(n_x - 1)$-dimensional tangent plane, then the controllability boundary and the target boundary must be tangent at points where they meet, and $\boldsymbol{\lambda}_f$ must be normal or "transverse" to $\partial\mathcal{X}$.

A Controllability Minimum Principle

Our results for determining trajectories in the boundary of the controllable may now be summarized.

Controllability Minimum Principle: *Let* $\mathbf{u}(\mathbf{x})$ *be an admissible control law that generates a trajectory* $\mathbf{x}(t)$ *satisfying*

$$\dot{\mathbf{x}} = \mathbf{f}[\mathbf{x}, \mathbf{u}(\mathbf{x})], \qquad \mathbf{u}(\mathbf{x}) \in \mathcal{U}, \tag{6.2-43}$$

with $\mathbf{x}(t)$ *lying in the boundary* ∂C *of the controllable set* C to X *for some time interval* $0 \leq t \leq t_f$. *Then there exists a nonzero continuous* n_x*-dimensional outward normal vector* $\boldsymbol{\lambda}(t)$ *to* ∂C *such that, for all* $t \in [0, t_f]$,

$$0 = \mathcal{H}[\mathbf{x}(t), \mathbf{u}[\mathbf{x}(t)], \boldsymbol{\lambda}(t)] = \min_{\mathbf{u} \in \mathcal{U}} \mathcal{H}[\mathbf{x}(t), \mathbf{u}, \boldsymbol{\lambda}(t)], \tag{6.2-44}$$

with $\boldsymbol{\lambda}(t)$ *satisfying the adjoint equations*

$$\begin{aligned} \dot{\boldsymbol{\lambda}}^{\mathrm{T}} &= -\frac{\partial \mathcal{H}}{\partial \mathbf{x}} - \frac{\partial \mathcal{H}}{\partial \mathbf{u}} \frac{\partial \mathbf{u}}{\partial \mathbf{x}} \\ &= -\frac{\partial \mathcal{H}}{\partial \mathbf{x}} \qquad \text{for } \mathcal{U} = \{\mathbf{u} | \mathbf{h}(\mathbf{u}) \geq \mathbf{0}\}, \end{aligned} \tag{6.2-45}$$

where the function $\mathcal{H}(\cdot)$ is defined by

$$\mathcal{H}(\mathbf{x}, \mathbf{u}, \boldsymbol{\lambda}) \triangleq \boldsymbol{\lambda}^{\mathrm{T}} \mathbf{f}(\mathbf{x}, \mathbf{u}). \tag{6.2-46}$$

If $\mathbf{x}(t)$ *reaches the target* $X = \{\mathbf{x} \in R^{n_x} | \mathbf{g}(\mathbf{x}) = \mathbf{0}\}$ *at time* t_f, *with* $\mathbf{x}_f = \mathbf{x}(t_f)$ *and* $\boldsymbol{\lambda}_f = \boldsymbol{\lambda}(t_f)$, *then there exists a nonzero vector* $\boldsymbol{\rho} = [\rho_1 \ \cdots \ \rho_{n_g}]^{\mathrm{T}}$ *such that the following target transversality conditions hold*:

$$\mathbf{0} = \mathbf{g}(\mathbf{x}_f) \tag{6.2-47}$$

$$\boldsymbol{\lambda}_f^{\mathrm{T}} = \boldsymbol{\rho}^{\mathrm{T}} \frac{\partial \mathbf{g}(\mathbf{x}_f)}{\partial \mathbf{x}}. \tag{6.2-48}$$

These terminal conditions also apply at the boundary of targets defined by inequalities, $X = \{\mathbf{x} \in R^{n_x} | \mathbf{g}(\mathbf{x}) \leq \mathbf{0}\}$, *with the added information that* $\boldsymbol{\rho} \geq \mathbf{0}$, *with* $\rho_j = 0$ *if* $g_j(\mathbf{x}_f) < 0$ *at* $\mathbf{x}_f \in \partial X$.

Example 6.2-2: Zermelo's Navigation Problem Consider again Zermelo's problem (Example 2.2-6), of a swimmer trying to reach a circular island in a wide river. The equations of motion are

$$\dot{x}_1 = -2 + \cos u$$
$$\dot{x}_2 = \sin u, \tag{6.2-49}$$

where **x** is the swimmer's position and u is his or her heading angle relative to the positive x_1-axis. The target is a unit circle at the origin, defined by

$$g(\mathbf{x}) = x_1^2 + x_2^2 - 1 = 0. \tag{6.2-50}$$

Ignoring, for the moment, our previous experience with this system, it is clear that a controllability boundary must exist, since $\dot{x}_1 < 0$ implies that the swimmer cannot swim in the positive x_1-direction, upstream against the flow of the river. We now use the controllability minimum principle and the target transversality condition to find a control that generates a trajectory in the controllability boundary.

The $\mathcal{H}$ function is given by

$$\mathcal{H} = \lambda_1(-2 + \cos u) + \lambda_2 \sin u \tag{6.2-51}$$

and the adjoint equations (6.2-45) take on a particularly simple form, given by

$$\dot{\lambda}_1 = -\frac{\partial \mathcal{H}}{\partial x_1} = 0$$

$$\dot{\lambda}_2 = -\frac{\partial \mathcal{H}}{\partial x_2} = 0.$$

Therefore

$$\boldsymbol{\lambda}(t) = \boldsymbol{\lambda}_f = \text{constant}. \tag{6.2-52}$$

The control that minimizes $\mathcal{H}$ must satisfy the necessary condition

$$0 = \frac{\partial \mathcal{H}}{\partial u} = -\lambda_1 \sin u + \lambda_2 \cos u,$$

which implies

$$\tan u = \frac{\lambda_2}{\lambda_1}.$$

To resolve the quadrant in which the swimmer's heading angle u lies, write this result more specifically as

$$\cos u = \beta \frac{\lambda_1}{\sqrt{\lambda_1^2 + \lambda_2^2}}, \qquad \sin u = \beta \frac{\lambda_2}{\sqrt{\lambda_1^2 + \lambda_2^2}}, \tag{6.2-53}$$

where β is either +1 or −1. To determine β, substitute (6.2-53) into the $\mathcal{H}$ function and simplify the result to get

$$\mathcal{H} = -2\lambda_1 + \beta \sqrt{\lambda_1^2 + \lambda_2^2}. \tag{6.2-54}$$

Thus to minimize $\mathcal{H}$ we use $\beta = -1$ and the control is given by

$$u = \tan^{-1} \left[\frac{-\lambda_2}{-\lambda_1} \right] = \text{constant}, \tag{6.2-55}$$

where the $\tan^{-1}(\cdot)$ is the two-argument arctangent [as in the Fortran ATAN2$(\cdot, \cdot)$ function].

The boundary conditions on $\mathbf{x}(t_f) = \mathbf{x}_f$ and $\boldsymbol{\lambda}(t_f) = \boldsymbol{\lambda}_f$ at a point where the controllability boundary meets the target are given by the target transversality conditions

$$0 = g(\mathbf{x}_f) = x_{1f}^2 + x_{2f}^2 - 1 \tag{6.2-56}$$

$$\lambda_{1f} = \rho \frac{\partial g(\mathbf{x}_f)}{\partial x_1} = 2\rho x_{1f} \tag{6.2-57}$$

$$\lambda_{2f} = \rho \frac{\partial g(\mathbf{x}_f)}{\partial x_2} = 2\rho x_{2f}. \tag{6.2-58}$$

Using these results to evaluate the $\mathcal{H}$ function (6.2-54) at the terminal point yields

$$0 = \mathcal{H}_f = -4\rho x_{1f} - 2|\rho|.$$

Therefore we have

$$x_{1f} = -\frac{1}{2}\,\text{sgn}(\rho)$$

$$x_{2f} = \pm\sqrt{1 - x_{1f}^2} = \pm\frac{\sqrt{3}}{2}$$

$$\lambda_{1f} = -|\rho|$$

$$\lambda_{2f} = \pm\rho\sqrt{3}.$$

Because the $\mathcal{H}$ function for the controllability minimum principle and the adjoint equations are **homogeneous** in $\boldsymbol{\lambda}$ [i.e., $\mathcal{H}(\mathbf{x}, \mathbf{u}, k\boldsymbol{\lambda}) = k\mathcal{H}(\mathbf{x}, \mathbf{u}, \boldsymbol{\lambda})$], we can choose any convenient positive scale factor for the adjoint vector $\boldsymbol{\lambda}$ (e.g., choose $|\rho| = 1$). Since we want $\boldsymbol{\lambda}$ to point outside of the controllable set, we choose $\rho = +1$. This yields the following two solutions for the terminal conditions:

$$\mathbf{x}_f = \begin{bmatrix} -\frac{1}{2} \\ \frac{\sqrt{3}}{2} \end{bmatrix}, \qquad \boldsymbol{\lambda}_f = \begin{bmatrix} -1 \\ \sqrt{3} \end{bmatrix}, \qquad u = -\frac{\pi}{3} \tag{6.2-59}$$

$$\mathbf{x}_f = \begin{bmatrix} -\frac{1}{2} \\ -\frac{\sqrt{3}}{2} \end{bmatrix}, \qquad \boldsymbol{\lambda}_f = \begin{bmatrix} -1 \\ -\sqrt{3} \end{bmatrix}, \qquad u = \frac{\pi}{3}. \tag{6.2-60}$$

For a constant control u the equations of motion (6.2-49) can be transformed to

$$\frac{dx_2}{dx_1} = \frac{\sin u}{-2 + \cos u}$$

and integrated, to yield

$$x_2 - x_{2f} = \frac{\sin u}{-2 + \cos u}\,(x_1 - x_{1f}). \tag{6.2-61}$$

Substituting from (6.2-59) and (6.2-60) into (6.2-61), we see that the boundary of the controllable set is given, in part, by

$$x_2 = \pm\left[\frac{1}{\sqrt{3}}\left(x_1 + \frac{1}{2}\right) + \frac{\sqrt{3}}{2}\right], \qquad x_1 \geq -\frac{1}{2}, \tag{6.2-62}$$

which corresponds to two straight lines, with $\pm 30°$ slopes, that intersect the target tangentially, as illustrated in Figure 6.2-5.

The remainder of the controllability boundary, not found by the controllability minimum principle, consists of the left-hand portion of the target circle between the two tangency points. Along this circular arc, whose points are in the target X and thus in the controllable set C, the state immediately leaves the controllable set, regardless of the control u. Thus these points are also in the boundary of the controllable set. This is an example of a case where the boundary of the controllable set is not composed entirely of trajectories in ∂C. Nevertheless, we have still been able to construct the most significant portion of the controllability boundary using the controllability minimum principle. ■

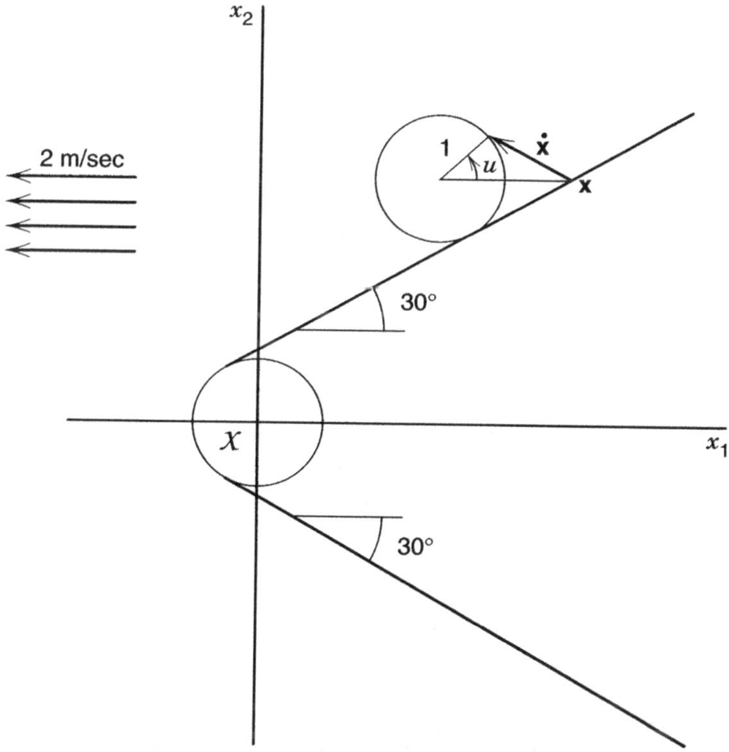

Figure 6.2-5 Controllable set for Zermelo's problem.

A Reachability Maximum Principle

The boundary of the reachable set $\partial\mathcal{R}$ from a state-space initial conditions set $\mathcal{X}$ is a semipermeable surface that trajectories cannot penetrate from the inside to the outside of $\mathcal{R}$. This boundary can be constructed by applying the controllability minimum principle to the retro-time system (6.1-2). However, for reference, we present a direct form of the necessary conditions using the forward-time system.

Reachability Maximum Principle: *Let* $\mathbf{u}(\mathbf{x})$ *be an admissible control law that generates a trajectory* $\mathbf{x}(t)$ *satisfying*

$$\dot{\mathbf{x}} = \mathbf{f}[\mathbf{x}, \mathbf{u}(\mathbf{x})], \qquad \mathbf{u}(\mathbf{x}) \in \mathcal{U}, \tag{6.2-63}$$

with $\mathbf{x}(t)$ *lying in the boundary* $\partial\mathcal{R}$ *of the reachable set* $\mathcal{R}$ *from* $\mathcal{X}$ *for some time interval* $0 \le t \le t_f$. *Then there exists a nonzero continuous* n_x*-dimensional outward normal vector* $\boldsymbol{\lambda}(t)$ *to* $\partial\mathcal{R}$ *such that, for all* $t \in [0, t_f]$,

$$0 = \mathcal{H}[\mathbf{x}(t), \mathbf{u}[\mathbf{x}(t)], \boldsymbol{\lambda}(t)] = \max_{\mathbf{u} \in \mathcal{U}} \mathcal{H}[\mathbf{x}(t), \mathbf{u}, \boldsymbol{\lambda}(t)], \tag{6.2-64}$$

with $\boldsymbol{\lambda}(t)$ *satisfying the adjoint equations*

$$\begin{aligned} \dot{\boldsymbol{\lambda}}^{\mathrm{T}} &= -\frac{\partial \mathcal{H}}{\partial \mathbf{x}} - \frac{\partial \mathcal{H}}{\partial \mathbf{u}} \frac{\partial \mathbf{u}}{\partial \mathbf{x}} \\ &= -\frac{\partial \mathcal{H}}{\partial \mathbf{x}} \qquad \text{for } \mathcal{U} = \{\mathbf{u} | \mathbf{h}(\mathbf{u}) \ge \mathbf{0}\}, \end{aligned} \tag{6.2-65}$$

where the function $\mathcal{H}(\cdot)$ *is defined by*

$$\mathcal{H}(\mathbf{x}, \mathbf{u}, \boldsymbol{\lambda}) \triangleq \boldsymbol{\lambda}^{\mathrm{T}} \mathbf{f}(\mathbf{x}, \mathbf{u}). \tag{6.2-66}$$

If $\mathbf{x}(t)$ *leaves the initial set* $\mathcal{X} = \{\mathbf{x} \in R^{n_x} | \mathbf{g}(\mathbf{x}) = \mathbf{0}\}$ *at time* $t = 0$, *with* $\mathbf{x}_i = \mathbf{x}(0)$ *and* $\boldsymbol{\lambda}_i = \boldsymbol{\lambda}(0)$, *then there exists a nonzero vector* $\boldsymbol{\rho} = [\rho_1 \ \cdots \ \rho_{n_g}]^{\mathrm{T}}$ *such that the following initial transversality conditions hold:*

$$\mathbf{0} = \mathbf{g}(\mathbf{x}_i) \tag{6.2-67}$$

$$\boldsymbol{\lambda}_i^{\mathrm{T}} = \boldsymbol{\rho}^{\mathrm{T}} \frac{\partial \mathbf{g}(\mathbf{x}_i)}{\partial \mathbf{x}}. \tag{6.2-68}$$

These initial conditions also apply at the boundary of initial sets defined by inequalities, $\mathcal{X} = \{\mathbf{x} \in R^{n_x} | \mathbf{g}(\mathbf{x}) \le \mathbf{0}\}$, *with the added information that* $\boldsymbol{\rho} \ge \mathbf{0}$, *with* $\rho_j = 0$ *if* $g_j(\mathbf{x}_i) < 0$ *at* $\mathbf{x}_i \in \partial\mathcal{X}$.

6.3 ADDITIONAL CONTROLLABLE SET RESULTS

Two important questions remain regarding use of the controllability minimum principle. Do boundary trajectories exist for a given problem? If so, how, in general, do we determine initial conditions for the system and adjoint equations? As we see below the first question has an easy answer. The second question is more difficult, and what we offer are a few "tricks" that can be used to answer it.

Existence of Boundary Trajectories

One case where boundary trajectories do exist is where the controllable set is closed [Grantham, 1973]. In this case the boundary of the controllable set intersects the target and the only way the system can be transferred to the target from a point on the boundary is by means of a trajectory that lies in the boundary. Since $\partial\mathcal{C} \cap \mathcal{X}$ is nonempty, the target transversality conditions (6.2-47) and (6.2-48) provide initial conditions for integrating the system and adjoint equations backward in time from the target with the control determined from the controllability minimum principle.

If the controllable set is not closed, we cannot automatically assume boundary trajectories exist. Even if they do, we cannot apply the transversality conditions. Other methods have to be used for finding initial conditions for the system and adjoint equations.

Example 6.3-1: No Trajectory in the Boundary of the Controllable Set Consider the scalar system

$$\dot{x} = x + u, \qquad |u| < 1.$$

The controllable set to the origin is $\mathcal{C} = \{x|\ |x| < 1\}$, with $\partial\mathcal{C}$ being the two points $x = \pm 1$. At these points there are trajectories in the boundary of the controllable set, namely, the equilibrium points corresponding to $u = \mp 1$. But these controls do not satisfy the constraints $|u| < 1$. For any admissible control, the trajectories from $x(0) = \pm 1$ yield $|x(t)| \rightarrow \infty$, which means that there are no trajectories in $\partial\mathcal{C}$ that can be generated by admissible controls. However, if the control constraint was $|u| \leq 1$, then there would be admissible controls that generate (equilibrium) trajectories in the boundary of the controllable set to the origin, $\mathcal{C} = \{x|\ |x| < 1\}$. ■

For most practical systems, however, trajectories do exist in the boundary of the controllable set.

Boundary Trajectory Theorem: *Consider a general system of the form*

$$\dot{\mathbf{x}} = \mathbf{f}(\mathbf{x}, \mathbf{u}), \qquad \mathbf{u} \in \mathcal{U} \subseteq R^{n_u}, \qquad (6.3\text{-}1)$$

where $\mathbf{x} \in R^{n_x}$, $\mathbf{f}(\cdot)$ *is* C^1, *and* $\mathcal{U}$ *is closed. Let* $C \subseteq R^{n_x}$ *be the controllable set to a closed target* $X \subset R^{n_x}$. *At every point* $\mathbf{x} \in \partial C - X$ *there exists an admissible control* $\mathbf{u}(\mathbf{x}) \in \mathcal{U}$ *that generates a trajectory* $\mathbf{x}(t) \in \partial C$ *for a nonzero time interval* $0 \le t \le T$.

We present a brief outline of a proof of this result. Let

$$\mathcal{F}(\mathbf{x}) = \{\dot{\mathbf{x}} | \dot{\mathbf{x}} = \alpha \mathbf{f}(\mathbf{x}, \mathbf{u}_1) + (1 - \alpha)\mathbf{f}(\mathbf{x}, \mathbf{u}_2), \alpha \in [0, 1], \mathbf{u}_1, \mathbf{u}_2 \in \mathcal{U}\} \tag{6.3-2}$$

denote the set of all possible Filippov velocity vectors at $\mathbf{x}$. This convex set is closed since $\mathbf{f}(\cdot)$ is continuous and $\mathcal{U}$ is closed. Thus, unlike Example 6.3-1, limiting velocities can be generated by admissible controls. Now consider a point $\mathbf{x}(0) \in \partial C - X$ and, by way of contradiction, suppose that for every admissible control $\mathbf{u}(\mathbf{x})$ the resulting trajectory $\mathbf{x}(t)$ immediately leaves ∂C, with a nonzero initial velocity $\dot{\mathbf{x}}(0) \in \mathcal{F}[\mathbf{x}(0)]$. From the properties of controllable sets, $\mathbf{x}(t)$ could not go to the interior of C, so it would have to go to the exterior of C. From the definition of a boundary point, every neighborhood of $\mathbf{x}(0) \in \partial C$ contains points $\hat{\mathbf{x}}$ in C. Since $\mathbf{x}(0)$ is exterior to X, we can pick a neighborhood small enough that the points $\hat{\mathbf{x}}$ are also exterior to X. Then by continuity of solutions with respect to initial conditions, every admissible control would generate trajectories $\hat{\mathbf{x}}(t)$ that transfer the state from these neighboring initial points $\hat{\mathbf{x}}(0) = \hat{\mathbf{x}} \in C - X$ to the exterior of C, but no trajectory could return to C, which contradicts $\hat{\mathbf{x}} \in C$ and completes the proof.

An alternate form of an existence result is proven in Gayek and Vincent [1988] for certain two-dimensional systems with a scalar bounded control. As a special case, consider a two-dimensional linear system

$$\dot{\mathbf{x}} = \mathbf{A}\mathbf{x} + \mathbf{b}u, \qquad \mathbf{x} \in R^2, \tag{6.3-3}$$

with a target at the origin and with a bounded scalar control. Suppose that:

(a) The Kalman controllability condition is satisfied, that is, $\text{rank}[\mathbf{b}, \mathbf{Ab}] = 2$;

(b) The eigenvalues of $\mathbf{A}$ have positive real parts; and

(c) $u = 0$ is an interior point of the bounded control constraint set

$$u \in \mathcal{U} = \{u | u_{\min} \le u \le u_{\max}\} \tag{6.3-4}$$

with $-\infty < u_{\min} < 0 < u_{\max} < \infty$.

Then through every point in ∂C there exists a trajectory $\mathbf{x}(t)$ in ∂C. For details of the proof, see Gayek and Vincent [1988].

Open Controllable Sets

If the controllable set is open, there are no transversality conditions to provide initial conditions. In fact, generally, not even one point on ∂C would be known. In certain cases, however, appropriate boundary conditions for open controllable sets can be found by using equilibrium points or the boundary attraction property discussed below.

Example 6.3-2: Linear Systems with Open Controllable Sets Consider a constrained linear system

$$\dot{\mathbf{x}} = \mathbf{Ax} + \mathbf{Bu}, \qquad \mathbf{u} \in \mathcal{U}, \tag{6.3-5}$$

in which the Kalman controllability condition

$$\text{rank}[\mathbf{B}, \mathbf{AB}, \ \ldots, \ \mathbf{A}^{n_x - 1}\mathbf{B}] = n_x \tag{6.3-6}$$

is satisfied and $\mathbf{u} = \mathbf{0}$ is an interior point of the control constraint set, that is,

$$\mathbf{0} \in \overset{\circ}{\mathcal{U}}. \tag{6.3-7}$$

Then the controllable set C to the origin is an open set. In addition, if $\mathcal{U}$ is bounded and any of the eigenvalues of $\mathbf{A}$ have positive real parts, then ∂C is nonempty. Furthermore, if all of the eigenvalues of $\mathbf{A}$ have positive real parts, then ∂C is bounded.

To verify these results, note that in the absence of control constraints (6.3-6) implies that a linear feedback control $\mathbf{u} = -\mathbf{Kx}$ can be designed to make the origin asymptotically stable. When the control constraints are invoked, condition (6.3-7) implies that the corresponding saturating control will be unsaturated sufficiently near the origin. In that region, the system behaves as if there were no control constraints and the Kalman conditions insure the existence of a control such that the state will actually reach the origin in finite time. Thus the origin is an interior point of the controllable set C. Since no trajectory from ∂C ever reaches the interior of C, we conclude that C is an open set.

Suppose one of the eigenvalues $\mu_j = \sigma_j + i\omega_j$ of $\mathbf{A}$ has a positive real part and, for simplicity, suppose the eigenvectors $\{\boldsymbol{\xi}_j\}$ are linearly independent. Then the coordinate transformation $\mathbf{z} = \mathbf{M}^{-1}\mathbf{x}$ transforms (6.3-5) to the decoupled form

$$\dot{z}_j = \mu_j z_j + \hat{\mathbf{b}}_j^{\mathrm{T}} \mathbf{u}, \qquad j = 1, \ \ldots, \ n_x,$$

where $\hat{\mathbf{b}}_j^{\mathrm{T}}$ is the jth row of $\hat{\mathbf{B}} = \mathbf{M}^{-1}\mathbf{B}$. From $|z_j|^2 = z_j^* z_j$, where $(\cdot)^*$ denotes the complex conjugate, we have

$$\frac{d|z_j|^2}{dt} = z_j^* \dot{z}_j + \dot{z}_j^* z_j$$
$$= (\mu_j + \mu_j^*) z_j^* z_j + z_j^* \hat{\mathbf{b}}_j^{\mathrm{T}} \mathbf{u} + \mathbf{u}^{\mathrm{T}} \hat{\mathbf{b}}_j^* z_j$$
$$= 2\sigma_j |z_j|^2 + 2 z_j^* \hat{\mathbf{b}}_j^{\mathrm{T}} \mathbf{u}.$$

If $\mathbf{u}$ is bounded, then the first term dominates the second for sufficiently large $|z_j|$. Then for $\sigma_j = Re(\mu_j) > 0$ and $|z_j(0)|$ sufficiently large, we have $|z_j(t)| \to \infty$. Thus if one or more of the eigenvalues of $\mathbf{A}$ has a positive real part, then for any bounded admissible control $\mathbf{u}(\mathbf{x})$ there is an initial state $\mathbf{x}(0)$ for which $\|\mathbf{x}(t)\| \to \infty$. Thus $\mathbf{x}(0)$ is exterior to the controllable set $\mathcal{C}$ to the origin, which implies $\partial\mathcal{C} \neq \emptyset$.

For the case where all of the eigenvalues of $\mathbf{A}$ have positive real parts, consider the rate of change of a symmetric positive definite function $V(\mathbf{x}) = \mathbf{x}^{\mathrm{T}}\mathbf{P}\mathbf{x}$, given by

$$\dot{V}(\mathbf{x}, \mathbf{u}) = \mathbf{x}^{\mathrm{T}}\mathbf{P}\dot{\mathbf{x}} + \dot{\mathbf{x}}^{\mathrm{T}}\mathbf{P}\mathbf{x}$$
$$= \mathbf{x}^{\mathrm{T}}(\mathbf{P}\mathbf{A} + \mathbf{A}^{\mathrm{T}}\mathbf{P})\mathbf{x} + 2\mathbf{x}^{\mathrm{T}}\mathbf{P}\mathbf{B}\mathbf{u}.$$

If the eigenvalues $\mu = \sigma + i\omega$ of $\mathbf{A}$ all have positive real parts, then, from Lyapunov's lemma (applied to $-\mathbf{A}$ with $\mathbf{Q} = \mathbf{I}$), there exists a positive-definite matrix $\mathbf{P}$ such that $\mathbf{P}\mathbf{A} + \mathbf{A}^{\mathrm{T}}\mathbf{P} = \mathbf{I}$, so that

$$\dot{V}(\mathbf{x}, \mathbf{u}) = \mathbf{x}^{\mathrm{T}}\mathbf{x} + 2\mathbf{x}^{\mathrm{T}}\mathbf{P}\mathbf{B}\mathbf{u}. \tag{6.3-8}$$

Thus for a bounded control $\mathbf{u}(\mathbf{x}) \in \mathcal{U}$ the first term in (6.3-8) dominates the second term as $\|\mathbf{x}\| \to \infty$. Hence for sufficiently large c no trajectory can penetrate the region $V(\mathbf{x}) \leq c$ from outside to inside. Thus the controllable set to the origin is bounded. ■

Equilibrium Points in a Controllability Boundary

One trick for finding a point on $\partial\mathcal{C}$ is available if the controllability minimum principle produces an equilibrium point in the boundary of the controllable set. In this case

$$\dot{\mathbf{x}} = \mathbf{f}(\mathbf{x}, \mathbf{u}) = \mathbf{0} \tag{6.3-9}$$

so that the condition

$$0 = \mathcal{H}(\mathbf{x}, \mathbf{u}, \boldsymbol{\lambda}) = \boldsymbol{\lambda}^{\mathrm{T}}\mathbf{f}(\mathbf{x}, \mathbf{u}) \tag{6.3-10}$$

is satisfied automatically. This information can often be used to find a point in the boundary of the controllable set.

Example 6.3-3: Linearized Inverted Pendulum As in Example 2.2-1, consider the system

$$\begin{aligned} \dot{x}_1 &= x_2 \\ \dot{x}_2 &= x_1 + u, \end{aligned} \qquad (6.3\text{-}11)$$

where the control is constrained by $|u| \le 1$ and the target is the origin. As we learned from Example 6.3-2, since the Kalman controllability condition (6.3-6) is satisfied, the origin is interior to the controllable set and, since the system is not stable, a controllability boundary will exist.

Even though the target transversality conditions do not apply, we can still apply the main part of the controllability minimum principle. For this system the $\mathcal{H}$ function is given by

$$\begin{aligned} \mathcal{H} &= \lambda_1 \dot{x}_1 + \lambda_2 \dot{x}_2 \\ &= \lambda_1 x_2 + \lambda_2 (x_1 + u). \end{aligned}$$

The adjoint equations (6.2-45) yield

$$\begin{aligned} \dot{\lambda}_1 &= -\frac{\partial \mathcal{H}}{\partial x_1} = -\lambda_2 \\ \dot{\lambda}_2 &= -\frac{\partial \mathcal{H}}{\partial x_2} = -\lambda_1. \end{aligned} \qquad (6.3\text{-}12)$$

To determine the control we note that u appears linearly in $\mathcal{H}$, which we can rewrite as

$$\mathcal{H} = \lambda_1 x_2 + \lambda_2 x_1 + \sigma u, \qquad (6.3\text{-}13)$$

where

$$\sigma(\mathbf{x}, \boldsymbol{\lambda}) \triangleq \frac{\partial \mathcal{H}}{\partial u} = \lambda_2 \qquad (6.3\text{-}14)$$

is a **switching function** for the control. If $\sigma \neq 0$, then $\partial \mathcal{H}/\partial u \neq 0$ and u cannot be interior to the constraints $|u| \le 1$. Hence the control that minimizes $\mathcal{H}$ is given by

$$u = -\text{sgn}(\sigma), \tag{6.3-15}$$

which switches between ±1 whenever $\sigma(t) \triangleq \sigma[\mathbf{x}(t), \boldsymbol{\lambda}(t)]$ changes sign.

Whenever the control variable appears linearly in $\mathcal{H}$, an important special case may occur, in which $\sigma(t) \equiv 0$ for a nonzero time interval (instead of $\sigma(t)$ simply passing through zero and changing sign at some switching time). This case is called **singular control,** since the value of the $\mathcal{H}$ function becomes independent of the control. Singular control is discussed in more detail in Chapter 8. For our current example singular control does not occur. To verify this, suppose that $\sigma(t) \equiv 0$. Then $\lambda_2(t) \equiv 0$ and the adjoint equations (6.3-12) imply $\lambda_1(t) \equiv 0$. But $\boldsymbol{\lambda} = \mathbf{0}$ is not allowed by the controllability minimum principle.

The orthogonality condition $\mathcal{H} = 0$ is satisfied if $\mathbf{f}(\mathbf{x}, \mathbf{u}) = \mathbf{0}$, and the corresponding equilibrium trajectory $\mathbf{x}(t) = \hat{\mathbf{x}} =$ constant is a valid candidate for a trajectory in the boundary of the controllable set. Since (6.3-15) implies $u = \pm 1$, the candidate equilibrium points are given by

$$\begin{aligned} \hat{\mathbf{x}} &= [1 \quad 0]^{\mathrm{T}}, & u &= -1 \\ \hat{\mathbf{x}} &= [-1 \quad 0]^{\mathrm{T}}, & u &= 1. \end{aligned} \tag{6.3-16}$$

These equilibrium points are in fact points in the boundary of the controllable set, for this example, but in addition there are also nonequilibrium trajectories in the boundary of the controllable set that approach these two points asymptotically. To find these other trajectories we can examine constant control ($u = 1$ or $u = -1$) solutions to the state equations that approach these equilibrium points asymptotically. The state equations can be written as

$$\frac{dx_2}{dx_1} = \frac{x_1 + u}{x_2},$$

which we rearrange as

$$x_2\, dx_2 = (x_1 + u)dx_1.$$

For $u =$ constant this expression can be integrated, using the corresponding equilibrium conditions in (6.3-16) to evaluate the constant of integration, yielding

$$(x_1 + u)^2 - x_2^2 = 0. \tag{6.3-17}$$

As illustrated in Figure 6.3-1, the constant control trajectories represented by (6.3-17) are hyperbolas and form saddle-point equilibriums at $\hat{\mathbf{x}} = [\mp 1 \quad 0]^{\mathrm{T}}$, corresponding to $u = \pm 1$. Trajectories using the control $u = +1$ are shown dashed and trajectories using $u = -1$ are shown dotted, except for the trajectories that approach these two equilibrium points, which are shown solid. The solid tra-

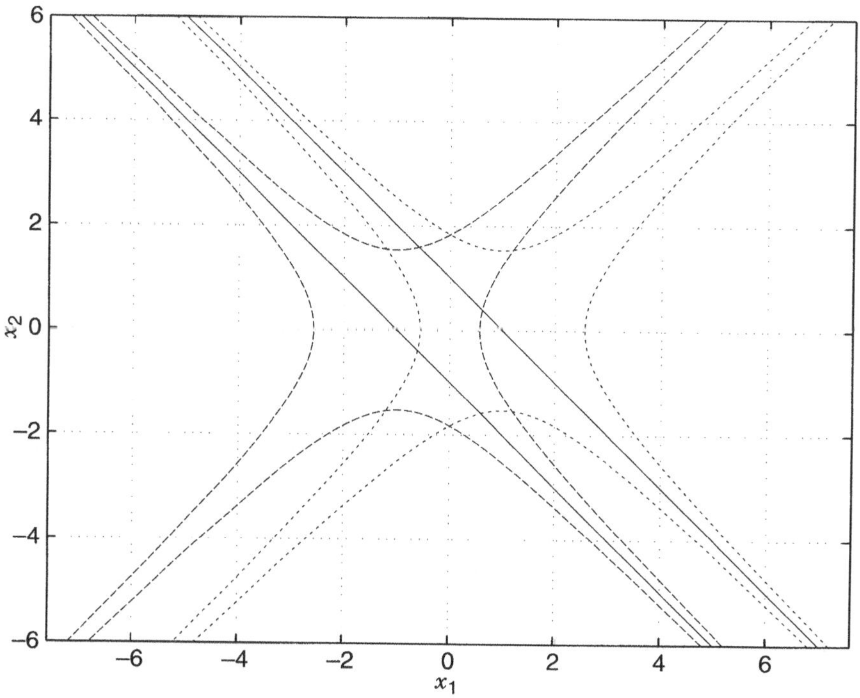

Figure 6.3-1 Controllability boundary for a linearized inverted pendulum.

jectories form the boundaries of the controllable set and are given by the two straight lines

$$x_2 = -x_1 - u, \qquad u = \pm 1.$$

This suggests a procedure for determining the boundary trajectories numerically when we have equilibrium points on ∂C. In this case the boundary trajectories could be obtained numerically by choosing an initial point very close to the equilibrium points and integrating the system equations of motion backward in time (to get a stable trajectory) under appropriate boundary control. We cannot start exactly on an equilibrium point since the resultant trajectory would remain there. For example, if we choose as an initial point $x_1 = 1$, $x_2 = \varepsilon$, where ε is a small positive number, and integrate the system equations backward with $u = -1$, a close approximation to the upper right-hand portion of ∂C is obtained. Choosing ε a small negative number generates a close approximation to the lower right-hand portion of ∂C. A similar procedure can be used to generate a close approximation to the left-hand portion of ∂C. ■

Boundary Attraction

There is another trick we can use that allows us to either find points on the boundary of the controllable set or in some cases to actually approximate the boundary with a trajectory that approaches the boundary. This trick is given to us by an attractive property associated with using boundary control in the neighborhood of the boundary of reachable sets. This property was proven in Gayek and Vincent [1985] for systems with a stable linear part. They showed that the boundary of the reachable set from the origin would be asymptotically stable under open-loop control $\mathbf{u}^*(t)$ obtained from the reachability maximum principle. In particular, for a system

$$\dot{\mathbf{x}} = \mathbf{A}\mathbf{x} + \mathbf{B}\mathbf{u} \tag{6.3-18}$$

with $\mathbf{u} \in \mathcal{U}$, if $\mathcal{U}$ is compact (closed and bounded), if all of the eigenvalues of $\mathbf{A}$ have negative real parts, and if there exists a trajectory $\mathbf{x}^*(t)$ in the boundary $\partial\mathcal{R}$ of the reachable set $\mathcal{R}$ from a point, generated by $\mathbf{u}^*(t) \in \mathcal{U}$, then $\mathbf{x}^*(t)$ is globally asymptotically stable.

We take a different approach here, because we deal with closed-loop controls and, as a consequence, we cannot generally expect asymptotic stability. For example, consider a limit cycle or a reachability boundary in two dimensions. We know that, at least for differentiable closed-loop controls $\mathbf{u}(\mathbf{x})$, no periodic solution could be asymptotically stable for the resulting autonomous system $\dot{\mathbf{x}} = \mathbf{f}[\mathbf{x}, \mathbf{u}(\mathbf{x})]$. In addition, we would also like to examine constrained linear systems in which some eigenvalues have positive real parts and others do not. The results that we present here are new, so we will be precise both in the statement and the proof of our results.

Boundary Attraction Theorem: *Let C be the controllable set to a closed target X for the system*

$$\dot{\mathbf{x}} = \mathbf{f}(\mathbf{x}, \mathbf{u}), \qquad \mathbf{u} \in \mathcal{U}, \tag{6.3-19}$$

where $\mathbf{f}(\cdot)$ is C^1, and let $\mathbf{u}(\mathbf{x})$ be an admissible control that: (a) transfers every state in C to the target X; (b) generates trajectories $\mathbf{x}(t)$, $-\infty < t < \infty$, that are unique forward and backward in time; and (c) generates any trajectories $\mathbf{x}^*(t) \in \partial C$. Suppose the following conditions are satisfied:

(i) *The controllable set C is open, with X in the interior $\mathring{C}$ of C.*

(ii) *$\partial C \neq \varnothing$ is composed of trajectories $\mathbf{x}^*(t)$, $-\infty < t < \infty$, and has a unique unit outward normal $\gamma(t)$ along $\mathbf{x}^*(t)$.*

(iii) *Along $\mathbf{x}^*(t)$*

$$0 < \psi(t) \triangleq \boldsymbol{\gamma}^{\mathrm{T}}(t)[\hat{\mathbf{A}}(t) + \hat{\mathbf{A}}^{\mathrm{T}}(t)]\boldsymbol{\gamma}(t) \tag{6.3-20}$$

where

$$\hat{\mathbf{A}}(t) = \frac{\partial \mathbf{f}}{\partial \mathbf{x}} + \frac{\partial \mathbf{f}}{\partial \mathbf{u}} \frac{\partial \mathbf{u}}{\partial \mathbf{x}} \tag{6.3-21}$$

is evaluated along $\mathbf{x} = \mathbf{x}^*(t), \mathbf{u} = \mathbf{u}[\mathbf{x}^*(t)]$.

Then there exists a neighborhood $\mathcal{N}$ *of* ∂C *such that for every* $\mathbf{x}(0) \in \mathcal{N} \cap \mathring{C}$ *the trajectory* $\mathbf{x}(t)$, $-\infty < t < \infty$, *generated by* $\mathbf{u}(\mathbf{x})$ *approaches* ∂C *as* $t \to -\infty$.

The implications of the boundary attraction theorem are that we do not need to know a precise point $\mathbf{x}^*(0) \in \partial C$; if the conditions (i)–(iii) are satisfied, we can integrate backward in time and a neighboring trajectory $\mathbf{x}(t)$ will approach ∂C.

To establish the boundary attraction theorem, we begin by noting that ∂C does not intersect the target since C is open and X is closed, with $X \subset \mathring{C}$. More specifically, every sufficiently small open neighborhood $\mathcal{N}$ of ∂C has the property that $\mathcal{N} \cap \mathring{C}$ is exterior to X. It follows that no forward-time ($t \geq 0$) trajectory $\mathbf{x}^*(t)$ in ∂C can be stable, either Lyapunov stable or orbitally stable. Otherwise, neighboring trajectories $\mathbf{x}(t)$ starting sufficiently inside $\mathcal{N} \cap \mathring{C}$ as generated by $\mathbf{u}(\mathbf{x})$ would be kept inside $\mathcal{N} \cap \mathring{C}$ for all $t \geq 0$ and therefore never reach X. However, we will show that backward in time, every retro-time trajectory in ∂C is in fact asymptotically orbitally stable.

Let $\mathbf{x}^*(t)$, $-\infty < t < \infty$, be any trajectory in ∂C, generated by $\mathbf{u}(\mathbf{x})$. By hypothesis, ∂C has a nonzero outward normal $\boldsymbol{\lambda}(t)$ along $\mathbf{x}^*(t)$, which satisfies the adjoint equations

$$\dot{\boldsymbol{\lambda}}^{\mathrm{T}} = -\boldsymbol{\lambda}^{\mathrm{T}}\hat{\mathbf{A}}(t), \tag{6.3-22}$$

where $\hat{\mathbf{A}}(t)$ is given by (6.3-21). Choose a vector $\boldsymbol{\eta}(0)$ such that

$$\boldsymbol{\lambda}^{\mathrm{T}}(0)\boldsymbol{\eta}(0) < 0. \tag{6.3-23}$$

Let $\mathbf{x}(t)$, $-\infty < t < \infty$, be a perturbed trajectory generated by the same control $\mathbf{u}(\mathbf{x})$, with

$$\mathbf{x}(0) = \mathbf{x}^*(0) + \alpha\boldsymbol{\eta}(0, \alpha) \in \mathcal{N} \cap \mathring{C}, \tag{6.3-24}$$

where $\boldsymbol{\eta}(0, \alpha) \to \boldsymbol{\eta}(0)$ as $\alpha \to 0$. Then from Taylor's theorem

$$\mathbf{x}(t) = \mathbf{x}^*(t) + \alpha\boldsymbol{\eta}(t) + \mathbf{O}(\alpha^2), \tag{6.3-25}$$

where $\boldsymbol{\eta}(t)$ is the corresponding solution to the state perturbation equations

$$\dot{\boldsymbol{\eta}} = \hat{\mathbf{A}}(t)\boldsymbol{\eta}. \tag{6.3-26}$$

As we have seen previously, $\boldsymbol{\lambda}^{\mathrm{T}}(t)\boldsymbol{\eta}(t) = \text{constant}$, except possibly across switching surfaces for $\mathbf{u}(\mathbf{x})$.

Since the eigenvalues for the adjoint equations are the negative of the eigenvalues for the state perturbation equations, it frequently happens that $\|\boldsymbol{\eta}(t)\| \to 0$ and $\|\boldsymbol{\lambda}(t)\| \to \infty$, or vice versa, while keeping $\boldsymbol{\lambda}^{\mathrm{T}}(t)\boldsymbol{\eta}(t) = \text{constant}$. We avoid this problem by calculating only the direction of $\boldsymbol{\lambda}(t)$. We can scale it to a constant length by defining a **normalized adjoint vector**

$$\gamma(t) = k\,\frac{\boldsymbol{\lambda}(t)}{\|\boldsymbol{\lambda}(t)\|}, \qquad k > 0, \tag{6.3-27}$$

and using the **normalized adjoint equations**

$$\dot{\boldsymbol{\gamma}}^{\mathrm{T}} = -\boldsymbol{\gamma}^{\mathrm{T}}\hat{\mathbf{A}}(t)\left[\mathbf{I} - \frac{\boldsymbol{\gamma}\boldsymbol{\gamma}^{\mathrm{T}}}{\|\boldsymbol{\gamma}\|^2}\right]. \tag{6.3-28}$$

These equations were developed in Lee [1991] and Lee et al. [1994] for long-term optimal control problems involving chaotic systems. In particular, the controllability minimum principle holds for $\mathbf{x}^*(t)$ and $\mathbf{u}(\mathbf{x})$ with $\boldsymbol{\lambda}$ replaced by $\boldsymbol{\gamma}$ in the $\mathcal{H}$ function.

Along $\mathbf{x}^*(t)$ we have $\|\boldsymbol{\gamma}(t)\| \equiv k$ and $\gamma(t)$ can be made a **unit outward normal** by setting $k = 1$. It follows that

$$\begin{aligned}
\frac{d}{dt}(\boldsymbol{\gamma}^{\mathrm{T}}\boldsymbol{\eta}) &= \dot{\boldsymbol{\gamma}}^{\mathrm{T}}\boldsymbol{\eta} + \boldsymbol{\gamma}^{\mathrm{T}}\dot{\boldsymbol{\eta}} \\
&= -\boldsymbol{\gamma}^{\mathrm{T}}\hat{\mathbf{A}}(t)\left[\mathbf{I} - \frac{\boldsymbol{\gamma}\boldsymbol{\gamma}^{\mathrm{T}}}{\|\boldsymbol{\gamma}\|^2}\right]\boldsymbol{\eta} + \boldsymbol{\gamma}^{\mathrm{T}}\hat{\mathbf{A}}(t)\boldsymbol{\eta} \\
&= \frac{1}{\|\boldsymbol{\gamma}\|^2}(\boldsymbol{\gamma}^{\mathrm{T}}\hat{\mathbf{A}}(t)\boldsymbol{\gamma})(\boldsymbol{\gamma}^{\mathrm{T}}\boldsymbol{\eta}) \\
&= \frac{\psi(t)}{2\|\boldsymbol{\gamma}\|^2}\boldsymbol{\gamma}^{\mathrm{T}}\boldsymbol{\eta},
\end{aligned} \tag{6.3-29}$$

where $\psi(t)$ is defined by (6.3-20). Thus

$$\boldsymbol{\gamma}^{\mathrm{T}}(t)\boldsymbol{\eta}(t) = \boldsymbol{\gamma}^{\mathrm{T}}(0)\boldsymbol{\eta}(0)e^{\beta(t)}, \tag{6.3-30}$$

where

$$\beta(t) = \frac{1}{2k^2} \int_0^t \psi(\tau)\, d\tau. \tag{6.3-31}$$

The results in (6.3-29)–(6.3-31) also hold in the case of jump changes in $\boldsymbol{\eta}(t)$ across a switching surface for a discontinuous control $\mathbf{u}(\mathbf{x})$. In particular, from (6.2-15), $\boldsymbol{\gamma}^{\mathrm{T}}\boldsymbol{\eta}_+ = \boldsymbol{\gamma}^{\mathrm{T}}\boldsymbol{\eta}_-$ since $\dot{\mathbf{x}}_+^*$ and $\dot{\mathbf{x}}_-^*$ are orthogonal to $\boldsymbol{\gamma}$.

For $\psi(t) > 0$ we have $\beta(t) < 0$ for $t < 0$ and $\beta(t) \to -\infty$ as $t \to -\infty$, which implies $\boldsymbol{\gamma}^{\mathrm{T}}(t)\boldsymbol{\eta}(t) \to 0$ as $t \to -\infty$; that is, $\boldsymbol{\eta}(t)$ becomes tangent to $\partial\mathcal{C}$ as $t \to -\infty$. Thus, in the limit as $\alpha \to 0$ in (6.3-25), we have $\mathbf{x}(t) \to \partial\mathcal{C}$.

6.4 DETERMINATION OF CONTROLLABLE SETS

In this section we use boundary attraction to obtain a controllable or reachable set for some linear systems under bounded control. As a general rule we find the controllable set for (6.3-18) by constructing the reachable set for the retro-system. This will be done by integrating the retro-system under boundary control in the neighborhood of the boundary. We have previously noted that, if all of the eigenvalues of $\mathbf{A}$ have positive real parts, then we are guaranteed that the reachable set for the retro-system will have attractive boundaries under open-loop control $\mathbf{u}^*(t)$. However, if this condition is not satisfied, the boundary attraction theorem can be used as illustrated in Example 6.4-3.

Two-Dimensional Systems

Example 6.4-1: Constrained Linear Systems with Complex Eigenvalues Consider the problem of determining the boundary of the controllable set to the origin for the system $\dot{\mathbf{x}} = \mathbf{A}\mathbf{x} + \mathbf{B}u$ given by

$$\begin{bmatrix} \dot{x}_1 \\ \dot{x}_2 \end{bmatrix} = \begin{bmatrix} -1 & -1 \\ 3 & 2 \end{bmatrix} \begin{bmatrix} x_1 \\ x_2 \end{bmatrix} + \begin{bmatrix} -1 \\ -1 \end{bmatrix} u$$

with $|u| \le \frac{1}{2}$. In this case the eigenvalues for the $\mathbf{A}$ matrix are $\mu = (1 \pm i\sqrt{3}/2)$. Since these eigenvalues have positive real parts, the reachable set from the origin for the retro-time system will have attractive boundaries.

To find the controllable set boundaries, we first find boundary control using the reachability maximum principle for the retro-time system. In this case,

$$\mathcal{H} = \lambda_1(x_1 + x_2) - \lambda_2(3x_1 + 2x_2) + (\lambda_1 + \lambda_2)u.$$

It follows that $\sigma = \lambda_1 + \lambda_2$ is a switching surface for boundary control. There is no possibility for $\sigma \equiv 0$, since that would imply, using the adjoint equations, that $\lambda_1 = \lambda_2 = 0$, which the reachability maximum principle does not allow. Thus, $u(\mathbf{x}) = \pm\frac{1}{2}$ will be piecewise constant.

At a switch point, $\sigma = 0$, we still have $\mathcal{H} = 0$ and these two conditions together require that

$$4x_1 + 3x_2 = 0$$

in order for there to exist nonzero solutions for λ_1 and λ_2. This equation defines a state-space switching surface for the control u. Let

$$\sigma_s = 4x_1 + 3x_2.$$

Assuming that the reachable set is convex, both σ and σ_s will be positive in the first quadrant, which in turn implies that σ and σ_s must be positive and negative together. It then follows from the reachability maximum principle that boundary control is given by

$$u = \begin{cases} \frac{1}{2} & \text{if } \sigma_s \geq 0 \\ -\frac{1}{2} & \text{if } \sigma_s < 0, \end{cases}$$

where we have resolved what control to use on the switching surface by setting $u = +\frac{1}{2}$ when $\sigma_s = 0$.

Since the reachability boundary is attractive, we need only start the retro-time system "near" this boundary and integrate the retro-time system forward in time to generate a trajectory that will approach the boundary. Since the constant control equilibrium points generated by $|u| = \frac{1}{2}$ will be inside the controllable set to the origin, these represent convenient initial conditions for integrating the retro-time system. In this case $u = +\frac{1}{2}$ will produce the equilibrium point $x_1 = \frac{3}{2}$, $x_2 = -2$, which lies on $\sigma_s = 0$. If we start the retro-time system at this point and integrate forward in time using the above control law, the resulting trajectory will approach a limit cycle. This limit cycle defines the boundary of the controllable set for the original forward system. Construction of this limit cycle is left as Exercise 6-14. ■

Example 6.4-2: Resonance Versus Reachable Set The system

$$\begin{bmatrix} \dot{x}_1 \\ \dot{x}_2 \end{bmatrix} = \begin{bmatrix} 0 & 1 \\ -1 & -1 \end{bmatrix} \begin{bmatrix} x_1 \\ x_2 \end{bmatrix} + \begin{bmatrix} 0 \\ 1 \end{bmatrix} u$$

with $|u| \leq u_{\max}$ corresponds to a damped linear oscillator

$$\ddot{y} + 2\zeta\omega_n\dot{y} + \omega_n^2 = \omega_n^2 u$$

with natural frequency $\omega_n = 1$ and damping ratio $\zeta = \frac{1}{2}$. Driving the system at its "resonant frequency" [Grantham and Vincent, 1993, p. 158],

$$u(t) = u_{\max} \sin \omega t, \qquad \omega = \omega_n\sqrt{1 - 2\zeta^2} = \frac{1}{\sqrt{2}},$$

produces the largest amplitude in y, that is, ω is the most "destabilizing" frequency when using a sine input. The question to be examined here is whether this control input will drive the system to the boundary of the reachable set from the origin.

The eigenvalues for the **A** matrix are $\mu = (-1+i\sqrt{3}/2)$. Since these eigenvalues have negative real parts we know that the reachable set for the above system will have attractive boundaries, and we should be able to find these boundaries as in the previous example by integrating the system equations forward in time under boundary control. The $\mathcal{H}$ function for this example may be written as

$$\mathcal{H} = \lambda_1 x_2 - \lambda_2(x_1 + x_2) + \lambda_2 u.$$

It follows that $\sigma = \lambda_2$ is a switching function for boundary control with no possibility for $\sigma \equiv 0$. Using the same arguments as in the previous example, $\sigma = 0$ and $\mathcal{H} = 0$ imply the state-space switching curve

$$\sigma_s = x_2,$$

where σ and σ_s must be positive and negative together. It then follows from the reachability maximum principle that boundary control is given by

$$u = \begin{cases} +u_{\max} & \text{if } \sigma_s \geq 0 \\ -u_{\max} & \text{if } \sigma_s < 0, \end{cases}$$

where we have resolved what control to use on the switching surface by setting $u = +u_{\max}$ when $\sigma_s = 0$. The equilibrium point $x_1 = u_{\max}$, $x_2 = 0$, corresponding to the control $u = u_{\max}$, lies on $\sigma_s = 0$. When we start the system at this point and integrate the system forward in time using the above control law, the resulting trajectory approaches a limit cycle that defines the boundary of the reachable set. Construction of this limit cycle and calculation of the resonant values of y (which turn out to lie inside the reachable set) are left as Exercise 6-15. ■

Example 6.4-3: Positive and Negative Real Eigenvalues Consider again Example 6.3-3, with the system given by

$$\begin{bmatrix} \dot{x}_1 \\ \dot{x}_2 \end{bmatrix} = \begin{bmatrix} 0 & 1 \\ 1 & 0 \end{bmatrix} \begin{bmatrix} x_1 \\ x_2 \end{bmatrix} + \begin{bmatrix} 0 \\ 1 \end{bmatrix} u, \qquad |u| \leq 1 \qquad (6.4\text{-}1)$$

with a target at the origin. In this case the eigenvalues for the **A** matrix are given by $\mu_1 = 1$, $\mu_2 = -1$. We cannot immediately conclude that the controllable set will have attractive boundaries for $t \to -\infty$. However, we can still use the boundary attraction theorem to find the controllable set for this problem, by trying to satisfy (6.3-20).

In Example 6.3-3 the controllability minimum principle yielded the piecewise constant control

$$u = -\text{sgn}(\sigma), \tag{6.4-2}$$

where $\sigma = \partial \mathcal{H}/\partial u = \boldsymbol{\lambda}^T \mathbf{b} = \lambda_2$ and

$$\mathcal{H} = \boldsymbol{\lambda}^T \mathbf{A}\mathbf{x} + \boldsymbol{\lambda}^T \mathbf{b} u = \lambda_1 x_2 + \lambda_2 x_1 + \sigma u.$$

Since **A** is already symmetric define

$$\psi = \boldsymbol{\lambda}^T \mathbf{A} \boldsymbol{\lambda}.$$

The eigenvalues and eigenvectors of **A** are

$$\mu_1 = 1 \qquad \boldsymbol{\xi}_1 = \begin{bmatrix} 1 \\ 1 \end{bmatrix}$$

$$\mu_2 = -1 \qquad \boldsymbol{\xi}_2 = \begin{bmatrix} 1 \\ -1 \end{bmatrix}.$$

Since the eigenvectors are linearly independent, we write $\boldsymbol{\gamma}(t)$ as

$$\boldsymbol{\gamma}(t) = \alpha_1(t)\boldsymbol{\xi}_1 + \alpha_2(t)\boldsymbol{\xi}_2.$$

Then we have

$$\begin{aligned} \psi(t) &= [\alpha_1 \boldsymbol{\xi}_1^T + \alpha_2 \boldsymbol{\xi}_2^T]\mathbf{A}[\alpha_1 \boldsymbol{\xi}_1 + \alpha_2 \boldsymbol{\xi}_2] \\ &= \alpha_1^2 \mu_1 \|\boldsymbol{\xi}_1\|^2 + \alpha_2^2 \mu_2 \|\boldsymbol{\xi}_2\|^2 + \alpha_1 \alpha_2 (\mu_1 + \mu_2) \boldsymbol{\xi}_1^T \boldsymbol{\xi}_2 \\ &= 2(\alpha_1^2 - \alpha_2^2), \end{aligned} \tag{6.4-3}$$

which can be made positive for all t by choosing $\alpha_2(t) = 0$ and $\alpha_1 =$ constant $\neq 0$. The resulting solution to the adjoint equations

$$\dot{\boldsymbol{\lambda}}^T = -\boldsymbol{\lambda}^T \mathbf{A}, \tag{6.4-4}$$

with $\boldsymbol{\lambda}(0) = \alpha_1 \boldsymbol{\xi}_1 = \alpha_1 [1 \quad 1]^{\mathrm{T}}$, is

$$\boldsymbol{\lambda}(t) = \alpha_1 e^{-\mu_1 t} \boldsymbol{\xi}_1 = e^{-\mu_1 t} \boldsymbol{\lambda}(0) = \alpha_1 e^{-t} \begin{bmatrix} 1 \\ 1 \end{bmatrix}. \tag{6.4-5}$$

Since $\boldsymbol{\lambda}(t) = e^{-\mu_1 t} \boldsymbol{\lambda}(0)$ is to be an outward normal to ∂C, the candidates for a controllability boundary are the hyperplanes perpendicular to $\boldsymbol{\lambda}(0)$; for this example lines perpendicular to $\boldsymbol{\xi}_1 = [1 \quad 1]^{\mathrm{T}}$.

Now consider the orthogonality condition

$$0 \equiv \min_{u \in \mathcal{U}} \mathcal{H} = \lambda_1 x_2 + \lambda_2 x_1 - |\lambda_2|. \tag{6.4-6}$$

The candidate controllability boundaries are determined by

$$0 = \boldsymbol{\lambda}^{\mathrm{T}} d\mathbf{x} = \alpha_1 e^{-t}(dx_1 + dx_2),$$

which yields

$$x_1 + x_2 = c,$$

where $\mathrm{sgn}(c) = \mathrm{sgn}(\alpha_1)$ for $\boldsymbol{\lambda}$ to be an outward normal. Substituting this result into (6.4-6) yields

$$\begin{aligned} 0 = \mathcal{H} &= \alpha_1 e^{-t}[x_2 + x_1 - \mathrm{sgn}(\alpha_1)] \\ &= \alpha_1 e^{-t}[c - x_1 + x_1 - \mathrm{sgn}(\alpha_1)] \\ &= |\alpha_1| e^{-t}(|c| - 1). \end{aligned}$$

The resulting surfaces,

$$x_1 + x_2 = \mathrm{sgn}(\alpha_1) = \pm 1,$$

are in fact the boundaries of the controllable set, as we have seen previously in Examples 2.2-1 and 6.3-3.

From the boundary attraction theorem these boundaries can also be generated by using trajectories that asymptotically approach the controllable set boundary backward in time. For $\boldsymbol{\lambda}(0) = \alpha_1 \boldsymbol{\xi}_1 = \alpha_1 [1 \quad 1]^{\mathrm{T}}$, $\alpha_1 \neq 0$, we integrate the state equations (6.4-1) and the adjoint equations (6.4-4) backward in time, using the control law (6.4-2). One can easily verify, as in Example 6.3-3, that for $\mathbf{x}(0)$ sufficiently close to the appropriate boundary line of $C = \{\mathbf{x} \mid |x_1 + x_2| < 1\}$—indeed for any $\mathbf{x}(0)$ for this example—the resulting trajectory $\mathbf{x}(t)$ approaches ∂C as $t \to -\infty$. In particular, the solution for $\boldsymbol{\lambda}(t)$ is given by (6.4-5) and the switching function $\sigma = \lambda_2(t)$ does not change sign. Thus $u(\mathbf{x})$ is a constant, given by

$u(\mathbf{x}) = -\text{sgn}(\alpha_1)$. The resulting trajectories have been illustrated previously in Figure 6.3-1. For $\alpha_1 > 0$ the $u = -1$ trajectories approach the right-hand boundary in Figure 6.3-1 as $t \to -\infty$ and for $\alpha_1 < 0$ the $u = 1$ retro-time trajectories approach the left-hand boundary. ■

Higher-Dimensional Systems

Using the boundary attraction property, it may be possible to learn important information about controllable and reachable sets in higher-dimensional spaces. For example, suppose that a two-dimensional state-space projection of the controllable set is of interest. By running trajectories that approach ∂C and plotting the results in the two dimensions of interest, a good approximation of the projection may be obtained. However, there is no guarantee of this. The results obtained will depend on how skillfully the initial points are chosen.

A more direct approach is given by Vincent and Wu [1990]. The idea is to obtain both an underestimate and an overestimate to the projection of the controllable set. If both the underestimate and the overestimate agree, then the exact projection has been obtained. Otherwise, the intersection of these two sets gives a measure of how well the projection is known. The method is easily applied to linear systems, and we illustrate the procedure with the following example.

Example 6.4-4: Maximum Energy A simple spring k and mass m system subject to a control force f is modeled by

$$m\ddot{y} + ky = f,$$

where y is the displacement of the mass from equilibrium. Suppose that $k/m = 6\ \text{sec}^{-2}$ and a proportional-integral-derivative control loop of the form

$$f = m\left[r - \left(5y + 6\int y\,dt + 6\dot{y}\right)\right]$$

is applied to the system so that the controlled system is given by

$$\frac{d^3y}{dt^3} + 6\ddot{y} + 11\dot{y} + 6y = \dot{r},$$

where $\dot{r}$ is a command input subject to $|\dot{r}| \leq 1$.

Assume that the system starts at the zero energy state. We need to determine the maximum possible energy that the command input can impart to the system. Theoretically, this problem can be solved by finding those points in the three-dimensional state space (position–velocity–acceleration) that are reachable from the origin under the bounded command input and then finding where on this reachable set the energy ($E = \frac{1}{2}m\dot{y}^2 + \frac{1}{2}ky^2$) is maximized. Note that, since the energy is of interest, it is sufficient to have knowledge of only the projection of the reachable set onto the position-velocity space.

Toward this end, consider finding the controllable set for the retro-time system

$$\frac{d^3y}{d\tau^3} - 6\ddot{y} + 11\dot{y} - 6y = u,$$

where $u = -\dot{r}$, τ is time-to-go, and the dots now denote derivatives with respect to τ. The retro-time system has eigenvalues $\mu_1 = 1$, $\mu_2 = 2$, and $\mu_3 = 3$. This, and our interest in the two-dimensional energy of the system, suggests breaking the third-order system into two other systems, a first-order system with an input u and an output w that feeds directly into one second-order system with an output y. In order to improve accuracy, we make the first-order system as fast as possible by constructing it with the eigenvalue $\mu = 3$. The first-order system is given by

$$\begin{aligned} \dot{z}_1 &= 3z_1 + u \\ w &= z_1, \end{aligned} \tag{6.4-7}$$

where z_1 is the state of this system driven by the control input u and w is the output. The second-order system is given by

$$\begin{aligned} \dot{p}_1 &= p_2 \\ \dot{p}_2 &= -2p_1 + 3p_2 + w \\ y &= p_1, \end{aligned} \tag{6.4-8}$$

where p_1 and p_2 are the two state variables, w is the input, and y is the output. It is easy to check that the first-order system and the second-order system, taken together, are equivalent to the above retro-time system. Also, the state space of interest is $p_1 = y$ and $p_2 = \dot{y}$.

The overestimate for the retro-system controllable set is found by first finding the controllable set from the origin for (6.4-7), which is clearly given by $-\frac{1}{3} < z_1 < \frac{1}{3}$. The output w from this system has the same range. We now find the controllable set for (6.4-8) subject to $|w| \le \frac{1}{3}$. This must be an overestimate of the projection of the controllable set onto the $(y, \dot{y})$ space for the retro-time system, since we are allowing w to change discontinuously. This is the reason we want the first system to be fast, so that z_1 will respond as quickly as possible to u.

To find an underestimate for the retro-system controllable set, one can simply find the domain of attraction obtained from (6.4-8) when a stabilizing feedback control (e.g., $u = -kz_1$ with $k > 3$) is used in (6.4-7).

If the underestimate and the overestimate differ, then a lower bound to the maximum total energy is obtained by finding where the total energy is maximized on the underestimate, and an upper bound to the maximum total energy is found by finding where the total energy is maximized on the overestimate. Details of solving this specific example problem are left for Exercise 6-16. ■

EXERCISES

6-1. Consider the system

$$\dot{x}_1 = u - 2$$
$$\dot{x}_2 = u$$

with control constraint $|u| \leq 1$ and target set

$$x_1^2 + x_2^2 - 1 = 0.$$

(a) In polar coordinates (r, θ) determine the θ range(s) corresponding to the "usable part" of the target, that is, where trajectories can reach the target $[\mathcal{H}(\mathbf{x}, u, \boldsymbol{\lambda}) \leq 0$ for some admissible u, with $\boldsymbol{\lambda}$ as an outward normal to the target$]$.

(b) Determine the controllable set to the target.

(c) Determine a feedback control $u(\mathbf{x})$ that will transfer every state in the controllable set to the target.

6-2. We have previously determined the controllable set to the origin for the system

$$\dot{x}_1 = x_2$$
$$\dot{x}_2 = x_1 + u$$

where $|u| \leq 1$. Use this information to determine the reachable and maneuverable sets for this system.

6-3. Plot the boundary of the reachable set from the origin for the system

$$\begin{bmatrix} \dot{x}_1 \\ \dot{x}_2 \end{bmatrix} = \begin{bmatrix} 0 & 1 \\ -2 & -3 \end{bmatrix} \begin{bmatrix} x_1 \\ x_2 \end{bmatrix} + \begin{bmatrix} 0 \\ 1 \end{bmatrix} u, \qquad |u| \leq 1.$$

6-4. Plot the boundary of the reachable set from the origin for the system

$$\begin{bmatrix} \dot{x}_1 \\ \dot{x}_2 \end{bmatrix} = \begin{bmatrix} 0 & 1 \\ -4 & -3 \end{bmatrix} \begin{bmatrix} x_1 \\ x_2 \end{bmatrix} + \begin{bmatrix} 0 \\ 1 \end{bmatrix} u, \qquad |u| \leq 1.$$

6-5. Given the system

$$\ddot{y} + 2\zeta\omega_n\dot{y} + \omega_n^2 y = u$$

determine the maximum acceleration the system can experience under the bounded control $|u| \leq 1$. Let $\omega_n^2 = 4$, $2\zeta\omega_n = 2$. *Hint:* Determine the boundaries of the reachable set for

$$\dot{x}_1 = x_2$$
$$\dot{x}_2 = -\omega_n^2 x_1 - 2\zeta\omega_n x_2 + u.$$

Then determine where

$$A = -\omega_n^2 x_1 - 2\zeta\omega_n x_2 + u$$

takes on a maximum on $\partial\mathcal{R}$.

6-6. Plot the boundary of the controllable set to the origin for the system

$$\dot{x}_1 = -x_2 - 2u$$
$$\dot{x}_2 = 2x_1 + 3x_2 + 3u$$

where the control is bounded by $1 \leq u \leq 1$.

6-7. Plot the boundary of the controllable set to the origin for the system

$$\dot{x}_1 = -x_2$$
$$\dot{x}_2 = x_1 + x_2 - u$$

where the control is bounded by $-1 \leq u \leq 1$.

6-8. Plot the boundary of the controllable set to $(x_1, x_2) = (1, 0)$ for the system

$$\dot{x}_1 = x_2$$
$$\dot{x}_2 = -x_1^2 + u$$

subject to $|u| \leq 1$.

6-9. A system is given by

$$\dot{x}_1 = x_2$$
$$\dot{x}_2 = -\sin x_1 - x_2 + u$$

where $|u| \le 0.5$. Determine the control law that will maintain the system on the boundary of the reachable set. Plot the boundary of the reachable set.

6-10. For Zermelo's problem of the form

$$\dot{x}_1 = 3 + 2\cos u$$
$$\dot{x}_2 = 2\sin u,$$

determine the control u that will drive the system along the boundary of the controllable set to the target

$$g = x_1^2 + x_2^2 - 1 = 0.$$

6-11. Consider the system

$$\dot{x}_1 = -x_2 - x_1 r(r^2 - u)^2$$
$$\dot{x}_2 = x_1 - x_2 r(r^2 - u)^2$$

with $1 \le u \le 2$, where $r = \sqrt{x_1^2 + x_2^2}$. Use the controllability minimum principle to determine the controllable set C to a small $r \le \varepsilon$ neighborhood of the origin. Determine a control law $u(\mathbf{x})$ that will transfer every state in C to the target.

6-12. Determine the controllable set for Example 2.3-4.

6-13. Goddard's Problem (Challenging!): [Leitmann, 1957; Ardema, 1979; Seywald and Cliff, 1993.] A variable-thrust sounding rocket ascends vertically through a constant-density atmosphere. After its fuel is exhausted it coasts on until it reaches its maximum height $h_{\max}$. Determine $h_{\max}$ and plot the altitude, speed, and thrust as functions of time for two cases:

(a) Analytically, for $C_D = 0$.

(b) Numerically, for $C_D = 0.05$.

Hint: Use the reachability maximum principle, the boundary attraction theorem, and reasonable initial guesses $\boldsymbol{\lambda}(0)$ for a $\boldsymbol{\lambda}(t)$ that will approach an outward normal to the boundary of the reachable set. Assuming a flat earth, no wind, constant gravity and density, and vertical flight, the system is governed by the equations of motion

$$\dot{h} = V$$

$$\dot{V} = \frac{T - D}{m} - g$$

$$\dot{m} = -\frac{T}{V_e}$$

$$0 \le T \le T_{\max},$$

where

$$D = C_D \tfrac{1}{2} \rho A V^2$$

and

$$T = \begin{cases} \in [0, T_{\max}] & \text{if } m > m_R \\ 0 & \text{if } m \le m_R, \end{cases}$$

with initial conditions

$$h(0) = 0$$
$$V(0) = 0$$
$$m(0) = m_R + m_F,$$

where h is the altitude (m), V is the vertical speed of the rocket (m/sec), m is the combined mass (kg) of the rocket and fuel, T is the thrust (N = kg·m/sec^2), with upper bound $T_{\max} = 15{,}000$ N, $g = 9.81$ m/sec^2 is the acceleration of gravity, C_D is the drag coefficient, $\rho = 1.22$ kg/m^3 is the density, $A = 0.16$ m^2 is the cross-sectional area of the rocket, and V_e = 1,700 m/sec is the relative exhaust velocity. The mass of the rocket is $m_R = 150$ kg and the initial fuel mass is $m_F = 300$ kg.

6-14. Plot the controllable set boundary for Example 6.4-1.

6-15. Plot the reachable set boundary for Example 6.4-2 as well as the resonant values of y under the sine input. Assume that $u_{\max} = 2$.

6-16. Carry out the details of solving Example 6.4-4. Plot both the underestimate and the overestimate on the same figure. Can you improve your underestimate? What is the upper bound to the maximum total energy in the system?

7

OPTIMAL CONTROL SYSTEMS

7.1 OPTIMAL CONTROL PROBLEMS

So far our approach to designing control systems has focused on stability and controllability, with a look at function minimizing controls for accomplishing these objectives. For many applications, finding a control that will drive the system to a target or finding a control that will maintain the system at a stable equilibrium point are sufficient objectives. However, for other applications it is not enough just to achieve these objectives. In addition, the control law may be required to optimize an overall performance measure. For example, in space applications, where there is a large cost associated with providing power to the controller, one might strive to minimize total power requirements in addition to maintaining stability and satisfying controllability requirements.

Performance Measures and Optimal Control Laws

Control algorithms that optimize some performance measure for a dynamic system are called **optimal control** laws. The performance objective may be to minimize cost, minimize time, minimize fuel, maximize profit, and so on. Fortunately there is a unified theory for optimal control systems that encompasses a very broad class of performance measures.

The cumulative cost function, whose value depends on the choice for the feedback control function $\mathbf{u}(\mathbf{x})$, is called a **functional** since the cost is a "function of a function." Finding a control function $\mathbf{u}(\mathbf{x})$ to minimize a cost functional could be posed as a problem in the calculus of variations, introduced in Section 3.6. However, in what follows, we do not take the classical calculus of variations approach. Rather we take an approach based on modern optimal control theory. In particular, we will use the geometric approach pioneered by Austin Blaquière and George Leitmann [Blaquière and Leitmann, 1967; Leitmann, 1981]. This approach provides a more natural formulation in terms of control notation and also avoids some difficulties the classical calculus of variations approach has with discontinuous controls. However, there does exist an interesting historical connection between the classical calculus of variations approach and the modern optimal control approach. A good reference is Pesch and Bulirsch [1994], which is well worth the effort for the interested reader.

If one or more control laws exist that will drive a system to a target, it is reasonable to assume that there must exist an optimal control that will do the job. However, we are not guaranteed *a priori* that an optimal control will exist for any given problem, and therefore, in what follows, we assume existence in order to develop necessary conditions for finding an optimal control.

The Integral Cost Problem in Optimal Control

There are many ways to set up an optimal control problem for a dynamical system, depending on the assumptions made regarding cost or performance criteria, constraint sets, and target sets. We seek to provide a formulation for optimal control problems that will cover most problems of interest, yet not be so complex as to obscure the main features of dynamical optimal processes.

Consider the **optimal control problem** of minimizing an integral "cost" functional associated with transferring the state from a given initial point to a specified target set. Specifically, we seek a control $\mathbf{u}(\mathbf{x}) \in \mathcal{U} \subseteq R^{n_u}$ that will minimize the cost integral

$$J[\mathbf{u}(\cdot)] = \int_0^{t_f} f_0(\mathbf{x}, \mathbf{u})\, dt \tag{7.1-1}$$

while transferring the state $\mathbf{x}(t) \in R^{n_x}$ of the dynamical system

$$\dot{\mathbf{x}} = \mathbf{f}(\mathbf{x}, \mathbf{u}) \tag{7.1-2}$$

from a specified initial point

$$\mathbf{x}(0) = \mathbf{x}_i \tag{7.1-3}$$

to any final state $\mathbf{x}(t_f)$ in a **terminal set** $\mathcal{X} \subset R^{n_x}$,

$$\mathbf{x}(t_f) \in \mathcal{X} = \{\mathbf{x} \in R^{n_x} \,|\, \mathbf{g}(\mathbf{x}) = \mathbf{0}\}. \tag{7.1-4}$$

The final time t_f is generally unspecified in this problem formulation. It is determined as the first time that $\mathbf{x}(t)$ reaches $\mathcal{X}$. At each instant of time $0 \leq t \leq t_f$ the control vector $\mathbf{u}$ is constrained to be an element of a specified **control constraint set** $\mathcal{U}$, of the form

$$\mathcal{U} = \{\mathbf{u} \in R^{n_u} | \mathbf{h}(\mathbf{u}) \geq \mathbf{0}\}. \tag{7.1-5}$$

In the above description $\mathbf{x} = [x_1 \cdots x_{n_x}]^{\mathrm{T}}$ is an n_x-dimensional state vector and $\mathbf{u} = [u_1 \cdots u_{n_u}]^{\mathrm{T}}$ is an n_u-dimensional control vector. The scalar-valued function $f_0(\mathbf{x}, \mathbf{u})$ and each component of the vector functions $\mathbf{f}(\mathbf{x}, \mathbf{u}) = [f_1 \ \cdots \ f_{n_x}]^{\mathrm{T}}$, $\mathbf{g}(\mathbf{x}) = [g_1 \ \cdots \ g_{n_g}]^{\mathrm{T}}$, and $\mathbf{h}(\mathbf{u}) = [h_1 \ \cdots \ h_{n_h}]^{\mathrm{T}}$ are assumed to be continuous and to have continuous partial derivatives with respect to their arguments. In addition, we assume that the gradient vectors $\partial g_i(\mathbf{x})/\partial \mathbf{x}$, $i = 1, \ldots, n_g$, are linearly independent and that the gradient vectors $\partial h_i(\mathbf{u})/\partial \mathbf{u}$, $i = 1, \ldots, n_h$, are linearly independent. The control function $\mathbf{u}(\mathbf{x})$ may be discontinuous. However, we only consider trajectories $\mathbf{x}(t)$ for which $\mathbf{u}[\mathbf{x}(t)]$ and $\partial \mathbf{u}[\mathbf{x}(t)]/\partial \mathbf{x}$ are at worst piecewise continuous.

It will be convenient to convert the cost integral to an equivalent differential equation. We do so by introducing an additional quantity called the **accumulated cost**, designated by the scalar x_0 and given by the differential equation

$$\dot{x}_0 = f_0(\mathbf{x}, \mathbf{u}). \tag{7.1-6}$$

Note that $f_0(\mathbf{x}, \mathbf{u})$ does not depend on x_0. That is, the current rate at which cost accumulates is assumed to depend only on the current state and control. It does not depend on the cost already accumulated. Since the initial value of the cost variable x_0 would have no effect on determining an optimal control, we choose $x_0(0) = 0$.

With $x_0(0) = 0$, the accumulated cost variable $x_0(t)$ is the cost incurred while driving the system from an initial state-space point $\mathbf{x}(0)$ to some other point $\mathbf{x}(t)$, at a time $t \leq t_f$, along the resultant trajectory to the terminal set $\mathcal{X}$ by using a particular control law. Thus our objective is to minimize $x_0(t_f)$.

Including the accumulated cost x_0 together with the state $\mathbf{x}$, we form an **augmented state** vector $\mathbf{X}^{\mathrm{T}} = [x_0 \quad \mathbf{x}^{\mathrm{T}}]$. The augmented state $\mathbf{X}(t)$, plotted in the (n_x+1)-dimensional augmented state space, is called an **augmented trajectory**. The augmented state is governed by the differential equations

$$\dot{\mathbf{X}} = \mathbf{F}(\mathbf{X}, \mathbf{u}), \tag{7.1-7}$$

where

$$\mathbf{F}(\mathbf{X}, \mathbf{u}) \triangleq \begin{bmatrix} f_0(\mathbf{x}, \mathbf{u}) \\ \mathbf{f}(\mathbf{x}, \mathbf{u}) \end{bmatrix}. \tag{7.1-8}$$

Since we have introduced an additional differential equation to our system of state equations, it is necessary to extend our definition of an admissible control. Recall from Section 2.1 that an **admissible control $\mathbf{u}(\mathbf{x})$** is any piecewise continuous and piecewise differentiable closed-loop control law that (a) yields a unique forward time solution to (7.1-2) from any initial state-space point of interest, and (b) satisfies $\mathbf{u} \in \mathcal{U}$ for any point along a trajectory generated by $\mathbf{u}(\mathbf{x})$. We expand this definition here in order to make it applicable to the augmented state space and to satisfy the termination requirement (7.1-4). In particular, an admissible control $\mathbf{u}(\mathbf{x})$ is said to be a **feasible control** at the point $\mathbf{x}(0)$ if it generates a unique solution to (7.1-7), that is, to (7.1-2) and (7.1-6), such that $\mathbf{x}(t_f) \in \mathcal{X}$. Thus a feasible control generates a unique solution from every point of the controllable set to $\mathcal{X}$, and the resulting trajectory reaches the target.

We seek a feasible control $\mathbf{u}(\mathbf{x})$ that minimizes the cost integral (7.1-1), that is, that minimizes $x_0(t_f)$. Such a control is called an **optimal control**. Since an optimal control must be sought among the class of feasible controls, controllability as discussed in Chapter 6 becomes a prerequisite for optimization.

The total cost is a function of the initial state $\mathbf{x}(0)$ and the feasible control function $\mathbf{u}(\mathbf{x})$, and it can be designated by the **cost functional** $V(\cdot)$ defined by

$$V[\mathbf{x}(0), \mathbf{u}(\mathbf{x})] \triangleq \int_0^{t_f} f_0[\mathbf{x}, \mathbf{u}(\mathbf{x})]\, dt. \qquad (7.1\text{-}9)$$

Our problem is to find a feasible control $\mathbf{u}(\mathbf{x})$ such that V takes on a minimum. A feasible control $\mathbf{u}^*(\mathbf{x})$ is an **optimal control** (global minimum control) for V at $\mathbf{x}(0)$ if and only if

$$V[\mathbf{x}(0), \mathbf{u}^*(\mathbf{x})] \le V[\mathbf{x}(0), \mathbf{u}(\mathbf{x})] \qquad (7.1\text{-}10)$$

for all feasible controls $\mathbf{u}(\mathbf{x})$. Along an optimal trajectory generated by $\mathbf{u}^*(\mathbf{x})$ we may consider any point $\mathbf{x} = \mathbf{x}(t)$ as an initial point. This property of optimal control [Leitmann, 1981, p. 86] is also referred to as the principle of optimality [Bellman and Dreyfus, 1964, p. 15]. It follows from the fact that (7.1-7) is autonomous. As a consequence, V may be thought of as a function of $\mathbf{x}$ only, using the definition

$$V^*(\mathbf{x}) \triangleq \int_t^{t_f} f_0[\mathbf{x}, \mathbf{u}^*(\mathbf{x})]\, dt. \qquad (7.1\text{-}11)$$

In this form $V^*(\mathbf{x})$ is known as the **optimal return function**. It is the **minimum cost-to-go** from any point on an optimal trajectory to the terminal set $\mathcal{X}$.

7.2 PONTRYAGIN'S MINIMUM PRINCIPLE

Suppose that an optimal trajectory, generated by a feasible and optimal control $\mathbf{u}^*(\mathbf{x})$, has been determined from an initial point $\mathbf{x}_i$ to the terminal set $\mathcal{X}$. Let $\mathcal{S}$ be a state-space curve through $\mathbf{x}_i$ containing initial conditions that all have the same minimum cost-to-go to $\mathcal{X}$. Consider the augmented state-space trajectories from each point of $\mathcal{S}$, as illustrated in Figure 7.2-1. All such trajectories define a surface Σ. Consider now some point designated by a on the augmented trajectory starting from $\mathbf{X}_i^{\mathrm{T}} = [0 \quad \mathbf{x}_i^{\mathrm{T}}]$. If a nonoptimal control were to be applied at point a, clearly the augmented trajectory could not go below the surface Σ at this point, in the direction of decreased cost [Leitmann, 1981, p. 93]. In particular, the surface Σ is a **semipermeable surface** and no trajectory $\mathbf{X}(t)$ in augmented state-cost space can penetrate Σ from above (higher cost) to below (lower cost).

This property of an optimal Σ surface can be stated more precisely. Let the vector $\mathbf{N} = [n_0 \quad n_1 \quad \cdots \quad n_{n_x}]^{\mathrm{T}}$ be an **upward normal** to Σ at any point along an optimal trajectory. The normal vector $\mathbf{N}$ is directed upward toward increasing cost as illustrated in Figure 7.2-1. It follows that $n_0 \geq 0$. Note that $n_0 = 0$ could correspond to a situation where a portion of the surface Σ is perpendicular to the state space. However, more commonly, the surface Σ has an edge, as shown in Figure 7.2-1, and an optimal trajectory that runs along the edge of the surface Σ can do so with either $n_0 > 0$ ($\mathbf{N}$ normal to Σ) or $n_0 = 0$ ($\mathbf{N}$ normal to the edge and in the plane of Σ). This corresponds to a situation where using nonoptimal

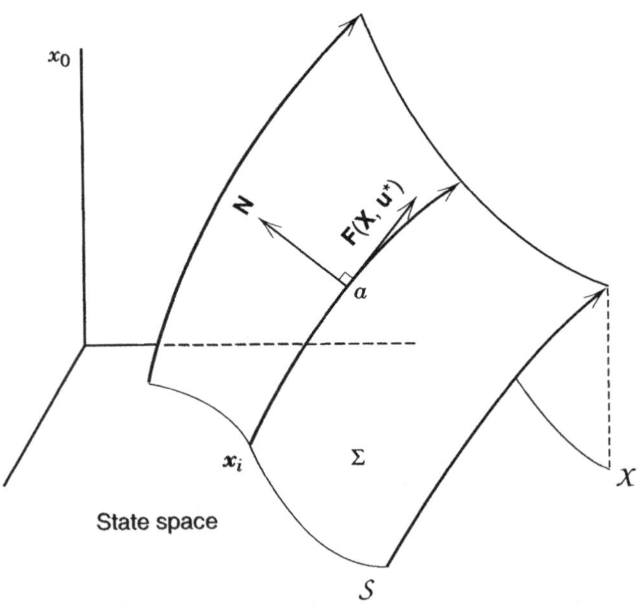

Figure 7.2-1 A Σ surface in augmented state space.

control would result in a trajectory which misses the target. This is a special case that we discuss in detail in Section 8.3. For the remainder of this section we assume that $n_0 > 0$.

As a semipermeable surface, the optimal Σ surface has the property that, for any $\mathbf{X} \in \Sigma$,

$$0 = \mathbf{N}^T\mathbf{F}(\mathbf{X}, \mathbf{u}^*) \leq \mathbf{N}^T\mathbf{F}(\mathbf{X}, \mathbf{u}) \tag{7.2-1}$$

for all $\mathbf{u} \in \mathcal{U}$, where $\mathbf{u}^*$ is the value of the control determined from $\mathbf{u}^*(\mathbf{x})$ at $\mathbf{x}$. This property is a minimizing condition. At every point of an optimal trajectory the dot product between the upward normal vector $\mathbf{N}$ to the Σ surface and the augmented velocity vector $\mathbf{F}$ must take on a minimum with respect to the control $\mathbf{u} \in \mathcal{U}$, with zero being the minimum value for the dot product. This property, which is analogous to the semipermeability condition for the controllability minimum principle, follows from the fact that trajectories starting on Σ cannot go below Σ, since this would correspond to a total cost less than the minimum cost. The zero condition of (7.2-1) simply reflects the fact that optimal trajectories that start on Σ stay on it.

Consider an **augmented adjoint vector** $\mathbf{\Lambda}^T = [\lambda_0 \quad \boldsymbol{\lambda}^T] = [\lambda_0 \quad \lambda_1 \quad \cdots \quad \lambda_{n_x}]$, where the λ_0 component of $\mathbf{\Lambda}$ corresponds to the x_0 component of the augmented state vector $\mathbf{X}$ and the **adjoint vector** $\boldsymbol{\lambda}$ corresponds to the state vector $\mathbf{x}$ for the dynamic system (7.1-2). From the results in Chapter 6, if $\mathbf{\Lambda}(0)$ is an upward normal to Σ at $\mathbf{X}(0)$ and if $\mathbf{\Lambda}(t)$ is defined as the solution to the augmented adjoint differential equations

$$\dot{\mathbf{\Lambda}}^T = -\mathbf{\Lambda}^T \left[\frac{\partial \mathbf{F}}{\partial \mathbf{X}} + \frac{\partial \mathbf{F}}{\partial \mathbf{u}} \frac{\partial \mathbf{u}}{\partial \mathbf{X}} \right], \tag{7.2-2}$$

then $\mathbf{\Lambda}(t)$ will be an upward normal to Σ at $\mathbf{X}(t)$. The initial condition for $\mathbf{\Lambda}$ must be chosen such that $\mathbf{\Lambda}(0)$ is in the direction of $\mathbf{N}(0)$. Thus $\lambda_0 \geq 0$.

Notice that, since $\mathbf{F} = \mathbf{F}[\mathbf{X}, \mathbf{u}(\mathbf{x})]$ is not a function of the x_0 component of $\mathbf{X}$, equation (7.2-2) in partitioned form,

$$[\dot{\lambda}_0 \quad \dot{\boldsymbol{\lambda}}^T] = -[\lambda_0 \quad \boldsymbol{\lambda}^T] \begin{bmatrix} \dfrac{\partial f_0}{\partial x_0} + \dfrac{\partial f_0}{\partial \mathbf{u}} \dfrac{\partial \mathbf{u}}{\partial x_0} & \dfrac{\partial f_0}{\partial \mathbf{x}} + \dfrac{\partial f_0}{\partial \mathbf{u}} \dfrac{\partial \mathbf{u}}{\partial \mathbf{x}} \\ \dfrac{\partial \mathbf{f}}{\partial x_0} + \dfrac{\partial \mathbf{f}}{\partial \mathbf{u}} \dfrac{\partial \mathbf{u}}{\partial x_0} & \dfrac{\partial \mathbf{f}}{\partial \mathbf{x}} + \dfrac{\partial \mathbf{f}}{\partial \mathbf{u}} \dfrac{\partial \mathbf{u}}{\partial \mathbf{x}} \end{bmatrix},$$

can be written equivalently as

$$\dot{\lambda}_0 = 0 \Rightarrow \lambda_0 = \text{constant} \geq 0 \tag{7.2-3}$$

$$\dot{\boldsymbol{\lambda}}^{\mathrm{T}} = -\boldsymbol{\Lambda}^{\mathrm{T}} \left[\frac{\partial \mathbf{F}}{\partial \mathbf{x}} + \frac{\partial \mathbf{F}}{\partial \mathbf{u}} \frac{\partial \mathbf{u}}{\partial \mathbf{x}} \right]. \tag{7.2-4}$$

Furthermore, note that (7.2-2) is **homogeneous** (of degree one) in $\boldsymbol{\Lambda}$. That is, if $\boldsymbol{\Lambda}(t)$ is a solution, then so is $k\boldsymbol{\Lambda}(t)$ for $k \geq 0$. Thus we can scale $\boldsymbol{\Lambda}(t)$ in any convenient fashion. It will be convenient to consider separately two cases, depending on whether $\lambda_0 = 0$ or $\lambda_0 > 0$. The "**abnormal control**" case, where $\lambda_0 = 0$, reduces to the same problem for which the controllability minimum principle was developed. We consider this case further in Section 8.3. For the "**normal control**" case, where $\lambda_0 > 0$, the vector $\boldsymbol{\Lambda}(t)$ will be scaled by taking $\lambda_0 = 1$.

For reference we mention an alternate scaling approach [Lee, 1991], which keeps $\|\boldsymbol{\Lambda}(t)\| = 1$. This approach, used in Section 6.3, has the advantage that it eliminates computational problems associated with the fact that for $\lambda_0 = 1$ we usually get $\|\boldsymbol{\Lambda}(t)\| \to \infty$ as $t \to \infty$. This is because the adjoint and state-perturbation equations have exactly opposite stability characteristics, that is, their "eigenvalues" (Lyapunov exponents) are negatives of each other. The disadvantage, mathematically, is that the resulting scaled adjoint equations [see (6.3-28)] would no longer be linear in $\boldsymbol{\Lambda}(t)$. This type of scaling is not required mathematically and is only necessary computationally for very large t_f. To keep things simple, we use the $\lambda_0 = 1$ scaling here.

There exists a relationship between the vector $\boldsymbol{\Lambda}$ and the optimal return function defined by (7.1-11). If the optimal return function (minimum cost-to-go) were known, then the Σ surface could be defined by

$$V^*(\mathbf{x}) + x_0 = c, \tag{7.2-5}$$

as illustrated in Figure 7.2-2, where c is the minimum cost associated with driving the system from $\mathcal{S}$ to the terminal set $\mathcal{X}$. Since $\boldsymbol{\Lambda}$ is in the direction of the upward normal to the Σ surface, it follows that we could scale $\boldsymbol{\Lambda}$ so that

$$\boldsymbol{\Lambda} = k \operatorname{grad}[x_0 + V^*(\mathbf{x})]$$

where $k > 0$ is any suitable constant. If $\lambda_0 > 0$ and we take $k = 1$, we obtain

$$[\lambda_0 \quad \lambda_1 \quad \cdots \quad \lambda_{n_x}] = \left[1 \quad \frac{\partial V^*}{\partial x_1} \quad \cdots \quad \frac{\partial V^*}{\partial x_{n_x}} \right].$$

From this result we conclude that

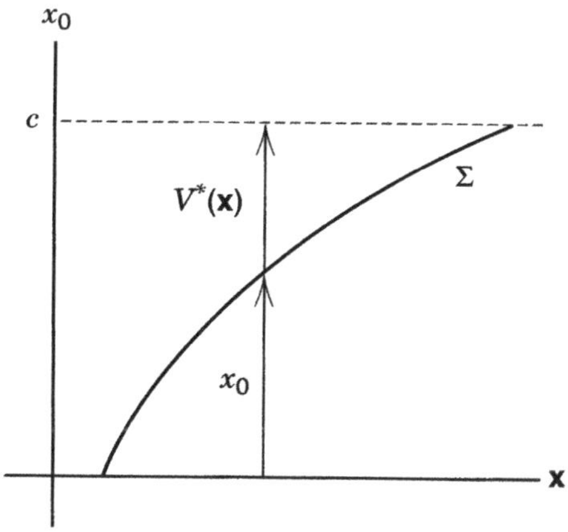

Figure 7.2-2 Total cost equals the accumulated cost plus the optimal return cost.

$$\lambda_0 = 1 \tag{7.2-6}$$

and

$$\boldsymbol{\lambda}^{\mathrm{T}} = \frac{\partial V^*(\mathbf{x})}{\partial \mathbf{x}}. \tag{7.2-7}$$

In other words, for the normal case with $\lambda_0 = 1$, the adjoint vector is the gradient of the optimal return function (provided that the gradient exists). This observation, along with some assumptions about the terminal set, allows us to obtain conditions for the final value of the adjoint vector.

Transversality Conditions

Assume that the terminal set is given by a set of equality conditions of the form

$$\mathcal{X} = \{\mathbf{x} \in R^{n_x} | \mathbf{g}(\mathbf{x}) = \mathbf{0}\}, \tag{7.2-8}$$

where $\mathbf{g} = [g_1 \ \cdots \ g_{n_g}]^{\mathrm{T}}$ is a continuous n_g-dimensional vector function of the state vector $\mathbf{x}$ with continuous partial derivatives with respect to $\mathbf{x}$. If we further assume that the gradient vectors $\partial g_i/\partial \mathbf{x}$, $i = 1, \ldots, n_g$, are linearly independent at every point of $\mathcal{X}$, then the tangent vectors to $\mathcal{X}$ at any point of $\mathcal{X}$ are exactly those vectors $\mathbf{e}$ that satisfy [Vincent and Grantham, 1981]

$$\frac{\partial \mathbf{g}(\mathbf{x})}{\partial \mathbf{x}} \mathbf{e} = \mathbf{0}. \tag{7.2-9}$$

In other words, the **tangent set** to $\mathcal{X}$ is defined by

$$\mathcal{T} = \left\{ \mathbf{e} \in R^{n_x} \,\middle|\, \frac{\partial \mathbf{g}(\mathbf{x})}{\partial \mathbf{x}} \mathbf{e} = \mathbf{0} \right\}. \tag{7.2-10}$$

Since the cost-to-go is zero on the terminal set, the terminal set $\mathcal{X}$ is a surface of constant $V^*(\mathbf{x})$. Thus the gradient of the minimum cost-to-go on the terminal set must be normal to the tangent set $\mathcal{T}$ to $\mathcal{X}$, which in turn implies that the adjoint vector is also normal to the tangent set $\mathcal{T}$; that is,

$$\frac{\partial V^*}{\partial \mathbf{x}} \mathbf{e} = 0 \Rightarrow \boldsymbol{\lambda}^{\mathrm{T}} \mathbf{e} = 0 \tag{7.2-11}$$

for all $\mathbf{e} \in \mathcal{T}$. The implication of (7.2-9) and (7.2-11) is that the vector $\boldsymbol{\lambda}$ is a linear combination of the row vectors in the matrix $\partial \mathbf{g}/\partial \mathbf{x}$ [Vincent and Grantham, 1981]. Thus at the terminal time t_f, with $\mathbf{x}_f \triangleq \mathbf{x}(t_f)$ and $\boldsymbol{\lambda}_f \triangleq \boldsymbol{\lambda}(t_f)$, we have

$$\boldsymbol{\lambda}_f^{\mathrm{T}} = \boldsymbol{\rho}^{\mathrm{T}} \frac{\partial \mathbf{g}(\mathbf{x}_f)}{\partial \mathbf{x}}, \tag{7.2-12}$$

where $\boldsymbol{\rho} = [\rho_1 \ \cdots \ \rho_{n_g}]^{\mathrm{T}}$ is an n_g-dimensional vector. This condition is known as the **transversality condition**. The information supplied by the transversality condition provides just enough additional information to solve the optimization problem.

Information on $\boldsymbol{\lambda}$, determined from (7.2-12), is at the final time, whereas information on $\mathbf{x}$ is generally known at the initial time. Thus solving the adjoint equations together with the system equations requires solving a **two-point boundary value problem**.

The Optimal Control H Function

In order to avoid any confusion with the previous development of the controllability minimum principle in Chapter 6, we continue to designate the dot product between the adjoint vector $\boldsymbol{\lambda}$ and the state-space velocity vector by $\mathcal{H}(\mathbf{x}, \mathbf{u}, \boldsymbol{\lambda}) = \boldsymbol{\lambda}^{\mathrm{T}} \mathbf{f}(\mathbf{x}, \mathbf{u})$ and define the **optimal control H function** as

$$\begin{aligned} H(\mathbf{x}, \mathbf{u}, \boldsymbol{\lambda}, \lambda_0) &\triangleq \lambda_0 f_0(\mathbf{x}, \mathbf{u}) + \boldsymbol{\lambda}^{\mathrm{T}} \mathbf{f}(\mathbf{x}, \mathbf{u}) \\ &= \lambda_0 f_0(\mathbf{x}, \mathbf{u}) + \mathcal{H}(\mathbf{x}, \mathbf{u}, \boldsymbol{\lambda}) \\ &= \boldsymbol{\Lambda}^{\mathrm{T}} \mathbf{F}(\mathbf{x}, \mathbf{u}). \end{aligned} \tag{7.2-13}$$

So far we can conclude that, if $\mathbf{u}^*(\mathbf{x})$ is a normal optimal control for the integral-cost optimal control problem defined in Section 7.1, then there must exist a nonzero continuous vector function $\boldsymbol{\Lambda}^{\mathrm{T}}(t) = [\lambda_0 \quad \boldsymbol{\lambda}^{\mathrm{T}}(t)]$, with constant $\lambda_0 = 1$ (or 0 in the "abnormal" case), and a constant multiplier vector $\boldsymbol{\rho} = [\rho_1 \ \cdots \ \rho_{n_g}]^{\mathrm{T}}$ where, from (7.2-4), $\boldsymbol{\lambda} = [\lambda_1 \ \cdots \ \lambda_{n_x}]^{\mathrm{T}}$ is a solution to

$$\dot{\boldsymbol{\lambda}}^{\mathrm{T}} = -\frac{\partial H}{\partial \mathbf{x}} - \frac{\partial H}{\partial \mathbf{u}} \frac{\partial \mathbf{u}}{\partial \mathbf{x}} \tag{7.2-14}$$

satisfying

$$\boldsymbol{\lambda}^{\mathrm{T}}(t_f) = \boldsymbol{\rho}^{\mathrm{T}} \frac{\partial \mathbf{g}[\mathbf{x}(t_f)]}{\partial \mathbf{x}}, \tag{7.2-15}$$

such that

$$\min_{\mathbf{u} \in \mathcal{U}} H(\mathbf{x}, \mathbf{u}, \boldsymbol{\lambda}, \lambda_0) = H(\mathbf{x}, \mathbf{u}^*, \boldsymbol{\lambda}, \lambda_0) = 0 \tag{7.2-16}$$

for every point $\mathbf{x}$ of an optimal trajectory generated by $\mathbf{u}^*(\mathbf{x})$, where $\mathbf{u}^*$ denotes the value of the control determined from the optimal control $\mathbf{u}^*(\mathbf{x})$ at $\mathbf{x}$.

The reason for introducing the H function definition (7.2-13) is that it allows us to focus on the main feature, that the function H must take on a global minimum with respect to $\mathbf{u}$ at every point of an optimal trajectory. Thus the optimal control approach requires us to solve a parameter optimization problem at each point along an optimal trajectory. This differs from the calculus of variations approach (see Section 3.6), where we must find a complete path function $\mathbf{q}(\cdot)$ to extremize an integral cost functional $J[\mathbf{q}(\cdot)]$. Furthermore, instead of the calculus of variations stationary conditions, $\delta J = 0$ (for which a local maximum, minimum, or inflection point are all valid candidate solutions), we have the much stronger necessary condition that H must take on a global minimum with respect to $\mathbf{u} \in \mathcal{U}$.

Minimizing the H Function

For a given value of $\boldsymbol{\lambda}$ at some point $\mathbf{x}$ along an optimal trajectory, H is a function of $\mathbf{u}$ only. We wish to minimize this function with respect to $\mathbf{u}$ subject to the restriction that $\mathbf{u} \in \mathcal{U}$. Necessary conditions for solving this type of problem are available from **parameter optimization theory** previously developed in Chapter 3. Here we require that the set $\mathcal{U}$ must be defined in terms of the algebraic relations (7.1-5). In a more general setting, **state-dependent control constraints** of the form

$$\mathcal{U}(\mathbf{x}) = \{\mathbf{u} \in R^{n_u} \,|\, \mathbf{h}(\mathbf{x}, \mathbf{u}) \geq \mathbf{0}\}, \tag{7.2-17}$$

could be considered where the constraint set $\mathcal{U}$ is a function of $\mathbf{x}$. However, as shown below, by restricting our analysis to **state-independent control constraints** of the form (7.1-5) the second term on the right-hand side of (7.2-14) is identically zero. For those situations where this is not the case, use of the results obtained so far would be difficult to apply. This is because the feedback control law, which is required to satisfy (7.2-14)–(7.2-16), could not be evaluated from these equations unless the control law were known!

We now further restrict the functions used to define the constraint set. The functions $\mathbf{h}(\mathbf{u})$ are further restricted to that class that will produce a **regular** set for $\mathcal{U}$. The set $\mathcal{U}$ will be regular provided that at every point of $\mathcal{U}$ the tangent vectors to $\mathcal{U}$ are exactly those vectors $\mathbf{e}$ that satisfy the condition

$$\frac{\partial h_j}{\partial \mathbf{u}} \mathbf{e} \geq 0 \tag{7.2-18}$$

for all $\mathbf{h}$ components $h_j(\mathbf{u})$ such that $h_j(\mathbf{u}) = 0$. This requirement is usually met, especially for the type of constraint functions usually employed in optimal control problems. Having imposed the regularity condition on $\mathcal{U}$ we may now state the following necessary conditions for minimizing $H(\mathbf{x}, \mathbf{u}, \boldsymbol{\lambda}, \lambda_0)$ with respect to $\mathbf{u}$ (see Section 3.2):

Minimization Necessary Conditions: *Let* $\mathbf{u}^*$ *be a regular point of* $\mathcal{U}$. *If* H *takes on a minimum with respect to* $\mathbf{u}$ *at* $\mathbf{u}^*$, *then Lagrange multipliers* $\boldsymbol{\gamma} = [\gamma_1 \ \cdots \ \gamma_{n_h}]^{\mathrm{T}}$ *exist such that*

$$\left.\frac{\partial L}{\partial \mathbf{u}}\right|_{\mathbf{u}=\mathbf{u}^*} = \mathbf{0}^{\mathrm{T}} \tag{7.2-19}$$

$$\mathbf{h}(\mathbf{u}^*) \geq \mathbf{0} \tag{7.2-20}$$

$$\boldsymbol{\gamma}^{\mathrm{T}}\mathbf{h}(\mathbf{u}^*) = 0 \tag{7.2-21}$$

$$\boldsymbol{\gamma} \geq \mathbf{0} \tag{7.2-22}$$

where

$$L(\mathbf{x}, \mathbf{u}, \boldsymbol{\lambda}, \lambda_0, \boldsymbol{\gamma}) \triangleq H(\mathbf{x}, \mathbf{u}, \boldsymbol{\lambda}, \lambda_0) - \boldsymbol{\lambda}^{\mathrm{T}}\mathbf{h}(\mathbf{u}). \tag{7.2-23}$$

The Minimum Principle

For state-independent control constraint sets of the form (7.1-5), the necessary conditions developed in the previous section for minimizing the H function also allow us to greatly simplify the adjoint equations (7.2-14).

It follows from (7.2-19) that

$$\frac{\partial H}{\partial \mathbf{u}} = \boldsymbol{\gamma}^{\mathrm{T}} \frac{\partial \mathbf{h}}{\partial \mathbf{u}}. \tag{7.2-24}$$

Thus the second term on the right-hand side of (7.2-14) may be written as

$$\frac{\partial H}{\partial \mathbf{u}} \frac{\partial \mathbf{u}}{\partial \mathbf{x}} = \boldsymbol{\gamma}^{\mathrm{T}} \frac{\partial \mathbf{h}}{\partial \mathbf{u}} \frac{\partial \mathbf{u}}{\partial \mathbf{x}} = \boldsymbol{\gamma}^{\mathrm{T}} \frac{d\mathbf{h}}{d\mathbf{x}}. \tag{7.2-25}$$

From (7.2-17) and (7.2-21)–(7.2-22), $\boldsymbol{\gamma}^{\mathrm{T}}\mathbf{h}(\mathbf{u})$ has a stationary minimum at $\mathbf{u} = \mathbf{u}^*(\mathbf{x})$. It then follows that the right-hand side of (7.2-25) is zero. Thus the second term on the right-hand side of (7.2-14) is zero. This result is extremely important because unless $[\partial H/\partial \mathbf{u}]\partial \mathbf{u}/\partial \mathbf{x} = \mathbf{0}^{\mathrm{T}}$ it is not possible to solve for $\boldsymbol{\lambda}$ from (7.2-14), since $\partial \mathbf{u}(\mathbf{x})/\partial \mathbf{x}$ is not known *a priori*. There are some situations in which the term on the left-hand side of (7.2-25) would be zero, or at least partially calculable, even if the control constraints $\mathbf{h}(\cdot)$ were functions of both $\mathbf{x}$ and $\mathbf{u}$, but in general, this case is difficult to handle, as we see in Section 7.7.

The results obtained so far can now be summarized in terms of an optimal control minimum principle [Pontryagin et al., 1962]:

Optimal Control Minimum Principle: *If* $\mathbf{u}^*(\mathbf{x}) \in \mathcal{U}$ *is an optimal control for the integral cost optimal control problem (7.1-1)–(7.1-5), then there must exist a continuous and piecewise differentiable vector function* $\boldsymbol{\lambda}(t) = [\lambda_1 \ \cdots \ \lambda_{n_x}]^{\mathrm{T}}$, *a constant* $\lambda_0 \geq 0$ [= 1 *or* 0], *with* $(\lambda_0, \boldsymbol{\lambda}) \neq \mathbf{0}$, *and a constant multiplier vector* $\boldsymbol{\rho} = [\rho_1 \ \cdots \ \rho_{n_x}]^{\mathrm{T}}$ *such that* $\boldsymbol{\lambda}(t)$ *satisfies the adjoint equations*

$$\dot{\boldsymbol{\lambda}}^{\mathrm{T}} = -\left.\frac{\partial H}{\partial \mathbf{x}}\right|_{\mathbf{u}=\mathbf{u}^*}, \tag{7.2-26}$$

such that, at the terminal set

$$\mathbf{g}[\mathbf{x}(t_f)] = \mathbf{0}, \tag{7.2-27}$$

$\boldsymbol{\lambda}(t_f)$ *satisfies the transversality conditions*

$$\boldsymbol{\lambda}^{\mathrm{T}}(t_f) = \boldsymbol{\rho}^{\mathrm{T}} \left.\frac{\partial \mathbf{g}}{\partial \mathbf{x}}\right|_{\mathbf{x}=\mathbf{x}(t_f)}, \tag{7.2-28}$$

and such that H *takes on a global minimum value with respect to* $\mathbf{u} \in \mathcal{U}$ *at every point* $\mathbf{x}$ *along the trajectory* $\mathbf{x}(t)$ *generated by* $\mathbf{u}^*(\mathbf{x})$. *Furthermore the minimum value of* H *at every such point is zero:*

$$0 = H[\mathbf{x}(t), \mathbf{u}^*[\mathbf{x}(t)], \boldsymbol{\lambda}(t), \lambda_0] = \min_{\mathbf{u} \in \mathcal{U}} H[\mathbf{x}(t), \mathbf{u}, \boldsymbol{\lambda}(t), \lambda_0], \tag{7.2-29}$$

where

$$H(\mathbf{x}, \mathbf{u}, \boldsymbol{\lambda}, \lambda_0) \triangleq \lambda_0 f_0(\mathbf{x}, \mathbf{u}) + \boldsymbol{\lambda}^{\mathrm{T}} \mathbf{f}(\mathbf{x}, \mathbf{u}). \tag{7.2-30}$$

Note that the nonlinear minimization necessary conditions given by (7.2-19)–(7.2-23) can be used to minimize H as a function of $\mathbf{u}$. If our objective is to maximize the cost functional (7.1-1), then there is a corresponding **optimal control maximum principle** that can be used. Its proof parallels the above development, and it may be stated in exactly the same way as the minimum principle except that every reference to "minimum" or "min" in the minimum principle is replaced by "maximum" or "max." Alternately we can continue to use the minimum principle by replacing f_0 by $-f_0$.

The minimum principle itself is a necessary condition for optimal control. Any solution satisfying the minimum principle (or maximum principle) is called an **extremal**. The actual use of the minimum principle in conjunction with the above necessary conditions is best illustrated through its application to examples. The examples in the following section serve to illustrate the general procedure.

An alternate computational approach to optimal control, based on the methods of Chapter 3, is also possible. By using control parametrization, the optimal control problem may be reduced to a mathematical programming problem and hence is solvable by existing efficient optimization software packages. See Teo, et al. [1991] for details on how to implement this procedure.

7.3 SOME APPLICATIONS OF THE MINIMUM PRINCIPLE

Given an initial point and a target, direct application of the necessary conditions provided by the minimum principle results in an open-loop solution for the control $\mathbf{u}^*(t) = \mathbf{u}^*[\mathbf{x}(t), \boldsymbol{\lambda}(t)]$. While open-loop solutions are of interest in some applications, most control applications will require us to provide the optimal solution in terms of feedback control $\mathbf{u}^*(\mathbf{x})$. As a general rule we seek a complete **synthesis** of the solution, that is, we find the optimal solution from every feasible initial point to the target and then use this result to consolidate the closed-loop control. Unless otherwise specified, we assume the normal control case by choosing $\lambda_0 = 1$.

Example 7.3-1: Shortest Distance Consider the problem of determining the shortest distance from a point to a terminal set in a two-dimensional Euclidean space. We first solve this problem with the terminal set a fixed point, and then solve it again for the more general case. That shortest-distance paths in a plane are straight lines will come as no surprise. But some of the variations on this problem will prove interesting, and help to illustrate the general procedure for solving optimal control problems.

Let (x_1, x_2) denote the position of a point that moves at a constant speed, but in any direction that we choose. The shortest-distance problem is to minimize the total distance traveled

$$s = \int ds$$

in moving from the initial point to the final point, where

$$ds = \sqrt{(dx_1)^2 + (dx_2)^2}$$

is the differential arc length along the path.

To state the problem in a particularly simple form usable by the minimum principle, let t be arc length s. Then the system "dynamics" can be written as

$$\begin{aligned} \dot{x}_0 &= 1 \\ \dot{x}_1 &= \cos u \\ \dot{x}_2 &= \sin u, \end{aligned} \tag{7.3-1}$$

where u is the current direction of motion, counterclockwise from the positive x_1-axis. Since there are no constraints on u, the constraint set is the entire one-dimensional space $\mathcal{U} = R^1$.

Shortest-Distance to a Point For the case of a fixed point target, at $\mathbf{x}_f = (x_{1f}, x_{2f})$, the target is described by

$$\begin{aligned} g_1(\mathbf{x}) &= x_1 - x_{1f} = 0 \\ g_2(\mathbf{x}) &= x_2 - x_{2f} = 0. \end{aligned} \tag{7.3-2}$$

To solve this problem using the minimum principle, we first form the H function

$$H = 1 + \lambda_1 \cos u + \lambda_2 \sin u.$$

Since u is unconstrained, a necessary condition for minimizing H is given by

$$0 = \frac{\partial H}{\partial u} = -\lambda_1 \sin u + \lambda_2 \cos u,$$

yielding

$$\tan u = \frac{\lambda_2}{\lambda_1}. \tag{7.3-3}$$

To determine which quadrant corresponds to the control that actually does minimize H, consider the **second-order necessary condition** [Vincent and Grantham, 1981, p. 48]

$$0 \le \frac{\partial^2 H}{\partial u^2} = -\lambda_1 \cos u - \lambda_2 \sin u.$$

From (7.3-3)

$$\sin u = \beta \frac{\lambda_2}{\sqrt{\lambda_1^2 + \lambda_2^2}} \qquad \cos u = \beta \frac{\lambda_1}{\sqrt{\lambda_1^2 + \lambda_2^2}}, \tag{7.3-4}$$

where $\beta = \pm 1$. Thus

$$\frac{\partial^2 H}{\partial u^2} = -\beta \frac{\lambda_1^2 + \lambda_2^2}{\sqrt{\lambda_1^2 + \lambda_2^2}} = -\beta \|\boldsymbol{\lambda}\|.$$

Since $H = 0$ implies $\boldsymbol{\lambda} \neq \mathbf{0}$, the second-order condition is satisfied with $\beta = -1$.

This example illustrates that a candidate for optimal control will, in general, depend on the adjoint variables. The adjoint variables are obtained from (7.2-26), which for this problem reduces to the trivial differential equations

$$\dot{\lambda}_1 = 0$$
$$\dot{\lambda}_2 = 0.$$

Thus the adjoint vector is constant. In particular we may write

$$\lambda_1(t) = \lambda_{1f} \triangleq \lambda_1(t_f)$$
$$\lambda_2(t) = \lambda_{2f} \triangleq \lambda_2(t_f).$$

Thus the optimal control candidate is constant control and is given by

$$u^* = \tan^{-1}\left[\frac{-\lambda_{2f}}{-\lambda_{1f}}\right]. \tag{7.3-5}$$

While this result guarantees that H takes on a minimum, this in itself does not represent a sufficient condition from which we can conclude that a minimum-time (shortest-path, in this case) trajectory has been found. The minimum principle is only a necessary condition. However, in this case, constant control, which results in straight line trajectories, is indeed optimal.

In general the final values for the adjoint variables are determined from the transversality conditions. From (7.2-28) and (7.3-2) we obtain

$$\lambda_{1f} = \rho_1 \frac{\partial g_1(\mathbf{x}_f)}{\partial x_1} + \rho_2 \frac{\partial g_2(\mathbf{x}_f)}{\partial x_1} = \rho_1$$

$$\lambda_{2f} = \rho_1 \frac{\partial g_1(\mathbf{x}_f)}{\partial x_2} + \rho_2 \frac{\partial g_2(\mathbf{x}_f)}{\partial x_2} = \rho_2.$$

In this case no useful information is obtained from the transversality conditions since one set of unknown constants is replaced by another set of unknown constants. The reason for this situation is that the terminal set is a fixed point. Thus the tangent plane to the target is not unique and any vector $\boldsymbol{\lambda}_f$ is an "outward normal" to the target.

Using the optimal control law (7.3-4) with $\beta = -1$ and evaluating the H function at the final time, we have

$$0 = H_f = 1 - \sqrt{\lambda_{1f}^2 + \lambda_{2f}^2}.$$

Thus $\boldsymbol{\lambda}_f$ can be any unit vector, which we could represent in polar coordinates by

$$\lambda_{1f} = \cos \varphi$$

$$\lambda_{2f} = \sin \varphi.$$

For each choice of φ these results determine corresponding values for λ_{1f} and λ_{2f}, which in turn determine the control u^*. Then we integrate the system equations backward in time from the terminal point $\mathbf{x}_f$. Those values of λ_{1f} and λ_{2f} for which the state-space trajectory passes through the initial point will satisfy all of the necessary conditions. We use backwards integration, in part, because it generates trajectories that satisfy the terminal conditions on both the state and adjoint vectors.

An alternate forward integration approach, starting from an initial state $\mathbf{x}(0)$ (called the "shooting method") can also be used. It consists of iteratively guessing an initial value for the adjoint vector $\boldsymbol{\lambda}(0)$ and then integrating the state and adjoint equations forward in time, trying to hit the target while satisfying the corresponding terminal transversality conditions on $\boldsymbol{\lambda}(t_f)$.

The backwards integration process is illustrated in Figure 7.3-1. When applied to the specific problem of finding the shortest path from a given initial point to a given fixed final point, it is the same as the shooting method. However, more commonly, our objective would be to find the shortest path from any feasible initial point to the fixed final point and the backward integration procedure just outlined solves that problem as well! If we can find a control law $\mathbf{u}^*(\mathbf{x})$ that minimizes the cost from every feasible point to the terminal set, then a complete **synthesis** of the optimal solutions as provided by the minimum principle is said to have been done. Our objective will always be to form a complete synthesis. This synthesis is relatively simple to perform for most one- and two-dimensional problems. However, unless they have some special structure, a complete synthesis for higher-dimensional systems may be very difficult to obtain.

Another way of stating the synthesis problem is to note that, whether forward or backward integration is used, direct application of the minimum principle yields the control $\mathbf{u}^*$ as a function $\mathbf{u}^*(\mathbf{x}, \boldsymbol{\lambda})$ of the state and adjoint vectors. These, in turn, are obtained as solutions $\mathbf{x}(t)$ and $\boldsymbol{\lambda}(t)$ to the corresponding differential equations, usually by means of a numerical integration procedure. Thus along an optimal trajectory the minimum principle yields the control in an **open-loop** form $\mathbf{u}^*(t) = \mathbf{u}^*[\mathbf{x}(t), \boldsymbol{\lambda}(t)]$. The synthesis problem is to determine the optimal closed-loop **feedback** control $\mathbf{u}^*(\mathbf{x})$, by examining all optimal trajectories that reach the target. Backwards integration is particularly well-suited to this task, since it only proudces trajectories that terminate at the target set.

Because of the simplicity of the differential equations for this example, we do not need to depend on a numerical synthesis for this problem. Instead, we

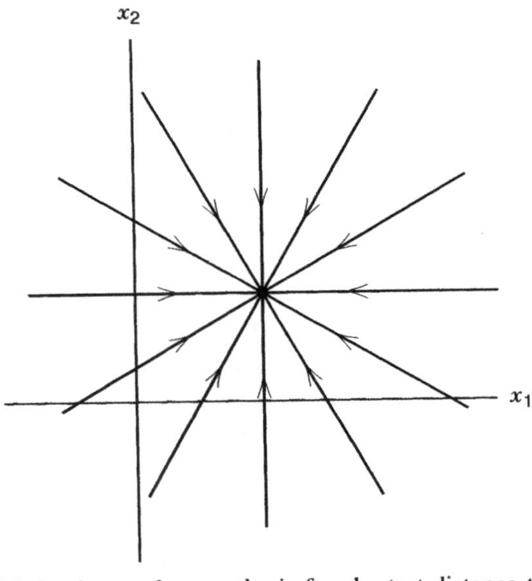

Figure 7.3-1 A complete synthesis for shortest distance to a point.

can synthesize an optimal feedback control analytically. Let u^* be the control obtained by some choice for $\lambda_{1f} = \rho_1$ and $\lambda_{2f} = \rho_2$. Integrating the system equations backward in time yields

$$x_0 = c - \tau$$
$$x_1 = x_{1f} - \tau \cos u^*$$
$$x_2 = x_{2f} - \tau \sin u^*,$$

where τ is the time (distance) to go from $\mathbf{x}$ to the target and c is the minimum cost (shortest distance) from the initial point to the final point. Combining the latter two equations we obtain

$$(x_1 - x_{1f})^2 + (x_2 - x_{2f})^2 = \tau^2.$$

It follows from (7.2-5) that the optimal return function is given by

$$V^*(\mathbf{x}) = \sqrt{(x_1 - x_{1f})^2 + (x_2 - x_{2f})^2}. \tag{7.3-6}$$

The optimal return function is constant on circles centered about $\mathbf{x}_f$. From (7.2-7) we deduce that, at points where $V^*(\mathbf{x})$ is differentiable (i.e., for $\mathbf{x} \neq \mathbf{x}_f$), the components of the constant adjoint vector, in terms of their final values, are given by

$$\begin{aligned} \lambda_{1f} &= \frac{x_1 - x_{1f}}{V^*(\mathbf{x})} \\ \lambda_{2f} &= \frac{x_2 - x_{2f}}{V^*(\mathbf{x})}. \end{aligned} \tag{7.3-7}$$

Then from (7.3-5) we obtain the synthesized closed-loop optimal control

$$u^*(\mathbf{x}) = \tan^{-1}\left(\frac{x_{2f} - x_2}{x_{1f} - x_1}\right). \tag{7.3-8}$$

Under this control law, using the four quadrant $\tan^{-1}(\cdot, \cdot)$ function, every point in the state space can be driven to the target along the shortest path.

Shortest Distance to a Curve Consider now the same problem except with the terminal set a curve given by the relation

$$g(\mathbf{x}) = 0, \tag{7.3-9}$$

where $g(\cdot)$ is a scalar function of $\mathbf{x}$, with nonzero gradient $\partial g(\mathbf{x}_f)/\partial \mathbf{x}$ at the terminal point $\mathbf{x}_f$. The transversality condition (7.2-28) for this problem yields

$$\lambda_{1f} = \rho \frac{\partial g(\mathbf{x}_f)}{\partial x_1} \qquad\qquad (7.3\text{-}10)$$
$$\lambda_{2f} = \rho \frac{\partial g(\mathbf{x}_f)}{\partial x_2}.$$

From $H = 0$ and (7.3-4) with $\beta = -1$, we have

$$0 = H_f = 1 - \sqrt{\lambda_{1f}^2 + \lambda_{2f}^2} = 1 - |\rho| \left\| \frac{\partial g(\mathbf{x}_f)}{\partial \mathbf{x}} \right\|.$$

Thus

$$\rho = \pm \frac{1}{\|\partial g(\mathbf{x}_f)/\partial \mathbf{x}\|}. \qquad\qquad (7.3\text{-}11)$$

Although we can get analytic solutions for this example optimal control problem, we describe the general **backwards-numerical-integration procedure** required for such solutions: Pick a point $\mathbf{x}_f$ on $\mathbf{g}(\mathbf{x}) = \mathbf{0}$ and then pick $\boldsymbol{\rho}$ so that $\boldsymbol{\lambda}_f$ from the transversality conditions, along with $\mathbf{u}^*{}_f = \mathbf{u}^*(\mathbf{x}_f)$ from the optimal control law, satisfies $H_f = 0$. Then integrate the state and adjoint equations backwards in time, recomputing the required optimal control at each integration step, yielding a retro-time trajectory. Repeat the process with other $\mathbf{x}_f$ and/or $\boldsymbol{\rho}$ values, until either the retro-time trajectory passes through a given initial state, or until all retro-time trajectories have been generated, for a complete synthesis of the solution.

We can solve our shortest-distance problem using a little geometry. It follows from (7.3-5) and (7.3-10) that the control for this case satisfies

$$\tan u^* = \frac{\partial g(\mathbf{x}_f)/\partial x_2}{\partial g(\mathbf{x}_f)/\partial x_1}, \qquad\qquad (7.3\text{-}12)$$

where the quadrant for u^* is as yet unresolved. We note from Figure 7.3-2 that the tangent of the angle θ that the gradient $\partial g(\mathbf{x}_f)/\partial \mathbf{x}$ makes with respect to the x_1-axis is equal to the ratio $\tan u^*$ given by (7.3-12). Thus either $u^* = \theta$ or $u^* = \theta \pm \pi$. From the equations of motion (7.3-1), the angle the trajectory makes with respect to the x_1-axis satisfies

$$\tan u^* = \frac{\dot{x}_2}{\dot{x}_1}.$$

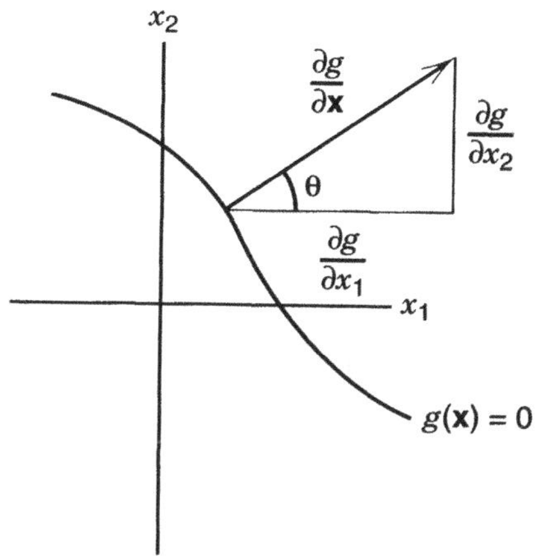

Figure 7.3-2 The gradient to the terminal set.

Since u^* is a constant it follows from (7.3-12) that, for this problem, the trajectory $\mathbf{x}(t)$ must be a straight line that intersects the target **transversely**, that is, parallel to the gradient vector $\partial g(\mathbf{x}_f)/\partial \mathbf{x}$ and perpendicular to the target. If the initial point $\mathbf{x}(0)$ is on the opposite side of the curve $g(\mathbf{x}) = 0$ from the direction of the gradient $\partial g(\mathbf{x}_f)/\partial \mathbf{x}$, then $\mathbf{x}(t)$ moves in the direction of the gradient, with $u^* = \theta$. Conversely, if the initial state and the gradient are on the same side of the terminal set, then $u^* = \theta \pm \pi$. This information allows us to determine the solution to the optimal control problem.

STRAIGHT LINE TARGET For example, suppose that the initial point is the origin and the target set is the straight line give by

$$g(\mathbf{x}) = x_1 + x_2 - 10 = 0.$$

If we think of integrating the system equations backward in time, from various points along the target, using the control law $u^* = \theta$, then the trajectories will be normal to the target. As illustrated in Figure 7.3-3, one need only find a point on the target set such that a trajectory intersects the origin. The solution in this case is given by $u^* = \pi/4$, which the reader can easily verify using (7.3-12).

A complete synthesis of the solution is illustrated in Figure 7.3-4. In this case, since $\partial g/\partial \mathbf{x} = [1 \quad 1]$ is constant, the closed-loop feedback control law, from $\tan u^* = 1$, is given by

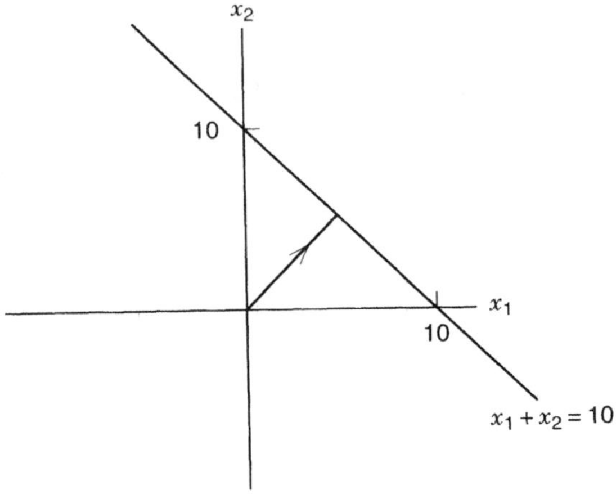

Figure 7.3-3 A trajectory from the origin that is normal to the terminal set.

$$u^*(\mathbf{x}) = \begin{cases} \dfrac{\pi}{4} & \text{if } g(\mathbf{x}) < 0 \\ -\dfrac{3\pi}{4} & \text{if } g(\mathbf{x}) > 0. \end{cases}$$

PARABOLIC TARGET If the target set is more complicated, then a complete synthesis of the solution is also more complicated. Consider the case shown in Figure 7.3-5, where the target set is a parabola given by

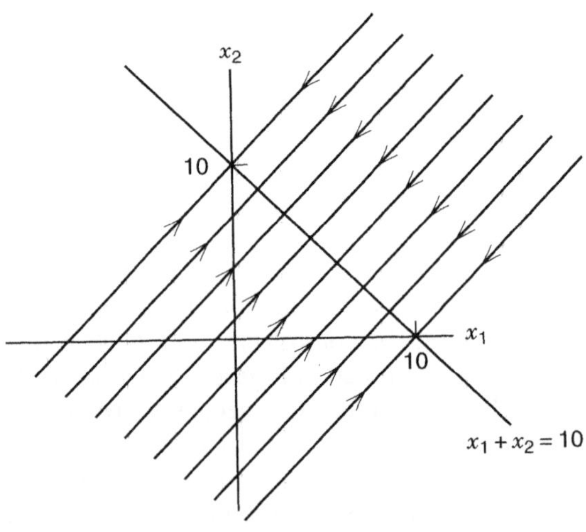

Figure 7.3-4 A complete synthesis for the shortest distance to a line.

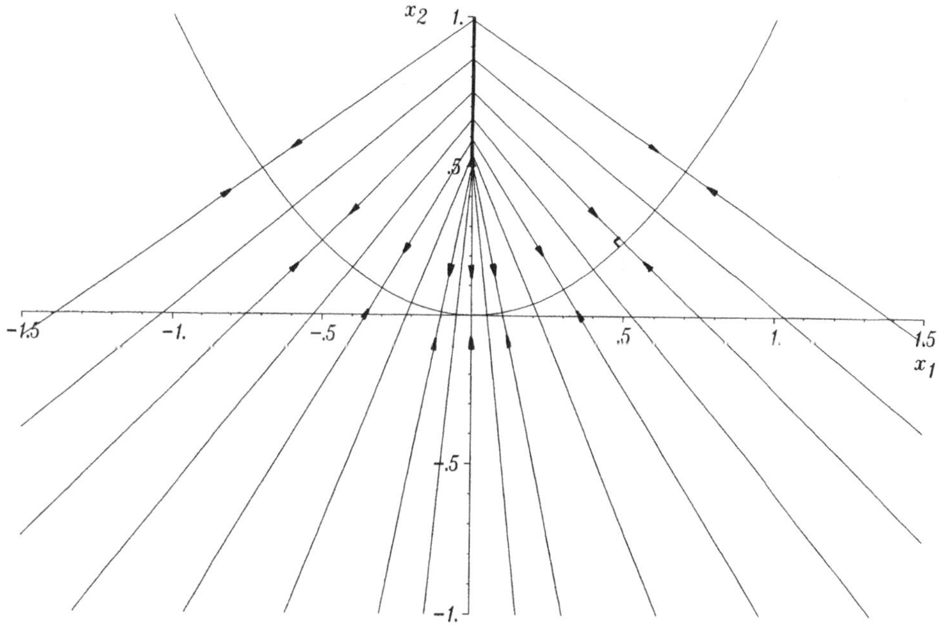

Figure 7.3-5 A case with intersecting extremals.

$$0 = g(\mathbf{x}) = x_1^2 - x_2.$$

In this case, trajectories generated by backward integration intersect above the parabola. Each trajectory satisfies the optimal control necessary conditions and hence is an extremal. Additional information about the problem must be used to select an appropriate control algorithm in the case of **intersecting extremals**. Typically there are points of intersection where two or more trajectories have the same minimum cost.

Consider a pair of intersecting retro-time trajectories above the parabola, corresponding to $\mathbf{x}_f = (\pm a, a^2)$, $a > 0$. We use the subscripts r and l to denote quantities associated with the right-hand and left-hand trajectories, respectively. For the right-hand trajectory,

$$u_r^* = -\tan^{-1}\left(\frac{1}{2a}\right), \qquad \cos u_r^* = \frac{2a}{\sqrt{1+4a^2}}, \qquad \sin u_r^* = \frac{-1}{\sqrt{1+4a^2}}$$

and

$$x_{1r} = \frac{-2a\tau_r}{\sqrt{1+4a^2}} + a$$

$$x_{2r} = \frac{\tau_r}{\sqrt{1+4a^2}} + a^2.$$

For the left-hand trajectory,

$$u_l^* = \pi - \tan^{-1}\left(\frac{1}{2a}\right), \qquad \cos u_l^* = \frac{-2a}{\sqrt{1+4a^2}}, \qquad \sin u_l^* = \frac{-1}{\sqrt{1+4a^2}}$$

and

$$x_{1l} = \frac{2a\tau_l}{\sqrt{1+4a^2}} - a$$

$$x_{2l} = \frac{\tau_l}{\sqrt{1+4a^2}} + a^2.$$

These two trajectories intersect on the positive x_2-axis at

$$x_1 = 0, \qquad x_2 = \tfrac{1}{2} + a^2$$

with

$$\tau_r = \tau_l = \tfrac{1}{2}\sqrt{1+4a^2}.$$

It is clear that the backwards extremals should not cross the x_2-axis. Forward in time from a point on the positive x_2-axis, the shortest distance to the parabola is the same for either extremal, whether we select the extremal that moves to the right or the left. Thus the positive x_2-axis, with $x_2 > \frac{1}{2}$, is a **dispersal surface**, from which optimal trajectories depart on either side and have the same minimum cost from the starting point where they intersect.

Note that there is one point on the x_2-axis, at $x_2 = \frac{1}{2}$ corresponding to $a = 0$, where the two optimal trajectories coalesce to a single trajectory. Below this point there is only one optimal trajectory to the parabola, namely along the x_2-axis. Thus the dispersal surface ends at the point $\mathbf{x} = (0, \frac{1}{2})$. Note that this point is *not* the focus of the parabola, which is located at $\mathbf{x} = (0, \frac{1}{4})$. ■

Example 7.3-2: Zermelo's Problem Consider again a slightly different version of Zermelo's problem previously discussed in Examples 2.2-6, 5.3-1, and

5.3-2. Here we reverse the direction of the river and the swimmer is attempting to reach a point target. Assume that the river moves with a uniform current of 2 km/hr and that the swimmer swims at 1 km/hr. The dynamics are given by

$$\dot{x}_1 = 2 + \cos u$$
$$\dot{x}_2 = \sin u,$$

where u is the direction the swimmer can swim relative to the river velocity. The river flows from left to right, as illustrated in Figure 7.3-6. Suppose that the swimmer wishes to reach some other point in the river in minimum time. Thus he or she wishes to minimize

$$\int_0^{t_f} dt,$$

so that

$$\dot{x}_0 = 1.$$

Since there are no constraints on u, the control constraint set is the entire real axis; that is, $\mathcal{U} = R^1$. The terminal set is a fixed point $\mathbf{x}_f$ given by

$$g_1(\mathbf{x}) = x_1 - x_{1f} = 0$$
$$g_2(\mathbf{x}) = x_2 - x_{2f} = 0.$$

For this problem it is clear that not all initial points can be driven to the terminal set, since the swimmer cannot make headway upstream. We determine

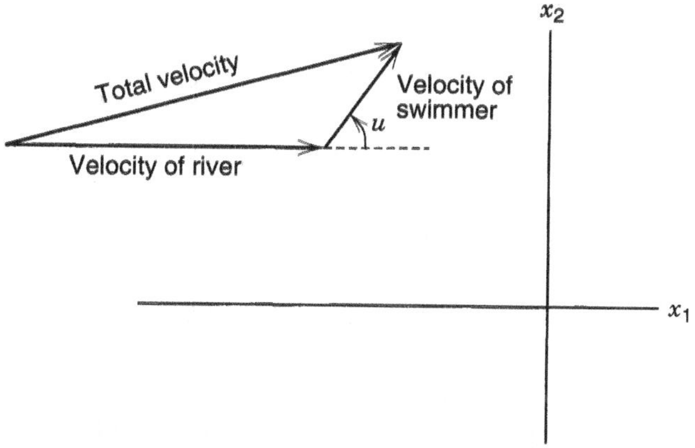

Figure 7.3-6 The dynamics of Zermelo's navigation problem.

the controllable set to the target as a by-product of performing the synthesis of the optimal control $u^*(\mathbf{x})$.

To solve the optimal control problem we first form

$$H = \lambda_0 + \lambda_1(2 + \cos u) + \lambda_2 \sin u,$$

where $\lambda_0 = 1$. The necessary condition for minimizing H is given by

$$0 = \frac{\partial H}{\partial u} = -\lambda_1 \sin u + \lambda_2 \cos u,$$

which yields the minimizing control

$$u^* = -\tan^{-1}\left(\frac{\lambda_2}{\lambda_1}\right), \tag{7.3-13}$$

with the adjoint variables determined from

$$\dot{\lambda}_1 = 0$$
$$\dot{\lambda}_2 = 0.$$

Since the endpoint $\mathbf{x}(t_f) = \mathbf{x}_f$ is fixed, and since the transversality conditions (7.2-28) yield no useful information, we arrive at the result that λ_1 and λ_2 are unknown constants. Thus, from (7.3-13), extremal control is constant control. A complete synthesis for this problem is obtained by integrating backward from the terminal point using constant control. Figure 7.3-7 illustrates typical trajectories obtained by such a process. All possible trajectories lie in a cone, as shown, and hence the cone represents the controllable set to the terminal point illustrated. The angle that the boundary of the cone makes with the x_1-axis can be easily calculated. We note that the slope of any trajectory is given by

$$\frac{\dot{x}_2}{\dot{x}_1} = \frac{\sin u}{2 + \cos u} \tag{7.3-14}$$

and that the maximum (minimum) value this slope can have occurs when

$$\cos u = -\tfrac{1}{2} \Rightarrow u = \pm 120°.$$

Substituting this value into (7.3-14) yields a maximum slope of ±30°.

Integrating the system equations backward in time under constant optimal control u^* yields

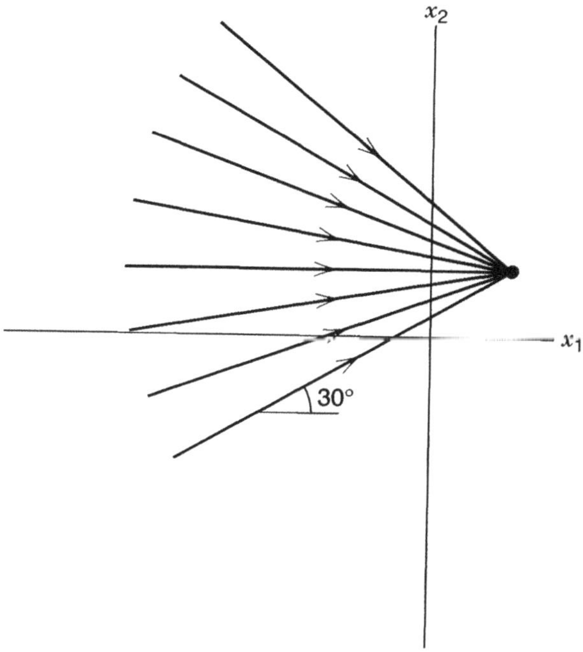

Figure 7.3-7 Minimum-time trajectories for Zermelo's navigation problem.

$$x_0 = c - \tau$$
$$x_1 = x_{1f} - 2\tau - \tau \cos u^*$$
$$x_2 = x_{2f} - \tau \sin u^*,$$

where $\tau = t_f - t \geq 0$ is the time-to-go. Combining the latter two equations yields

$$\psi(\mathbf{x}, \tau) \triangleq (x_1 - x_{1f} + 2\tau)^2 + (x_2 - x_{2f})^2 - \tau^2 = 0. \tag{7.3-15}$$

Thus lines of constant minimum time-to-go are the solid portions of the circles illustrated in Figure 7.3-8. The dashed portions of the circles correspond to maximum-time trajectories to the target.

The **envelope** of these constant time-to-go circles can be obtained directly from (7.3-15). As we move from one circle to the next, $\psi \equiv 0$, so that

$$0 = d\psi = \frac{\partial \psi}{\partial \mathbf{x}} d\mathbf{x} + \frac{\partial \psi}{\partial \tau} d\tau.$$

Along an envelope, however, constant-τ curves are mutually tangent and $\partial\psi/\partial\mathbf{x}$ is perpendicular to the envelope; thus

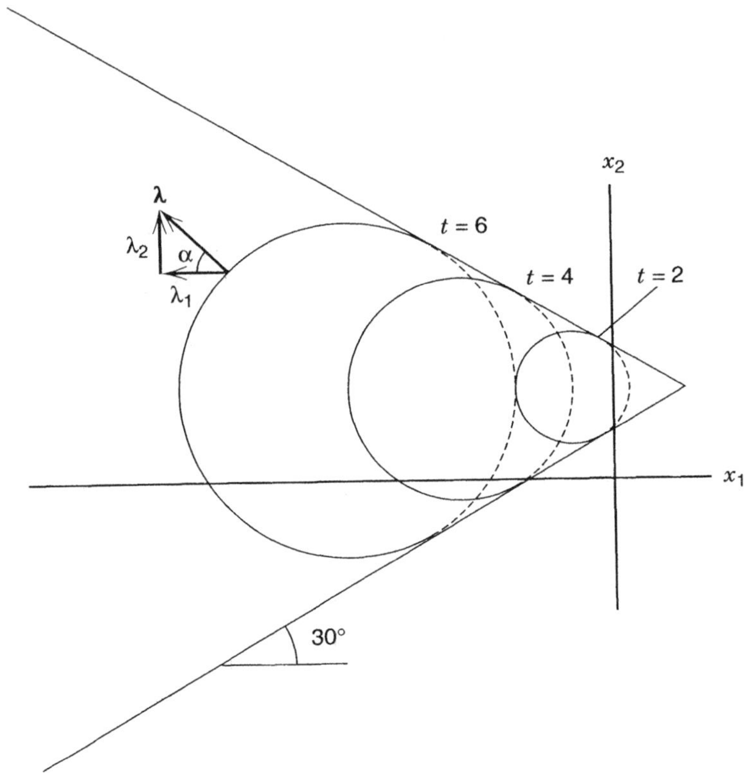

Figure 7.3-8 Minimum-time contours for Zermelo's navigation problem.

$$\frac{\partial \psi}{\partial \mathbf{x}} \, d\mathbf{x} = 0.$$

Thus we conclude that $\partial \psi / \partial \tau = 0$ along an envelope. Taking the partial derivative of the left-hand side of (7.3-15) with respect to τ, setting the result equal to zero, and solving for τ yields

$$\tau = -\frac{2}{3} \, (x_1 - x_{1f}).$$

Thus along the envelope we have

$$\cos u^* = \frac{1}{\tau}(-x_1 + x_{1f} - 2\tau) = -\frac{1}{2}$$

$$\sin u^* = \pm\frac{\sqrt{3}}{2}.$$

Substituting these results into (7.3-14) yields

$$\frac{\dot{x}_2}{\dot{x}_1} = \pm\frac{1}{\sqrt{3}},$$

which implies a slope of ±30°. As expected, the constant time-to-go circles are tangent to the boundary of the controllable set. At the tangency points the minimum time-to-go equals the maximum time-to-go, and the swimmer has no freedom of choice if he or she wishes to reach the target point $\mathbf{x}_f$.

Since

$$V^* = c - x_0 = \tau,$$

we can determine the optimal return function V^* by solving for τ from (7.3-15). Expanding (7.3-15), we obtain

$$3\tau^2 + 4\tau(x_1 - x_{1f}) + (x_1 - x_{1f})^2 + (x_2 - x_{2f})^2 = 0.$$

Thus the minimum-time-to-go optimal return function is

$$V^*(\mathbf{x}) = \tau - -\frac{2}{3}(x_1 - x_{1f}) + \frac{1}{3}\sqrt{(x_1 - x_{1f})^2 - 3(x_2 - x_{2f})^2},$$

where

$$x_1 - x_{1f} \leq 0, \qquad \text{and} \qquad |x_2 - x_{2f}| \leq \frac{1}{\sqrt{3}}|x_1 - x_{1f}|.$$

The optimal feedback control law $u^*(\mathbf{x})$ can be calculated from this function using (7.3-13) and (7.2-7). Since $\boldsymbol{\lambda} = \text{grad}\, V^*(\mathbf{x})$ is normal to lines of constant $V^*(\mathbf{x})$, it follows from (7.3-5) that, if α is the angle that $\boldsymbol{\lambda}$ makes to the negative x_1-direction at a point $\mathbf{x}$, then

$$\tan u^* = -\tan\alpha,$$

as illustrated in Figure 7.3-8. That is,

$$u^*(\mathbf{x}) = -\tan^{-1}\left(\frac{\partial V^*/\partial x_2}{\partial V^*/\partial x_1}\right)$$

and the swimmer swims opposite to grad $V^*(\mathbf{x})$. ■

Example 7.3-3: Juggernaut In India there is a huge statue, called Jagannath, of Vishnu the lord of the world. Each year in a Hindu ceremony, hundreds of people collectively move this huge statue from one place to another. Over the years they have learned precisely how to apply their collective force to place this statue in motion and then to bring the statue to rest in the desired location. From this ceremony comes the word "juggernaut," for a massive inexorable object that crushes anything in its path.

For one-dimensional motion in the absence of friction, the equation of motion is that of a mass m acted upon by a bounded force:

$$m\ddot{y} = F, \qquad |F| \leq F_{\max},$$

where y is the displacement from the desired position and F is the applied force in the y direction. In terms of scaled state-space variables

$$(x_1, x_2) = \left(\frac{y}{F_{\max}/m}, \frac{\dot{y}}{F_{\max}/m}\right),$$

with $u = F/F_{\max}$, the equations of motion are

$$\begin{aligned} \dot{x}_1 &= x_2 \\ \dot{x}_2 &= u, \end{aligned} \tag{7.3-16}$$

with

$$|u| \leq u_{\max} = 1. \tag{7.3-17}$$

We wish to transfer the state to the origin in minimum time; that is, we seek a feedback control $u(\mathbf{x})$ to minimize

$$J[u(\cdot)] = \int_0^{t_f} dt$$

subject to the state equations (7.3-16) and the control constraint (7.3-17), where t_f is determined by the terminal conditions

$$\mathbf{g}(\mathbf{x}_f) = \begin{bmatrix} x_{1f} \\ x_{2f} \end{bmatrix} = \begin{bmatrix} 0 \\ 0 \end{bmatrix}$$

at $\mathbf{x}(t_f) = \mathbf{x}_f$.

A solution to this minimum-time problem has already been presented in Example 2.1-1, to illustrate switching surfaces. Here we solve the problem using the minimum principle to generate the optimal trajectories shown in Figure 7.3-9.

The H function is

$$\begin{aligned} H &= \lambda_0 f_0 + \boldsymbol{\lambda}^{\mathrm{T}}\dot{\mathbf{x}} \\ &= \lambda_0 + \lambda_1 x_2 + \lambda_2 u. \end{aligned} \tag{7.3-18}$$

The adjoint equations are

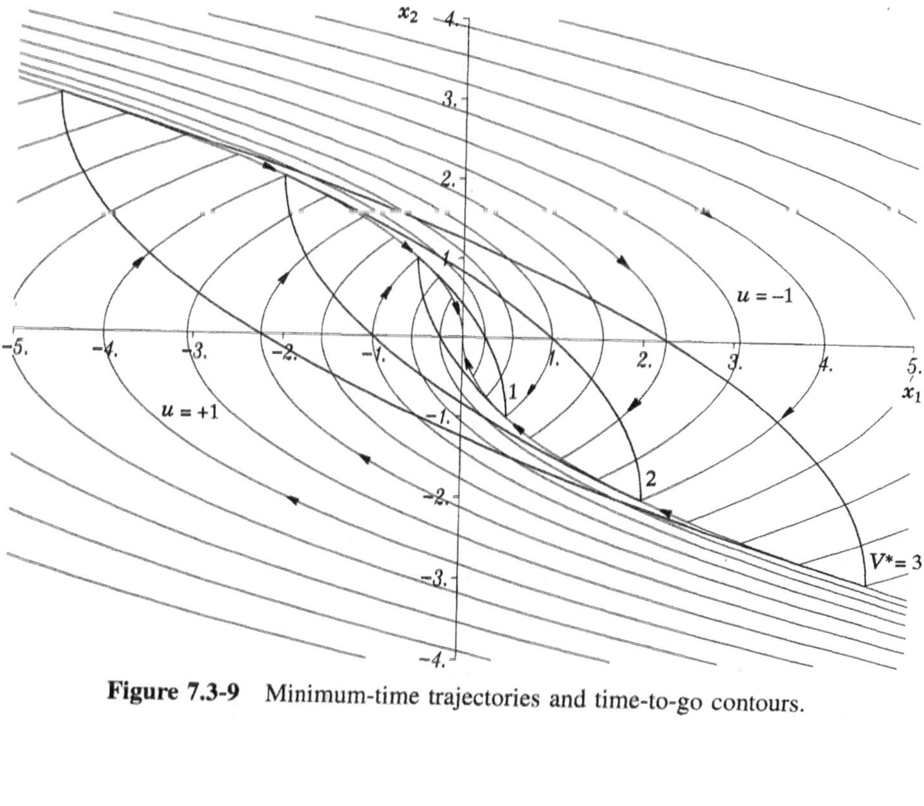

Figure 7.3-9 Minimum-time trajectories and time-to-go contours.

$$\dot{\lambda}_1 = -\frac{\partial H}{\partial x_1} = 0$$
$$\dot{\lambda}_2 = -\frac{\partial H}{\partial x_2} = -\lambda_1, \tag{7.3-19}$$

with the terminal transversality conditions

$$\boldsymbol{\lambda}^{\mathrm{T}} = \boldsymbol{\rho}^{\mathrm{T}} \frac{\partial \mathbf{g}(\mathbf{x}_f)}{\partial \mathbf{x}} = \boldsymbol{\rho}^{\mathrm{T}} \tag{7.3-20}$$

at $\mathbf{g}(\mathbf{x}_f) = \mathbf{0}$. Since $\mathbf{x}_f = \mathbf{0}$ is a fixed point the transversality conditions (7.3-20) yield no useful information; λ_{1f} and λ_{2f} are undetermined constants.

From the minimum principle

$$0 = \min_{|u| \le 1} H = \min_{|u| \le 1} \{\lambda_0 + \lambda_1 x_2 + \lambda_2 u\}$$

we obtain the optimal control

$$u^* = -\text{sgn}(\sigma), \tag{7.3-21}$$

where

$$\sigma(t) \triangleq \frac{\partial H}{\partial u} = \lambda_2(t) \tag{7.3-22}$$

is not identically zero for a nonzero time interval. To verify this note that $0 = \sigma = \lambda_2$ and $0 = \dot{\sigma} = \dot{\lambda}_2 = -\lambda_1$ would require $\boldsymbol{\lambda}(t) = \mathbf{0}$. Then $H = 0$ would require $(\lambda_0, \boldsymbol{\lambda}) = \mathbf{0}$, which is not allowed by the minimum principle. Substituting the optimal control into H, with $\lambda_0 = 1$, yields

$$H = 1 + \lambda_1 x_2 - |\lambda_2|.$$

To generate optimal trajectories we integrate the state and adjoint equations backwards from the terminal point at the origin. From $H_f = 0$ we have $\lambda_{2f} = \pm 1$, with corresponding $u_f^* = \mp 1$. Pick either of the two cases, say $\lambda_{2f} = -1$ with $u_f^* = 1$, and pick a value for λ_{1f} (to generate a trajectory in the upper right-hand side of Figure 7.3-9). In terms of the time-to-go $\tau = t_f - t$ the retro-time adjoint equations are

$$\frac{d\lambda_1}{d\tau} = 0$$

$$\frac{d\lambda_2}{d\tau} = \lambda_1,$$

which yields

$$\lambda_1 = \lambda_{1f}$$

$$\lambda_2 = \lambda_{1f}\tau + \lambda_{2f}.$$

Note that, if λ_{1f} and λ_{2f} have opposite signs, then the switching function $\sigma(\tau) = \lambda_2(\tau)$ will change sign, causing the control $u^*(\tau)$ to switch, at a switching time

$$\tau_s = -\frac{\lambda_{2f}}{\lambda_{1f}}.$$

Otherwise, the switching function does not change sign, and the control $u^*(\tau)$ stays constant at $u^*(\tau) = u_f^*$. For the $\lambda_{2f} = -1$ choose $\lambda_{1f} > 0$ so we do get a switch. The resulting control is

$$u^*(\tau) = \begin{cases} 1 & \text{for } 0 \le \tau < \tau_s = 1/\lambda_{1f} \\ -1 & \text{for } \tau \ge \tau_s. \end{cases}$$

Starting from the origin and using this control to integrate the retro-time state equations

$$\frac{dx_1}{d\tau} = -x_2$$

$$\frac{dx_2}{d\tau} = -u$$

yields

$$x_2(\tau) = \begin{cases} -u_f^*\tau & \text{for } 0 \le \tau \le \tau_s \\ (\tau - \tau_s)u_f^* + x_{2s} & \text{for } \tau \ge \tau_s, \end{cases} \tag{7.3-23}$$

so that

$$x_1(\tau) = \begin{cases} u_f^* \dfrac{\tau^2}{2} & \text{for } 0 \le \tau \le \tau_s \\ -\frac{1}{2}(\tau - \tau_s)^2 u_f^* - (\tau - \tau_s)x_{2s} + x_{1s} & \text{for } \tau \ge \tau_s, \end{cases} \tag{7.3-24}$$

where the switching point is

$$\begin{aligned} x_{1s} &= u_f^* \frac{\tau_s^2}{2} \\ x_{2s} &= -u_f^* \tau_s. \end{aligned} \tag{7.3-25}$$

For $\lambda_{2f} = -1$ and $u_f^* = 1$, we generate all of the upper trajectories in Figure 7.3-9 by varying $\lambda_{1f} > 0$ and thus the retro switching time $\tau_s = 1/\lambda_{1f}$. For the other case, $\lambda_{2f} = 1$ and $u_f^* = -1$, we generate the lower trajectories in Figure 7.3-9 by varying $\lambda_{1f} < 0$, with $\tau_s = -1/\lambda_{1f}$.

Since the control is $u^* = 1$ below the switching surface in Figure 7.3-9 and $u^* = -1$ above it, we can synthesize the closed-loop control simply by determining the switching surface $\sigma(\mathbf{x}) = 0$. By eliminating τ_s in (7.3-25) and using $u_f^* = \pm 1$, the results can be written as

$$u^*(\mathbf{x}) = \begin{cases} -\text{sgn}[\sigma(\mathbf{x})] & \text{if } \sigma(\mathbf{x}) \ne 0 \\ -\text{sgn}(x_2) & \text{if } \sigma(\mathbf{x}) = 0 \text{ and } \mathbf{x} \ne \mathbf{0} \\ 0 & \text{if } \mathbf{x} = \mathbf{0}, \end{cases} \tag{7.3-26}$$

where the switching function (two half-parabolas, in the second and fourth quadrants) can be written as

$$\sigma(\mathbf{x}) = \begin{cases} x_2 + \sqrt{2x_1} & \text{if } x_1 \ge 0 \\ x_2 - \sqrt{2|x_1|} & \text{if } x_1 \le 0. \end{cases} \tag{7.3-27}$$

To fully complete the synthesis, the minimum time-to-go can be determined by solving for τ using (7.3-23)–(7.3-25). For $0 \le \tau \le \tau_s$ equations (7.3-23) and (7.3-24) yield

$$\tau = -\frac{x_2}{u_f^*} \qquad \text{on} \qquad x_1 = -\frac{x_2^2}{2u_f^*}, \qquad \frac{x_2}{u_f^*} \le 0. \tag{7.3-28}$$

For $\tau \ge \tau_s$, using x_{2s} from (7.3-25) in (7.3-23) yields

$$\tau_s = \frac{\tau}{2} - \frac{x_2}{2u_f^*}.$$

Using this result to compute x_{1s} in (7.3-25) and then substituting the results for τ_s and x_{1s} into (7.3-24) and simplifying the result yields

$$x_1 = \frac{u_f^*}{2}(a^2 + 2ab - b^2) = \frac{u_f^*}{2}[(a+b)^2 - 2b^2],$$

where

$$a = \frac{1}{2}\left(\tau - \frac{x_2}{u_f^*}\right)$$

$$b = \frac{1}{2}\left(\tau + \frac{x_2}{u_f^*}\right).$$

Thus

$$x_1 = \frac{u_f^*}{2}\left[\tau^2 - \frac{1}{2}\left(\tau + \frac{x_2}{u_f^*}\right)^2\right].$$

Solving for τ, accounting for the two cases $u_f^* = \pm 1$, and incorporating (7.3-28), we find the minimum cost-to-go to be

$$V^*(\mathbf{x}) = \tau = \begin{cases} x_2 + \sqrt{2x_2^2 + 4x_1} & \text{if } \sigma(\mathbf{x}) > 0 \\ -x_2 + \sqrt{2x_2^2 - 4x_1} & \text{if } \sigma(\mathbf{x}) < 0 \\ |x_2| & \text{if } \sigma(\mathbf{x}) = 0, \end{cases}$$

where $\sigma(\mathbf{x})$ is given by (7.3-27). Curves of constant $V^*(\mathbf{x})$ are plotted in Figure 7.3-9. One can verify that at any point $\mathbf{x}$ where $\sigma(\mathbf{x}) \neq 0$:

$$\boldsymbol{\lambda}^T = \frac{\partial V^*(\mathbf{x})}{\partial \mathbf{x}}. \qquad (7.3\text{-}29)$$

But on $\sigma(\mathbf{x}) = 0$ this result does not hold, since (a) the gradient $\partial V^*(\mathbf{x})/\partial \mathbf{x}$ is discontinuous across $\sigma(\mathbf{x}) = 0$, and (b) $\boldsymbol{\lambda}(t)$ is multivalued on $\sigma(\mathbf{x}) = 0$. We leave the details as an exercise for the reader (Exercise 7-3). ■

Example 7.3-4: Missile Guidance Problem We see in Chapter 9 that the missile guidance problem would more properly be formulated as a two-player optimal control problem, called a pursuit-evasion differential game. However, in the situation of a nonmaneuvering evader a very important guidance law can be obtained using the optimal control approach.

Consider the two-dimensional pursuit-evasion situation depicted in Figure 7.3-10. The nonrotating (z_1, z_2) coordinate system is attached to the missile (the pursuer), which is moving with a constant speed S_u at an angle z_3 with respect to the z_1-axis. The missile can change direction by controlling its turning rate u. The evader is the target, which is moving with constant speed S_v at an angle z_4 with respect to the z_1-axis. The target can change direction by controlling its turning rate v. For now we assume that v is a known function $v(t)$. The **line-of-sight** angle γ is the angle from the z_2-axis to a line drawn between the missile and the target. No bounds are placed on the controls u and v.

From Figure 7.3-10 the equations of motion are given by

$$\begin{aligned} \dot{z}_1 &= S_v \cos z_4 - S_u \cos z_3 \\ \dot{z}_2 &= S_v \sin z_4 - S_u \sin z_3 \\ \dot{z}_3 &= u \\ \dot{z}_4 &= v. \end{aligned} \tag{7.3-30}$$

In terms of the complementary angles

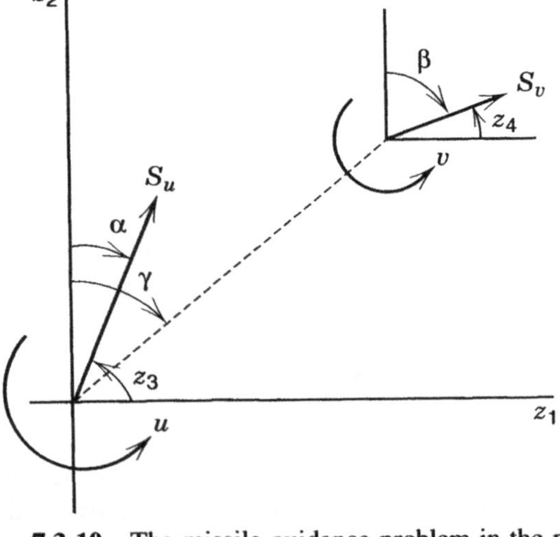

Figure 7.3-10 The missile guidance problem in the plane.

$$z_3 = \frac{\pi}{2} - \alpha$$
$$z_4 = \frac{\pi}{2} - \beta, \tag{7.3-31}$$

we have

$$\dot{z}_1 = S_v \cos\left(\frac{\pi}{2} - \beta\right) - S_u \cos\left(\frac{\pi}{2} - \alpha\right) = S_v \sin\beta - S_u \sin\alpha$$
$$\dot{z}_2 = S_v \sin\left(\frac{\pi}{2} - \beta\right) - S_u \sin\left(\frac{\pi}{2} - \alpha\right) = S_v \cos\beta - S_u \cos\alpha$$
$$\dot{\alpha} = -u$$
$$\dot{\beta} = -v.$$

Assume that either (a) α is small and β is small (tail-chase), or (b) α is small and $\beta = \pi + \bar{\beta}$ where $\bar{\beta}$ is small (head-to-head). If we now make small angle assumptions, then

$$\dot{z}_1 = S_v\beta - S_u\alpha \qquad \text{or} \qquad \dot{z}_1 = -S_v\bar{\beta} - S_u\alpha \Rightarrow \dot{z}_1 = \pm S_v\beta - S_u\alpha$$
$$\dot{z}_2 = \pm S_v - S_u$$
$$\dot{\alpha} = -u \tag{7.3-32}$$
$$\dot{\beta} = -v,$$

with β now being a small angle measured either from 0 (for the "+" tail-chase case) or π (for the "−" head-on case). Equations (7.3-32) are equivalent to

$$\ddot{z}_1 = \mp S_v v + S_u u$$
$$\dot{z}_2 = \pm S_v - S_u.$$

If we define the normal acceleration of the missile, normal acceleration of the target, and closing velocity as

$$A_{M_n} = S_u u$$
$$A_{T_n} = \pm S_v v$$
$$V_c = \pm S_v - S_u,$$

the equations become

$$\ddot{z}_1 = A_{M_n} - A_{T_n}$$
$$\dot{z}_2 = V_c.$$

In terms of the state variables

$$x_1 = z_1$$
$$x_2 = \dot{z}_1$$
$$x_3 = z_2,$$

the system may be written as

$$\begin{aligned} \dot{x}_1 &= x_2 \\ \dot{x}_2 &= A_{M_n} - A_{T_n} \\ \dot{x}_3 &= V_c. \end{aligned} \tag{7.3-33}$$

Note that the first two equations are the same as the juggernaut equations, except that now the control u in the juggernaut equations has been replaced by the accelerations of both the pursuer and the evader. The two systems also differ in dimension. Here x_1 corresponds to horizontal range, x_2 to horizontal velocity, and x_3 to vertical range.

Line of Sight The line of sight angle γ is defined by

$$\tan\gamma \triangleq \frac{z_1}{z_2}.$$

We are interested in the rate of change of line of sight, obtained from the time derivative

$$\dot{\gamma}\sec^2\gamma = \frac{z_2\dot{z}_1 - z_1\dot{z}_2}{z_2^2} = \frac{z_2[S_v\cos z_4 - S_u\cos z_3] - z_1[S_v\sin z_4 - S_u\sin z_3]}{z_2^2}.$$

Noting that $\sec\gamma = r/z_2$ where $r = \sqrt{z_1^2 + z_2^2}$ and using the definitions (7.3-31), we obtain

$$\dot{\gamma} = \frac{z_2[S_v\cos(\pi/2 - \beta) - S_u\cos(\pi/2 - \alpha)] - z_1[S_v\sin(\pi/2 - \beta) - S_u\sin(\pi/2 - \alpha)]}{r^2}.$$

We now make the following small angle assumptions: α is small and β is small, with β measured either from 0 (tail-chase) or π (head-to-head). We also assume z_1 is small and replace r with z_2 to obtain

$$\dot{\gamma} = \frac{\pm S_v \beta - S_u \alpha}{z_2} - \frac{(\pm S_v - S_u) z_1}{z_2^2}.$$

Now from the first equation of (7.3-32) and from the definitions for x_1, x_2, x_3, and V_c, we obtain

$$\dot{\gamma} = \frac{x_2}{x_3} - \frac{x_1}{x_3^2} V_c. \tag{7.3-34}$$

Optimal Guidance The pursuer's objective is to hit the evader, as defined by the terminal conditions

$$x_1(t_f) = 0$$
$$x_3(t_f) = 0,$$

using a minimum amount of control action as given by the cost function

$$x_0 = \tfrac{1}{2} \int_0^{t_f} A_{M_n}^2 \, dt.$$

A quadratic cost is chosen since A_{M_n} may be either positive or negative. This "penalty" on the use of control by the pursuer is imposed, in part, because we are not modeling any bounds on the pursuer's control.

The H function for this problem, with $\lambda_0 = 1$, is given by

$$H = \tfrac{1}{2} A_{M_n}^2 + \lambda_1 x_2 + \lambda_2 (A_{M_n} - A_{T_n}) + \lambda_3 V_c.$$

Since the control variable is unconstrained, a necessary condition for minimizing H is obtained by taking the partial derivative of H with respect to A_{M_n} and setting the result equal to zero:

$$\frac{\partial H}{\partial A_{M_n}} = A_{M_n} + \lambda_2 = 0 \Rightarrow A_{M_n} = -\lambda_2.$$

From the adjoint equations we obtain

$$\dot{\lambda}_1 = 0 \Rightarrow \lambda_1(t) = \lambda_1(t_f) = \rho_1$$
$$\dot{\lambda}_2 = -\lambda_1 \text{ where } \lambda_2(t_f) = 0$$
$$\dot{\lambda}_3 = 0 \Rightarrow \lambda_3(t) = \lambda_3(t_f) = \rho_2$$

where the transversality conditions have been used to obtain the right-hand relationships. We integrate the second adjoint equation backward in time using $\tau \triangleq t_f - t$ to obtain

$$\lambda_2 = \rho_1 \tau.$$

Thus

$$A_{M_n} = -\rho_1 \tau. \tag{7.3-35}$$

This result gives an open-loop solution for the control in terms of an as-yet-unknown constant ρ_1. We can determine the constant and also find the closed-loop control law by integrating the state equations backward in time using this open-loop control.

Substituting (7.3-35) into the linear state equations (7.3-33) and integrating backward, we obtain

$$x_3 = -V_c \tau$$

$$x_2 = x_2(t_f) + \frac{\rho_1}{2}\tau^2 + \int_0^\tau A_{T_n}\, d\tau$$

$$x_1 = -x_2(t_f)\tau - \frac{\rho_1}{6}\tau^3 - \int_0^\tau \left(\int_0^\tau A_{T_n}\, d\tau \right) d\tau.$$

Using the x_2 equation to solve for $x_2(t_f)$ and substituting into the x_1 equation yields

$$x_1 = -x_2 \tau + \frac{\rho_1}{2}\tau^3 + \tau \int_0^\tau A_{T_n}\, d\tau - \frac{\rho_1}{6}\tau^3 - \int_0^\tau \left(\int_0^\tau A_{T_n}\, d\tau \right) d\tau.$$

Using (7.3-35) to eliminate ρ_1 and simplifying yields

$$x_1 = -x_2 \tau - \frac{1}{3}\tau^2 A_{M_n} + \tau \int_0^\tau A_{T_n}\, d\tau - \int_0^\tau \left(\int_0^\tau A_{T_n}\, d\tau \right) d\tau.$$

Solving for the missile acceleration yields

$$A_{M_n} = 3\left[-\frac{x_2}{\tau} - \frac{x_1}{\tau^2} + \frac{1}{\tau}\int_0^\tau A_{T_n}\, d\tau - \frac{1}{\tau^2}\int_0^\tau \left(\int_0^\tau A_{T_n}\, d\tau \right) d\tau \right].$$

Using the x_3 equation to eliminate τ yields the closed-loop optimal guidance law

$$A_{M_n} = 3\left[\frac{x_2}{x_3} V_c - \frac{x_1}{x_3^2} V_c^2 + \frac{1}{\tau}\int_0^\tau A_{T_n}\, d\tau - \frac{1}{\tau^2}\int_0^\tau \left(\int_0^\tau A_{T_n}\, d\tau\right) d\tau\right].$$

Implementation of this control law would appear to be difficult, since we must not only know the closing velocity V_c, target acceleration A_{T_n}, and the time-to-go τ (which can usually be measured or estimated) but all three state variables as well. However, the first two terms in brackets are just $V_c\dot{\gamma}$, which means that only $\dot{\gamma}$ need be measured rather than the three state variables. Factoring out V_c it follows from (7.3-34) that the missile acceleration is given by

$$A_{M_n} = 3V_c\left[\dot{\gamma} + \frac{1}{V_c\tau}\int_0^\tau A_{T_n}\, d\tau - \frac{1}{V_c\tau^2}\int_0^\tau \left(\int_0^\tau A_{T_n}\, d\tau\right) d\tau\right]. \quad (7.3\text{-}36)$$

While this guidance law is applicable for any known target acceleration, there are two cases of particular importance.

Constant Target Velocity Setting $A_{T_n} = 0$, and the **navigation gain** $N = 3$, we obtain

$$A_{M_n} = N V_c \dot{\gamma}. \quad (7.3\text{-}37)$$

This control law is known as **proportional navigation**. That is, the control acceleration ($= S_u u$) is proportional to the rate of change of the line-of-sight angle. This control law is deeply rooted in intuitive notion. As we show in Section 9.7, two vehicles moving with constant velocity on a collision course will have $\dot{\gamma} = 0$. Therefore, in a missile guidance setting, if $\dot{\gamma} \neq 0$, it is reasonable to change the missile's heading so that $\gamma \rightarrow 0$.

Proportional navigation has been used for many years. Apparently early pirates successfully used this algorithm. It is still basic to the guidance systems of modern missiles. Variations on proportional navigation often involve using different values for the navigation gain N. Recall that $A_{M_n} = S_u u$ and $V_c = \pm S_v - S_u$. It then follows that for stationary targets ($S_v = 0$) the above control law simplifies to

$$u = -N\dot{\gamma}.$$

Constant Target Acceleration Setting $A_{T_n} = \overline{A}_{T_n} =$ constant we obtain from (7.3-36)

$$A_{M_n} = N V_c \left(\dot{\gamma} + \frac{\overline{A}_{T_n}}{V_c} - \frac{\overline{A}_{T_n}}{2V_c}\right) = N V_c \dot{\gamma} + \frac{N}{2}\overline{A}_{T_n}.$$

This control law is known as **augmented proportional navigation** and will produce better results than proportional navigation when used against constant accelerating targets, provided that a good estimate for $\overline{A}_{T_n}$ is known.

It should be noted that the above guidance laws are applicable only to a tail-chase or head-to-head situation. The small angles assumption is not just a device to simplify computations, it is a fundamental premise. If the assumption is violated, then extremely high values may occur for the pursuer's turning rate u. Equation (7.3-34) gives an indication of this. If x_3 approaches zero ($z_2 = 0$ as the pursuer and evader fly by each other), then $\dot{\gamma}$ approaches infinity!

As an example of how this might occur in a real-world application, suppose that the evader is a low-altitude cruise missile and the pursuer is a surface-to-air missile. In Figure 7.3-10, let the z_2-axis be the ground and let z_1 be altitude, and suppose that the evader's velocity vector is directed in the negative z_2-direction. If the cruise missile is approaching the surface-to-air launcher from very far away and the interceptor is launched in the direction of the evader, then initially γ will be small. But as the interceptor approaches the cruise missile the line of sight will change very rapidly and if the surface-to-air missiles does not have a very high "max-g" turning-rate capability the evader could fly past the pursuer. ■

7.4 SINGULAR OPTIMAL CONTROLS

It is not always possible to solve for the control directly using the necessary condition (7.2-29). This will happen whenever a component $\partial H/\partial u_j$ of the function $\partial H/\partial \mathbf{u}$ does not contain any components of the control vector $\mathbf{u}$. This is known as the **singular control** case, where the determinant of $\partial^2 H/\partial \mathbf{u}^2$ is zero, and it will occur whenever a control variable appears linearly in the H function. In a singular control situation, additional necessary conditions must be used in order to solve for the control [Johnson and Gibson, 1963; Goh, 1966; Kelley et al., 1967]. However, the additional required necessary conditions are easily generated for most problems where singular control occurs. We illustrate the procedure by way of example.

Example 7.4-1: Maximum Yield Harvest The **logistics equation**, given by

$$\dot{x}_1 = \frac{r}{K}\, x_1(K - x_1),$$

is often used to describe the growth of a single species population system, such as fish in a lake. The variable x_1 represents some appropriate measure of the population size (numbers, mass, etc.), the parameter r is called the **intrinsic growth rate**, and the parameter K is called the **carrying capacity**. The quantities r, K, and x_1 are all positive. The solution to this differential equation with

x_1 initially small gives an "S" shaped growth curve with time, with exponential growth at small populations and then a leveling off as the population nears K.

We can "harvest" this system by subtracting a term ux_1 from the right-hand side. In this case, u stands for the amount of harvesting **effort**. The product of harvesting effort times population size gives the rate at which the species is harvested. The total harvest over a time period $0 \le t \le T$ is given by

$$\int_0^T ux_1 \, dt.$$

Our objective is to harvest this system such that at the end of a fixed time interval the maximum amount has been harvested [Goh, 1969; Cliff and Vincent, 1973].

To state this as a minimum-cost optimal control problem, we minimize the negative of the total harvest integral. In addition, since the final time is specified, we must introduce time as a second state variable $x_2 = t$. The corresponding cost and state equations can be written as

$$\dot{x}_0 = -ux_1$$

$$\dot{x}_1 = \frac{r}{K} x_1(K - x_1) - ux_1$$

$$\dot{x}_2 = 1.$$

The harvesting effort is limited between $u = 0$ (no harvest) and $u = u_{max}$ (maximum-effort harvest). Thus the control set $\mathcal{U}$ is defined by the two inequality constraints

$$h_1 = u \ge 0$$
$$h_2 = u_{max} - u \ge 0.$$

The terminal set is defined in terms of the single equality constraint

$$g(\mathbf{x}) = x_2(t_f) - T = 0,$$

where T is the specified time interval.

To solve this problem, we first form the H function (with $\lambda_0 = 1$) as

$$H = -ux_1 + \lambda_1 \left[\frac{r}{K} x_1(K - x_1) - ux_1 \right] + \lambda_2.$$

Since u is constrained, necessary conditions for minimizing H are obtained from (7.2-19)–(7.2-23). For this problem the $L(\cdot)$ function is given by

$$L = H - \gamma_1 u - \gamma_2(u_{\max} - u).$$

Setting $\partial L/\partial u$ equal to zero yields

$$0 = \frac{\partial L}{\partial u} = -x_1(1 + \lambda_1) - \gamma_1 + \gamma_2.$$

Setting

$$\sigma \triangleq \frac{\partial H}{\partial u} = -x_1(1 + \lambda_1)$$

and utilizing (7.2-20)–(7.2-22), we have three cases involving the Lagrange multipliers $\gamma_1 \geq 0$ and $\gamma_2 \geq 0$, which indicate whether one, or the other, or neither of the control inequalities is active. Note that since the two control constraints, $u = u_{\max}$ and $u = 0$, cannot both be active, we will have either $\gamma_1 = 0$ or $\gamma_2 = 0$ or both.

If $\sigma < 0$, then $\partial L/\partial u = 0$ requires $\gamma_1 = 0$ and $\gamma_2 > 0$, which implies $u^* = u_{\max}$. Conversely, if $\sigma > 0$, then $\partial L/\partial u = 0$ requires $\gamma_1 > 0$ and $\gamma_2 = 0$, yielding $u^* = 0$. If $\sigma = 0$, then $\partial L/\partial u = 0$ requires $\gamma_1 = \gamma_2 = 0$ and yields no direct information for the optimal control u^*. So far we can conclude that

$$u^* = \begin{cases} u_{\max} & \text{if } \sigma < 0 \\ ? & \text{if } \sigma \equiv 0 \\ 0 & \text{if } \sigma > 0. \end{cases}$$

The quantity σ is known as a **switching function**, defined by

$$\sigma(\mathbf{x}, \boldsymbol{\lambda}, \lambda_0) \triangleq \frac{\partial H(\mathbf{x}, u, \boldsymbol{\lambda}, \lambda_0)}{\partial u} \tag{7.4-1}$$

for the case where $H(\cdot)$ is linear in u. Clearly, for this example, the control program is determined by the switching function, which in turn is a function of both x_1 and λ_1. Additional conditions are needed to handle the case where $\sigma(t) \triangleq \sigma[\mathbf{x}(t), \boldsymbol{\lambda}(t)] \equiv 0$ on some nonzero time interval. This case is called **singular control**, because the control u ceases to affect the value of H directly.

The adjoint variables are obtained from (7.2-26), and for this problem are given by the differential equations

$$\dot{\lambda}_1 = -\frac{\partial H}{\partial x_1} = u - \lambda_1 \left(r - \frac{2r}{K} x_1 - u \right)$$

$$\dot{\lambda}_2 = -\frac{\partial H}{\partial x_2} = 0.$$

It follows that λ_2 is a constant. The transversality conditions (7.2-28) yield the results

$$\lambda_1(t_f) = \rho \frac{\partial g(\mathbf{x}_f)}{\partial x_1} = 0$$

$$\lambda_2(t_f) = \rho \frac{\partial g(\mathbf{x}_f)}{\partial x_2} = \rho.$$

No useful information is obtained from the second transversality condition. However, we have sufficient information to solve the problem.

We must first resolve the question of whether it is possible to have $\sigma \equiv 0$. In order for $\sigma \equiv 0$, all time derivatives of σ must also be zero. The first time derivative of σ is given by

$$\dot{\sigma} = -\dot{x}_1(1 + \lambda_1) - x_1\dot{\lambda}_1.$$

Substituting for $\dot{x}_1$ and $\dot{\lambda}_1$ yields

$$\dot{\sigma} = \frac{r}{K} x_1[-K + x_1 - \lambda_1 x_1] = \frac{r}{K} x_1[-K + 2x_1 + \sigma],$$

which reduces to

$$\dot{\sigma} = \frac{r}{K} x_1[2x_1 - K]$$

for $\sigma = 0$. Setting this expression equal to zero implies that

$$x_1 = \frac{K}{2}$$

under singular control. The control that will maintain this state is obtained by differentiating σ again, yielding

$$\ddot{\sigma} = \frac{r}{K} \dot{x}_1[4x_1 - K + \sigma] + \frac{r}{K} x_1\dot{\sigma}.$$

For $\sigma = \dot{\sigma} = 0$ and $x_1 = K/2$ we have

$$\ddot{\sigma} = \frac{rK}{2}\left[\frac{r}{2} - u\right].$$

Setting this expression equal to zero yields

$$u = \frac{r}{2}.$$

If a singular control exists, it may or may not be part of an optimal control sequence. A **necessary condition for a singular control** [Johnson and Gibson, 1963; Kelley et al., 1967] to minimize the H function is that

$$(-1)^n \frac{\partial}{\partial u}\left(\frac{d^{2n}}{dt^{2n}}\left[\frac{\partial H}{\partial u}\right]\right) \geq 0, \tag{7.4-2}$$

where, during the repeated differentiation of the switching function $\sigma = \partial H/\partial u$, the control (which will always cancel out of odd numbered derivatives) first appears in the $(2n)$th derivative. For our example, $n = 1$ and the necessary condition (7.4-2) is satisfied.

The fact that we have obtained a singular control does not necessarily imply that singular control will enter every optimal solution. For example, let $r = K = 1$. Figure 7.4-1 illustrates the complete synthesis of this problem

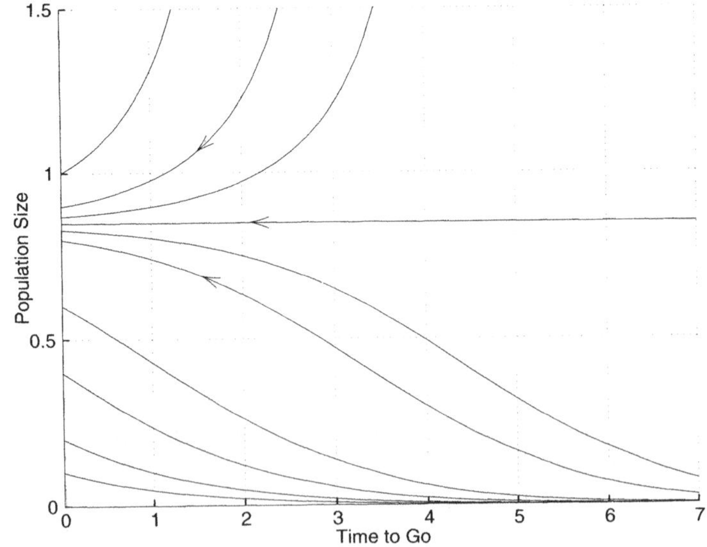

Figure 7.4-1 Optimal harvesting for $r = K = 1$ and $u_{max} = 0.15$.

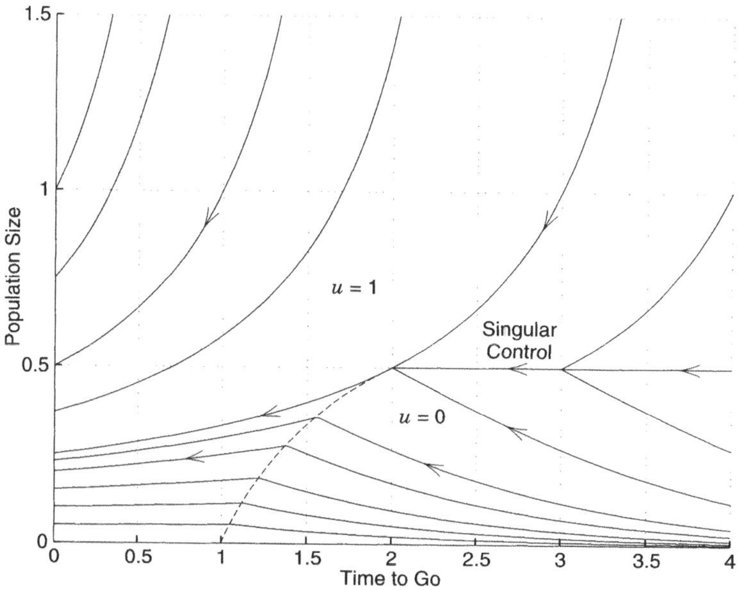

Figure 7.4-2 Optimal harvesting for $r = K = u_{max} = 1$.

when $u_{max} = 0.15$. In this case singular control is not possible, since the singular control $u = 0.5$ does not satisfy the constraints $0 \leq u \leq u_{max}$. All extremals in this case are composed of maximum control arcs ($u = 0.15$) only.

Figure 7.4-2, for $r = K = u_{max} = 1$, illustrates a case where singular control is required to complete the synthesis of the solution. Even in this case the singular solution would not be used from all starting conditions.

The solutions plotted in Figures 7.4-1 and 7.4-2 are obtained by integrating the system equations and adjoint equations backward in time from the terminal manifold. By choosing various points on the terminal set to start from, initial conditions (in the retro-time sense) for the state equations are known. The initial condition for λ_1 is determined from the transversality condition $\lambda_{1f} = 0$, and the initial condition for λ_2 is determined from $H_f = 0$. Note that the terminal control is always maximum control. This follows from the fact that at the terminal time the switching function is negative.

The singular trajectory in Figure 7.4-2 is found numerically by choosing $x_1(0)$, integrating backwards until the switching condition $\sigma = 0$ occurs, and then iteratively adjusting $x_1(0)$ via some root-finding scheme until both $\sigma = 0$ and $\dot{\sigma} = 0$ occur at the same time. Backwards integration is then continued by using the singular control $u = 0.5$ for various time intervals, followed by a switch to either $u = 0$ or $u = u_{max}$ to fill the state space with trajectories. ■

7.5 NONSTANDARD PROBLEMS

Often a problem may be formulated in such a way that it is not in the standard form introduced in Section 7.1. For example, time may be an explicit part of the problem statement, as in the fixed time interval problem considered in Example 7.4-1. This problem was not in standard form since time t explicitly appeared in the formulation. As illustrated by the example, this particular situation can be easily dealt with by defining time as a state variable, for example, $x_{n_x} = t$, with the differential equation $\dot{x}_{n_x} = 1$.

Our procedure for handling other nonstandard problems is the same: We convert nonstandard problems into standard ones by appropriate manipulation. Actually, a surprising number of nonstandard problems can be converted in this way.

Terminal Cost Problems

In many applications the most natural way of formulating an optimal control problem may be to state the cost as a function of the terminal state, rather than as an integral cost problem.

The **terminal cost problem** is to minimize a function $\psi(\mathbf{x})$ evaluated at the terminal state $\mathbf{x}_f \triangleq \mathbf{x}(t_f)$. This problem can be converted to the standard integral cost form by noting that

$$x_0(t_f) \triangleq \psi(\mathbf{x}_f) - \psi[\mathbf{x}(0)] = \int_0^{t_f} \frac{d\psi[\mathbf{x}(t)]}{dt}\, dt.$$

Thus the accumulated cost $x_0(t)$ obeys the differential equation

$$\dot{x}_0 = \frac{\partial \psi(\mathbf{x})}{\partial \mathbf{x}} \mathbf{f}(\mathbf{x}, \mathbf{u}).$$

Example 7.5-1: A Terminal Cost Problem In this example [Leitmann, 1981a, p. 206] the following sum is to be minimized:

$$t_f + \alpha x_1^2(t_f)$$

subject to the system dynamics

$$\dot{x}_1 = x_2$$
$$\dot{x}_2 = u,$$

where

$$|u| \le 1$$

and the terminal set is defined by

$$x_2(t_f) = 0.$$

If we define

$$x_0 = t + \alpha x_1^2 + \text{constant},$$

with $x_0(0) = 0$, then the accumulated cost for this problem obeys the equation

$$\dot{x}_0 = 1 + 2\alpha x_1 x_2.$$

The problem is now in the form of the integral cost optimal control problem.■

The Problem of Bolza

If an optimal control problem has both an integral cost term and an added terminal cost, we could convert the terminal cost portion to integral form, as in Example 7.5-1. However, we can also obtain necessary conditions that apply directly to the original terminal-plus-integral cost problem.

The **problem of Bolza**, in its most general form, is to minimize an initial + final + integral cost functional of the form

$$J[\mathbf{u}(\cdot)] = \theta(\mathbf{x}_i) + \psi(\mathbf{x}_f) + \int_0^{t_f} f_0(\mathbf{x}, \mathbf{u})\, dt \tag{7.5-1}$$

for a control system

$$\dot{\mathbf{x}} = \mathbf{f}(\mathbf{x}, \mathbf{u}), \tag{7.5-2}$$

subject to the control constraints

$$\mathbf{u} \in \mathcal{U} = \{\mathbf{u} | \mathbf{h}(\mathbf{u}) \ge \mathbf{0}\}, \tag{7.5-3}$$

with the terminal state $\mathbf{x}_f \triangleq \mathbf{x}(t_f)$, at an unspecified terminal time t_f, constrained by n_g terminal conditions

$$\mathbf{g}(\mathbf{x}_f) = \mathbf{0}, \tag{7.5-4}$$

and the initial state $\mathbf{x}_i \triangleq \mathbf{x}(0)$ constrained by n_m initial conditions

$$\mathbf{m}(\mathbf{x}_i) = \mathbf{0}. \tag{7.5-5}$$

A General Minimum Principle for the Bolza Problem For the problem of Bolza, Pontryagin's minimum principle yields the following necessary conditions:

$$0 = H[\mathbf{x}, \mathbf{u}(\mathbf{x}), \boldsymbol{\zeta}, \zeta_0] = \min_{\mathbf{u} \in \mathcal{U}} H(\mathbf{x}, \mathbf{u}, \boldsymbol{\zeta}, \zeta_0) \tag{7.5-6}$$

$$\dot{\boldsymbol{\zeta}}^{\mathrm{T}} = -\frac{\partial H}{\partial \mathbf{x}} \tag{7.5-7}$$

$$\mathbf{g}(\mathbf{x}_f) = \mathbf{0} \tag{7.5-8}$$

$$\mathbf{m}(\mathbf{x}_i) = \mathbf{0} \tag{7.5-9}$$

$$\boldsymbol{\zeta}_f^{\mathrm{T}} = \frac{\partial \Psi(\mathbf{x}_f, \boldsymbol{\rho})}{\partial \mathbf{x}} \tag{7.5-10}$$

$$\boldsymbol{\zeta}_i^{\mathrm{T}} = \frac{\partial \Theta(\mathbf{x}_i, \boldsymbol{\mu})}{\partial \mathbf{x}}, \tag{7.5-11}$$

where $\zeta_0 = \text{constant} \geq 0$, $\boldsymbol{\zeta}_f \triangleq \boldsymbol{\zeta}(t_f)$, $\boldsymbol{\zeta}_i \triangleq \boldsymbol{\zeta}(0)$, and

$$H(\mathbf{x}, \mathbf{u}, \boldsymbol{\zeta}, \zeta_0) \triangleq \zeta_0 f_0(\mathbf{x}, \mathbf{u}) + \boldsymbol{\zeta}^{\mathrm{T}}\mathbf{f}(\mathbf{x}, \mathbf{u}) \tag{7.5-12}$$

$$\Psi(\mathbf{x}, \boldsymbol{\rho}) \triangleq \psi(\mathbf{x}) + \boldsymbol{\rho}^{\mathrm{T}}\mathbf{g}(\mathbf{x}) \tag{7.5-13}$$

$$\Theta(\mathbf{x}, \boldsymbol{\mu}) \triangleq \theta(\mathbf{x}) + \boldsymbol{\mu}^{\mathrm{T}}\mathbf{m}(\mathbf{x}). \tag{7.5-14}$$

We have used ζ_0 and $\boldsymbol{\zeta}$ to denote the adjoint variables for the problem of Bolza in order to emphasize that they have different values than the adjoint variables of the standard minimum principle for the same problem converted to integral cost form.

Problems with an Integral Constraint

A problem may be nonstandard because of the inclusion of constraints on some quantity other than those defining the control constraint set. The particular form of constraint we consider is called an **isoperimetric constraint**, in which the value of some integral must satisfy a constraint.

Example 7.5-2: An Integral Constraint Suppose we wish to minimize the time of transfer to the origin for a system [Leitmann, 1981a, p. 208] defined by

$$\dot{x}_1 = x_2$$
$$\dot{x}_2 = u,$$

where u is constrained by the isoperimetric constraint

$$\int_0^{t_f} u^2\, dt \leq 10.$$

If we define a new state variable x_3 by the equation

$$\dot{x}_3 = u^2, \tag{7.5-15}$$

with $x_3(0) = 0$, then the integral constraint is equivalent to the terminal condition

$$x_3(t_f) - 10 \leq 0. \tag{7.5-16}$$

This inequality terminal condition in turn can be expressed as an equality in terms of an additional unknown parameter $x_4(t_f)$, yielding

$$x_3(t_f) - 10 + x_4^2(t_f) = 0. \tag{7.5-17}$$

The parameter $x_4(t_f)$, called a **slack variable**, is introduced into the system by defining it as a state variable with the trivial dynamics

$$\dot{x}_4 = 0. \tag{7.5-18}$$

The final value of x_4 is left unspecified and is to be determined as part of the optimization process.

The problem has now been reformulated as a standard one with four state variables and a new terminal set condition. ■

Optimal Parametric Controls

Consider an optimal control problem that involves finding not only an optimal control $\mathbf{u}(\mathbf{x}) \in \mathcal{U}$, but also a constant unconstrained optimal **parameter** vector $\mathbf{w} \in R^{n_w}$, to minimize

$$J = x_0(t_f) = \int_0^{t_f} f_0(\mathbf{x}, \mathbf{w}, \mathbf{u})\, dt$$

subject to the state equations

$$\dot{\mathbf{x}} = \mathbf{f}(\mathbf{x}, \mathbf{w}, \mathbf{u}),$$

where the terminal time t_f is unspecified, the initial state $\mathbf{x}(0)$ is given, and the terminal state $\mathbf{x}_f = \mathbf{x}(t_f)$ is required to satisfy

$$\mathbf{g}(\mathbf{x}_f) = \mathbf{0}.$$

If we were to apply the minimum principle, with $\mathbf{w}$ as another control vector along with $\mathbf{u}(\mathbf{x})$, to minimize H, the answer would generally be in the form $\mathbf{w} = \mathbf{w}[\mathbf{x}(t), \boldsymbol{\lambda}(t)]$; that is, $\mathbf{w}$ would not be constant along a trajectory $\mathbf{x}(t)$. To handle this difficulty, we treat $\mathbf{w}$ not as a control variable but as a state variable, by introducing the differential equations

$$\dot{\mathbf{w}} = \mathbf{0}. \tag{7.5-19}$$

However, since $\mathbf{w}$ is unknown, there are no *a priori* initial or terminal conditions imposed on $\mathbf{w}$.

Let

$$\begin{aligned} H(\mathbf{x}, \mathbf{w}, \mathbf{u}, \boldsymbol{\lambda}, \boldsymbol{\lambda}_w, \lambda_0) &= \lambda_0 \dot{x}_0 + \boldsymbol{\lambda}^{\mathrm{T}} \dot{\mathbf{x}} + \boldsymbol{\lambda}_w^{\mathrm{T}} \dot{\mathbf{w}} \\ &= \lambda_0 f_0(\mathbf{x}, \mathbf{w}, \mathbf{u}) + \boldsymbol{\lambda}^{\mathrm{T}} \mathbf{f}(\mathbf{x}, \mathbf{w}, \mathbf{u}). \end{aligned}$$

The minimization conditions for $\mathbf{u}(\mathbf{x})$, the adjoint equations and terminal transversality conditions on $\boldsymbol{\lambda}(t)$, and the terminal conditions on $\mathbf{x}(t_f)$ are the same as in the minimum principle. For $\mathbf{w}$ we have the additional adjoint equations

$$\dot{\boldsymbol{\lambda}}_w^{\mathrm{T}} = -\frac{\partial H}{\partial \mathbf{w}} - \frac{\partial H}{\partial \mathbf{u}} \frac{\partial \mathbf{u}}{\partial \mathbf{w}}, \tag{7.5-20}$$

where the second term, as usual, is zero if the control constraints $\mathbf{u} \in \mathcal{U}$ do not depend on $\mathbf{w}$. Since the initial and final values for the state variable $\mathbf{w}$ are unspecified, we have the initial and final transversality conditions

$$\begin{aligned} \boldsymbol{\lambda}_w(0) &= \mathbf{0} \\ \boldsymbol{\lambda}_w(t_f) &= \mathbf{0}. \end{aligned} \tag{7.5-21}$$

Thus the parameter optimization part of the problem is to find a constant parameter vector $\mathbf{w} \in R^{n_w}$ so that the two-point boundary value problem of (7.5-20) and (7.5-21) is satisfied, which is equivalent to

$$\int_0^{t_f} \left[\frac{\partial H}{\partial w_j} + \frac{\partial H}{\partial \mathbf{u}} \frac{\partial \mathbf{u}}{\partial w_j} \right] dt = 0, \qquad j = 1, \ldots, n_w. \tag{7.5-22}$$

Note that the resulting optimal parameter $\mathbf{w}^*$ will, in general, depend on the initial and final states. However, the parameters are constant along any given optimal control trajectory.

Example 7.5-3: Integral-Square Error In Example 3.3-2 we considered the problem of minimizing the integral-square error

$$J = \int_0^{t_f} (\hat{\theta} - x_1)^2 \, dt \tag{7.5-23}$$

for the system

$$\begin{aligned} \dot{x}_1 &= x_2 \\ \dot{x}_2 &= -\omega_n^2 x_1 - 2\zeta\omega_n x_2 + \hat{\theta}, \end{aligned}$$

where $\hat{\theta}$, $\mathbf{x}(0) = \mathbf{0}$, $\omega_n = 1$, and $t_f = T \to \infty$ are given, and ζ is a parameter to be chosen, with $0 < \zeta < 1$.

To apply the minimum principle, consider the system $\dot{\mathbf{x}} = \mathbf{A}\mathbf{x} + \mathbf{b}$ given by

$$\begin{bmatrix} \dot{x}_1 \\ \dot{x}_2 \end{bmatrix} = \begin{bmatrix} 0 & 1 \\ -1 & -2\zeta \end{bmatrix} \begin{bmatrix} x_1 \\ x_2 \end{bmatrix} + \begin{bmatrix} 0 \\ \hat{\theta} \end{bmatrix}, \tag{7.5-24}$$

with an additional differential equation $\dot{t} = 1$ because of the fixed terminal time $t_f = T$. For $w = \zeta$ and

$$H = (\hat{\theta} - x_1)^2 + \lambda_1 x_2 + \lambda_2(\hat{\theta} - x_1 - 2\zeta x_2) + \lambda_t \tag{7.5-25}$$

condition (7.5-22) yields

$$\int_0^T \frac{\partial H}{\partial \zeta} \, dt = -2 \int_0^T \lambda_2 x_2 \, dt = 0, \tag{7.5-26}$$

where

$$\dot{\lambda}_1 = -\frac{\partial H}{\partial x_1} = 2(\hat{\theta} - x_1) + \lambda_2$$

$$\dot{\lambda}_2 = -\frac{\partial H}{\partial x_2} = -\lambda_1 + 2\zeta\lambda_2 \qquad (7.5\text{-}27)$$

$$\dot{\lambda}_t = -\frac{\partial H}{\partial t} = 0.$$

Even though the optimization is only with respect to a constant parameter ζ, the minimum principle procedure requires the integration of both the state and adjoint equations and the solution of a two-point boundary value problem, involving the terminal transversality conditions on the adjoint variables and the integral condition (7.5-26). For a fixed finite terminal time, given by the target

$$g(x_{1f}, x_{2f}, t_f) = t_f - T = 0,$$

the transversality conditions would yield $\lambda_{1f} = \lambda_{2f} = 0$, with λ_{tf} otherwise undetermined except for the condition $H_f = 0$. However, for the infinite-time problem, where the target $t_f = \infty$ is never actually achieved, no boundary conditions can be imposed on the adjoint variables, except $H = 0$ and an implicit requirement that the cost integral and the parametric integral (7.5-26) exist, that is are finite, as $t_f = T \rightarrow \infty$.

Since the state and adjoint equations are linear in this example, they can be integrated analytically. However, the development that follows masks, somewhat, a laborious amount of algebra, carried out with the aid of the Maple symbolic algebra program.

The state equations (7.5-24) have the eigenvalues and eigenvectors

$$\mu_1 = -\zeta + i\omega, \qquad \xi_1 = \begin{bmatrix} 1 \\ -\zeta + i\omega \end{bmatrix}$$

$$\mu_2 = -\zeta - i\omega, \qquad \xi_2 = \begin{bmatrix} 1 \\ -\zeta - i\omega \end{bmatrix},$$

and the **state transition matrix**

$$\mathbf{\Phi}(t) = e^{\mathbf{A}t} = \frac{e^{-\zeta t}}{\omega} \begin{bmatrix} \omega \cos \omega t + \zeta \sin \omega t & \sin \omega t \\ -\sin \omega t & \omega \cos \omega t - \zeta \sin \omega t \end{bmatrix},$$

where $\omega = \sqrt{1 - \zeta^2}$. The solution to the state equations is

$$\mathbf{x}(t) = \mathbf{\Phi}(t)\mathbf{x}(0) + \int_0^t \mathbf{\Phi}(t-\tau)\mathbf{b}\, d\tau$$

and for $\mathbf{x}(0) = \mathbf{0}$ and $\mathbf{b} = [0 \quad \hat{\theta}]^{\mathrm{T}}$ we get

$$\mathbf{x}(t) = \begin{bmatrix} \hat{\theta} \\ 0 \end{bmatrix} + \frac{\hat{\theta}}{\omega} e^{-\zeta t} \begin{bmatrix} -\omega \cos \omega t - \zeta \sin \omega t \\ \sin \omega t \end{bmatrix}. \tag{7.5-28}$$

For $\zeta > 0$ the cost integral exists as $t_f \rightarrow \infty$ since the state equations are asymptotically stable to $\mathbf{x}(t) \rightarrow (\hat{\theta}, 0)$.

Ignoring the constant λ_t and using $\boldsymbol{\lambda} = [\lambda_1 \quad \lambda_2]^{\mathrm{T}}$, the adjoint equations (7.5-27) are of the form

$$\dot{\boldsymbol{\lambda}} = -\mathbf{A}^{\mathrm{T}}\boldsymbol{\lambda} + \mathbf{c}(t), \qquad \mathbf{c}(t) = \begin{bmatrix} 2[\hat{\theta} - x_1(t)] \\ 0 \end{bmatrix}, \tag{7.5-29}$$

with the eigenvalues and eigenvectors

$$\mu_3 = \zeta + i\omega, \qquad \boldsymbol{\xi}_3 = \begin{bmatrix} 1 \\ \zeta + i\omega \end{bmatrix}$$

$$\mu_4 = \zeta - i\omega, \qquad \boldsymbol{\xi}_4 = \begin{bmatrix} 1 \\ \zeta - i\omega \end{bmatrix}$$

and the **adjoint transition matrix**

$$\mathbf{\Phi}_\lambda(t) = e^{-\mathbf{A}^{\mathrm{T}}t} = [\mathbf{\Phi}^{-1}(t)]^{\mathrm{T}} = \mathbf{\Phi}^{\mathrm{T}}(-t) \tag{7.5-30}$$

given by

$$\mathbf{\Phi}_\lambda(t) = \frac{e^{\zeta t}}{\omega} \begin{bmatrix} \omega \cos \omega t - \zeta \sin \omega t & \sin \omega t \\ -\sin \omega t & \omega \cos \omega t + \zeta \sin \omega t \end{bmatrix}.$$

The solution to the adjoint equations is

$$\boldsymbol{\lambda}(t) = \mathbf{\Phi}_\lambda(t)\boldsymbol{\lambda}(0) + \mathbf{\Phi}_\lambda(t) \int_0^t \mathbf{\Phi}_\lambda(-\tau)\mathbf{c}(\tau)\, d\tau. \tag{7.5-31}$$

Using $\mathbf{c}(t)$ from (7.5-29), $\hat{\theta} - x_1(t)$ from (7.5-28), and $\omega^2 = 1 - \zeta^2$, we have

$$\boldsymbol{\lambda}(t) = \boldsymbol{\Phi}_\lambda(t) \begin{bmatrix} \lambda_{10} + \alpha_1(t) \\ \lambda_{20} + \alpha_2(t) \end{bmatrix},$$

where $\boldsymbol{\lambda}(0) = [\lambda_{10} \quad \lambda_{20}]^{\mathrm{T}}$ and

$$\alpha_1(t) = \frac{-\hat{\theta}}{2\zeta\omega} [e^{-2\zeta t}(\zeta^2(3 - 4\zeta^2)\cos 2\omega t + \zeta\omega(4\zeta^2 - 1)\sin 2\omega t + 1) - \omega^2(4\zeta^2 + 1)]$$

$$\alpha_2(t) = \frac{-\hat{\theta}}{2\omega} [e^{-2\zeta t}((1 - 2\zeta^2)\cos 2\omega t + 2\zeta\omega \sin 2\omega t + 1) - 2\omega^2].$$

Thus

$$\lambda_1(t) = \frac{e^{\zeta t}}{\omega} \{[\lambda_{10} + \alpha_1(t)](\omega \cos \omega t - \zeta \sin \omega t) + [\lambda_{20} + \alpha_2(t)] \sin \omega t\} \tag{7.5-32}$$

$$\lambda_2(t) = \frac{e^{\zeta t}}{\omega} \{-[\lambda_{10} + \alpha_1(t)] \sin \omega t + [\lambda_{20} + \alpha_2(t)](\omega \cos \omega t + \zeta \sin \omega t)\},$$

which yields

$$\begin{aligned} \lambda_2 x_2 = {} & \frac{\hat{\theta} \sin \omega t}{\omega^2} \{-\lambda_{10} \sin \omega t + \lambda_{20}(\omega \cos \omega t + \zeta \sin \omega t)\} \\ & + \frac{\hat{\theta}^2 \sin^2 \omega t}{2\omega^3\zeta} \{e^{-2\zeta t}[\zeta^2(3 - 4\zeta^2)\cos 2\omega t + \zeta\omega(4\zeta^2 - 1)\sin 2\omega t + 1] \\ & - \omega^2(4\zeta^2 + 1)\} \\ & - \frac{\hat{\theta}^2 \sin \omega t}{2\omega^3} \{e^{-2\zeta t}[(1 - 2\zeta^2)\cos 2\omega t + 2\zeta\omega \sin 2\omega t + 1] - 2\omega^2\} \\ & \cdot (\omega \cos \omega t + \zeta \sin \omega t). \end{aligned}$$

Integrating this result, we get

$$\frac{\omega^2}{\hat{\theta}} \int_0^t \lambda_2 x_2 \, dt = \left(\frac{1}{2} \lambda_{10} + \frac{\hat{\theta}\omega(1+2\zeta^2)}{4\zeta} - \frac{1}{2} \lambda_{20}\zeta \right) \left(-t + \frac{1}{2} \frac{\sin 2\omega t}{\omega} \right)$$

$$+ \left(\frac{1}{2} \lambda_{20}\omega + \frac{1}{2} \hat{\theta}\omega^2 \right) \frac{1 - \cos 2\omega t}{2\omega}$$

$$+ \frac{\hat{\theta}(3\zeta^2 - 2\zeta^4 - 1)}{8\zeta\omega} (\zeta + e^{-2\zeta t}[\omega \sin 2\omega t - \zeta \cos 2\omega t])$$

$$+ \frac{\hat{\theta}}{4} (\zeta^2 - 1)(\omega - e^{-2\zeta t}[\omega \cos 2\omega t + \zeta \sin 2\omega t])$$

$$+ \frac{\hat{\theta}}{8\zeta^2\omega} (2\zeta^4 - 3\zeta^2 + 1)(1 - e^{-2\zeta t}).$$

In the limit as $t \rightarrow \infty$ we have

$$\frac{\omega^2}{\hat{\theta}} \int_0^\infty \lambda_2 x_2 \, dt = \left(\frac{1}{2} \lambda_{10} + \frac{\hat{\theta}\omega(1+2\zeta^2)}{4\zeta} - \frac{1}{2} \lambda_{20}\zeta \right) \left(-t + \frac{1}{2} \frac{\sin 2\omega t}{\omega} \right)$$

$$+ (\lambda_{20} + \hat{\theta}\omega) \frac{1 - \cos 2\omega t}{4}$$

$$+ \frac{\hat{\theta}\omega}{8(\zeta^2 + \omega^2)\zeta^2} (4\zeta^4 - 5\zeta^2 + 1).$$

In order for this result to be zero it is necessary that

$$\lambda_2(0) = -\hat{\theta}\omega$$

$$\lambda_1(0) = \lambda_2(0)\zeta - \frac{\theta\omega(1+2\zeta^2)}{2\zeta}$$

and

$$0 = 4\zeta^4 - 5\zeta^2 + 1,$$

which has the roots

$$\zeta = \pm 1, \; \pm \tfrac{1}{2}.$$

For $0 < \zeta < 1$ we have

$$\zeta^* = \tfrac{1}{2},$$

which agrees with the minimum integral-square error result obtained previously, using parameter optimization methods in Example 3.3-2. Comparing the two examples we find that the minimum principle procedure is more laborious. ■

The minimum principle is applicable to the general case in which both an optimal control $\mathbf{u}^*(\mathbf{x})$ and an optimal parameter $\mathbf{w}^*$ are to be found. For the parameter-only case (i.e., where the class of controls is restricted to constant controls) an alternate set of necessary conditions, akin to the minimum principle but treating $\mathbf{w}$ directly as a parametric control rather than indirectly as a state variable, is presented in Vincent and Grantham [1981]. These conditions, while equivalent to those provided by the minimum principle [Vincent and Lee, 1983], will generally be easier to solve.

7.6 DYNAMIC PROGRAMMING

In order to solve optimal control problems using Pontryagin's minimum principle, we have used backwards integration. From a point $\mathbf{x}_f$ on the terminal surface we determine the terminal adjoint vector $\boldsymbol{\lambda}_f$ from the transversality conditions and the terminal optimal control $\mathbf{u}_f^*$ from the minimum principle. The state and adjoint equations are then integrated backwards in time, using at each instant of time the optimal control calculated from the minimum principle. In this way we generate an optimal trajectory $\mathbf{x}(t)$ and its associated adjoint vector $\boldsymbol{\lambda}(t)$. Repeating this backwards integration process from each point on the terminal surface (or using different $\boldsymbol{\lambda}_f$ values for the fixed point case), we generate optimal trajectories that fill the state space.

While this approach typically yields an open-loop optimal control $\mathbf{u}^*(t) = \mathbf{u}^*[\mathbf{x}(t), \boldsymbol{\lambda}(t)]$ for any given trajectory, if we keep track of the control values along all optimal trajectories, we can synthesize the optimal closed-loop feedback control $\mathbf{u}^*(\mathbf{x})$. In certain types of problems, involving "bang-bang" or "on-off" switching controls, this synthesis problem is quite tractable, particularly if the switching surfaces $\sigma(\mathbf{x}) = 0$ can be determined. But in general the synthesis process can be quite formidable.

In this section we examine a direct approach for determining the optimal feedback control $\mathbf{u}^*(\mathbf{x})$. The approach is called **dynamic programming** [Bellman, 1957; Hestenes, 1966, p. 286]. It is essentially backwards integration, but designed to construct the optimal return, or minimum cost-to-go, function $V^*(\mathbf{x})$ and the corresponding feedback control $\mathbf{u}^*(\mathbf{x})$ directly.

Hamilton–Jacobi–Bellman Equation

For the general problem of Bolza the cost-to-go is

$$V = \psi[\mathbf{x}(t_f)] + \int_t^{t_f} f_0(\mathbf{x}, \mathbf{u})\, dt \tag{7.6-1}$$

and the minimum cost-to-go from the current state $\mathbf{x} = \mathbf{x}(t)$ to the final state $\mathbf{x}(t_f)$ using the optimal feedback control $\mathbf{u}^*(\mathbf{x})$ is

$$V^*(\mathbf{x}) = \psi[\mathbf{x}(t_f)] + \int_t^{t_f} f_0[\mathbf{x}, \mathbf{u}^*(\mathbf{x})]\, dt. \tag{7.6-2}$$

Bellman's **principle of optimality** is the underlying postulate used for a direct construction of $V^*(\mathbf{x})$. This principle instructs us to use optimal control for the remainder of a trajectory regardless of how the system arrived at the current state. Applying this principle backwards in time from the terminal surface will yield the minimum cost-to-go $V^*(\mathbf{x})$ and the corresponding optimal feedback control $\mathbf{u}^*(\mathbf{x})$.

Suppose that at a point $\mathbf{x}$ a nonoptimal control $\mathbf{u} \in \mathcal{U}$ is employed for a short time Δt, yielding a new state $\mathbf{x} + \mathbf{f}(\mathbf{x}, \mathbf{u})\Delta t$. From this new state onward, we revert to the optimal control. Then the total cost-to-go from $\mathbf{x}$ to the target is

$$V = f_0(\mathbf{x}, \mathbf{u})\, \Delta t + V^*[\mathbf{x} + \mathbf{f}(\mathbf{x}, \mathbf{u})\Delta t].$$

Thus the minimum cost-to-go from $\mathbf{x}$ is

$$V^*(\mathbf{x}) = \min_{\mathbf{u} \in \mathcal{U}} \{f_0(\mathbf{x}, \mathbf{u})\, \Delta t + V^*[\mathbf{x} + \mathbf{f}(\mathbf{x}, \mathbf{u})\Delta t]\}. \tag{7.6-3}$$

Assuming $V^*(\mathbf{x})$ is C^2 and expanding the last term in a Taylor series, we have

$$0 = \min_{\mathbf{u} \in \mathcal{U}} \{f_0(\mathbf{x}, \mathbf{u}) + \frac{\partial V^*}{\partial \mathbf{x}}\, \mathbf{f}(\mathbf{x}, \mathbf{u})\}.$$

In other words, the optimal return function $V^*(\mathbf{x})$ satisfies the **Hamilton–Jacobi–Bellman** partial differential equation

$$0 = \min_{\mathbf{u} \in \mathcal{U}} \left\{H\left(\mathbf{x}, \mathbf{u}, \frac{\partial V^*}{\partial \mathbf{x}}\right)\right\}, \tag{7.6-4}$$

with the boundary condition (at the terminal surface) given by

$$V^*(\mathbf{x}) = \psi(\mathbf{x}) \qquad \text{on} \qquad \mathbf{g}(\mathbf{x}) = 0, \tag{7.6-5}$$

where $H(\mathbf{x}, \mathbf{u}, \partial V^*/\partial \mathbf{x}) = f_0(\mathbf{x}, \mathbf{u}) + (\partial V^*/\partial \mathbf{x})\mathbf{f}(\mathbf{x}, \mathbf{u})$.

The advantage of this approach is that we obtain $\mathbf{u}(\mathbf{x})$ and $V^*(\mathbf{x})$ directly. The disadvantage is that we must solve a partial differential equation to do so. Thus analytical solutions using this approach will be rare. One notable exception is provided by the following important example.

Linear-Quadratic Optimal Control

For a linear system

$$\dot{\mathbf{x}} = \mathbf{A}(t)\mathbf{x} + \mathbf{B}(t)\mathbf{u} \tag{7.6-6}$$

with no control constraints, we wish to design a feedback controller $\mathbf{u}(\mathbf{x}, t)$ so that $\mathbf{x}(t) \to \mathbf{0}$. For a fixed terminal time $t_f = T$ consider the problem of determining $\mathbf{u}(\mathbf{x}, t)$ to minimize

$$J = \tfrac{1}{2}\mathbf{x}_f^{\mathrm{T}}\mathbf{S}_f\mathbf{x}_f + \tfrac{1}{2}\int_0^{t_f} [\mathbf{x}^{\mathrm{T}}\mathbf{Q}(t)\mathbf{x} + \mathbf{u}^{\mathrm{T}}\mathbf{R}(t)\mathbf{u}]\, dt, \tag{7.6-7}$$

where $\mathbf{S}_f(t_f) \geq 0$, $\mathbf{R}(t) > 0$, and $\mathbf{Q}(t) \geq 0$ are symmetric matrices of appropriate sizes. The terminal portion of this cost is a penalty on $\mathbf{x}_f \neq \mathbf{0}$, the first term in the integral is a penalty on $\mathbf{x}(t) \neq \mathbf{0}$, and the second term in the integral is a penalty on the use of control. Since this is a fixed-time problem we append a differential equation $\dot{t} = 1$ and a terminal condition

$$t_f - T = 0.$$

In terms of the expanded state $[\mathbf{x}^{\mathrm{T}} \quad t]$ the H function is

$$H\left(\mathbf{x}, t, \mathbf{u}, \frac{\partial V^*}{\partial \mathbf{x}}, \frac{\partial V^*}{\partial t}\right) = \tfrac{1}{2}[\mathbf{x}^{\mathrm{T}}\mathbf{Q}(t)\mathbf{x} + \mathbf{u}^{\mathrm{T}}\mathbf{R}(t)\mathbf{u}] + \frac{\partial V^*}{\partial \mathbf{x}}[\mathbf{A}(t)\mathbf{x} + \mathbf{B}(t)\mathbf{u}] + \frac{\partial V^*}{\partial t}, \tag{7.6-8}$$

where

$$V^*(\mathbf{x}, t) = \min_{\mathbf{u}(\cdot)} \left\{\tfrac{1}{2}\mathbf{x}_f^{\mathrm{T}}\mathbf{S}_f\mathbf{x}_f + \tfrac{1}{2}\int_t^{t_f} [\mathbf{x}^{\mathrm{T}}\mathbf{Q}(t)\mathbf{x} + \mathbf{u}^{\mathrm{T}}\mathbf{R}(t)\mathbf{u}]\, dt\right\} \tag{7.6-9}$$

is the minimum cost-to-go. From the Hamilton–Jacobi–Bellman equation the optimal control $\mathbf{u}^*(\mathbf{x}, t)$ and the optimal return function $V^*(\mathbf{x}, t)$ are determined by solving

$$0 = \min_{\mathbf{u}} \left\{ H\left(\mathbf{x}, t, \mathbf{u}, \frac{\partial V^*}{\partial \mathbf{x}}, \frac{\partial V^*}{\partial t} \right) \right\} \tag{7.6-10}$$

with the boundary conditions

$$V^*(\mathbf{x}_f, t_f) = \tfrac{1}{2}\mathbf{x}_f^{\mathrm{T}}\mathbf{S}_f\mathbf{x}_f \tag{7.6-11}$$

at $t_f = T$. Minimizing H with respect to $\mathbf{u}$ yields

$$\mathbf{0}^{\mathrm{T}} = \frac{\partial H}{\partial \mathbf{u}} = \mathbf{u}^{*\mathrm{T}}\mathbf{R}(t) + \frac{\partial V^*}{\partial \mathbf{x}}\mathbf{B}(t),$$

or

$$\mathbf{u}^*(\mathbf{x}, t) = -\mathbf{R}^{-1}(t)\mathbf{B}^{\mathrm{T}}(t)\left[\frac{\partial V^*}{\partial \mathbf{x}}\right]^{\mathrm{T}}. \tag{7.6-12}$$

Then the Hamilton–Jacobi–Bellman equation (7.6-10) becomes

$$0 = \tfrac{1}{2}\mathbf{x}^{\mathrm{T}}\mathbf{Q}(t)\mathbf{x} + \frac{\partial V^*}{\partial \mathbf{x}}\mathbf{A}(t)\mathbf{x} - \frac{1}{2}\frac{\partial V^*}{\partial \mathbf{x}}\mathbf{B}(t)\mathbf{R}^{-1}(t)\mathbf{B}^{\mathrm{T}}(t)\left[\frac{\partial V^*}{\partial \mathbf{x}}\right]^{\mathrm{T}} + \frac{\partial V^*}{\partial t}. \tag{7.6-13}$$

In view of the symmetry of both (7.6-13) and the boundary condition (7.6-11), we try a solution of the form

$$V^*(\mathbf{x}, t) = \tfrac{1}{2}\mathbf{x}^{\mathrm{T}}\mathbf{S}(t)\mathbf{x}, \tag{7.6-14}$$

where $\mathbf{S}(t)$ is symmetric and

$$\mathbf{S}(t_f) = \mathbf{S}_f. \tag{7.6-15}$$

With

$$\frac{\partial V^*}{\partial t} = \tfrac{1}{2}\mathbf{x}^{\mathrm{T}}\dot{\mathbf{S}}\mathbf{x}$$

and

$$\frac{\partial V^*}{\partial \mathbf{x}} = \mathbf{x}^T\mathbf{S}$$

we find that (7.6-13) becomes

$$0 = \tfrac{1}{2}\mathbf{x}^T\mathbf{Q}(t)\mathbf{x} + \mathbf{x}^T\mathbf{S}\mathbf{A}(t)\mathbf{x} - \tfrac{1}{2}\mathbf{x}^T\mathbf{S}\mathbf{B}(t)\mathbf{R}^{-1}(t)\mathbf{B}^T(t)\mathbf{S}\mathbf{x} + \tfrac{1}{2}\mathbf{x}^T\dot{\mathbf{S}}\mathbf{x}.$$

Since each term is a scalar, and thus equal to its transpose, we have

$$0 = \tfrac{1}{2}\mathbf{x}^T\{\mathbf{Q}(t) + \mathbf{S}\mathbf{A}(t) + \mathbf{A}^T(t)\mathbf{S} - \mathbf{S}\mathbf{B}(t)\mathbf{R}^{-1}(t)\mathbf{B}^T(t)\mathbf{S} + \dot{\mathbf{S}}\}\mathbf{x}.$$

Since this equation must hold for all $\mathbf{x}$ we find that $\mathbf{S}(t)$ must satisfy the **matrix Riccati equation**

$$\mathbf{0} = \mathbf{Q}(t) + \mathbf{S}\mathbf{A}(t) + \mathbf{A}^T(t)\mathbf{S} - \mathbf{S}\mathbf{B}(t)\mathbf{R}^{-1}(t)\mathbf{B}^T(t)\mathbf{S} + \dot{\mathbf{S}} \tag{7.6-16}$$

with the boundary condition (7.6-15). After solving this quadratic differential equation for the matrix $\mathbf{S}(t)$, the optimal feedback control from (7.6-12) becomes

$$\mathbf{u}^*(\mathbf{x}, t) = -\mathbf{K}(t)\mathbf{x}, \tag{7.6-17}$$

where

$$\mathbf{K}(t) = \mathbf{R}^{-1}(t)\mathbf{B}^T(t)\mathbf{S}(t). \tag{7.6-18}$$

Linear-Quadratic Regulator

For the constant-coefficient linear system

$$\dot{\mathbf{x}} = \mathbf{A}\mathbf{x} + \mathbf{B}\mathbf{u} \tag{7.6-19}$$

we seek a control $\mathbf{u}(\mathbf{x})$ so that $\mathbf{x}(t) \to \mathbf{0}$ as $t \to \infty$. Choose constant symmetric positive definite matrices $\mathbf{R}$ and $\mathbf{Q}$ and consider the problem of minimizing

$$J = \tfrac{1}{2}\int_0^\infty [\mathbf{x}^T\mathbf{Q}\mathbf{x} + \mathbf{u}^T\mathbf{R}\mathbf{u}]\, dt. \tag{7.6-20}$$

If this infinite-time optimal control problem has a solution, then in order for the integral in (7.6-20) to exist we will have $\mathbf{x}(t) \to \mathbf{0}$ and $\mathbf{u}^*[\mathbf{x}(t)] \to \mathbf{0}$ as $t \to \infty$, since $\mathbf{R}$ and $\mathbf{Q}$ are positive definite. Also, the optimal return function

$$V^*(\mathbf{x}) = \min_{\mathbf{u}(\cdot)} \tfrac{1}{2} \int_t^{\infty} [\mathbf{x}^T\mathbf{Q}\mathbf{x} + \mathbf{u}^T\mathbf{R}\mathbf{u}]\, dt \tag{7.6-21}$$

cannot depend on t, since $t_f - t_1 = t_f - t_2$ for all t_1, t_2 as $t_f \to \infty$. Thus $\partial V^*/\partial t = 0$. Also note that, if $\mathbf{x} \equiv \mathbf{0}$, then $\mathbf{u}^* \equiv \mathbf{0}$. Thus we require $V^*(\mathbf{x}) = 0$ at $\mathbf{x} = \mathbf{0}$.

Following the same development as in the fixed-time nonautonomous case, we find that

$$\mathbf{u}^*(\mathbf{x}) = -\mathbf{K}\mathbf{x}, \tag{7.6-22}$$

where

$$\mathbf{K} = \mathbf{R}^{-1}\mathbf{B}^T\mathbf{S} \tag{7.6-23}$$

and $\mathbf{S}$ is the positive definite symmetric solution to the (quadratic) matrix **algebraic Riccati equation**

$$\mathbf{0} = \mathbf{Q} + \mathbf{S}\mathbf{A} + \mathbf{A}^T\mathbf{S} - \mathbf{S}\mathbf{B}\mathbf{R}^{-1}\mathbf{B}^T\mathbf{S}. \tag{7.6-24}$$

Note that

$$V^*(\mathbf{x}) = \tfrac{1}{2}\mathbf{x}^T\mathbf{S}\mathbf{x}$$

is a Lyapunov function establishing global asymptotic stability of the origin, since $V^*(\mathbf{x})$ is positive definite and

$$0 = H = \tfrac{1}{2}\,[\mathbf{x}^T\mathbf{Q}\mathbf{x} + \mathbf{u}^{*T}\mathbf{R}\mathbf{u}^*] + \frac{\partial V^*(\mathbf{x})}{\partial \mathbf{x}}\dot{\mathbf{x}}$$

implies

$$\frac{dV^*[\mathbf{x}(t)]}{dt} < 0 \qquad \text{for } \mathbf{x}(t) \neq 0.$$

This stability result may also be obtained in cases where $\mathbf{Q} \geq 0$, but nonzero, as we illustrate in the next example.

Except for the quadratic term the algebraic Riccati equation (7.6-24) is the same as the matrix Lyapunov equation in Chapter 4, for the stability of autonomous linear systems. This observation leads to an iterative technique [Kleinman, 1968; Rothschild and Jameson, 1970] for solving the matrix algebraic Riccati equation in terms of quadratic improvements to solutions to the matrix Lyapunov equation. The technique is essentially Newton's method applied to (7.6-24).

Example 7.6-1: Algebraic Riccati Equation Consider the problem of minimizing

$$J = \frac{1}{2}\int_0^\infty (x_1^2 + u^2)\,dt$$

for the system

$$\dot{x}_1 = x_2$$
$$\dot{x}_2 = x_1 + u.$$

For

$$\mathbf{R} = 1, \qquad \mathbf{Q} = \begin{bmatrix} 1 & 0 \\ 0 & 0 \end{bmatrix}, \qquad \mathbf{A} = \begin{bmatrix} 0 & 1 \\ 1 & 0 \end{bmatrix},$$
$$\mathbf{B} = \begin{bmatrix} 0 \\ 1 \end{bmatrix}, \qquad \mathbf{S} = \begin{bmatrix} s_{11} & s_{12} \\ s_{12} & s_{22} \end{bmatrix}$$

the algebraic Riccati equation (7.6-24) yields

$$0 = 1 + 2s_{12} - s_{12}^2$$
$$0 = s_{11} + s_{22}(1 - s_{12})$$
$$0 = 2s_{12} - s_{22}^2,$$

and the positive definite solution

$$\mathbf{S} = \begin{bmatrix} 3.108 & 2.414 \\ 2.414 & 2.197 \end{bmatrix}.$$

The optimal control is

$$u^*(\mathbf{x}) = -\mathbf{R}^{-1}\mathbf{B}^{\mathrm{T}}\mathbf{S}\mathbf{x} = -[2.414 \quad 2.197]\begin{bmatrix} x_1 \\ x_2 \end{bmatrix}.$$

For the optimal return function

$$V^*(\mathbf{x}) = \tfrac{1}{2}\mathbf{x}^T\mathbf{S}\mathbf{x}$$

we have

$$\frac{dV^*[\mathbf{x}(t)]}{dt} = -\tfrac{1}{2}\,[\mathbf{x}^T\mathbf{Q}\mathbf{x} + \mathbf{u}^{*T}\mathbf{R}\mathbf{u}^*]$$

$$= -\tfrac{1}{2}\mathbf{x}^T \begin{bmatrix} 6.829 & 5.305 \\ 5.305 & 4.828 \end{bmatrix} \mathbf{x} < 0 \qquad \text{for } \mathbf{x} \neq \mathbf{0}.$$

Thus the origin is globally asymptotically stable. ■

7.7 STATE CONSTRAINTS

In this section we consider the case where the control constraints are (explicitly or implicitly) state dependent [Hestenes, 1966, p. 352; Leitmann, 1981a, p. 167]. We consider two cases. The first deals with explicit **state-dependent control constraints**, that is, $\mathbf{u} \in \mathcal{U}(\mathbf{x})$. The second, implicit, case deals with control constraints $\mathbf{u} \in \mathcal{U}$ that are not explicit functions of the state $\mathbf{x}$, but the motion of the system is subject to **state constraints** $\mathbf{x}(t) \in X$, where $X \subset R^{n_x}$ is a specified state constraint set. This latter case can induce implicit constraints on a control $\mathbf{u}(\mathbf{x})$ in order to keep a trajectory $\mathbf{x}(t)$ from violating the state constraints.

State-Dependent Control Constraints

Here we consider trajectories that are interior to any state constraint region X, with explicit state-dependent control constraints of the form

$$\mathbf{u} \in \mathcal{U}(\mathbf{x}) = \{\mathbf{u} \in R^{n_u} \,|\, \mathbf{h}(\mathbf{x}, \mathbf{u}) \geq \mathbf{0}\}, \qquad (7.7\text{-}1)$$

where the components of $\mathbf{h} = [h_1 \cdots h_{n_h}]^T$ are continuous and continuously differentiable in their arguments. Let $\mathbf{u}^*(\mathbf{x})$ be an optimal control with corresponding trajectory $\mathbf{x}(t)$, $0 \leq t \leq t_f$. Suppose that during some time interval $t_e \leq t \leq t_l$ certain of the inequalities in (7.7-1) are active as equalities. We assume that none of the active constraints are pure state constraints. That is, if there are constraints of the form $h_j(\mathbf{x}) \geq 0$, then

$$h_j[\mathbf{x}(t)] > 0 \qquad \text{for } t_e \leq t \leq t_l. \qquad (7.7\text{-}2)$$

We also assume that $\mathbf{u}^*[\mathbf{x}(t)] \in \mathcal{U}[\mathbf{x}(t)]$ is a **regular point** [Vincent and Grantham, 1981, p. 35] of the constraint set $\mathcal{U}[\mathbf{x}(t)]$ along $\mathbf{x}(t)$. In other words, the tangent vectors to $\mathcal{U}[\mathbf{x}(t)]$ are exactly those vectors $\mathbf{e} \in R^{n_u}$ such that

$$\frac{\partial h_j}{\partial \mathbf{u}}\,\mathbf{e} \geq 0$$

for all $\mathbf{h}$ components such that $h_j(\mathbf{x}(t), \mathbf{u}^*[\mathbf{x}(t)]) = 0$.

The necessary conditions for $\mathbf{u}^*(\mathbf{x})$ to minimize

$$H(\mathbf{x}, \mathbf{u}, \boldsymbol{\lambda}, \lambda_0) = \lambda_0 f_0(\mathbf{x}, \mathbf{u}) + \boldsymbol{\lambda}^{\mathrm{T}}\mathbf{f}(\mathbf{x}, \mathbf{u})$$

along $\mathbf{x}(t)$, $t_e \leq t \leq t_l$, is that [Vincent and Grantham, 1981, p. 53] there exists a Lagrange multiplier vector $\boldsymbol{\gamma}(t) \in R^{n_h}$ such that

$$\mathbf{0}^{\mathrm{T}} = \frac{\partial H(\mathbf{x}, \mathbf{u}^*, \boldsymbol{\lambda}(t), \lambda_0)}{\partial \mathbf{u}} - \boldsymbol{\gamma}^{\mathrm{T}}(t)\frac{\partial \mathbf{h}(\mathbf{x}, \mathbf{u}^*)}{\partial \mathbf{u}} \tag{7.7-3}$$

$$\mathbf{0} \leq \boldsymbol{\gamma}(t) \tag{7.7-4}$$

$$\mathbf{0} = \boldsymbol{\gamma}^{\mathrm{T}}(t)\mathbf{h}(\mathbf{x}, \mathbf{u}^*), \tag{7.7-5}$$

where $\mathbf{u}^* = \mathbf{u}^*(\mathbf{x})$, $\mathbf{x} = \mathbf{x}(t)$, and (7.7-5) implies $\gamma_j = 0$ if $h_j > 0$. Thus the adjoint equations (7.2-14) can be written as

$$\dot{\boldsymbol{\lambda}}^{\mathrm{T}} = -\frac{\partial H}{\partial \mathbf{x}} - \frac{\partial H}{\partial \mathbf{u}}\frac{\partial \mathbf{u}}{\partial \mathbf{x}} = -\frac{\partial H}{\partial \mathbf{x}} - \boldsymbol{\gamma}^{\mathrm{T}}(t)\frac{\partial \mathbf{h}}{\partial \mathbf{u}}\frac{\partial \mathbf{u}}{\partial \mathbf{x}}, \tag{7.7-6}$$

where the partial derivatives are evaluated at $\mathbf{x} = \mathbf{x}(t)$, $\mathbf{u}^* = \mathbf{u}^*[\mathbf{x}(t)]$.

Now, for this $\boldsymbol{\gamma}(t)$, consider the function $\psi(\mathbf{x}, \mathbf{u}) \triangleq \boldsymbol{\gamma}(t)^{\mathrm{T}}\mathbf{h}(\mathbf{x}, \mathbf{u})$. Along $\mathbf{x}(t)$ we have $\boldsymbol{\gamma}^{\mathrm{T}}(t)\mathbf{h}[\mathbf{x}, \mathbf{u}^*(\mathbf{x})] = 0$. Suppose we make an arbitrary small change in $\mathbf{x}$, to $\mathbf{x} + \Delta\mathbf{x}$. The set of active constraints in (7.7-1) may change, but we take $\Delta\mathbf{x}$ sufficiently small so that any inactive pure state constraints (7.7-2) remain inactive. From (7.7-1) and (7.7-4), $\boldsymbol{\gamma}^{\mathrm{T}}(t)\mathbf{h}[\mathbf{x} + \Delta\mathbf{x}, \mathbf{u}^*(\mathbf{x} + \Delta\mathbf{x})] \geq 0$. Thus $\mathbf{u}^*(\mathbf{x})$ yields a stationary minimum for $\psi(\mathbf{x}, \mathbf{u})$, from which we have

$$\mathbf{0}^{\mathrm{T}} = \boldsymbol{\gamma}^{\mathrm{T}}(t)\left[\frac{\partial \mathbf{h}}{\partial \mathbf{x}} + \frac{\partial \mathbf{h}}{\partial \mathbf{u}}\frac{\partial \mathbf{u}}{\partial \mathbf{x}}\right]. \tag{7.7-7}$$

Thus

$$\dot{\boldsymbol{\lambda}}^{\mathrm{T}} = -\frac{\partial H}{\partial \mathbf{x}} + \boldsymbol{\gamma}^{\mathrm{T}}(t)\frac{\partial \mathbf{h}}{\partial \mathbf{x}}. \tag{7.7-8}$$

Extending these results to the entire optimal trajectory $\mathbf{x}(t)$, $t \in [0, t_f]$, we have:

Minimum Principle with State-Dependent Control Constraints: *If* $\mathbf{u}^*(\mathbf{x})$ *is an optimal control with corresponding trajectory* $\mathbf{x}(t)$ *interior to any state constraint region for the problem of minimizing*

$$\int_0^{t_f} f_0(\mathbf{x}, \mathbf{u})\, dt \tag{7.7-9}$$

subject to

$$\dot{\mathbf{x}} = \mathbf{f}(\mathbf{x}, \mathbf{u}) \tag{7.7-10}$$

with state-dependent control constraints

$$\mathbf{u} \in \mathcal{U}(\mathbf{x}) = \{\mathbf{u} \mid \mathbf{h}(\mathbf{x}, \mathbf{u}) \geq \mathbf{0}\}, \tag{7.7-11}$$

a specified initial state $\mathbf{x}(0)$, *and the terminal conditions*

$$\mathbf{g}(\mathbf{x}_f) = \mathbf{0} \tag{7.7-12}$$

at $\mathbf{x}_f = \mathbf{x}(t_f)$, *then there exists a constant* λ_0 [= 1 *or* 0], *a continuous vector* $\boldsymbol{\lambda}(t) \in R^{n_x}$, *a vector* $\boldsymbol{\gamma}(t) \in R^{n_h}$, *and a constant vector* $\boldsymbol{\rho} \in R^{n_g}$ *such that* $(\lambda_0, \boldsymbol{\lambda}) \neq \mathbf{0}$ *and*

$$0 = H[\mathbf{x}(t), \mathbf{u}^*, \boldsymbol{\lambda}(t), \lambda_0] = \min_{\mathbf{u} \in \mathcal{U}(\mathbf{x})} H[\mathbf{x}(t), \mathbf{u}, \boldsymbol{\lambda}(t), \lambda_0] \tag{7.7-13}$$

$$\frac{\partial L}{\partial \mathbf{u}} = \mathbf{0}^{\mathrm{T}} \tag{7.7-14}$$

$$\mathbf{h}[\mathbf{x}, \mathbf{u}^*(\mathbf{x})] \geq \mathbf{0} \tag{7.7-15}$$

$$\boldsymbol{\gamma}^{\mathrm{T}}\mathbf{h}[\mathbf{x}, \mathbf{u}^*(\mathbf{x})] = 0 \tag{7.7-16}$$

$$\boldsymbol{\gamma} \geq \mathbf{0}, \tag{7.7-17}$$

where the partial derivatives are evaluated at $\mathbf{x} = \mathbf{x}(t)$, $\mathbf{u} = \mathbf{u}^*(\mathbf{x})$, *with* $\boldsymbol{\lambda}(t)$ *satisfying the modified adjoint equations*

$$\dot{\boldsymbol{\lambda}}^{\mathrm{T}} = -\frac{\partial L}{\partial \mathbf{x}} \tag{7.7-18}$$

and the terminal transversality conditions

$$\boldsymbol{\lambda}_f^{\mathrm{T}} = \boldsymbol{\rho}^{\mathrm{T}} \frac{\partial \mathbf{g}(\mathbf{x}_f)}{\partial \mathbf{x}}, \tag{7.7-19}$$

where

$$H(\mathbf{x}, \mathbf{u}, \boldsymbol{\lambda}, \lambda_0) \triangleq \lambda_0 f_0(\mathbf{x}, \mathbf{u}) + \boldsymbol{\lambda}^{\mathrm{T}}\mathbf{f}(\mathbf{x}, \mathbf{u}) \tag{7.7-20}$$

$$L(\mathbf{x}, \mathbf{u}, \boldsymbol{\lambda}, \lambda_0, \boldsymbol{\gamma}) \triangleq H(\mathbf{x}, \mathbf{u}, \boldsymbol{\lambda}, \lambda_0) - \boldsymbol{\gamma}^{\mathrm{T}}\mathbf{h}(\mathbf{x}, \mathbf{u}). \tag{7.7-21}$$

Note that the vector $\boldsymbol{\gamma}(t)$ may be discontinuous along $\mathbf{x}(t)$, since the set of active inequalities may change at various times. However, both H and $\boldsymbol{\lambda}$ are continuous. For an approach that does not employ the Lagrange multiplier vector, see Leitmann [1981, pp. 167–174].

Example 7.7-1: Power-Limited Regulator Consider a variation of the juggernaut problem (Example 7.3-3) in which the people moving the statue are limited in power as well as in the force that they can apply. Specifically, we wish to transfer the state to the origin in minimum time for the system

$$\dot{x}_1 = x_2$$
$$\dot{x}_2 = u,$$

where the force per unit mass u is limited in magnitude

$$|u| \le u_{\max}$$

and the actuator that produces the force is limited in power

$$|ux_2| \le p.$$

Thus we have state-dependent control constraints $u \in \mathcal{U}(\mathbf{x})$ defined by $\mathbf{h}(\mathbf{x}, u) \ge \mathbf{0}$, where $\mathbf{h} = [h_1 \ \cdots \ h_4]^{\mathrm{T}}$ and

$$0 \le h_1 = p - ux_2$$
$$0 \le h_2 = p + ux_2$$
$$0 \le h_3 = u_{\max} - u$$
$$0 \le h_4 = u_{\max} + u.$$

The H function is

$$H = \lambda_0 + \lambda_1 x_2 + \lambda_2 u,$$

where $\lambda_0 = 1$ or 0. In terms of the switching function $\sigma(t) = \partial H/\partial u = \lambda_2(t)$, the control that minimizes H is given by

$$u^* = -\hat{u}\,\mathrm{sgn}(\sigma), \tag{7.7-22}$$

where

$$\hat{u}(\mathbf{x}) = \min\left\{u_{\max}, \frac{p}{|x_2|}\right\} = \begin{cases} u_{\max} & \text{if } |x_2| \le \hat{x}_2 \\ \dfrac{p}{|x_2|} & \text{if } |x_2| \ge \hat{x}_2 \end{cases} \tag{7.7-23}$$

and

$$\hat{x}_2 = \frac{p}{u_{\max}}. \tag{7.7-24}$$

It can be shown that the singular control case, where $\sigma(t) \equiv 0$ on a nonzero time interval, does not occur since it would require $(\lambda_0, \lambda_1, \lambda_2) = \mathbf{0}$.

In terms of the Lagrangian

$$\begin{aligned} L &= H - \boldsymbol{\gamma}^{\mathrm{T}}\mathbf{h} \\ &= \lambda_0 + \lambda_1 x_2 + \lambda_2 u - \gamma_1(p - ux_2) - \gamma_2(p + ux_2) \\ &\quad - \gamma_3(u_{\max} - u) - \gamma_4(u_{\max} + u), \end{aligned}$$

the control u^* satisfies the necessary conditions (7.7-14)–(7.7-17) for minimizing H and allows us to write the adjoint equations as

$$\begin{aligned} \dot{\lambda}_1 &= -\frac{\partial L}{\partial x_1} = 0 \\ \dot{\lambda}_2 &= -\frac{\partial L}{\partial x_2} = -\lambda_1 + (\gamma_2 - \gamma_1)u^*. \end{aligned}$$

For $|x_2| \le \hat{x}_2$, where $|u^*|$ is limited by $u_{\max}$, we have $(\gamma_2 - \gamma_1)u^* = 0$,

$$u^* = -u_{\max}\, \mathrm{sgn}(\lambda_2), \tag{7.7-25}$$

and the adjoint equations reduce to

$$\begin{aligned} \dot{\lambda}_1 &= 0 \\ \dot{\lambda}_2 &= -\lambda_1. \end{aligned} \tag{7.7-26}$$

For $|x_2| > \hat{x}_2$, where $|u^*|$ is power limited, we have

$$0 = \frac{\partial L}{\partial u} = \lambda_2 - (\gamma_2 - \gamma_1)x_2,$$

$$u^* = -\frac{p}{|x_2|}\,\text{sgn}(\lambda_2), \tag{7.7-27}$$

and

$$\dot{\lambda}_2 = -\lambda_1 + \lambda_2\,\frac{u^*}{x_2} = -\lambda_1 - \lambda_2\,\text{sgn}(\lambda_2)\,\frac{p}{x_2|x_2|}$$

$$= -\lambda_1 + \text{sgn}(x_2)|\lambda_2|\,\frac{p}{x_2^2}.$$

Thus for $|x_2| > \hat{x}_2$ the adjoint equations reduce to

$$\begin{aligned} \dot{\lambda}_1 &= 0 \\ \dot{\lambda}_2 &= -\lambda_1 + \text{sgn}(x_2)|\lambda_2|\,\frac{p}{x_2^2}. \end{aligned} \tag{7.7-28}$$

Figure 7.7-1 shows the minimum-time trajectories for this example with $u_{\max} = p = 1$. These trajectories were obtained by backwards integration of the state and adjoint equations from $\mathbf{x}_f = \mathbf{0}$ with $\lambda_0 = 1$, $\lambda_{2f} = \pm 1$ (from $H_f = 0$), and various choices for λ_{1f}. The optimal feedback control $u^*(\mathbf{x})$, indicated in the figure, is power limited for $|x_2| \geq 1$. ■

State Constraints

Consider now the case where we have a pure state constraint [Hestenes, 1966, p. 352; Bryson and Ho, 1975, p. 117]

$$\mathbf{x} \in X = \{\mathbf{x} \in R^{n_x} \,|\, s(\mathbf{x}) \geq 0\}, \tag{7.7-29}$$

with state-independent control constraints

$$\mathbf{u} \in \mathcal{U} = \{\mathbf{u} \in R^{n_u} \,|\, \mathbf{h}(\mathbf{u}) \geq \mathbf{0}\}, \tag{7.7-30}$$

where $\mathbf{h} = [h_1 \cdots h_{n_h}]^T$ is continuous and continuously differentiable and $s(\mathbf{x})$ is continuously differentiable through as many orders as required for the following analysis. We also assume that $\mathbf{x} = \mathbf{x}(t)$ and $\mathbf{u} = \mathbf{u}^*(\mathbf{x})$ are regular points of X and $\mathcal{U}$, respectively.

If $s[\mathbf{x}(t)] = 0$ for a nonzero time interval $t \in [t_e, t_l)$, then so are the time derivatives of $s[\mathbf{x}(t)]$. At $\mathbf{x} = \mathbf{x}(t)$ let

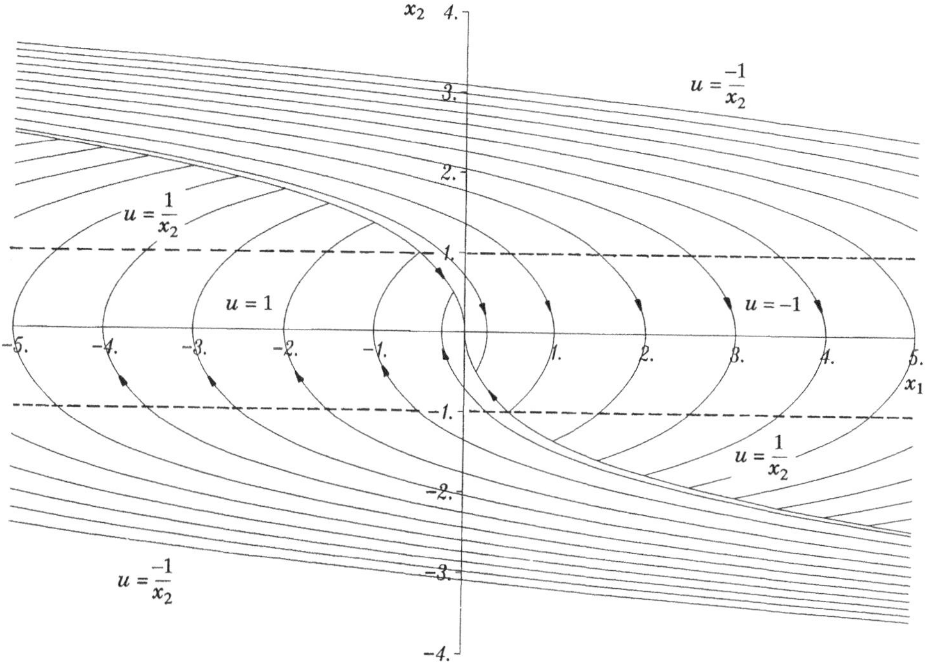

Figure 7.7-1 Minimum-time trajectories for a power-limited regulator.

$$\mathbf{S}(\mathbf{x}, \mathbf{u}) = \begin{bmatrix} s(\mathbf{x}) \\ \dot{s}(\mathbf{x}) \\ \ddot{s}(\mathbf{x}) \\ \vdots \\ s^{(q)}(\mathbf{x}, \mathbf{u}) \end{bmatrix}, \tag{7.7-31}$$

where $s^{(q)} \triangleq d^q s/dt^q$ and $q \geq 1$ is the lowest order derivative in which one or more of the control variables appears explicitly. The state constraint is called a qth-order **state constraint**. In general, q may be either odd or even for a state constraint, in contrast to singular control where the control only appears in an even order derivative of a switching function.

In the following we use the subscripts e and l to denote points where a forward-time trajectory enters or leaves a state constraint boundary, respectively. Specifically, let $\mathbf{x}_e = \mathbf{x}(t_e)$ be a point where $\mathbf{x}(t)$ reaches the state constraint boundary and begins to travel along it for a nonzero time interval $[t_e, t_l)$. Then $\mathbf{x}_e$ must be such that

$$0 = s(\mathbf{x}_e) = \dot{s}(\mathbf{x}_e) = \cdots = s^{(q-1)}(\mathbf{x}_e) \tag{7.7-32}$$

and $\mathbf{u}^*(\mathbf{x}_e)$ must be such that

$$0 = s^{(q)}[\mathbf{x}_e, \mathbf{u}^*(\mathbf{x}_e)]. \tag{7.7-33}$$

Then $\dot{\mathbf{x}}^*(t_e)$ is tangent to $s(\mathbf{x}) = 0$. Indeed, $\mathbf{x}(t)$ will remain on $s(\mathbf{x}) = 0$ provided

$$0 \equiv s^{(q)}(\mathbf{x}(t), \mathbf{u}^*[\mathbf{x}(t)]) \qquad \text{for } t \in [t_e, t_l).$$

In particular, the state constraint

$$s[\mathbf{x}(t)] \geq 0$$

is equivalent to a requirement that

$$s^{(q)}(\mathbf{x}, \mathbf{u}) \geq 0 \qquad \text{for all } \mathbf{u} \in \mathcal{U} \tag{7.7-34}$$

and

$$s^{(q)}[\mathbf{x}, \mathbf{u}^*(\mathbf{x})] = 0 \tag{7.7-35}$$

at each point $\mathbf{x} = \mathbf{x}(t)$, $t \in [t_e, t_l)$.

The qth-order state inequality condition (7.7-34) along with the control variable inequalities (7.7-30) are, for the most part, analogous to the case of state-dependent control constraints. The state constraint problem can be handled by adding (7.7-35) to the list of active inequalities, provided conditions (7.7-32) are satisfied at $\mathbf{x}_e = \mathbf{x}(t_e)$. However, there is a key difference. Because $s(\mathbf{x}) \geq 0$ restricts the state, we are not able to make an arbitrary small move $\Delta\mathbf{x}$, leading to a stationary condition (7.7-7). As a consequence the analogous development of the adjoint equations will stop at (7.7-6) since we cannot transform the adjoint equations to the form (7.7-8). However, in this case the term involving $\partial\mathbf{u}/\partial\mathbf{x}$ can be calculated, because at least part of $\mathbf{u}(\mathbf{x})$ is determined by the state constraint condition (7.7-35).

The adjoint equations are

$$\dot{\boldsymbol{\lambda}}^{\mathrm{T}} = -\frac{\partial H}{\partial \mathbf{x}} - \frac{\partial H}{\partial \mathbf{u}}\frac{\partial \mathbf{u}}{\partial \mathbf{x}}, \tag{7.7-36}$$

where the second term is zero away from the state constraint boundary since the control constraints $\mathbf{h}(\mathbf{u}) \geq \mathbf{0}$ are state independent. The problem is to calculate the second term in (7.7-36) when the trajectory is on the state constraint boundary.

If $\mathbf{u}^*(\mathbf{x})$ is an optimal control with corresponding trajectory $\mathbf{x}(t)$ satisfying a qth-order state constraint $s[\mathbf{x}(t)] \geq 0$, and if $s[\mathbf{x}(t)] = 0$ for $t \in [t_e, t_l]$, then the necessary condition for minimizing H subject to the constraints (7.7-30) and (7.7-34) are that there exist a vector $\boldsymbol{\gamma}(t) \in R^{n_h}$ and a scalar $\mu(t) \geq 0$ such that, for $t \in [t_e, t_l)$,

$$\mathbf{0}^{\mathrm{T}} = \frac{\partial H}{\partial \mathbf{u}} - \boldsymbol{\gamma}^{\mathrm{T}} \frac{\partial \mathbf{h}}{\partial \mathbf{u}} - \mu \frac{\partial s^{(q)}}{\partial \mathbf{u}} \tag{7.7-37}$$

$$\mathbf{h}(\mathbf{u}) \geq \mathbf{0} \tag{7.7-38}$$

$$\boldsymbol{\gamma} \geq \mathbf{0} \tag{7.7-39}$$

$$\boldsymbol{\gamma}^{\mathrm{T}}\mathbf{h}(\mathbf{u}) = 0 \tag{7.7-40}$$

$$s^{(q)}(\mathbf{x}, \mathbf{u}) \geq 0 \tag{7.7-41}$$

$$\mu s^{(q)}(\mathbf{x}, \mathbf{u}) = 0, \tag{7.7-42}$$

with

$$\mathbf{S}[\mathbf{x}, \mathbf{u}^*(\mathbf{x})] = \mathbf{0} \qquad \text{on } s[\mathbf{x}(t)] = 0,\ t \in [t_e, t_l) \tag{7.7-43}$$

$$\mu(t) = 0 \qquad \text{for } s[\mathbf{x}(t)] > 0, \tag{7.7-44}$$

where $\mathbf{S}(\mathbf{x}, \mathbf{u})$ is defined by (7.7-31), and all quantities are evaluated at $\mathbf{x} = \mathbf{x}(t)$, $\mathbf{u} = \mathbf{u}^* = \mathbf{u}^*(\mathbf{x})$. From (7.7-37)

$$\frac{\partial H}{\partial \mathbf{u}} \frac{\partial \mathbf{u}^*}{\partial \mathbf{x}} = \boldsymbol{\gamma}^{\mathrm{T}} \frac{\partial \mathbf{h}}{\partial \mathbf{u}} \frac{\partial \mathbf{u}^*}{\partial \mathbf{x}} + \mu \frac{\partial s^{(q)}}{\partial \mathbf{u}} \frac{\partial \mathbf{u}^*}{\partial \mathbf{x}}.$$

Thus, with the quantities $\boldsymbol{\gamma}$ and μ fixed, consider the function

$$\psi(\mathbf{x}, \mathbf{u}) \triangleq \boldsymbol{\gamma}^{\mathrm{T}}\mathbf{h}(\mathbf{u}) + \mu s^{(q)}(\mathbf{x}, \mathbf{u}).$$

With $\mathbf{u} = \mathbf{u}^*(\mathbf{x})$ we have $\psi[\mathbf{x}, \mathbf{u}^*(\mathbf{x})] = 0$ at $\mathbf{x} \in \partial X$. If we make a small move, to $\mathbf{x} + \Delta\mathbf{x} \in X$, we have $\psi[\mathbf{x} + \Delta\mathbf{x}, \mathbf{u}^*(\mathbf{x} + \Delta\mathbf{x})] \geq 0$. Therefore, $\psi[\mathbf{x}, \mathbf{u}^*(\mathbf{x})]$ has a local minimum at $\mathbf{x}$ subject to the constraint $s(\mathbf{x}) \geq 0$. Thus there exists a Lagrange multiplier $\alpha \geq 0$, with $\alpha s(\mathbf{x}) = 0$, such that

$$\begin{aligned} \mathbf{0}^{\mathrm{T}} &= \frac{d\psi[\mathbf{x}, \mathbf{u}^*(\mathbf{x})]}{d\mathbf{x}} - \alpha \frac{\partial s(\mathbf{x})}{\partial \mathbf{x}} \\ &= \left[\boldsymbol{\gamma}^{\mathrm{T}} \frac{\partial \mathbf{h}}{\partial \mathbf{u}} + \mu \frac{\partial s^{(q)}}{\partial \mathbf{u}}\right] \frac{\partial \mathbf{u}^*}{\partial \mathbf{x}} + \mu \frac{\partial s^{(q)}}{\partial \mathbf{x}} - \alpha \frac{\partial s(\mathbf{x})}{\partial \mathbf{x}} \end{aligned} \tag{7.7-45}$$

Hence

$$\frac{\partial H}{\partial \mathbf{u}} \frac{\partial \mathbf{u}^*}{\partial \mathbf{x}} = -\mu \frac{\partial s^{(q)}}{\partial \mathbf{x}} + \alpha \frac{\partial s(\mathbf{x})}{\partial \mathbf{x}}$$

and the adjoint equations become

$$\dot{\boldsymbol{\lambda}}^{\mathrm{T}} = -\frac{\partial H}{\partial \mathbf{x}} + \mu \frac{\partial s^{(q)}}{\partial \mathbf{x}} - \alpha \frac{\partial s(\mathbf{x})}{\partial \mathbf{x}},$$

where μ, $\alpha \geq 0$, with $\mu = \alpha = 0$ for $s(\mathbf{x}) > 0$.

In terms of

$$L(\mathbf{x}, \mathbf{u}, \boldsymbol{\lambda}, \lambda_0, \boldsymbol{\gamma}, \mu, \alpha) \triangleq H(\mathbf{x}, \mathbf{u}, \boldsymbol{\lambda}, \lambda_0) - \boldsymbol{\gamma}^{\mathrm{T}}\mathbf{h}(\mathbf{u}) - \mu s^{(q)}(\mathbf{x}, \mathbf{u}) + \alpha s(\mathbf{x}), \quad (7.7\text{-}46)$$

the modified adjoint equations can be written as

$$\dot{\boldsymbol{\lambda}}^{\mathrm{T}} = -\frac{\partial L}{\partial \mathbf{x}} \quad (7.7\text{-}47)$$

and the minimization necessary condition (7.7-37) can be written as

$$\mathbf{0}^{\mathrm{T}} = \frac{\partial L}{\partial \mathbf{u}}. \quad (7.7\text{-}48)$$

Note that if we had used $\psi[\mathbf{x}, \mathbf{u}^*(\mathbf{x})] = 0$ in a neighborhood of $s[\mathbf{x}(t)] = 0$, $t \in (t_e, t_l)$, yielding

$$\mathbf{0}^{\mathrm{T}} = \frac{d\psi[\mathbf{x}, \mathbf{u}^*(\mathbf{x})]}{d\mathbf{x}} = \mu \frac{\partial s^{(q)}}{\partial \mathbf{x}} + \left[\boldsymbol{\gamma}^{\mathrm{T}} \frac{\partial \mathbf{h}}{\partial \mathbf{u}} + \mu \frac{\partial s^{(q)}}{\partial \mathbf{u}}\right] \frac{\partial \mathbf{u}^*}{\partial \mathbf{x}},$$

which is equivalent to $\alpha = 0$, the resulting adjoint equations would have been the same as in Bryson and Ho [1975, p. 117] but without their jump conditions on $\boldsymbol{\lambda}(t_l)$. On the other hand, for a first-order constraint the adjoint equations (7.7-47) and $\alpha > 0$ differ from those in Hestenes [1966, p. 352], which are the same as Bryson and Ho [1975, p. 117] for a first-order constraint, but with no jump conditions. But both of these references deal with open-loop controls $\mathbf{u}(t)$ instead of closed-loop controls $\mathbf{u}(\mathbf{x})$. Perhaps this is part of the source of the differences in the three results. Our result in (7.7-47) should, although we have not reproven it, give $H = 0$ and $\boldsymbol{\lambda}^{\mathrm{T}} = \partial V^*(\mathbf{x})/\partial \mathbf{x}$ without requiring discontinuous changes in $\boldsymbol{\lambda}(t)$ at $t = t_l$ or $t = t_e$.

Example 7.7-2: Avoidance Region Consider the problem of transferring the state to the origin in minimum time (shortest distance) for the system

$$\dot{x}_1 = \cos u$$
$$\dot{x}_2 = \sin u \tag{7.7-49}$$

while avoiding a region consisting of the interior of a unit disk centered at $(x_1, x_2) = (2, 0)$, as illustrated in Figure 7.7-2. In other words, the state $\mathbf{x}(t)$ must satisfy the inequality constraint

$$s(\mathbf{x}) = (x_1 - 2)^2 + x_2^2 - 1 \geq 0. \tag{7.7-50}$$

This is a first-order constraint since

$$\dot{s}(\mathbf{x}, u) = 2(x_1 - 2)\cos u + 2x_2 \sin u \tag{7.7-51}$$

contains the control u explicitly.

The H function, with $\lambda_0 = 1$, is

$$H = 1 + \lambda_1 \cos u + \lambda_2 \sin u \tag{7.7-52}$$

and the L function is

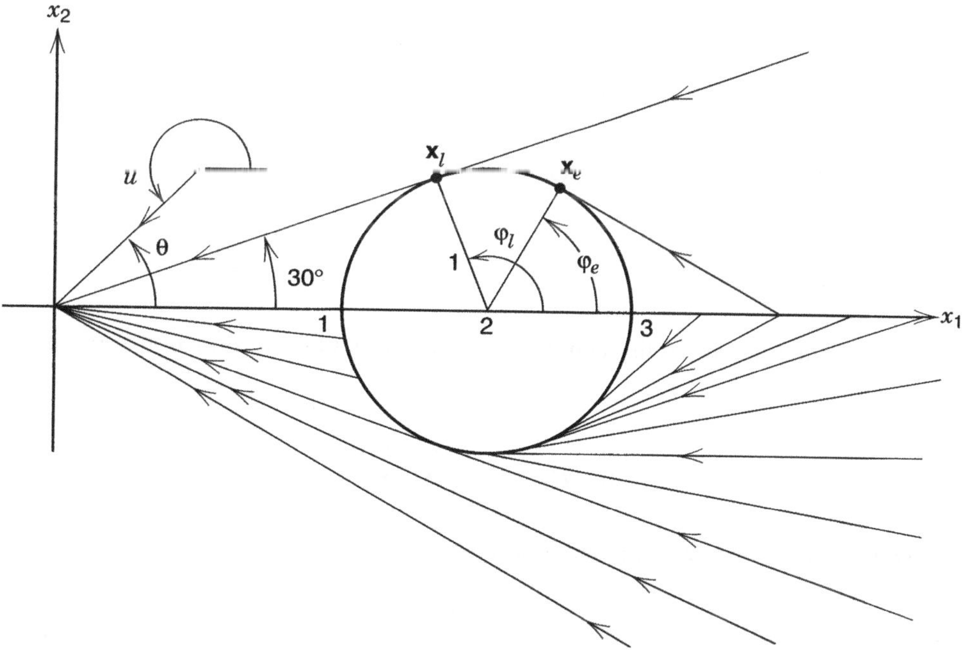

Figure 7.7-2 Minimum-time trajectories with an avoidance region.

$$L = H - \mu\dot{s} + \alpha s$$
$$= 1 + \lambda_1 \cos u + \lambda_2 \sin u - \mu[2(x_1 - 2)\cos u + 2x_2 \sin u]$$
$$+ \alpha[(x_1 - 2)^2 + x_2^2 - 1]. \tag{7.7-53}$$

The optimal control that minimizes H, subject to $\dot{s}(\mathbf{x}, u) \geq 0$ if $s(\mathbf{x}) = 0$, must satisfy

$$0 = \frac{\partial L}{\partial u}$$
$$= -\lambda_1 \sin u + \lambda_2 \cos u - 2\mu[-(x_1 - 2)\sin u + x_2 \cos u],$$

where $\mu \geq 0$ with $\mu = 0$ if $s(\mathbf{x}) > 0$. This necessary condition yields

$$\tan u^* = \frac{\lambda_2 - 2\mu x_2}{\lambda_1 - 2\mu(x_1 - 2)}. \tag{7.7-54}$$

The adjoint equations are

$$\begin{aligned} \dot{\lambda}_1 &= -\frac{\partial L}{\partial x_1} = 2\mu \cos u^* - 2\alpha(x_1 - 2) \\ \dot{\lambda}_2 &= -\frac{\partial L}{\partial x_2} = 2\mu \sin u^* - 2\alpha x_2, \end{aligned} \tag{7.7-55}$$

where $\mu = \alpha = 0$ for $s(\mathbf{x}) > 0$.

Off the State Constraint Off the state constraint, for $\mu = 0$, we have

$$\cos u^* = \frac{-\lambda_1}{\sqrt{\lambda_1^2 + \lambda_2^2}}, \qquad \sin u^* = \frac{-\lambda_2}{\sqrt{\lambda_1^2 + \lambda_2^2}}. \tag{7.7-56}$$

The terminal transversality conditions at $\mathbf{x}_f = \mathbf{0}$ yield no information about $\boldsymbol{\lambda}_f$, except that

$$0 = H_f = 1 - \sqrt{\lambda_{1f}^2 + \lambda_{2f}^2}.$$

Thus we can write $\boldsymbol{\lambda}_f$ in polar coordinates as

$$\begin{aligned} \lambda_{1f} &= \cos\theta \\ \lambda_{2f} &= \sin\theta. \end{aligned} \tag{7.7-57}$$

In terms of the time-to-go $\tau = t_f - t$, integration of the state and adjoint equations backward in time from $\mathbf{x}_f = \mathbf{0}$, with $\mu = \alpha = 0$, yields

$$\lambda_1(\tau) = \cos\theta$$
$$\lambda_2(\tau) = \sin\theta, \tag{7.7-58}$$

with

$$u^*(\tau) = \theta + \pi \tag{7.7-59}$$

and

$$x_1(\tau) = \tau\cos\theta$$
$$x_2(\tau) = \tau\sin\theta. \tag{7.7-60}$$

Thus the minimum-time trajectories are straight lines with slope

$$\frac{dx_2}{dx_1} = \tan\theta$$

and the minimum time-to-go is

$$\tau = \sqrt{x_1^2 + x_2^2}. \tag{7.7-61}$$

As illustrated in Figure 7.7-2, one of three things happen to the retro-time trajectories. If $|\theta| > 30°$, then $\mathbf{x}(\tau)$ never intersects the state constraint boundary $s(\mathbf{x}) = 0$. For $|\theta| < 30°$ retro time trajectories intersect the state constraint at a point

$$x_1(\tau) = 2 + \cos\varphi$$
$$x_2(\tau) = \sin\varphi, \tag{7.7-62}$$

where $120° < \varphi < 240°$, such that

$$0 > \frac{ds[\mathbf{x}(\tau)]}{d\tau} = 2(x_1 - 2)\cos\theta + 2x_2\sin\theta = 2\cos(\varphi - \theta),$$

since

$$\tau = \sqrt{x_1^2 + x_2^2} < \sqrt{3}$$

and the law of cosines

$$2^2 = 1^2 + \tau^2 - 2(1)\tau \cos(\pi - \theta - [\pi - \varphi])$$

yields

$$\cos(\varphi - \theta) = \frac{\tau^2 - 3}{2\tau} < 0.$$

Thus further retro-time integration yields trajectories that penetrate the state constraint. For this case the retro-time optimal trajectories simply stop when they reach the state constraint.

On the State Constraint The third case, where $\theta = \pm 30°$, yields retro-time trajectories that are tangent to $s(\mathbf{x}) = 0$ at $\mathbf{x} = (3/2, \pm\sqrt{3}/2)$ with $\tau = \tau_l = \sqrt{3}$ and $\varphi = \varphi_l = \pm 2\pi/3$ in (7.7-62). Further retro-time integration, with no change in the adjoint equations, causes the trajectories to continue on, back into the $s(\mathbf{x}) > 0$ region. This leaves a "shadow" to the right of the state constraint in Figure 7.7-2, which is empty of trajectories. To fill in this region the adjoint equations must be changed at τ_l to cause trajectories that travel on the state constraint boundary $s(\mathbf{x}) = 0$ for a while, say, until $\tau = \tau_e$, before re-entering the $s(\mathbf{x}) > 0$ region.

Consider a trajectory that runs along the top portion of the state constraint in Figure 7.7-2. Symmetric results occur for the bottom portion. From (7.7-51), along $s(\mathbf{x}) = 0$ with $\dot{s}(\mathbf{x}, u) = 0$, which corresponds to $\alpha = 0$ from (7.7-45), we get

$$u^*(\mathbf{x}) = \tan^{-1}\left[\frac{x_1 - 2}{-x_2}\right], \tag{7.7-63}$$

or, since $(x_1 - 2)^2 + x_2^2 = 1$,

$$\cos u^* = -x_2, \qquad \sin u^* = x_1 - 2. \tag{7.7-64}$$

Thus the retro-time state equations are

$$\frac{dx_1}{d\tau} = -\cos u^* = x_2$$

$$\frac{dx_2}{d\tau} = -\sin u^* = 2 - x_1,$$

with $\mathbf{x}(\tau_l) = \mathbf{x}_l = [3/2 \quad \sqrt{3}/2]^{\mathrm{T}}$. Since the motion is along a circular state constraint, the solution can be written in polar form as

$$x_1 = 2 + \cos\varphi$$
$$x_2 = \sin\varphi, \tag{7.7-65}$$

where, for unit speed on a unit circle, $R\,d\varphi = -V\,d\tau$ yields

$$\tau = \tau_l + (\varphi_l - \varphi) \tag{7.7-66}$$

with $\tau_l = \sqrt{3}$ and, on the top portion of the state constraint, $\varphi_l = 2\pi/3$. From (7.7-64) and (7.7-65) we also note that

$$u^* = \varphi + \frac{\pi}{2} \tag{7.7-67}$$

and

$$\cos u^* = -\sin\varphi, \qquad \sin u^* = \cos\varphi. \tag{7.7-68}$$

From the necessary condition (7.7-53) for minimizing H subject to $\dot{s}(\mathbf{x}, u) \geq 0$, we have

$$\begin{aligned}\mu &= \frac{-\lambda_1 \sin u + \lambda_2 \cos u}{2[-(x_1 - 2)\sin u + x_2 \cos u]} \\ &= \frac{\lambda_1 \sin\varphi + \lambda_2 \cos\varphi}{2[(x_1 - 2)\sin\varphi + x_2 \cos\varphi]} \\ &= \frac{\lambda_1 \sin\varphi + \lambda_2 \cos\varphi}{2\sin 2\varphi}.\end{aligned} \tag{7.7-69}$$

We could use this result to form the adjoint equations (7.7-55). However, since we have a single control variable u we have been able to completely determine the control $u^*(\mathbf{x})$ along the state constraint, from $\dot{s}(\mathbf{x}, u) = 0$ on $s(\mathbf{x}) = 0$, so we can compute $\partial u^*(\mathbf{x})/\partial\mathbf{x}$ and then use the direct form of the adjoint equations (7.7-36). Later we examine the necessary condition $\mu \geq 0$.

From (7.7-63) we get

$$\frac{\partial u^*}{\partial \mathbf{x}} = \frac{1}{1 + ((x_1 - 2)/x_2)^2}\begin{bmatrix} \dfrac{1}{-x_2} & \dfrac{x_1 - 2}{x_2^2} \end{bmatrix} = [-x_2 \quad x_1 - 2],$$

and from (7.7-64)

$$\frac{\partial H}{\partial u}\frac{\partial u^*}{\partial \mathbf{x}} = [(-\lambda_1 \sin u^* + \lambda_2 \cos u^*) \qquad (-x_2)(-\lambda_1 \sin u^* + \lambda_2 \cos u^*)(x_1 - 2)]$$

$$= [\lambda_1 x_2(x_1 - 2) + \lambda_2 x_2^2 \qquad -\lambda_1(x_1 - 2)^2 - \lambda_2 x_2(x_1 - 2)].$$

Then the adjoint equations

$$\dot{\boldsymbol{\lambda}}^{\mathrm{T}} = -\frac{\partial H}{\partial \mathbf{x}} - \frac{\partial H}{\partial u}\frac{\partial u^*}{\partial \mathbf{x}}$$

become

$$\begin{aligned} \dot{\lambda}_1 &= -\lambda_1 x_2(x_1 - 2) - \lambda_2 x_2^2 \\ \dot{\lambda}_2 &= \lambda_1(x_1 - 2)^2 + \lambda_2 x_2(x_1 - 2). \end{aligned} \tag{7.7-70}$$

These equations can be integrated backward in time, with $\mathbf{x}(\tau)$ given by (7.7-65) and (7.7-66) and with $\boldsymbol{\lambda}(\tau_l) = [\cos\theta \quad \sin\theta]^{\mathrm{T}} = [\sqrt{3}/2 \quad 1/2]^{\mathrm{T}}$ on the top portion of the state constraint.

Using (7.7-65) and (7.7-66), with $d(\ )/dt = -d(\ )/d\tau = d(\ )/d\varphi$, the adjoint equations can be written as

$$\begin{aligned} \frac{d\lambda_1}{d\varphi} &= -(\lambda_1 \cos\varphi + \lambda_2 \sin\varphi)\sin\varphi \\ \frac{d\lambda_2}{d\varphi} &= \lambda_1 \cos^2\varphi + \lambda_2 \sin\varphi\cos\varphi. \end{aligned} \tag{7.7-71}$$

From

$$0 = H = 1 - \lambda_1 x_2 + \lambda_2(x_1 - 2) = 1 - \lambda_1 \sin\varphi + \lambda_2 \cos\varphi \tag{7.7-72}$$

we have

$$\lambda_2 = \frac{\lambda_1 \sin\varphi - 1}{\cos\varphi}. \tag{7.7-73}$$

Thus

$$\frac{d\lambda_1}{d\varphi} = -(\lambda_1 - \sin\varphi)\tan\varphi,$$

which has the solution

$$\lambda_1 = (C - \varphi)\cos\varphi + \sin\varphi.$$

At $\mathbf{x}(\tau_l)$, with $\varphi = \varphi_l = 2\pi/3$ and $\lambda_{1l} = \sqrt{3}/2$ on the top portion of the state constraint, we have

$$C = \varphi_l + \frac{\lambda_{1l} - \sin\varphi_l}{\cos\varphi_l} = \varphi_l.$$

Thus

$$\lambda_1 = (\varphi_l - \varphi)\cos\varphi + \sin\varphi. \qquad (7.7\text{-}74)$$

Instead of integrating the second adjoint equation directly, we can determine λ_2 from $H = 0$, since the adjoint equations, by construction, imply H = constant. From (7.7-73) and (7.7-74) we get

$$\lambda_2 = (\varphi_l - \varphi)\sin\varphi - \cos\varphi. \qquad (7.7\text{-}75)$$

From (7.7-69)

$$\begin{aligned}\mu &= \frac{\lambda_1 \sin\varphi + \lambda_2\cos\varphi}{2\sin 2\varphi}\\ &= \frac{[(\varphi_l - \varphi)\cos\varphi + \sin\varphi]\sin\varphi + [(\varphi_l - \varphi)\sin\varphi - \cos\varphi]\cos\varphi}{2\sin 2\varphi}\\ &= \tfrac{1}{2}(\varphi_l - \varphi - \cot 2\varphi).\end{aligned}$$

Note that the condition $\mu \geq 0$ is satisfied throughout the range $0 \leq \varphi \leq \varphi_l = 2\pi/3$ along the top portion of the trjaectory.

In the State Constraint Shadow To fill in the region to the right of the state constraint in Figure 7.7-2 we leave the state constraint, at $\varphi = \varphi_e$ (which we choose) and

$$\tau_e = \tau_l + \varphi_l - \varphi_e, \qquad (7.7\text{-}76)$$

by choosing $\mu(\tau) = 0$ for $\tau \geq \tau_e$. This discontinuous change in $\mu(\tau)$ causes a switch back to the normal adjoint equations and causes the retro-time trajectory $\mathbf{x}(\tau)$ to leave the state constraint tangentially and return to the $s(\mathbf{x}) > 0$ region, along a straight-line trajectory

$$\begin{aligned}x_1(\tau) &= (\tau - \tau_e)\cos u_e^* + x_{1e}\\ x_2(\tau) &= (\tau - \tau_e)\sin u_e^* + x_{2e},\end{aligned} \qquad (7.7\text{-}77)$$

where

$$x_{1e} = 2 + \cos \varphi_e \qquad x_{2e} = \sin \varphi_e \tag{7.7-78}$$

and

$$u_e^* = \varphi_e + \frac{\pi}{2}. \tag{7.7-79}$$

Along the retro-time trajectory, $u^*(\tau) = u_e^*$ since $\boldsymbol{\lambda}(\tau) = \boldsymbol{\lambda}_e$, where

$$\lambda_{1e} = (\varphi_l - \varphi_e)\cos \varphi_e + \sin \varphi_e \qquad \lambda_{2e} = (\varphi_l - \varphi_e)\sin \varphi_e - \cos \varphi_e. \tag{7.7-80}$$

The minimum time-to-go is

$$V^*(\mathbf{x}) = \tau = \tau_e + \sqrt{(x_1 - x_{1e})^2 + (x_2 - x_{2e})^2}, \tag{7.7-81}$$

where, from the top portion of the state constraint,

$$\tau_e = \sqrt{3} + \frac{2\pi}{3} - \varphi_e.$$

The retro-time trajectory continues until, if $\varphi_e < \pi/2$, it intersects at $x_2(\tau) = 0$ with its mirror image from the bottom portion of the state constraint, forming a dispersal surface along which the minimum time-to-go is the same for trajectories above and below the x_1-axis. ■

EXERCISES

7-1. Rework Example 7.3-2 with the terminal set specified by

(a) $g(\mathbf{x}) = x_1 + 3x_2 - 1 = 0$.

(b) $g(\mathbf{x}) = x_1^2 + x_2^2 - 1 = 0$.

7-2. Some pirates have buried a pot of gold on the shoreline of a circular island 1 mile in radius. From the center of the island, the treasure is 20° east of north. You are currently located 5 miles from the center of the island 135° east of north. If a 3 mile/hr current is flowing due east, determine the minimum time to get to the treasure, by sea, using a boat capable of traveling at 10 miles/hr relative to the water. *Hint:* Consider a two-part cost: terminal-plus-integral.

7-3. For a numerical solution of the minimum-time problem in Example 7.3-3:

(a) Using retro-time numerical integration of the state and adjoint equations with a *fixed time step* (e.g., fourth-order Runge–Kutta, but not variable time step routines such as LSODAR or MATLAB's ode23 or ode45 commands), numerically plot the minimum-time trajectories in Figure 7.3-9. Choose a plotting time step small enough so there are no visible "kinks" or straight-line segments in the trajectories.

(b) On the plot generated in (a), numerically plot the minimum time-to-go contours $V^*(\mathbf{x}) = c$, for $c = 1, 2, 3$.

(c) Consider the retro-time trajectory that passes through $\mathbf{x} = (x_1, x_2) = (1, 0)$. From the analytic solutions in Example 7.3-3, verify that $\boldsymbol{\lambda}^{\mathrm{T}} = \partial V^*/\partial \mathbf{x}$ at $\mathbf{x} = (1, 0)$. At the point $\mathbf{x}(\tau) = (1, 0)$ on the plot in (a), where τ is the time-to-go, plot (by hand) the vector $\boldsymbol{\lambda}(\tau)$ generated by numerical retro-time integration. Now, consider the switching point at $\hat{\mathbf{x}} = (\frac{1}{2}, -1)$. There are two trajectories that meet at this point, forward in time. For the corresponding numerical retro-time trajectories, at the point $\hat{\mathbf{x}}$ on the plot in (a) plot (by hand) the two values of $\boldsymbol{\lambda}(\tau)$.

(d) For the numerical retro-time trajectory in (a) that passes through $\mathbf{x} = (1, 0)$, numerically plot $H(\tau)$ versus τ, $0 \leq \tau \leq 3$, during retro-time integration. Discuss why $H(\tau)$ is not exactly equal to zero in this plot and discuss the cause of any sudden changes in $H(\tau)$. Experiment with different integration time-step sizes. Also, see what happens to $H(\tau)$ if you modify your program to introduce a coding error (sign error, wrong coefficient, etc.), so that the state and adjoint equations no longer correspond to each other.

7-4. In 1696 John Bernoulli propounded the famous brachistochrone problem. The solution of this problem led to the development of the calculus of variations by Euler, Lagrange, and others. The problem is to determine the path a mass on a frictionless wire should follow in order to minimize the time of descent, under gravity, between two points, starting from rest. Set up this problem for solution using the minimum principle and show that an extremal is given by an arc of a cycloid. (See Example 3.6-2, but do not use the calculus of variations formulation).

7-5. **Goh's Problem**[1]**:** Determine the feedback control law that will minimize the accumulated cost corresponding to

$$\dot{x}_0 = -u$$

[1]Dr. Bean San Goh (personal communication) originally formulated this problem as a lottery winner problem. Given that you have just won the lottery and have put the money in the bank, how should you withdraw the money to maximize your return? It is required that at the end of your life there should be enough money for the funeral, but in the meantime, you should live no more frugally than an assistant professor and no more luxuriously than a full professor. We are a little uncertain as to how Dr. Goh determines t_f.

in a given time interval t_f, with the system dynamics given by

$$\dot{x}_1 = 0.05x_1 - u.$$

The control is limited by $35 \le u \le 65$ and the final value of x_1 must satisfy

$$x_1(t_f) = 15.$$

7-6. For the missile guidance problem of Example 7.3-4 assume that S_u = 2000 m/sec, S_v = 3000 m/sec, $\alpha(0) = 0$, $\beta(0) = 176°$ at an initial range of 50,000 m. For the following two cases, use (7.3-30) to generate the plots of the intercept of the missile with the target. Also plot the total control used as a function of time, up to the point of intercept.

(a) $A_{T_n} = 0$.

(b) $A_{T_n} = 7g$, where $g = 9.81$ m/sec^2 is the acceleration of gravity.

7-7. Rework Example 7.4-1 with "rate harvesting" rather than "effort harvesting." In other words, the rate at which harvesting occurs should be of the form u, rather than ux_1. Fix the final population number at zero. Let $r = k = u_{\max} = 1$.

7-8. A prey (x_1) predator (x_2) system subject to harvesting of the prey is modeled by the system of equations

$$\begin{aligned} \dot{x}_1 &= x_1(1 - x_2) - ux_1 \\ \dot{x}_2 &= x_2(x_1 - 1), \end{aligned}$$

where $0 \le u \le u_{\max}$ is the harvesting effort. Determine the feedback control program that will drive this system from any point in the positive (x_1, x_2) space to the unharvested equilibrium point $x_1 = x_2 = 1$ while minimizing the cost given by

$$J = \int_0^{t_f} (x_1 + u)\, dt.$$

Consider two cases:

(a) $u_{\max} = 0.5$.

(b) $u_{\max} = 1$.

7-9. Solve Example 7.5-1 with $\alpha = 1$ and $\alpha = 0.1$.

7-10. The one-dimensional system

$$\dot{x}_1 = u$$

has an accumulated cost given by

$$\dot{x}_0 = x_1^2.$$

By plotting retro-time trajectories analogous to Figure 7.4-2, determine the optimal feedback control law that will drive the system to the point $x_1(t_f) = 3$ in a fixed time interval t_f. The control is bounded by $|u| \le 1$.

7-11. Consider a Bolza problem from a fixed initial point, that is, with no initial cost term $\theta(\mathbf{x}_i)$ and no initial conditions surface $\mathbf{m}(\mathbf{x}_i) = \mathbf{0}$. Show that the applicable necessary conditions in (7.5-6)–(7.5-14) for this problem of Bolza follow from Pontryagin's minimum principle, by converting the Bolza problem to an integral cost problem. Develop formulas relating the $H(\cdot)$ function and adjoint variables in the Bolza problem necessary conditions to those of the minimum principle applied to the integral cost representation of the problem of Bolza.

7-12. Solve Example 7.5-2.

7-13. The simplest problem in the calculus of variations is to minimize a cost functional of the form

$$\int_0^{t_f} \phi[y(t), \dot{y}(t), t]\, dt.$$

By letting $x_1 = y$, $x_2 = t$, and $u = \dot{y}$, show that the minimum principle can be used to derive the classical Euler necessary condition

$$\frac{d}{dt}\left[\frac{\partial \phi}{\partial \dot{y}}\right] - \frac{\partial \phi}{\partial y} = 0.$$

7-14. Dido's Problem[2]: Find the trajectory $\mathbf{x}(t)$ that will maximize

$$\int_0^{t_f} x_2 \cos u\, dt$$

while driving the system

[2]According to legend, when the Tyrian princess Dido landed on the North African shore of the Mediterranean sea, she met a local chieftain who offered her all the land she could enclose between the shoreline and a rope made of the hide of a cow. Upon this land she was to found Carthage.

$$\dot{x}_1 = \cos u$$
$$\dot{x}_2 = \sin u$$

from the origin to the terminal surface $x_2(t_f) = 0$, subject to the constraint $t_f = l$. Show that the trajectory is a particular semicircle.

7-15. An optimization problem is given by

$$x_0 = \int_0^T x_1^2 \, dt$$

$$\dot{x}_1 = u$$

where $|u| \leq 1$. Using the minimum principle, determine the optimal feedback control law which will drive the system to the point $x_1(T) = 2$ in a fixed time interval T. You are to obtain a complete syntheses of the optimal solution with a representative sketch of the optimal trajectories to the target. In particular, highlight the optimal trajectory starting from $x_1(0) = 2$, with $T = 6$.

7-16. Use dynamic programming to solve the problem of minimizing

$$\frac{1}{2} \int_0^{t_f} u^2 \, dt$$

subject to

$$\dot{x} = x^3 + u$$
$$x(t_f) = 0.$$

7-17. Consider the steel production problem from Example 5.3-6, of maximizing $S(t_f)$ for the system

$$\dot{M} = c u_1 S$$
$$\dot{S} = [(a - 1)u_2 - u_1]S$$

subject to the control constraints

$$u_1 \geq 0, \qquad u_2 \geq 0, \qquad u_1 + u_2 \leq 1, \qquad u_2 \leq \frac{b}{a}\frac{M}{S}.$$

For $t_f = 10$, $a = 2$, $b = 4$, and $c = 1$, restate the problem as a standard integral cost optimal control problem. Determine the optimal controls and plot the corresponding optimal trajectories $S(\tau)$ versus $M(\tau)$, $\tau = t_f - t$, from all feasible values of $M > 0$ and $S > 0$. In each region of the plot indicate the optimal feedback controls $u_1^*(S, M)$ and $u_2^*(S, M)$.

7-18. Determine $u(x)$, and the corresponding $x(t)$, to minimize

$$\int_0^{t_f} dt$$

subject to

$$\dot{x} = x + u$$
$$u^2 \le 4x^2$$
$$x(t_f) = 0,$$

with $x(0) = 1$.

8

OPTIMAL CONTROL DESIGN

8.1 LINEAR SYSTEMS WITH LINEAR COST

A linear constant-coefficient dynamical system is of the form

$$\dot{\mathbf{x}} = \mathbf{A}\mathbf{x} + \mathbf{B}\mathbf{u}, \tag{8.1-1}$$

where $\mathbf{x}$ is an $n_x \times 1$ state vector, $\mathbf{A}$ is an $n_x \times n_x$ constant matrix, $\mathbf{B}$ is an $n_x \times n_u$ constant matrix, and $\mathbf{u}$ is an $n_u \times 1$ control vector. A linear cost integral is of the form

$$\int_0^{t_f} (\mathbf{q}^{\mathrm{T}}\mathbf{x} + \mathbf{r}^{\mathrm{T}}\mathbf{u} + c)dt, \tag{8.1-2}$$

where $\mathbf{q}$ is an $n_x \times 1$ state coefficient vector, $\mathbf{r}$ is an $n_u \times 1$ control coefficient vector, and c is a scalar constant.

In what follows we initially assume that the control is a scalar constrained by simple inequalities of the form $u_{\min} \leq u \leq u_{\max}$; that is,

$$\begin{aligned} h_1 &= u_{\max} - u \geq 0 \\ h_2 &= u - u_{\min} \geq 0, \end{aligned} \tag{8.1-3}$$

where $u_{\max} \geq u_{\min}$ are constants. The final state is assumed to be fixed (e.g., the origin). With a fixed final state no information is obtained from the transversality conditions. However, the minimum principle is easily applied to determine the nature of optimal control.

The H function for this problem is given by

$$H = \lambda_0(\mathbf{q}^{\mathrm{T}}\mathbf{x} + ru + c) + \boldsymbol{\lambda}^{\mathrm{T}}(\mathbf{A}\mathbf{x} + \mathbf{b}u).$$

Alternately, this can be written as

$$H = \lambda_0(\mathbf{q}^{\mathrm{T}}\mathbf{x} + c) + \boldsymbol{\lambda}^{\mathrm{T}}\mathbf{A}\mathbf{x} + \sigma u$$

where

$$\sigma = \frac{\partial H}{\partial u} = \lambda_0 r + \boldsymbol{\lambda}^{\mathrm{T}}\mathbf{b} \tag{8.1-4}$$

is the **switching function** for this system.

A necessary condition for minimizing H subject to the constraints on u is obtained by first forming the L function

$$L = H - \gamma_1(u_{\max} - u) - \gamma_2(u - u_{\min})$$

and then setting the derivative of L with respect to u equal to zero:

$$0 = \frac{\partial L}{\partial u} = \sigma + \gamma_1 - \gamma_2.$$

From this result and the requirements that $\boldsymbol{\gamma}^{\mathrm{T}}\mathbf{h} = 0$ and $\boldsymbol{\gamma} \geq \mathbf{0}$, it follows that

$$u^* = \begin{cases} u_{\min} & \text{if } \sigma > 0 \\ ? & \text{if } \sigma \equiv 0 \\ u_{\max} & \text{if } \sigma < 0. \end{cases} \tag{8.1-5}$$

Thus the nature of the control is determined from the switching function. Except for the possibility of singular control (where $\sigma(t)$ is identically zero, as discussed in Section 7.4), the control must either be minimum control or maximum control, according to the sign of the switching function.

Time-Optimal Control

An important special case of linear cost is the **time-optimal control** problem. We obtain the necessary conditions for this case by setting the constants $\mathbf{q} = \mathbf{0}$, $r = 0$, and $c = 1$. The switching function is then given by

$$\sigma = \boldsymbol{\lambda}^{\mathrm{T}}\mathbf{b},$$

with the adjoint equations

$$\dot{\boldsymbol{\lambda}}^{\mathrm{T}} = -\boldsymbol{\lambda}^{\mathrm{T}}\mathbf{A}.$$

The possibility of singular control is examined by setting repeated derivatives of the switching function equal to zero. In other words, using the notation $\sigma^{(n)} = d^n\sigma/dt^n$, singular control requires

$$\begin{aligned}
\sigma &= \boldsymbol{\lambda}^{\mathrm{T}}\mathbf{b} \equiv \mathbf{0} \\
\dot{\sigma} &= \dot{\boldsymbol{\lambda}}^{\mathrm{T}}\mathbf{b} = -\boldsymbol{\lambda}^{\mathrm{T}}\mathbf{A}\mathbf{b} \equiv \mathbf{0} \\
\ddot{\sigma} &= -\dot{\boldsymbol{\lambda}}^{\mathrm{T}}\mathbf{A}\mathbf{b} = \boldsymbol{\lambda}^{\mathrm{T}}\mathbf{A}^2\mathbf{b} \equiv \mathbf{0} \\
&\vdots \\
\sigma^{(n_x-1)} &= (-1)^{n_x-1}\boldsymbol{\lambda}^{\mathrm{T}}\mathbf{A}^{n_x-1}\mathbf{b} \equiv \mathbf{0},
\end{aligned}$$

which yields a set of n_x homogeneous equations in $\boldsymbol{\lambda}$. This set of equations for σ and its derivatives imply the trivial solution $\boldsymbol{\lambda}(t) = \mathbf{0}$ if the vector coefficients of $\boldsymbol{\lambda}$ in each of these equations are linearly independent, that is, if the determinant

$$|\mathcal{P}| = |\mathbf{b}, \mathbf{A}\mathbf{b}, \ \ldots, \ \mathbf{A}^{n_x-1}\mathbf{b}| \neq 0.$$

A trivial solution for $\boldsymbol{\lambda}$ would require $H = \lambda_0 = 0$, resulting in $(\lambda_0, \boldsymbol{\lambda}) = \mathbf{0}$. This is not allowed by the minimum principle. Thus we conclude that, if $|\mathcal{P}| \neq 0$, then there can be no singular solution. If there can be no singular solution, then the control law must be **bang-bang**. That is, when $|\mathcal{P}| \neq 0$ the minimum-time control for the linear system (8.1-1) will be a combination of maximum control and minimum control, given by

$$u^* = \begin{cases} u_{\min} & \text{if } \sigma > 0 \\ u_{\min} \text{ or } u_{\max} & \text{if } \sigma = 0 \\ u_{\max} & \text{if } \sigma < 0. \end{cases} \tag{8.1-6}$$

We note that the matrix $\mathcal{P}$ is the controllability matrix discussed in Section 2.2. Thus another way of thinking of this result is that whenever the Kalman controllability condition is satisfied, minimum-time control for linear systems with bounded controls will be bang-bang.

The results obtained for a scalar control can be generalized to include a vector control $\mathbf{u}$. In this case the system equations are given by

$$\dot{\mathbf{x}} = \mathbf{A}\mathbf{x} + \mathbf{B}\mathbf{u}. \tag{8.1-7}$$

If each component of the control is constrained by inequalities of the form

$$u_{i_{\min}} \leq u_i \leq u_{i_{\max}} \tag{8.1-8}$$

for $i = 1, \ldots, n_u$, and if

$$|\mathcal{P}_i| = |\mathbf{b}_i, \mathbf{A}\mathbf{b}_i, \ \ldots, \ \mathbf{A}^{n_x - 1}\mathbf{b}_i| \neq 0, \tag{8.1-9}$$

where $\mathbf{b}_i$ is the ith column of $\mathbf{B}$, then minimum-time control u_i^* is bang-bang, given by

$$u_i^* = \begin{cases} u_{i_{\min}} & \text{if } \sigma_i > 0 \\ u_{i_{\min}} \text{ or } u_{i_{\max}} & \text{if } \sigma = 0 \\ u_{i_{\max}} & \text{if } \sigma_i < 0, \end{cases} \tag{8.1-10}$$

where

$$\sigma_i = \boldsymbol{\lambda}^{\mathrm{T}}\mathbf{b}_i. \tag{8.1-11}$$

Furthermore it can be shown [Leitmann, 1981a, p. 215] that *if the eigenvalues of* **A** *are real, then there can be at most* $n_x - 1$ *switches in* u_i^*.

Minimum-Time Control of Second-Order IO Systems

One system of particular interest is an input-output (IO) system of the form

$$\ddot{y} + 2\zeta\omega_n\dot{y} + \omega_n^2 y = q_0 u,$$

with $q_0 \neq 0$ and all coefficients real (although we allow imaginary ω_n, for $\omega_n^2 < 0$, provided ζ is also imaginary). When $\omega_n > 0$ and $\zeta > 0$ this equation models a spring-mass-dashpot subject to a control input. This system is studied frequently in linear control systems theory, but without any explicit constraints on the control. Here, the practical implication of control constraints are of particular interest, and we assume that u is constrained by

$$-1 \leq u \leq 1.$$

Let us now apply the previous bang-bang results to deduce the minimum-time control that will drive this system to the origin. Using standard methods, we first express the system in terms of state-space notation. With $\mathbf{x} = [y \quad \dot{y}]^{\mathrm{T}}$ we have the state equations

$$\dot{\mathbf{x}} = \mathbf{A}\mathbf{x} + \mathbf{b}u,$$

where

$$\mathbf{A} = \begin{bmatrix} 0 & 1 \\ -\omega_n^2 & -2\zeta\omega_n \end{bmatrix}, \qquad \mathbf{b} = \begin{bmatrix} 0 \\ q_0 \end{bmatrix}.$$

The problem of transferring the state of two-dimensional linear systems with bounded controls the origin in minimum time is called **Bushaw's problem.** However, Professor Bushaw says he prefers to be known for the solution, rather than the problem. His solutions to this class of problem [Bushaw, 1958] were developed prior to the advent of Pontryagin's minimum principle, using topological methods. We use the minimum principle to investigate minimum-time control for various cases based on the eigenvalues of **A**.

The Kalman controllability determinant for this system is given by

$$|\mathcal{P}| = -q_0^2,$$

so the minimum-time control will always be bang-bang for $q_0 \neq 0$. This is true even for higher-order input-output systems, since they will always have at least one state-space representation [Grantham and Vincent, 1993, p. 61] that will satisfy the controllability condition $|\mathcal{P}| \neq 0$, provided $q_0 \neq 0$. For the remainder of this section we take $q_0 = 1$.

The eigenvalues and eigenvectors for the **A** matrix are given by

$$\mu_{1,2} = -\zeta\omega_n \pm \sqrt{(\zeta\omega_n)^2 - \omega_n^2}, \qquad \xi_1 = \begin{bmatrix} 1 \\ \mu_1 \end{bmatrix}, \qquad \xi_2 = \begin{bmatrix} 1 \\ \mu_2 \end{bmatrix}.$$

The nature of minimum-time control will vary depending on these eigenvalues. There are two consequences to this dependency. If the eigenvalues are real, minimum-time control will involve at most one switch in the control, whereas if they are complex conjugates we have no *a priori* information on the maximum number of switches. The second consequence is: *if any of the eigenvalues have positive real part, then the controllable set to the origin will not be all of state space.*

Systems with Nonpositive Real Eigenvalues If $\zeta \geq 1$ with $\omega_n > 0$, the origin for the system under no control ($u = 0$) will be globally asymptotically stable, with negative real eigenvalues $\mu_2 < \mu_1 < 0$. More importantly, since the eigenvalues are real there will be at most one switch in the minimum-time control. Knowledge of this fact allows us to very quickly obtain a complete synthesis of the optimal solution since there will be no need to integrate the adjoint equations in order to determine the nature of the control. Instead we need only integrate the state equations backward from the origin with either $u = +1$ or

$u = -1$ until a specified backward **switching time** τ_s is reached and then the control is switched to $u = -1$ or $u = +1$. By selecting various switching times a complete synthesis can be obtained.

If $\omega_n = 0$, the origin for the uncontrolled system will be unstable, corresponding to a repeated zero eigenvalue $\mu_1 = \mu_2 = 0$. However, even in this case the entire state space is still controllable to the origin. The special case of $\omega_n = 0$ has been previously discussed in Example 2.1-1, and the minimum-time optimal control problem to the origin was worked in detail in Example 7.3-3.

Example 8.1-1: A Critically Damped System We obtain a model for a critically damped spring-mass-dashpot system by setting the undamped natural frequency $\omega_n = 1$ rad/sec and the damping ratio $\zeta = 1$. This results in repeated negative real eigenvalues $\mu_{1,2} = -1$.

The state equations for this system are

$$\dot{x}_1 = x_2$$
$$\dot{x}_2 = -x_1 - 2x_2 + u$$

with $|u| < 1$. Since the system is controllable and has negative real eigenvalues, it follows that the control force that will bring the mass to rest at zero displacement in minimum time is bang-bang with at most one switch in the control. This information allows us to immediately obtain the optimal control trajectories, as illustrated in Figure 8.1-1. The trajectories in this figure were obtained by integrating backward from the origin, with the control equal to $u = -1$, switching to $u = +1$ at a switching time $\tau_s \in \{0.4, 0.5, 0.55, 0.6, 1\}$, and then integrating backward from the origin, with the control equal to $u = +1$, switching to $u = -1$ at the same switching times. The two nonswitching trajectories (corresponding to $\tau_s = \infty$), which have a common point at the origin, constitute a switching arc σ (described below). This figure illustrates a complete synthesis of the optimal control solution $u^*(\mathbf{x})$ with the optimal control given by $u^* = -1$ whenever the state of the system is either above the switching arc or on the upper portion of σ, and $u^* = +1$ whenever the state of the system is either below the switching arc or on the lower portion of σ.

We can also obtain the optimal control solutions by direct application of the minimum principle. The H function in this case is

$$H = 1 + \lambda_1 x_2 + \lambda_2(u - x_1 - 2x_2),$$

yielding adjoint equations

$$\dot{\lambda}_1 = \lambda_2$$
$$\dot{\lambda}_2 = -\lambda_1 + 2\lambda_2.$$

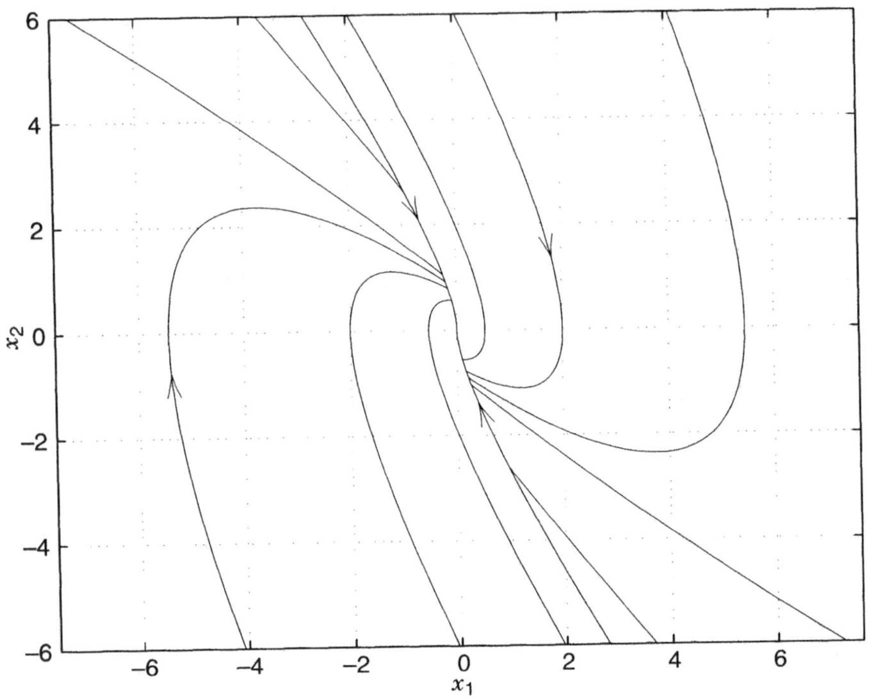

Figure 8.1-1 Minimum-time trajectories for $\omega_n = 1$, $\zeta = 1$.

Using the optimal control

$$u^* = -\text{sgn}(\lambda_2)$$

and

$$0 = H_f = 1 - |\lambda_{2f}|,$$

we have

$$\lambda_{2f} = \pm 1.$$

Integrating the adjoint equations backward in time we have

$$\lambda_1(\tau) = [\lambda_{1f} + (\lambda_{1f} - \lambda_{2f})\tau]e^{-\tau}$$
$$\lambda_2(\tau) = [\lambda_{2f} + (\lambda_{1f} - \lambda_{2f})\tau]e^{-\tau}.$$

For $\lambda_{1f}/\lambda_{2f} > 1$ the optimal control $u^*(\tau)$ will switch from $u^*(\tau) = u_f^* = -\text{sgn}(\lambda_{2f})$ to $u^*(\tau) = -u_f^*$ at a switching time, from $\lambda_2(\tau_s) = 0$, given by

$$\tau_s = \frac{1}{\lambda_{1f}/\lambda_{2f} - 1}.$$

For $\lambda_{1f}/\lambda_{2f} \leq 1$ the optimal control will not switch. Thus the optimal control will switch at most once, as expected.

Integrating the state equations backwards from the origin with $u^*(\tau) = u_f^* = \pm 1$, $0 \leq \tau < \tau_s$, yields

$$x_1(\tau) = u_f^*[1 + (\tau - 1)e^\tau]$$
$$x_2(\tau) = -u_f^* \tau e^\tau.$$

Switching control at $\tau = \tau_s$ and continuing to integrate the state equations backwards yields

$$x_1(\tau) = -u_f^* + [u_f^* + x_{1s} - (u_f^* + x_{1s} + x_{2s})(\tau - \tau_s)]e^{\tau - \tau_s}$$
$$x_2(\tau) = [x_{2s} + (u_f^* + x_{1s} + x_{2s})(\tau - \tau_s)]e^{\tau - \tau_s}$$

for $\tau \geq \tau_s$, where

$$x_{1s} = u_f^*[1 + (\tau_s - 1)e^{\tau_s}]$$
$$x_{2s} = -u_f^* \tau_s e^{\tau_s}.$$

The switching surface $\sigma(\mathbf{x}) = 0$ is defined parametrically by the points $\mathbf{x}(\tau_s)$, from which we obtain

$$x_1 = u_f^* - x_2 - u_f^* e^{\tau_s}.$$

Hence

$$\tau_s = \ln\left(1 - \frac{x_1 + x_2}{u_f^*}\right)$$

and

$$x_2 = -u_f^*\left(1 - \frac{x_1 + x_2}{u_f^*}\right)\ln\left(1 - \frac{x_1 + x_2}{u_f^*}\right),$$

which can be written as

$$\exp\left(\frac{x_2}{x_1 + x_2 - u_f^*}\right) = 1 - \frac{x_1 + x_2}{u_f^*}.$$

Accounting for the two cases $u_f^* = \pm 1$, along with $\tau_s \geq 0$, yields the switching curve

$$\sigma(\mathbf{x}) = \begin{cases} \exp\left(\dfrac{x_2}{x_1 + x_2 - 1}\right) - 1 + (x_1 + x_2) & \text{if } x_1 + x_2 \leq 0 \\ \exp\left(\dfrac{x_2}{x_1 + x_2 + 1}\right) - 1 - (x_1 + x_2) & \text{if } x_1 + x_2 \geq 0. \end{cases}$$

The resulting optimal feedback control is given by

$$u^*(\mathbf{x}) = \begin{cases} -\text{sgn}[\sigma(\mathbf{x})] & \text{if } \sigma(\mathbf{x}) \neq 0 \\ -\text{sgn}(x_2) & \text{if } \sigma(\mathbf{x}) = 0 \text{ and } \mathbf{x} \neq \mathbf{0} \\ 0 & \text{if } \mathbf{x} = \mathbf{0}. \end{cases}$$ ■

Systems with a Positive Real Eigenvalue For a two-dimensional system with a positive real eigenvalue, both eigenvalues will be real, so there will be at most one switch in the control. Again, we can generate a complete synthesis of the control by integrating the system equations backward from the origin with the control switching at a selected retro-time τ_s. For these systems, the controllable set to the origin will not be all of state space. This makes good physical sense since in the eigenvector direction associated with a positive eigenvalue, the unstable $\mathbf{A}\mathbf{x}$ term in the state equations will dominate the bounded $\mathbf{b}u$ term for sufficiently large $\mathbf{x}$. With both eigenvalues positive and distinct, the corresponding eigenvectors will be linearly independent. It follows that, in this case, dominance would occur in every direction from the origin. In fact, even if the eigenvalues are positive and equal, there will be a region outside of some neighborhood of the origin for which the system instability will be greater than any bounded stabilizing control.

If $\zeta \leq -1$ with $\omega_n > 0$ or if $\omega_n^2 < 0$ with $\zeta\omega_n$ real, the origin for the system under no control ($u = 0$) will be unstable with both eigenvalues real and at least one of them positive. Numerical integration results are illustrated in Figure 8.1-2 for the case when $\omega_n = 1$ and $\zeta = -1$. In this case, both of the eigenvalues

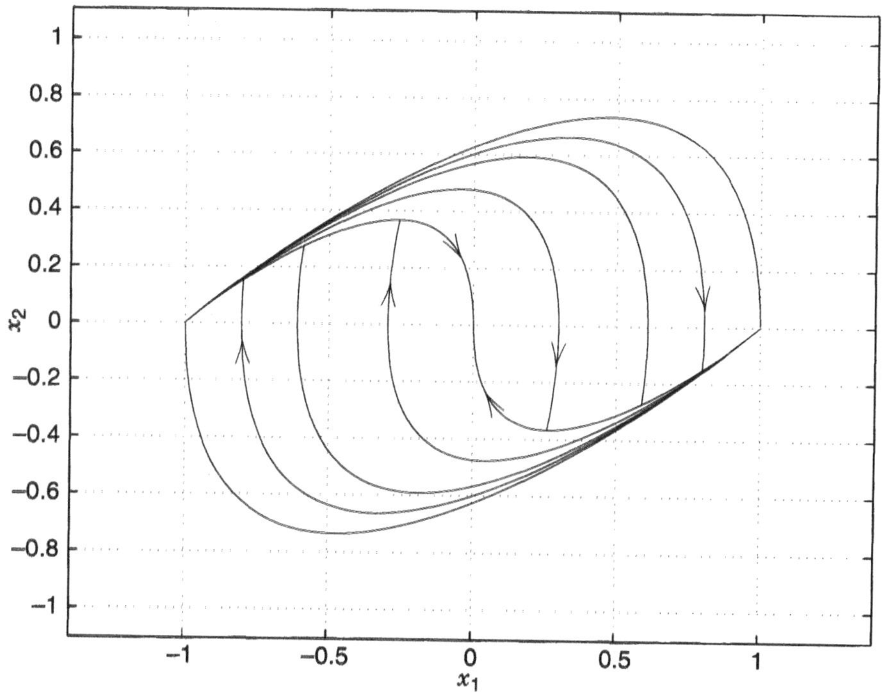

Figure 8.1-2 Minimum-time trajectories for $\omega_n = 1$, $\zeta = -1$.

are positive and equal ($\mu_1 = \mu_2 = 1$). In the retro-time sense all minimum-time trajectories approach one of two equilibrium points, corresponding to either $u = +1$ or $u = -1$. The equilibrium points lie on the boundary of the controllable set to the origin for this case.

Recall that no points that lie on the boundary of a controllable set are controllable to the target unless the target itself intersects the boundary. The boundary of the controllable set shown in Figure 8.1-2 represents just that region in state space where the control is able to "balance" the destabilizing effect of the system. Outside this region, it is not possible to reach the target (the origin in this case). Inside this region, the minimum-time controller makes one switch when a forward-time trajectory reaches the constant control arc through the origin. Note that close to the boundary of the controllable set, the minimum-time controller switches very near $x_2 = 0$. That is, in a neighborhood of the boundary, minimum-time control is very nearly the "boundary control" of Section 6.2. This result is true regardless of the performance criteria. Near a controllable set boundary, all optimal controllers will be the same. This observation can be used to produce suboptimal controllers, as will be demonstrated later.

Figure 8.1-3 illustrates minimum-time trajectories for another unstable system with real eigenvalues. In this case $\zeta = 0$ and $\omega_n^2 = -1$. Since one eigenvalue

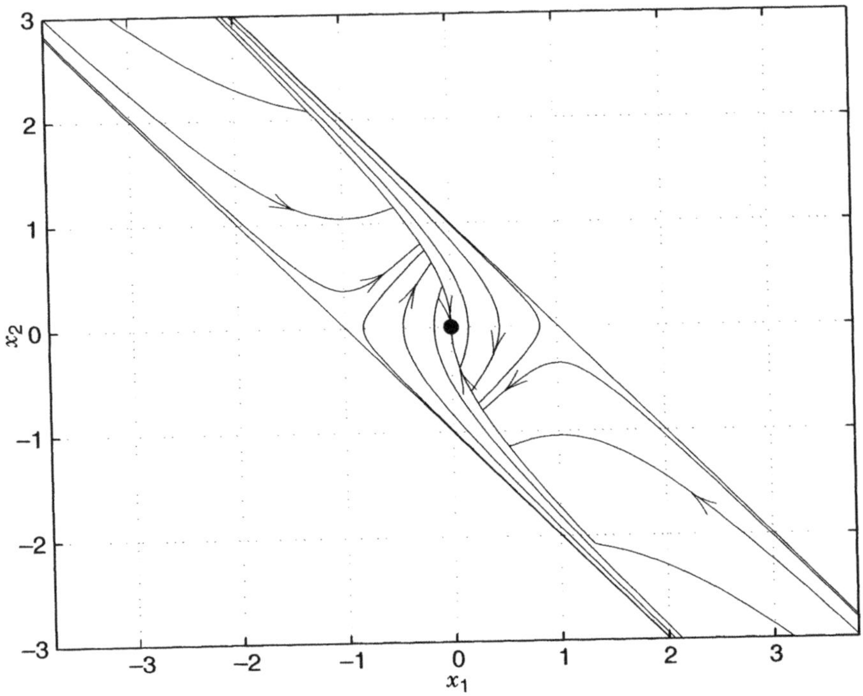

Figure 8.1-3 Minimum-time trajectories for $\omega_n^2 = -1$, $\zeta = 0$.

is positive and the other is negative, the controllable set is unbounded, but it is not all of the state space.

Stable Systems with Complex Eigenvalues If $0 \le \zeta < 1$ with $\omega_n > 0$, the origin for the state-space system under no control ($u = 0$) will be stable with complex eigenvalues and the controllable set to the origin is the entire state space. We now have no information as to the number of switches for a minimum-time controller, so it will be necessary to integrate the adjoint equations along with the system equations in order to determine the optimal control from (8.1-6). The adjoint equations in this case are given by

$$\begin{aligned} \dot{\lambda}_2 &= \omega_n^2 \lambda_2 \\ \dot{\lambda}_2 &= -\lambda_1 + 2\zeta\omega_n\lambda_2. \end{aligned} \tag{8.1-12}$$

Example 8.1-2: Undamped Harmonic Oscillator If $\zeta = 0$ and $\omega_n = 1$, we have a stable system with pure imaginary eigenvalues. This system corresponds to a spring-mass system with no damping whose dynamics are given by

$$\dot{x}_1 = x_2$$
$$\dot{x}_2 = -x_1 + u. \tag{8.1-13}$$

Since an analytical solution to this problem can be obtained easily, it is useful to again compare both the numerical and analytical approaches. Let us first consider the numerical approach.

Figure 8.1-4 illustrates the state-space trajectories obtained using a numerical integration. To produce these trajectories the adjoint and the system equations, (8.1-12) and (8.1-13), are integrated backward in time, from the origin, using the control law (8.1-10) with the switching function in this case given by

$$\sigma = \lambda_2.$$

The initial value of the control (in the retro-time sense) is determined by the initial value of λ_2. The retro-time state-space trajectories that depart from the origin in the fourth quadrant, corresponding to initially using $u = +1$, were generated using $\lambda_2(0) = \lambda_{2f} = -1$. Likewise, all trajectories that depart the origin in the second quadrant, corresponding to initially using $u = -1$, were generated by using $\lambda_2(0) = \lambda_{2f} = +1$.

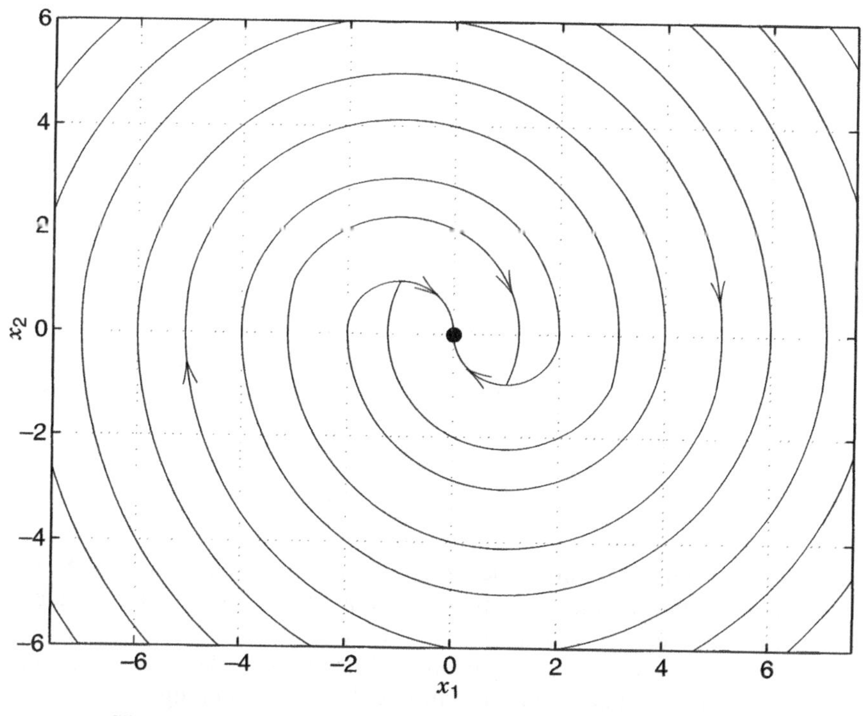

Figure 8.1-4 Minimum-time trajectories for $\omega_n = 1$, $\zeta = 0$.

The family of trajectories illustrated in Figure 8.1-4 was obtained by assigning $\lambda_1(0) = \lambda_{1f} \in [0, 20]$ for the trajectories that depart from the origin in the fourth quadrant and assigning $\lambda_1(0) = \lambda_{1f} \in [-20, 0]$ for the trajectories that depart from the origin in the second quadrant. Clearly $\lambda_1(0)$ dictates the switching sequence obtained. The first switch (in the retro-time sense), which occurs on one of the two arcs leaving the origin, is easy to see. Additional switching between $u = \pm 1$ continues to occur as the trajectories "spiral" out from the origin. In fact, a switch takes place before each crossing of the x_1-axis. Letting $\lambda_1(0) \rightarrow \infty$ simply moves the point of first switching to the x_1-axis. All trajectories between $\lambda_1(0) = \pm 20$ and $\lambda_1(0) \rightarrow \pm\infty$ are indistinguishable at this scale.

Note that in this numerical approach the switching surface $\sigma(\mathbf{x}) = 0$ may be obtained by plotting a point each time $\sigma(\tau) = \lambda_2(\tau) = 0$ along a trajectory $\mathbf{x}(\tau)$, for a family of such trajectories corresponding to $\lambda_1(0)$, and then "connecting the dots" to draw the switching surface. An alternate approach, based on a variational differential equation for motion along a switching surface, can be found in Fritchman [1985].

Consider now an analytical approach to our example minimum-time problem. We know that the control $u = -\text{sgn}(\sigma)$ is bang-bang, that is, $u = \pm 1$. However, we do not know where the switches will take place in the state space. Under the control $u = +1$ the system equations (8.1-13) can be integrated easily to yield

$$(x_1 - 1)^2 + x_2^2 = c_1$$

and under $u = -1$ we obtain

$$(x_1 + 1)^2 + x_2^2 = c_2,$$

where c_1 and c_2 are positive constants of integration. Thus under constant $u = \pm 1$ control the trajectories are circles about the points $(x_1, x_2) = (1, 0)$ or $(-1, 0)$.

The adjoint equations (8.1-12) are independent of the state and can be integrated directly to yield, for $\zeta = 0$ and $\omega_n = 1$, the forward time solutions

$$\lambda_1(t) = a\cos(t - \alpha)$$
$$\lambda_2(t) = -a\sin(t - \alpha),$$

where a and α are constants depending on the initial conditions for $\boldsymbol{\lambda}$. It follows that the value of the switching function $\sigma(t) = \lambda_2(t)$ along a trajectory is a sinusoidal function with period 2π. This means that u will switch from one extreme to the other every π seconds. Since it also takes π seconds for the system to traverse a half circle, it follows that the switching points must lie on semicircular arcs, shown dashed in Figure 8.1-5. Trajectories that enter the

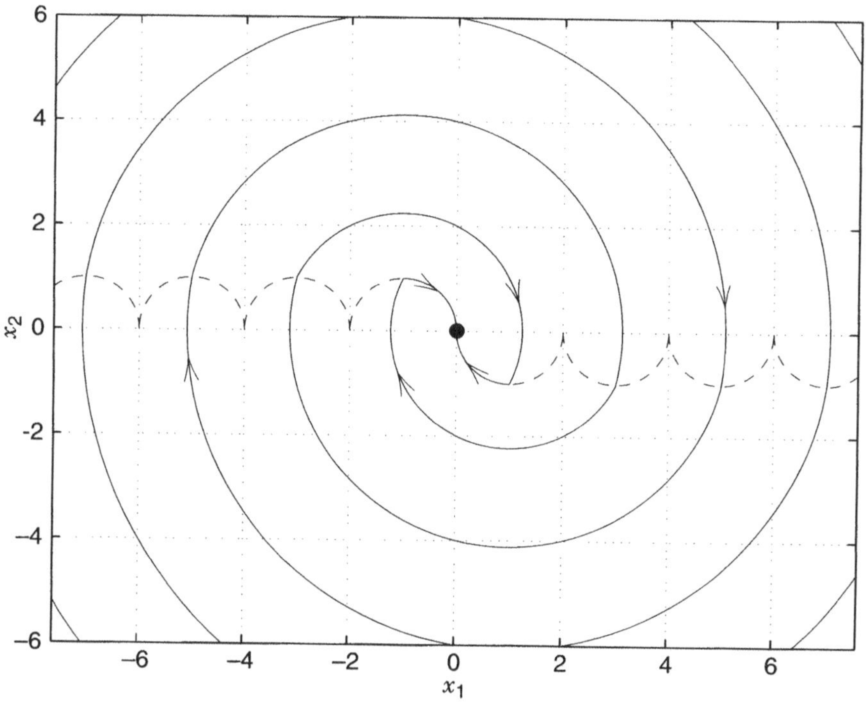

Figure 8.1-5 Switching arc segments.

origin from the second quadrant switch control on the 1st, 3rd, ... semicircle to the left and on the 2nd, 4th, ... semicircle to the right. Trajectories that enter the origin from the fourth quadrant switch control on the 1st, 3rd, ... semicircle to the right and on the 2nd, 4th, ... semicircle to the left. Since we now know the exact location of the switching surface, a feedback control algorithm can easily be written to implement minimum-time control to the origin.

From the point of view of exactly reaching the origin, the final switch is the most critical one. All of the other switches could take place anywhere in a neighborhood of the x_1-axis. Thus a simple suboptimal controller could be implemented by having each switch take place on the x_1-axis, except for the last switch, which must take place on the proper semicircular arc in order to bring the system in to the origin. This suboptimal control would be much easier to implement than the optimal one. ■

The above example is a special case of a stable system with pure imaginary eigenvalues. Consider now the same "spring-mass" system with damping. Figure 8.1-6 illustrates the numerical solutions for minimum-time trajectories for the damped harmonic oscillator with $\zeta = 0.2$ and $\omega_n = 1$. Note that the nature of the solutions is similar to the previous case. Even the same retro-time initial

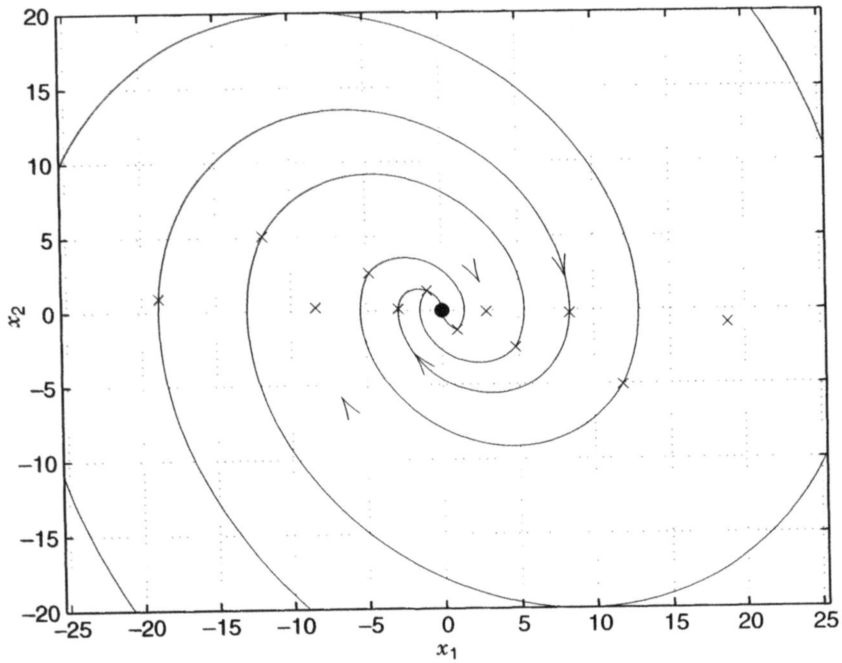

Figure 8.1-6 Minimum-time trajectories for $\omega_n = 1$, $\zeta = 0.2$.

condition values for λ_1 to λ_2 are used to generate the trajectories. The major difference is that now it would be more difficult to determine, analytically, the location of the control switches. These are marked by the X's in this figure. However, analytical solutions do exist [Lee and Marcus, 1967, pp. 42–47], for both this case and the next case.

Unstable Systems with Complex Eigenvalues If $-1 < \zeta < 0$ with $\omega_n > 0$, the origin for the system under no control ($u = 0$) will be unstable with complex eigenvalues, so the controllable set to the origin will be a bounded subset of state space. Again we have no information as to the number of switches for an optimal minimum-time controller. As in the previous case it is necessary to integrate the adjoint equations along with the system equations in order to determine the optimal control from (8.1-6).

As an example, consider the case where $\zeta = -0.5$ and $\omega_n = 1$. The adjoint equations are again given by (8.1-12), with the system equations given by

$$\dot{x}_1 = x_2$$
$$\dot{x}_2 = -x_1 + x_2 + u.$$

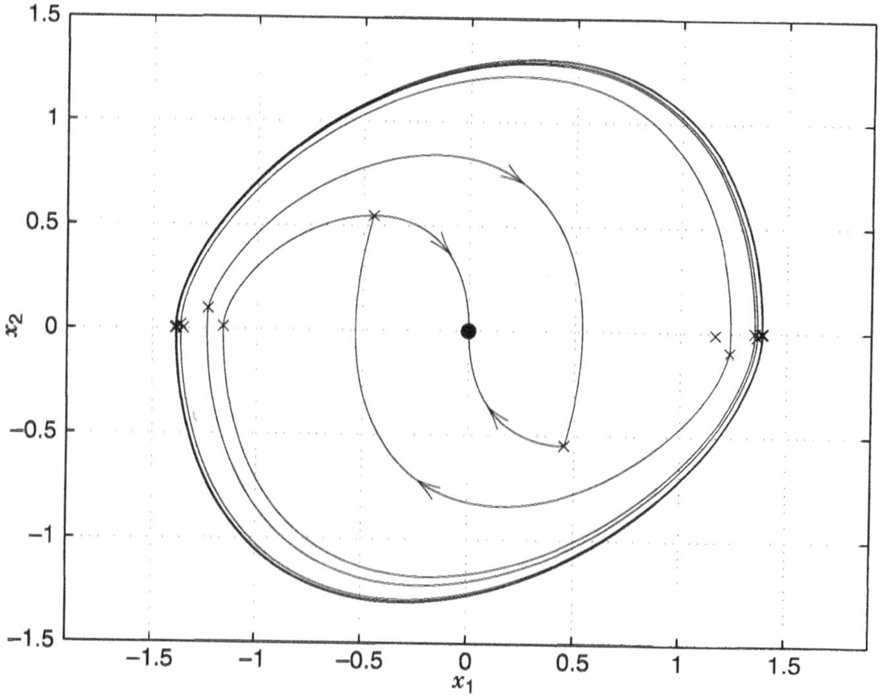

Figure 8.1-7 Minimum-time trajectories for $\omega_n = 1$, $\zeta = -0.5$.

A numerical solution to this problem is illustrated in Figure 8.1-7. The same backwards integration procedure as in Figure 8.1-2 was used to generate the solutions, using the pairs $(\lambda_{1f}, \lambda_{2f}) = (-1, 1)$, $(20, 1)$, $(1, -1)$, and $(-20, -1)$. Since the eigenvalues are all unstable, the controllable set is bounded. This time, however, with unstable complex eigenvalues, all system trajectories (when integrated in the retro-time sense) converge to a limit cycle. This limit cycle is, of course, the boundary of the controllable set, and the same boundary would be obtained no matter what performance criterion was used to obtain the system trajectories. Only the details on the interior of the controllable set will differ using different performance criteria.

Minimum-Time Control as a Feedback Controller

We have noted in the previous examples that, once a complete synthesis of a minimum-time problem has been obtained, it is then possible to formulate either an exact or approximate closed-loop feedback algorithm for the control. Such algorithms can be readily implemented using microprocessors. However, we should note that nowhere in the statement of the standard integral cost optimal control problem (Section 7.1) is stability an imposed requirement. It is only required that the system be driven to a target. It is not required that the system must remain at or near the target. Indeed for some of the nonlinear examples cited in Chapter 7 it is not even possible to keep the system at the target. Thus we should not expect *a priori* that minimum-time control or indeed any other optimal control will necessarily keep the system at the target.

How do nonlinear minimum-time control algorithms compare with the classical feedback controllers for linear systems? First of all one must decide exactly what it is that is being compared. Linear state-variable controllers [Grantham and Vincent, 1993] do not incorporate control bounds and are typically designed on the basis of a stability criteria, with no direct consideration of the time required to return the state to (a neighborhood of) the origin. Indeed, since such controllers are designed for asymptotic stability, they take an infinite time to reach the origin.

One major advantage of a closed-loop controller synthesized from an optimal control solution is that it will be effective throughout the controllable set. We have previously shown in Chapter 6 that the domain of attraction for controllers designed by other methods (e.g., LQR design) will generally be a subset of the controllable set. One disadvantage of closed-loop optimal controllers is that they often result in bang-bang control.

Suboptimal controllers effective throughout the controllable set can be designed by modifying bang-bang optimal closed-loop controllers so that they do not chatter in the neighborhood of the target. This can often be done by replacing a sign switching function (SGN) with a saturation function (SAT) in the neighborhood of the target. This method has the advantage of maintaining the optimal solution when the solution is "far" from the target and simulating the behavior of state feedback control when the solution is "near" the target.

Another approach to suboptimal control that avoids chatter can be found in methods that approximate optimal solutions using neural networks [Goh et al., 1993; Cottrell et al., 1996]. The idea is to obtain closed-loop control by training neural networks using open-loop optimal control solutions. Once the net is trained, one can feed the state (or other data, depending on how the net was trained) into the neural network and obtain a control as an output. A trained neural network smooths the corners of discontinuous functions so that pure bang-bang control will not be produced.

8.2 LINEAR SYSTEMS WITH QUADRATIC COST

An important class of optimal control problems involving linear systems is where the cost functional is quadratic in the control and state. We assume a linear system of the form

$$\dot{\mathbf{x}} = \mathbf{A}\mathbf{x} + \mathbf{B}\mathbf{u}, \tag{8.2-1}$$

where $\mathbf{x}$ is an $n_x \times 1$ state vector, $\mathbf{A}$ is an $n_x \times n_x$ constant matrix, $\mathbf{B}$ is an $n_x \times$

n_u constant matrix, and **u** is an $n_u \times 1$ control vector. In this case no bounds are assumed to exist on the controls. The final time is unspecified, and the desired final state is an asymptotically stable origin.

Linear System Quadratic Cost Regulator (LQR)

In Chapter 2 we introduced the LQR method for control design. Here we derive this result using the minimum principle. The cost functional for the **linear-quadratic regulator** problem is of the form

$$x_0 = \frac{1}{2}\int_0^{t_f} (\mathbf{x}^T\mathbf{Q}\mathbf{x} + \mathbf{u}^T\mathbf{R}\mathbf{u})dt, \tag{8.2-2}$$

where **Q** and **R** are $n_x \times n_x$ and $n_u \times n_u$ constant real symmetric matrices, respectively, with $\mathbf{R} > 0$ positive definite and $\mathbf{Q} \geq 0$ at least positive semidefinite. We have examined this problem previously, in Section 7.6, in terms of dynamic programming and the Hamilton–Jacobi–Bellman partial differential equation. Here we use the minimum principle in its standard form.

Note that, since the final time is unspecified, it may (and in fact does) happen that $t_f \to \infty$. In that case, since the integrand in (8.2-2) is nonnegative, a necessary condition for the integral to exist (i.e., be finite) is that the integrand must approach zero. That is, $\mathbf{u}(t) \to \mathbf{0}$ and $\mathbf{x}(t) \to \mathbf{0}$ (if $\mathbf{Q} > 0$). Thus if a solution to the optimal control problem exists and if $\mathbf{Q} > 0$, then the resulting control law must cause the state to approach the origin, as desired.

This cost functional has received a lot of attention from controls engineers. The reason for this is that it incorporates some of the qualitative features one seeks for a controller. The quadratic cost associated with the state represents a penalty for deviating from the nominal rest state ($\mathbf{x} = \mathbf{0}$), whereas the quadratic cost associated with the control represents a penalty for using control. This latter objective might be very important, for example, in space applications, where the cost of control is very expensive. In addition, the quadratic control cost allows for the formulation of a control problem without the need to explicitly account for control constraints. One can force the satisfaction of a control constraint by choosing the appropriate components of **R** sufficiently large. One other big advantage with this formulation is that we will be able to obtain the resulting optimal control law as a linear state-vector feedback controller.

The cost functional in (8.2-2) involves a weighted trade-off between the state penalty and the control penalty and, indeed, a trade-off between individual elements of the state and control vectors. From a design viewpoint there are no set rules for choosing the elements of the $\mathbf{R} > 0$ and $\mathbf{Q} \geq 0$ matrices. Rather than focusing on the "cost" aspect, involving "apples versus oranges" trade-offs, the performance measure given by (8.2-2) could simply be viewed as a device for producing a state feedback controller that will produce asymptotic stability for the origin.

The H function for this problem is given by

$$H = \lambda_0 \tfrac{1}{2}[\mathbf{x}^{\mathrm{T}}\mathbf{Q}\mathbf{x} + \mathbf{u}^{\mathrm{T}}\mathbf{R}\mathbf{u}] + \boldsymbol{\lambda}^{\mathrm{T}}[\mathbf{A}\mathbf{x} + \mathbf{B}\mathbf{u}].$$

A necessary condition for minimizing H with respect to $\mathbf{u}$ is obtained by taking the partial derivative of H with respect to $\mathbf{u}$ and setting the result equal to zero,

$$\frac{\partial H}{\partial \mathbf{u}} = \lambda_0 \mathbf{u}^{\mathrm{T}}\mathbf{R} + \boldsymbol{\lambda}^{\mathrm{T}}\mathbf{B} = \mathbf{0}^{\mathrm{T}},$$

to obtain (for $\lambda_0 = 1$)

$$\mathbf{u}^{\mathrm{T}} = -\boldsymbol{\lambda}^{\mathrm{T}}\mathbf{B}\mathbf{R}^{-1},$$

or equivalently

$$\mathbf{u} = -\mathbf{R}^{-1}\mathbf{B}^{\mathrm{T}}\boldsymbol{\lambda}. \tag{8.2-3}$$

We can now substitute this result into the H function. From the fact that $H = 0$ we obtain

$$H = \tfrac{1}{2}[\mathbf{x}^{\mathrm{T}}\mathbf{Q}\mathbf{x} + \boldsymbol{\lambda}^{\mathrm{T}}\mathbf{B}\mathbf{R}^{-1}\mathbf{B}^{\mathrm{T}}\boldsymbol{\lambda}] + \boldsymbol{\lambda}^{\mathrm{T}}\mathbf{A}\mathbf{x} - \boldsymbol{\lambda}^{\mathrm{T}}\mathbf{B}\mathbf{R}^{-1}\mathbf{B}^{\mathrm{T}}\boldsymbol{\lambda} = 0. \tag{8.2-4}$$

At any state position $\mathbf{x}$, we can think of this as an equation in $\boldsymbol{\lambda}$. The symmetry of the equation suggests assuming a solution of the form

$$\boldsymbol{\lambda} = \mathbf{S}\mathbf{x}, \tag{8.2-5}$$

where $\mathbf{S}$ is a constant symmetric $n_x \times n_x$ matrix. In order for this to be a solution to (8.2-4), it must satisfy $H = 0$. Substituting into the above expression and multiplying through by 2, we obtain

$$\mathbf{x}^{\mathrm{T}}[\mathbf{Q} + 2\mathbf{S}\mathbf{A} - \mathbf{S}\mathbf{B}\mathbf{R}^{-1}\mathbf{B}^{\mathrm{T}}\mathbf{S}]\mathbf{x} = 0.$$

Since this relation must be true for all $\mathbf{x}$ along a trajectory, the fact that the quadratic form must be zero implies that the symmetric part of the matrix in brackets must be zero. In other words, we can solve for $\mathbf{S}$ by finding a solution to the following quadratic matrix equation:

$$\mathbf{Q} + \mathbf{S}\mathbf{A} + [\mathbf{S}\mathbf{A}]^{\mathrm{T}} - \mathbf{S}\mathbf{B}\mathbf{R}^{-1}\mathbf{B}^{\mathrm{T}}\mathbf{S} = \mathbf{0}. \tag{8.2-6}$$

This equation is called the **algebraic Riccati equation** [Kailath, 1980, pp. 230–234].

If a solution for **S** exists from (8.2-6), then the control is given by

$$\mathbf{u} = -\mathbf{R}^{-1}\mathbf{B}^{\mathrm{T}}\mathbf{S}\mathbf{x}. \tag{8.2-7}$$

The control obtained by this procedure is a linear state-variable feedback control of the form

$$\mathbf{u} = -\mathbf{K}\mathbf{x},$$

where the **gain matrix K** is a constant determined by the matrices **R**, **B**, and **S**. From linear control systems theory we know that this type of controller can produce a closed-loop system with any desired eigenvalues (pole placement). In fact, choosing the cost weighting matrices **Q** and **R** is equivalent to selecting the eigenvalues for a closed-loop state-variable feedback control system. However, there is no general relationship available that directly connects the matrices **Q** and **R** to the resulting eigenvalues for the closed-loop system matrix $\mathbf{A} - \mathbf{BK}$.

In formulating the quadratic-cost optimal control problem it is assumed that **R** is positive definite. If $\mathbf{R} = \mathbf{0}$ there is no penalty on using the control and $\mathbf{R}^{-1}$ does not exist. In this case the control cannot be computed from (8.2-7), although the implication from (8.2-7) might be that the control is infinite ("high gain"), because of division by zero. The problem with the $\mathbf{R} = \mathbf{0}$ case is that H becomes linear in **u**, with a vector of switching functions given by $\boldsymbol{\sigma}^{\mathrm{T}} = \boldsymbol{\lambda}^{\mathrm{T}}\mathbf{B}$. If the Kalman controllability condition, (2.2-1) or (8.1-9), is satisfied, then singular solutions in this case are not possible and a minimum-cost solution will only exist if there are bounds on **u**. This observation strengthens the notion that the quadratic performance index is a convenient way to handle bounded control without explicitly including the bounds in the statement of the problem. Note however, that in the presence of control bounds, the control given by (8.2-7) will saturate for sufficiently large **x**.

Example 8.2-1: A Scalar System To illustrate the usefulness of the linear-quadratic control procedure, we examine a simple scalar problem using two different methods to obtain the solution. The problem is to minimize the cost

$$\frac{1}{2}\int_0^{t_f} u^2\, dt$$

with the system given by

$$\dot{x} = ax + bu,$$

where $a \neq 0$ and $b \neq 0$ and x and u are scalars. Our objective is to drive the system from the initial state $x(0)$ to the final state $x(t_f) \to 0$ while minimizing the square of the total control used.

Since this problem is an LQR problem, we first use those methods for solution. It follows from the cost function that $R = 1$, so the linear feedback control is given by

$$u = -bSx,$$

where S is the solution to

$$2Sa - b^2S^2 = 0.$$

We have two solutions for S. The first solution $S = 0$ yields the control $u^* = 0$, which is the optimal solution if $a < 0$. However if $a > 0$, zero control will not drive the system to the target. Assuming $S \neq 0$ and $a > 0$, the second solution is given by

$$S = \frac{2a}{b^2}$$

with the feedback control

$$u^*(x) = -\frac{2a}{b}\, x.$$

Compare this approach with an analysis using the minimum principle applied in the standard way to the same linear-quadratic optimal control problem. We first form the H function (with $\lambda_0 = 1$)

$$H = \tfrac{1}{2}u^2 + \lambda(ax + bu),$$

from which we obtain the adjoint equation

$$\dot{\lambda} = -a\lambda$$

and a necessary condition for minimizing H is that

$$u + \lambda b = 0. \tag{8.2-8}$$

Solving this equation for u and substituting the solution for λ from the adjoint equation, we obtain

$$u^*(t) = -b\lambda(0)e^{-at}. \tag{8.2-9}$$

As expected, following the usual procedures associated with the minimum principle, we obtain an open-loop solution for the control. In order to convert this to a closed-loop solution we must now integrate the system equation and eliminate time between the x solution and the u solution. Substituting (8.2-9) into the system equation we obtain

$$\dot{x} = ax - b^2\lambda(0)e^{-at}.$$

For $a \neq 0$ this differential equation has a solution of the form

$$x(t) = \left[x(0) - \frac{b^2\lambda(0)}{2a}\right]e^{at} + \frac{b^2\lambda(0)}{2a}\,e^{-at}.$$

An optimal solution must drive the system to the target. If $a < 0$, then we must have $\lambda(0) = 0$; otherwise $|x(t)| \to \infty$. This, in turn, yields the $u^* = 0$ solution. If $a > 0$, then the first term in brackets must be zero, which implies that

$$\lambda(0) = \frac{2a}{b^2}x(0).$$

Thus

$$x(t) = x(0)e^{-at}.$$

Substituting $\lambda(0)$ into (8.2-9) and comparing with the result for $x(t)$ just obtained we conclude that

$$u^*(x) = -\frac{2a}{b}\,x,$$

which is the same result we previously obtained.

There is also a "short cut" way to get these results from the minimum principle. From (8.2-8), we solve for $\lambda = -u/b$ and substitute into $H = 0$ to obtain

$$\tfrac{1}{2}u^2 - \frac{u}{b}\,(ax + bu) = 0.$$

The solutions $u^* = 0$ and $u^* = -(2a/b)x$ follow from this expression.

The Riccati equation was easy to solve for this one-dimensional example. For higher dimensional problems, one would almost always seek a numerical solution. As mentioned in Chapter 2, the MATLAB command lqr provides such a numerical solution. See Example 2.2-2 for application of the LQR method to a problem in six dimensions. ■

Two Important Additional Properties of S

Under optimal control, the quadratic form

$$V = \tfrac{1}{2}\mathbf{x}^{T}\mathbf{S}\mathbf{x}, \tag{8.2-10}$$

where **S** is a positive definite solution to algebraic Riccati equation (8.2-6), is both a Lyapunov function for the dynamical system (8.2-1) and the optimal return function as well. To see this, first note that the time derivative of $V[\mathbf{x}(t)]$ is given by

$$\dot{V} = \tfrac{1}{2}(\dot{\mathbf{x}}^{T}\mathbf{S}\mathbf{x} + \mathbf{x}^{T}\mathbf{S}\dot{\mathbf{x}}). \tag{8.2-11}$$

Substituting (8.2-7) into (8.2-1) yields

$$\dot{\mathbf{x}} = (\mathbf{A} - \mathbf{B}\mathbf{R}^{-1}\mathbf{B}^{T}\mathbf{S})\mathbf{x}$$

so that $\dot{\mathbf{V}}$ may be written as

$$\dot{V} = \tfrac{1}{2}\mathbf{x}^{T}(\mathbf{A}^{T} - \mathbf{S}\mathbf{B}\mathbf{R}^{-1}\mathbf{B}^{T})\mathbf{S}\mathbf{x} + \tfrac{1}{2}\mathbf{x}^{T}\mathbf{S}(\mathbf{A} - \mathbf{B}\mathbf{R}^{-1}\mathbf{B}^{T}\mathbf{S})\mathbf{x}.$$

Rearranging terms yields

$$\dot{V} = \tfrac{1}{2}\mathbf{x}^{T}(\mathbf{A}^{T}\mathbf{S} + \mathbf{S}\mathbf{A} - \mathbf{S}\mathbf{B}\mathbf{R}^{-1}\mathbf{B}^{T}\mathbf{S})\mathbf{x} - \mathbf{x}^{T}(\mathbf{S}\mathbf{B}\mathbf{R}^{-1}\mathbf{B}^{T}\mathbf{S})\mathbf{x}.$$

It follows from (8.2-6) that the first term on the right side of $\dot{V}$ can be replaced by $-\tfrac{1}{2}\mathbf{x}^{T}\mathbf{Q}\mathbf{x}$ and from (8.2-7) that the second term on the right side of $\dot{V}$ can be replaced by $-\tfrac{1}{2}\mathbf{u}^{T}\mathbf{R}\mathbf{u}$. Thus

$$\dot{V} = -\tfrac{1}{2}\mathbf{x}^{T}\mathbf{Q}\mathbf{x} - \tfrac{1}{2}\mathbf{u}^{T}\mathbf{R}\mathbf{u} < \mathbf{0}$$

for $\mathbf{x} \neq \mathbf{0}$, and we conclude that V is a Lyapunov function for the system. Furthermore, since $\dot{V} = -\dot{x}_0$, we may write

$$x_0 = -\int_0^{\infty} \dot{V}\,dt = -V\Big|_0^{\infty} = -\tfrac{1}{2}\mathbf{x}^{T}\mathbf{S}\mathbf{x}\Big|_0^{\infty}.$$

Since the controlled system is stable the upper limit is zero and we conclude

$$x_0 = \tfrac{1}{2}\mathbf{x}^{\mathrm{T}}(0)\mathbf{S}\mathbf{x}(\mathbf{0}) = V[\mathbf{x}(0)]. \tag{8.2-12}$$

Thus $V(\mathbf{x})$ is the optimal return function; that is, the ellipsoid $V(\mathbf{x}) = c^*$ defines all initial states with the same minimum cost c^*.

Control Saturation

While the results of this section assume that the control does not saturate, it is useful to investigate the applicability of these results under control saturation. Consider the scalar control case, assume that $|u| < u_{\max}$, and suppose that the Riccati equation (8.2-6) has been used to solve for a positive definite $\mathbf{S}$. We again examine $\dot{V}$ as given by (8.2-11), only now we evaluate this expression using (8.2-1), so that

$$\dot{V} = \tfrac{1}{2}\{\mathbf{x}^{\mathrm{T}}(\mathbf{A}^{\mathrm{T}}\mathbf{S} + \mathbf{S}\mathbf{A})\mathbf{x} + u\mathbf{B}^{\mathrm{T}}\mathbf{S}\mathbf{x} + \mathbf{x}^{\mathrm{T}}\mathbf{S}\mathbf{B}u\}.$$

Since $\mathbf{S}$ is a solution to the Riccati equation it follows that

$$\mathbf{A}^{\mathrm{T}}\mathbf{S} + \mathbf{S}\mathbf{A} = -\mathbf{Q} + \mathbf{S}\mathbf{B}R^{-1}\mathbf{B}^{\mathrm{T}}\mathbf{S}.$$

Thus

$$\dot{V} = \tfrac{1}{2}\{-\mathbf{x}^{\mathrm{T}}\mathbf{Q}\mathbf{x} + (\mathbf{x}^{\mathrm{T}}\mathbf{S}\mathbf{B})R^{-1}(\mathbf{B}^{\mathrm{T}}\mathbf{S}\mathbf{x}) + u\mathbf{B}^{\mathrm{T}}\mathbf{S}\mathbf{x} + \mathbf{x}^{\mathrm{T}}\mathbf{S}\mathbf{B}u\}. \tag{8.2-13}$$

Our objective is to make $\dot{V} < 0$. Clearly, whenever the control is unsaturated, we can guarantee this by taking

$$u = -R^{-1}\mathbf{B}^{\mathrm{T}}\mathbf{S}\mathbf{x}, \tag{8.2-14}$$

which is the same as LQR feedback control, yielding

$$\dot{V} = -\tfrac{1}{2}\mathbf{x}^{\mathrm{T}}(\mathbf{Q} + \mathbf{S}\mathbf{B}R^{-1}\mathbf{B}^{\mathrm{T}}\mathbf{S})\mathbf{x} < 0 \qquad \text{for } \mathbf{x} \neq \mathbf{0}.$$

However, if the control law (8.2-14) violates $|u| \le u_{\max}$, we still need to eliminate any positive contribution provided by the second term on the right-hand side of (8.2-13). The best we can do to minimize $\dot{V}$ when $|u| \ge u_{\max}$ is to take

$$u = -u_{\max}\,\mathrm{sgn}(\mathbf{B}^{\mathrm{T}}\mathbf{S}\mathbf{x}).$$

Since R is a scalar in this case, we can combine the two results into

$$u = -u_{\max}\ \mathrm{sat}\left(\frac{R^{-1}\mathbf{B}^{\mathrm{T}}\mathbf{S}\mathbf{x}}{u_{\max}}\right).$$

We conclude that this is equivalent to simply letting the LQR controller saturate. Note that a controller based on this saturating LQR design will not necessarily guarantee stability throughout the state space, even if the controllable set (under bounded control) is the entire state space. In other words, the resulting domain of attraction to the origin may not be all of the controllable set. See Example 2.3-3.

8.3 ABNORMAL ARCS

Abnormality and Controllability

In the development of the optimal control minimum principle, we chose $\lambda_0 = 1$ as part of a normalizing procedure. One important consequence of this assumption was the result that the adjoint vector $\boldsymbol{\lambda}$ is the gradient of the optimal return function. However, in the development of the theory prior to this step it was recognized that more generally $\lambda_0 \geq 0$. Thus we still need to consider the case of $\lambda_0 = 0$. This latter situation is called the **abnormal case**, since the quantity to be optimized vanishes completely from the problem! Any trajectories generated by the minimum principle using $\lambda_0 = 0$ are called **abnormal arcs**. Abnormal arcs can be determined using the same procedures as for normal arcs by simply using

$$H(\mathbf{x}, \mathbf{u}^*, \boldsymbol{\lambda}) = \mathcal{H}(\mathbf{x}, \mathbf{u}^*, \boldsymbol{\lambda}) = \boldsymbol{\lambda}^{\mathrm{T}}\mathbf{f}(\mathbf{x}, \mathbf{u}^*). \tag{8.3-1}$$

By comparing the controllability minimum principle of Section 6.2 and the optimal control minimum principle of Section 7.2, it is apparent that the controllability minimum principle corresponds to the abnormal optimal control case. In particular, if optimal control trajectories exist on the boundary of a controllable set, then those trajectories must be abnormal arcs. If we seek abnormal arcs as part of an optimal control problem we discover that they have an important role to play not only in helping to define the boundary of the controllable set, but as a part of the optimal control structure as well.

As we see below, abnormal arcs may exist that do not lie on the boundary of the controllable set. Nevertheless, such optimal trajectories have interesting **barrier** or semipermeability properties, which play key roles in the structure of optimal control.

Since the minimum-time H function (with $\lambda_0 = 1$) and the $\mathcal{H}$ function for abnormal control only differ by a constant,

$$H = 1 + \mathcal{H},$$

one might think that abnormal solutions would be the same as minimum-time solutions. In general, they are not. That constant can be very significant! In order to illustrate the new role that abnormal arcs can play in the structure of an optimal control solution, let us examine some abnormal arcs for a number of optimal control examples that we have previously considered.

Linear Systems

For the linear system

$$\dot{\mathbf{x}} = \mathbf{A}\mathbf{x} + \mathbf{B}\mathbf{u}$$

the Kalman controllability condition is given by

$$|\mathcal{P}| = |\mathbf{B}, \mathbf{AB}, \ \ldots, \ \mathbf{A}^{n_x - 1}\mathbf{B}| \neq 0.$$

We have seen previously in Example 6.2-1, in the absence of any control constraints, that if the controllability condition is satisfied, then the controllability minimum principle does not have a solution, which means there are no abnormal optimal control arcs. However, abnormal arcs can exist if the components of the control are constrained by inequalities of the form

$$u_{i_{\min}} \leq u_i \leq u_{i_{\max}}$$

for $i = 1, \ldots, n_u$, and we can expect the same type of switching phenomena as in the case of minimum-time control for constrained linear systems.

Second-Order IO Systems The second-order IO system discussed in Section 8.1 is given by the equations

$$\begin{aligned} \dot{x}_1 &= x_2 \\ \dot{x}_2 &= -\omega_n^2 x_1 - 2\zeta\omega_n x_2 + u, \end{aligned}$$

where

$$-1 \leq u \leq 1.$$

Since $|\mathcal{P}| \neq 0$ abnormal control must be bang-bang. The $\mathcal{H}$ function is given by

$$\mathcal{H} = \lambda_1 x_2 + \lambda_2(-\omega_n^2 x_1 - 2\zeta\omega_n x_2 + u) = 0.$$

The switching function for the bang-bang control is given by

$$\frac{\partial \mathcal{H}}{\partial u} = \lambda_2.$$

Whenever the switching function is zero, we also have $x_2 = 0$. This follows from the requirement that $\mathcal{H} = 0$ and $\boldsymbol{\lambda} \neq \mathbf{0}$ at any point in time. In fact $x_2 = 0$ will be a switching surface for abnormal arcs provided that λ_2 changes sign whenever x_2 changes sign. In this case the abnormal control, which minimizes $\mathcal{H}$, is given by

$$u = -\text{sgn}(\lambda_2) = \pm\,\text{sgn}(x_2) = \pm 1.$$

To generate abnormal trajectories we integrate the system equations backward in time from the origin using abnormal control. We can start the integration process with either $u = +1$ or $u = -1$. If the trajectory intersects the x_1-axis, then the control must be switched to the opposite sign. All such trajectories must in fact be abnormal arcs since under such a control law the $\mathcal{H}$ function, which is initially zero, will remain zero. This is easily demonstrated by examining $d\mathcal{H}/dt$. The derivative of $\mathcal{H}$ with respect to time is given by

$$\frac{d\mathcal{H}}{dt} = \dot{\boldsymbol{\lambda}}^{\mathrm{T}}\mathbf{f} + \boldsymbol{\lambda}^{\mathrm{T}}\frac{\partial \mathbf{f}}{\partial \mathbf{x}}\mathbf{f} + \boldsymbol{\lambda}^{\mathrm{T}}\frac{\partial \mathbf{f}}{\partial u}\,\dot{u}.$$

Since $\boldsymbol{\lambda}$ is determined from the adjoint equations, it follows that the first two terms on the right-hand side will cancel. The third term is $[\partial\mathcal{H}/\partial u]\dot{u}$ and it is also zero, since either $\partial\mathcal{H}/\partial u = 0$ or $u = \pm 1$ is constant. In particular, for the second-order IO system $d\mathcal{H}/dt$ reduces to

$$\frac{d\mathcal{H}}{dt} = \lambda_2 \dot{u}.$$

Since u is constant whenever $\lambda_2 \neq 0$, it follows that $d\mathcal{H}/dt = 0$.

We now reconsider some of the examples we discussed in Section 8.1 and discover that abnormal arcs that do not lie on the boundary of the controllable set happen to be optimal trajectories with interesting **barrier** properties. Knowledge of these barriers provides information about the structure of the optimal control.

Systems with Nonpositive Real Eigenvalues Figure 8.3-1 illustrates the abnormal optimal control arcs obtained for the case when $\omega_n = 0$. This is the juggernaut problem of Example 7.3-3. Although the system is unstable due to repeated zero eigenvalues, the entire state space is controllable to the origin,

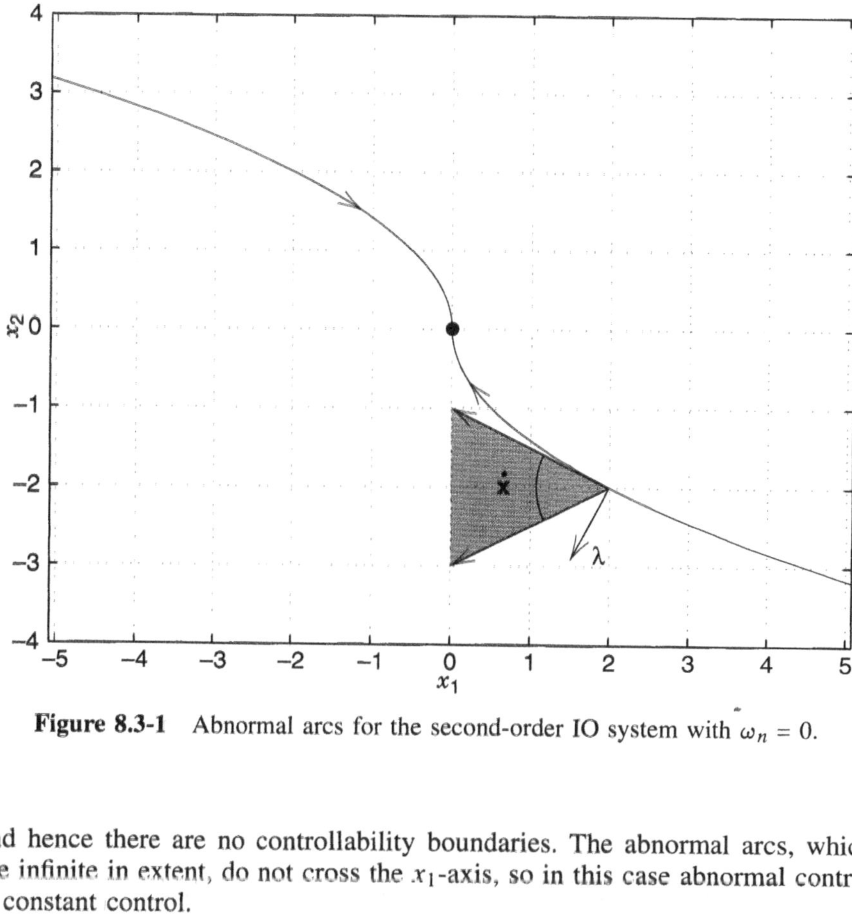

Figure 8.3-1 Abnormal arcs for the second-order IO system with $\omega_n = 0$.

and hence there are no controllability boundaries. The abnormal arcs, which are infinite in extent, do not cross the x_1-axis, so in this case abnormal control is constant control.

If we compare Figure 8.3-1 with Figure 7.3-9, we see that the abnormal arcs in Figure 8.3-1 are also minimum-time arcs. In fact they are the minimum-time switching arcs. Even though a boundary for the controllable set does not exist for this case, the abnormal arcs do correspond to a boundary of sorts. Some of the same conditions that exist on the boundary of a controllable set exist here. Namely, it is possible to drive the system to only one side of the abnormal arc. Consider, for example, the abnormal arc in the fourth quadrant in Figure 8.3-1. At any point along this arc it is only possible to drive the system in a direction below the arc, by changing control. The difference between this situation and the boundary of a controllable set is that here it is possible to eventually drive the system to points above the abnormal arc. However, in order to do this, one must take a devious route. In this case one would have to first encircle the origin before being able to return to a point above the abnormal arc. In this sense the abnormal arcs still represent a type of boundary, often referred to as a **barrier.**

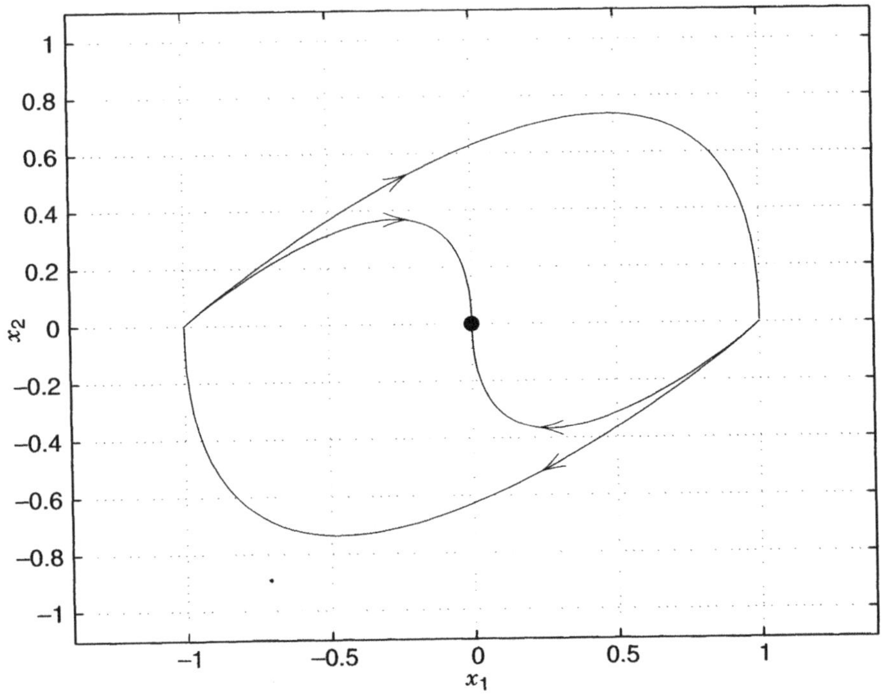

Figure 8.3-2 Abnormal arcs for the second-order IO system with $\omega_n = 1$, $\zeta = -1$.

Systems with a Positive Real Eigenvalue Figure 8.3-2 illustrates the abnormal control arcs for the specific case with $\zeta = -1$, $\omega_n = 1$. Since the eigenvalues are both positive the controllable set is bounded. In this case we have abnormal arcs that lie both on the boundary of the controllable set and on the interior of the controllable set. The abnormal arcs that lie on the interior of the controllable set are barriers in the same sense as described in the previous case. In the process of integrating the system equations backward from the origin under $u = +1$ or $u = -1$, the system trajectory will asymptotically approach the equilibrium points $(x_1, x_2) = (1, 0)$ or $(x_1, x_2) = (-1, 0)$, both of which lie on the switching surface $x_2 = 0$. Only in the limit as $t \to \infty$ does the abnormal arc obtained by integrating backward from the origin reach the switching surface. If we switch to opposite control at this point, the resulting trajectory from this point on lies on the boundary of the controllable set. Comparing this figure with Figure 8.1-2 we see again that abnormal arcs and minimum-time trajectories can coincide.

Systems with Complex Eigenvalues Figure 8.3-3 illustrates the abnormal arcs for the stable system with $\zeta = 0$ and $\omega_n = 1$, and Figure 8.3-4 illustrates the abnormal arcs for the unstable system with $\zeta = -0.5$ and $\omega_n = 1$. In the first case, the abnormal arcs are optimal trajectories that happen to switch on $x_2 = 0$. We cannot compare this result exactly with Figure 8.1-4, since that figure

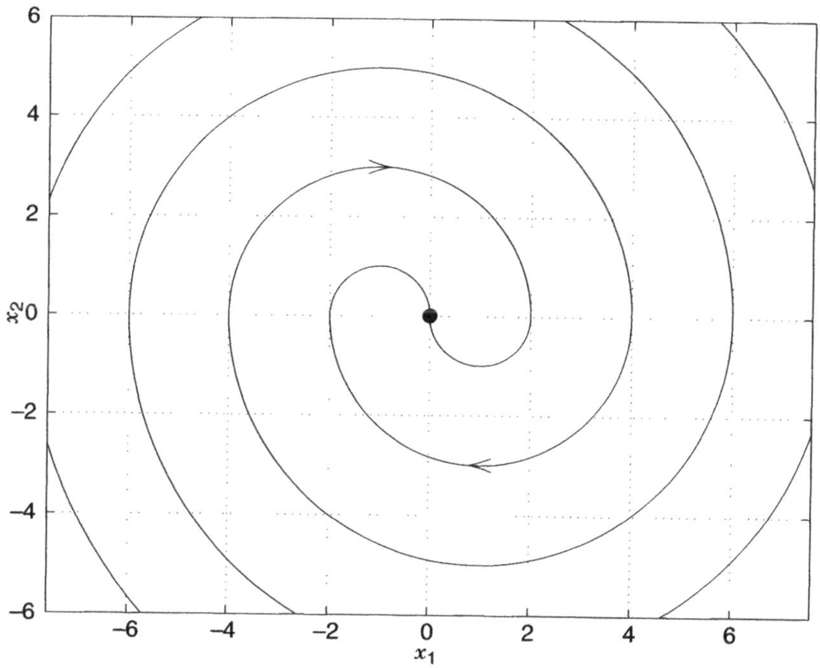

Figure 8.3-3 Abnormal arcs for the second-order IO system with $\omega_n = 1$, $\zeta = 0$.

was generated using finite values of $\lambda_1(0)$. Again the abnormal arcs are barriers since it is not possible to cross them on the way to the origin. For the second case, backward integration from the origin yields an abnormal arc that asymptotically approaches the boundary of the controllable set and, while not one of the trajectories shown in Figure 8.1-7, it corresponds to an optimal trajectory.

8.4 NONLINEAR SYSTEMS

Most of the optimal control systems that we have examined so far have either been linear systems with control constraints or nonlinear systems for which analytic solutions could be obtained for the state equations. When this is not the case, numerical methods must be employed.

Goddard's Problem

Consider a variable density version of Goddard's sounding rocket problem (see Exercise 6-13) [Tsiotras and Kelley, 1992]. The optimal control objective is to

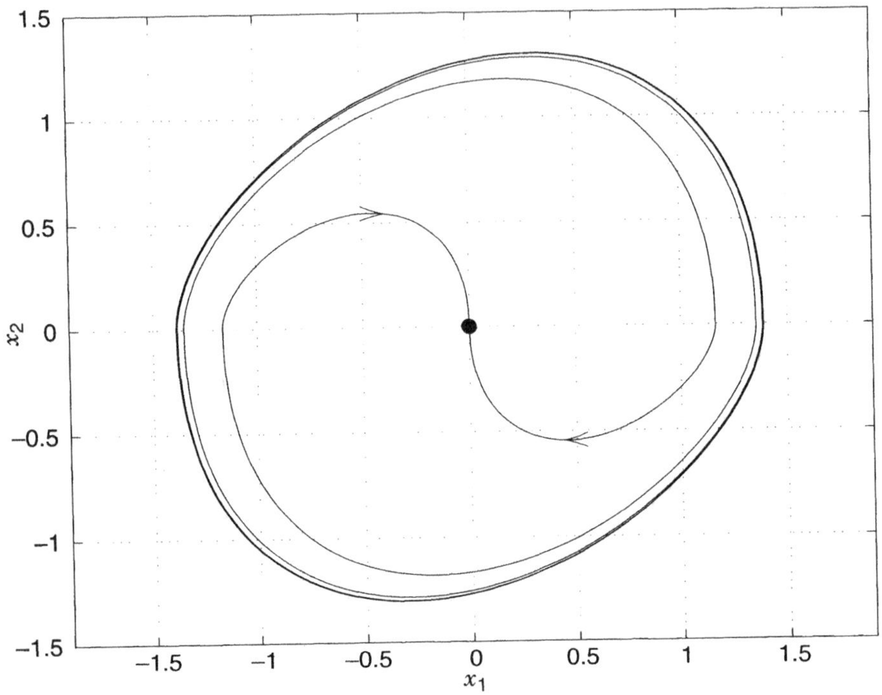

Figure 8.3-4 Abnormal arcs for the second-order IO system with $\omega_n = 1$, $\zeta = -0.5$.

maximize the total height (thrusting-plus-coasting) attained by a limited-fuel variable-thrust rocket ascending vertically through an atmosphere having an exponential density. The equations of motion are

$$\dot{h} = V$$
$$\dot{V} = \frac{T - D}{m} - g \tag{8.4-1}$$
$$\dot{m} = -\frac{T}{V_e},$$

with the initial conditions

$$h(0) = 0$$
$$V(0) = 0 \tag{8.4-2}$$
$$m(0) = m_R + m_F,$$

where $g = g(h)$ is the gravitational acceleration (e.g., constant or inverse square law), and

$$D = D(h, V) = C_D \tfrac{1}{2} \rho A V^2$$

is the drag, which we model as

$$D = C_D \tfrac{1}{2} \rho A V|V|, \tag{8.4-3}$$

so as to maintain direction information. The density $\rho = \rho(h)$ is modeled as

$$\rho = \rho_0 e^{-\beta h}. \tag{8.4-4}$$

The control variable is the upward thrust, $T = u$, which is related to the fuel flow rate and the exhaust velocity by

$$T = -\dot{m} V_e, \tag{8.4-5}$$

with

$$T \in \begin{cases} [0, T_{\max}] & \text{if } m > m_R \\ 0 & \text{if } m = m_R. \end{cases} \tag{8.4-6}$$

The quantities C_D, A, m_R, m_F, V_e, ρ_0, β, and $T_{\max}$ are all positive parameters (along with g for the constant gravity case).

The optimal control problem is to maximize

$$h(t_f) = \int_0^{t_f} V\,dt, \tag{8.4-7}$$

where the terminal conditions, corresponding to a final coasting phase until $\dot{h}(t_f) = 0$, are given by

$$0 = V(t_f) \tag{8.4-8}$$

$$0 = m(t_f) - m_R. \tag{8.4-9}$$

Note that the control constraints (8.4-6) are state dependent, since the maximum height will occur during the coasting phase, after the fuel has been exhausted. However, this will pose no difficulty since $u(\mathbf{x}) = 0$ during coasting not only allows us to calculate the $\partial u/\partial \mathbf{x}$ term needed in the adjoint equations, but also results in no change in the adjoint equations, since $\partial u/\partial \mathbf{x}$ will be zero.

Since we have a maximization problem, we use the corresponding optimal control maximum principle. Maximizing the H function

$$H = (\lambda_0 + \lambda_h)V + \lambda_V \left\{ \frac{T - D}{m} - g \right\} - \lambda_m \frac{T}{V_e} \tag{8.4-10}$$

yields

$$T^* = \begin{cases} T_{\max} & \text{if } \sigma > 0 \\ ? & \text{if } \sigma \equiv 0 \\ 0 & \text{if } \sigma < 0, \end{cases} \tag{8.4-11}$$

where the switching function is

$$\sigma(t) = \frac{\partial H}{\partial T} = \frac{\lambda_V}{m} - \frac{\lambda_m}{V_e}. \tag{8.4-12}$$

The adjoint equations are

$$\begin{aligned} \lambda_0 &= 1 \text{ or } 0 \\ \dot{\lambda}_h &= -\frac{\partial H}{\partial h} = \lambda_V \left(\frac{1}{m} \frac{\partial D}{\partial h} + \frac{\partial g}{\partial h} \right) \\ \dot{\lambda}_V &= -\frac{\partial H}{\partial V} = -(\lambda_0 + \lambda_h) + \frac{\lambda_V}{m} \frac{\partial D}{\partial V} \\ \dot{\lambda}_m &= -\frac{\partial H}{\partial m} = \lambda_V \frac{T - D}{m^2}, \end{aligned} \tag{8.4-13}$$

and the terminal transversality conditions yield

$$\lambda_{hf} = 0, \tag{8.4-14}$$

where λ_{Vf} and λ_{mf} are as yet undetermined. We have

$$T_f = 0 \tag{8.4-15}$$

from (8.4-6) and (8.4-9), so

$$0 > \sigma_f = \frac{\lambda_{Vf}}{m_f} - \frac{\lambda_{mf}}{V_e} \tag{8.4-16}$$

In addition,

$$0 = H_f = -\lambda_V g|_{t_f}$$

yields

$$\lambda_{Vf} = 0. \tag{8.4-17}$$

Thus

$$\lambda_{mf} > 0. \tag{8.4-18}$$

Singular Control The possibility of singular control, in which $\sigma(t) \equiv 0$ for a nonzero time interval, may play a central role in the optimal control solution. In particular, in an exponential atmosphere ($\beta > 0$) the "intuitive" solution of maximum thrust until burnout then coast may turn out not to be the optimal solution!

For singular control, $\sigma(t)$ and its derivatives must be zero. After a bit of algebra we have

$$0 = \sigma = \frac{\lambda_V}{m} - \frac{\lambda_m}{V_e} \tag{8.4-19}$$

$$0 = m\dot{\sigma} = -(\lambda_0 + \lambda_h) + \frac{\lambda_V}{m}\left(\frac{D}{V_e} + \frac{\partial D}{\partial V}\right) \tag{8.4-20}$$

$$0 = m^2\ddot{\sigma} = m\dot{\sigma}\left\{T + \frac{D}{V_e} + \frac{\partial D}{\partial V}\right\} + \frac{\lambda_V}{m}\{(T - D)\mathcal{A} - \mathcal{B}\} \tag{8.4-21}$$

where

$$\mathcal{A} = \frac{D}{V_e^2} + \frac{2}{V_e}\frac{\partial D}{\partial V} + \frac{\partial^2 D}{\partial V^2} \tag{8.4-22}$$

$$\mathcal{B} = \frac{\partial D}{\partial h} m\left(1 - \frac{V}{V_e}\right) - \frac{\partial g}{\partial h} m^2 - \frac{\partial^2 D}{\partial h\, \partial V} mV + \left(\frac{1}{V_e}\frac{\partial D}{\partial V} + \frac{\partial^2 D}{\partial V^2}\right) mg. \tag{8.4-23}$$

For the drag law (8.4-3) and (8.4-4) the required partial derivatives are given by

$$\frac{\partial D}{\partial V} = C_D \rho A |V|$$

$$\frac{\partial D}{\partial h} = -\beta \tfrac{1}{2} C_D \rho A V|V|$$

$$\frac{\partial^2 D}{\partial V^2} = C_D \rho A \operatorname{sgn}(V) \tag{8.4-24}$$

$$\frac{\partial^2 D}{\partial h\, \partial V} = -\beta C_D \rho A |V|.$$

On $\sigma = \dot{\sigma} \equiv 0$ the condition $\ddot{\sigma} \equiv 0$ requires either $\lambda_V \equiv 0$ or $(T-D)\mathcal{A} - \mathcal{B} \equiv 0$. Suppose $\lambda_V \equiv 0$. Then $\sigma \equiv 0$ would require $\lambda_m \equiv 0$, and $H \equiv 0$ would require $\lambda_0 + \lambda_h \equiv 0$. From the structure of H and the adjoint equations (8.4-13), $\lambda_0 + \lambda_h$ can be viewed as a single adjoint variable, say $\lambda_h' = \lambda_0 + \lambda_h$. We reject the $\lambda_V \equiv 0$ case since it would require $(\lambda_h', \lambda_V, \lambda_m) = (0, 0, 0)$, which is not allowed by the maximum principle. Thus the singular control is given by

$$T_{sing} = D + \frac{\mathcal{B}}{\mathcal{A}}. \tag{8.4-25}$$

The singular control necessary condition for the maximum principle is

$$0 \geq (-1)^q \frac{\partial}{\partial T} \left[\frac{d^{2q}}{dt^{2q}} \left(\frac{\partial H}{\partial T} \right) \right], \tag{8.4-26}$$

which for our first-order ($q = 1$) singularity yields

$$0 \leq \frac{\partial \ddot{\sigma}}{\partial T} = \frac{\lambda_V}{m^3} \mathcal{A}. \tag{8.4-27}$$

For the drag law (8.4-3) and (8.4-4), with $V > 0$, we have $\mathcal{A} > 0$. Thus (8.4-27) reduces to $\lambda_V \geq 0$ for $V > 0$.

Synthesis of Singular Surface We leave the numerical solution of Goddard's problem as an exercise, involving a very sensitive two-point boundary value problem. However, important analytical information about the singular surface is available.

In state space, the singular surface itself can be found from

$$0 = H|_{\sigma=0} = (\lambda_0 + \lambda_h)V - \frac{\lambda_V}{m}(D + mg)$$

$$= H|_{\sigma=\dot{\sigma}=0} = \frac{\lambda_V}{m}\left\{\left(\frac{D}{V_e} + \frac{\partial D}{\partial V}\right)V - D - mg\right\},$$

which yields

$$0 = S(h, V) = \left(\frac{D}{V_e} + \frac{\partial D}{\partial V}\right)V - D - mg. \tag{8.4-28}$$

For the draw law (8.4-3) this reduces to

$$0 = S(h, V) = \left(1 + \frac{V}{V_e}\right)D - mg. \tag{8.4-29}$$

Optimal Return Function as a Lyapunov Function

For the LQR problem, we found that the optimal return function was also a Lyapunov function, establishing the asymptotic stability of the origin. A corresponding result occurs for a large class of nonlinear systems.

Consider the problem of minimizing

$$\int_0^\infty f_0(\mathbf{x}, \mathbf{u})dt$$

for the system

$$\dot{\mathbf{x}} = \mathbf{f}(\mathbf{x}, \mathbf{u})$$

with control constraints $\mathbf{u} \in \mathcal{U}$ and a target at $\mathbf{x}_f = 0$. Assume that the target is contained in the interior of the controllable set and that optimal feedback control $\mathbf{u}^*(\mathbf{x})$ yields $f_0[\mathbf{x}, \mathbf{u}^*(\mathbf{x})] > 0$ for any state in the controllable set other than the target. In order for the cost integral to exist, with $f_0(\mathbf{x}, \mathbf{u}) \geq 0$, it is necessary that $f_0(\mathbf{x}(t), \mathbf{u}^*[\mathbf{x}(t)]) \to 0$ as $t \to \infty$ along any trajectory $\mathbf{x}(t)$ generated by $\mathbf{u}^*(\mathbf{x})$.

Let $V^*(\mathbf{x})$ be the minimum cost-to-go from $\mathbf{x}$ to the target and suppose that $V^*(\mathbf{x})$ is C^1. Since

$$V^*(\mathbf{x}) = \int_t^\infty f_0[\mathbf{x}, \mathbf{u}^*(\mathbf{x})]\, dt$$

is positive definite and, for $\lambda_0 = 1$,

$$0 = H = f_0[\mathbf{x}, \mathbf{u}^*(\mathbf{x})] + \frac{\partial V^*(\mathbf{x})}{\partial \mathbf{x}} \mathbf{f}[\mathbf{x}, \mathbf{u}^*(\mathbf{x})],$$

along optimal trajectories, we have

$$\frac{dV^*[\mathbf{x}(t)]}{dt} = -f_0[\mathbf{x}, \mathbf{u}^*(\mathbf{x})] < 0 \qquad \text{for } \mathbf{x}(t) \neq 0.$$

Thus $V^*(\mathbf{x})$ is a Lyapunov function and $\mathbf{x}(t) \to 0$. In other words, the origin is globally asymptotically stable throughout the controllable set.

Note that while the assumption $f_0[\mathbf{x}, \mathbf{u}^*(\mathbf{x})] > 0$ is often satisfied for optimal control problems of practical interest, it need not be satisfied for all problems. See Exercise 8-13 for an example where this assumption is not satisfied.

EXERCISES

8-1. Consider the linear dynamical system

$$\dot{x}_1 = x_2$$
$$\dot{x}_2 = x_1 + u,$$

where $|u| \leq 1$.

(a) Determine the nature of the minimum-time control from any point in the controllable set to the origin. *Hint:* First show that the solution is bang-bang with, at most, one switch. Then simply integrate backward from the origin switching to opposite control at various times.

(b) Draw a sketch of the controllable set and several representative minimum-time trajectories to the origin.

(c) Does a minimum-time trajectory exist from $x_1 = x_2 = 1$?

8-2. Determine the closed-loop control law that will minimize the time to drive the system

$$\dot{x}_1 = x_2$$
$$\dot{x}_2 = -x_2 + u$$
$$|u| \leq 1$$

from any point in state space to the origin. Use the hint from Problem 8-1a.

8-3. Given the system

$$\dot{x}_1 = x_2$$
$$\dot{x}_2 = u,$$

with $|u| \leq 1$, find the minimum-time control law that will drive the system from any point in state space to:

(a) The line

$$2x_1 + x_2 - 3 = 0.$$

(b) The circle

$$x_1^2 + x_2^2 - 1 = 0.$$

Hint: Again, for this problem, the solutions will be bang-bang with, at most, one switch. However synthesizing the solution on this information alone will be difficult. Use the minimum principle directly including the transversality conditions.

8-4. Show that the switching function for the system of Figure 8.1-5 can be expressed as

$$\lambda_2 = \sin(t + \alpha),$$

where

$$\tan \alpha - \frac{\lambda_2(0)}{\lambda_1(0)}.$$

8-5. A dynamical system is described by

$$\dot{x}_1 = |x_2|$$
$$\dot{x}_2 = u,$$

where $|u| \leq 1$. Determine the closed-loop control that will minimize the time from any initial point in the controllable set to the point $x_1(t_f) = 5$, $x_2(t_f) = 0$.

8-6. A certain control system is described by

$$\dot{x} = 3x + 2u.$$

Determine a linear state feedback control law for u that will drive the system from any initial state to the origin and minimize the accumulated cost of the form

$$x_0 = \int_0^{t_f} u^2 \, dt.$$

Solve this problem using:

(a) The Riccati equation.

(b) The minimum principle. Eliminate time between the solutions to the system and adjoint equations.

(c) The minimum principle. Eliminate λ between $H = 0$ and $\partial H / \partial u$.

8-7. Does an optimal control solution exist for Example 8.2-1 if $a = 0$? If not, discuss the nature of a suboptimal solution to this problem.

8-8. A system is described by

$$\begin{aligned} x_1 &= -x_1 - x_2 + u \\ x_2 &= 3x_1 + 2x_2 + u \\ |u| &\leq 0.5. \end{aligned}$$

(a) Determine the controllable set to the origin using abnormal arcs.

(b) Show that the switching curve for the abnormal arcs is given by

$$4x_1 + 3x_2 = 0.$$

(c) Determine the feedback control law for minimizing the time from any point in the controllable set to the origin.

8-9. A system is described by

$$\begin{aligned} x_1 &= -x_2 \\ x_2 &= 2x_1 + 3x_2 - u \\ |u| &\leq 1. \end{aligned}$$

(a) Determine the controllable set to the origin using abnormal arcs.

(b) Compare this controllable set with the set of points that can be transferred to the origin under a linear state variable feedback controller $u = -k_1x_1 - k_2x_2$. Since the control is limited, the state variable feedback controller must saturate at $u = 1$ or $u = -1$ for certain portions of the state space.

8-10. The leaf mass in a plant has been modeled by the equation

$$\dot{x} = RLx(1 - u)$$

where x = leaf mass, L = vegetative mass constant, R = photosynthetic constant, and u is the fraction of total plant growth due to photosynthesis that is allocated to seed production (i.e., $u = 0$ means all growth is to leaf and $u = 1$ means that all growth is to seed). Assume that the plant allocates resources in order to maximize seed production

$$x_0 = \int_0^{t_f} ux \; dt.$$

Determine the control program as a function of time that will maximize x_0 in the fixed time interval t_f.

8-11. The treatment of multiple myeloma using chemotherapy has been modeled using the following nondimensional form of the Gompertz equation [Swan and Vincent, 1977]:

$$\dot{x} = -x - \frac{pu}{1+u},$$

where x is the number of cancer cells in a patient, p is a retardation factor associated with treatment, and u is a control variable associated with the administration of drugs. Under this model, the smaller x is the more healthy the patient is. Determine an optimal closed-loop feedback control program $u^*(x)$ that will minimize the total amount of drugs used. That is, minimize

$$x_0 = \int_0^{t_f} u \; dt$$

during treatment over a fixed time interval, with $x(t_f) = 0$.

8-12. Challenging Problem! Numerically solve Goddard's problem (Section 8.4) for the constant gravity case, with the following parameter values: $g = 9.81$ m/sec^2, $C_D = 0.05$, $A = 0.16$ m^2, $V_e = 1700$ m/sec, $m_R = 150$ kg, $m_F = 300$ kg, $\rho_0 = 1.225$ kg/m^3, $\beta = 1.577 \times 10^{-4}$ m^{-1}. Plot the optimal trajectory in the (h, V) plane. Plot the optimal thrust history as a function of time.

8-13. Consider the problem of minimizing

$$\int_0^{t_f} \cos(\pi x)\, dt$$

subject to

$$\frac{dx}{dt} = -u$$

$$x(t_f) = 0$$

$$\frac{1}{4} \le u \le \frac{1}{2}.$$

In terms of the time-to-go, $\tau = t_f - t$, determine the optimal control $u(\tau)$ and the state $x(\tau)$. Accurately plot $u(\tau)$, $x(\tau)$, and $\cos(\pi x)$ versus τ for $0 \le \tau \le 11$. For the x range corresponding to $0 \le \tau \le 11$ plot the optimal return function $V(x)$ versus x and determine the optimal feedback control $u(x)$.

9

DIFFERENTIAL GAMES

9.1 INTRODUCTION

In this chapter we consider dynamical systems of the form

$$\dot{\mathbf{x}} = \mathbf{f}(\mathbf{x}, \mathbf{u}, \mathbf{v}) \tag{9.1-1}$$

in which the state $\mathbf{x} \in R^{n_x}$ evolves under the control of two input vectors $\mathbf{u} \in \mathcal{U} \subseteq R^{n_u}$ and $\mathbf{v} \in \mathcal{V} \subseteq R^{n_v}$. The inputs **u** and **v** are considered to be independent control inputs that, in general, are not under the command of a single controller. For example, consider the problem of driving a car using two drivers. One controls the steering wheel and the other the gas pedal and brake. Each in his or her own way controls the car, with the resultant motion being determined by both control actions.

Given that there are *two distinct controllers* involved and that each controller has a *different objective* for the dynamical system, the dynamical system is said to be a **differential game**. In order to emphasize the game-theoretic background of the material to follow, we use the term **player** rather than **controller**. This terminology underscores the different nature of the problem to be examined here. For example, the player controlling **v** might be an uncertain random input or **v** might be an input from a player with objectives directly opposed to those of the **u** player. The latter situation corresponds to a **worst-case** situation with respect to the **u** player.

In order to fix this idea more firmly in mind we begin by considering a game that attracted considerable attention in the development of differential games, starting in the 1960s. This game, called the **homicidal chauffeur game** [Isaacs, 1965], is a "simple" example from a large class of pursuit-evasion games, which includes the missile guidance problem introduced in Chapter 7.

The Homicidal Chauffeur Game

This game is a pursuit-evasion game in which a "homicidal chauffeur" (pursuer) tries to run over a slower but more maneuverable "pedestrian" (evader). The pursuer wants to cause a collision and the evader wants to avoid collision. We are particularly concerned with determining the locations from which the pursuer wins (that player can guarantee collision) or the evader wins (that player can guarantee avoidance).

In this game, as illustrated in Figure 9.1-1, there are two dynamical systems, each independently controlled by a separate operator. The first system is a constant-speed car of limited turning radius. The operator of the car controls u, the turning rate of the car. The second system is a pedestrian who also moves at constant speed and controls v, his instantaneous direction of travel.

Using a fixed coordinate system to describe the motion would result in two sets of differential equations. The first set would involve two equations for the pedestrian's position and the second set three equations for the car's position and heading. However, only two differential equations, rather than five, are needed to discribe the situation if a relative coordinate system is used. The relative motion of the pedestrian with respect to the car is obtained by using a

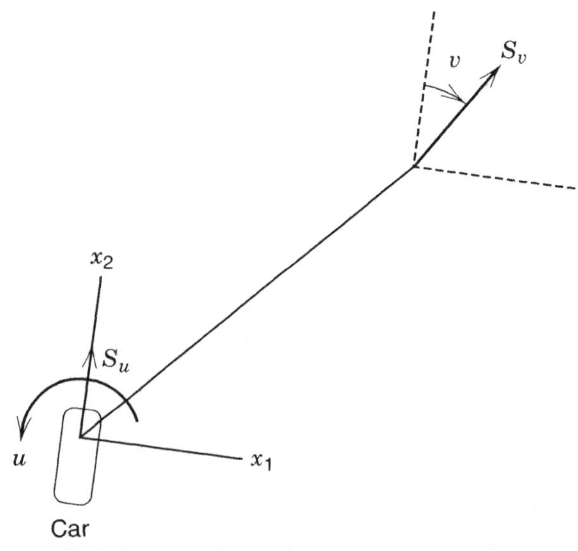

Figure 9.1-1 Geometry of the homicidal chauffeur game.

translating and rotating coordinate system attached to the car, as illustrated in Figure 9.1-1. The origin of the coordinate system moves with the car, and as the car turns so does the coordinate system, so that the x_2-axis is always aligned with the car's direction of motion. This is the same type of coordinate system that a fighter pilot or the captain of a destroyer would have when looking at a target on a radar screen.

If the turning rate u of the car were zero, so that it was going in a straight line, then the motion of the pedestrian with respect to the car would be given simply by

$$\dot{x}_1 = S_v \sin v$$
$$\dot{x}_2 = S_v \cos v - S_u, \tag{9.1-2}$$

where S_u is the car's speed and S_v is the pedestrian's speed, with $S_u > S_v$.

These equations are similar to the equations used to describe Zermelo's problem (Example 7.3-2), and indeed the same geometric interpretation can be given for the velocity $\dot{\mathbf{x}}$ of the system as a function of the control v. As illustrated in Figure 9.1-2*a*, the set of velocity vectors $\dot{\mathbf{x}}$ that can be produced via (9.1-2) is a circle of radius S_v, offset a distance S_u from $\mathbf{x}$. This circular geometry limits the possible directions for the velocity vector $\dot{\mathbf{x}}$ as a function of v. The maximum possible deviation angle α of the velocity vector $\dot{\mathbf{x}}$ from the x_1-direction, illustrated in Figure 9.1-2*a*, is independent of the state location. Therefore, for the case of a nonmaneuvering car, one can easily determine the region of state space within which the pedestrian cannot avoid the car. This region is illustrated in Figure 9.1-2*b*.

It is clear from Figure 9.1-2 that, when the car cannot maneuver, the pedestrian is vulnerable in only a relatively small region of space in front of the car. If we now allow the car to maneuver, this situation changes drastically.

If the car can maneuver, equations (9.1-2) must be modified in order to account for the rotating coordinate system attached to the car. In particular, from **Chasles' theorem**,

$$\left.\frac{d\mathbf{x}}{dt}\right|_{\text{rot}} = \frac{d\mathbf{x}}{dt} + \boldsymbol{\omega} \times \mathbf{x},$$

the cross product $\boldsymbol{\omega} \times \mathbf{x}$ must be added to the relative velocity from (9.1-2), where $\mathbf{x}$ is the position vector to the pedestrian and $\boldsymbol{\omega}$ is the angular velocity vector with which $\mathbf{x}$ rotates due to a rotation of the coordinate system. For the situation depicted in Figure 9.1-1, $\mathbf{x}$ rotates clockwise due to a counterclockwise rotation of the coordinate system, and the angular velocity vector $\boldsymbol{\omega}$ for $\mathbf{x}$ is directed into the figure and has magnitude u. The cross product $\boldsymbol{\omega} \times \mathbf{x}$ yields the components ux_2 in the x_1-direction and $-ux_2$ in the x_2-direction. Thus the equations of motion become

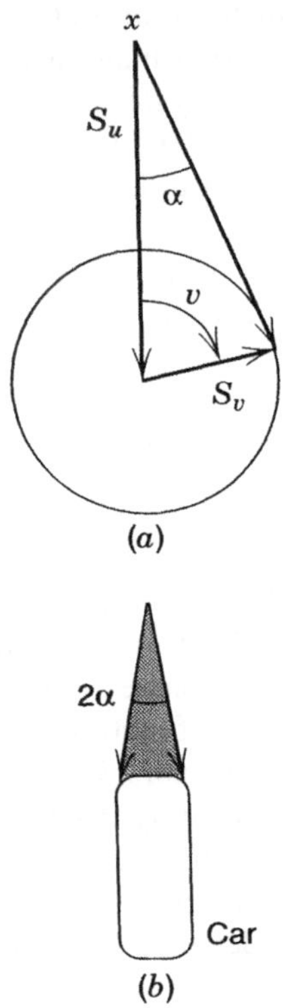

Figure 9.1-2 (*a*) The maximum deflection of the velocity vector when $u = 0$. (*b*) The pedestrian cannot avoid the car (with $u = 0$) when in the shaded region.

$$\begin{aligned} \dot{x}_1 &= S_v \sin v + ux_2 \\ \dot{x}_2 &= S_v \cos v - S_u - ux_1. \end{aligned} \tag{9.1-3}$$

Since the turning rate u for the car is bounded, we assume here that $|u| \leq u_{\max}$, with $u_{\max} > 0$. Note that both u and v must be specified before the motion of the pedestrian with respect to the car can be determined. In this way both players contribute to the ultimate motion of the (x_1, x_2) state space system. Qualitative game theory is now required in order to find the region of state space where the pedestrian cannot avoid the car.

9.2 QUALITATIVE GAMES

The class of differential games of interest here are dynamical systems of the form

$$\dot{\mathbf{x}} = \mathbf{f}(\mathbf{x}, \mathbf{u}, \mathbf{v}), \tag{9.2-1}$$

under the control of two players. The **u** player (**player 1**, or the **pursuer** in the pursuit-evasion game) controls $\mathbf{u} \in \mathcal{U}$, and the **v** player (**player 2**, or the **evader** in the pursuit-evasion game) controls $\mathbf{v} \in \mathcal{V}$. Assume that a **target set** $\mathcal{X}$ is specified and that the game ends the first time that $\mathbf{x} \in \mathcal{X}$. The **u** player wants to end the game, while the **v** player does not. The **qualitative game** is associated with the problem of determining the set of initial states from which the **u** player can *guarantee* to drive the system to the target no matter what control actions are taken by the **v** player to the contrary. Such a set of points in state space is called the **playable set** for the **u** player [Blaquière et al., 1969; Skowronski and Vincent, 1982]. If the **v** player's control did not appear in $\mathbf{f}(\cdot)$, then this would be a standard control problem and the controllable set to $\mathcal{X}$ for the **u** player could be found as in Chapter 6. However, in this two-player game setting, it is clear that the playable set will depend on the control capabilities of both players.

The Boundary of the Playable Set

We are particularly concerned with determining the boundary Σ of the playable set, which we assume is a smooth $(n_x - 1)$-dimensional surface with a unique tangent plane at each point in Σ. In Chapter 6, for the case of a single controller, we used some properties of the controllable set to develop necessary conditions for the control **u** to drive the system along a trajectory in the boundary of the controllable set. In this section we take a similar approach in seeking necessary conditions for controls **u** and **v** that will drive the system along the boundary of the playable set.

A basic property of the boundary of the playable set is that for each player there will be a "one-sided" **semipermeability** property associated with the velocity vectors generated by unilateral control actions of the other player. In other words, if both players try as hard as possible to achieve their own goals, then on the boundary of the playable set neither player can unilaterally pull the state to his or her own side of the boundary. Assuming that controls $\mathbf{u}^*(\mathbf{x})$ and $\mathbf{v}^*(\mathbf{x})$ exist such that the system can be driven along the boundary of the playable set, the resulting boundary trajectory represents a delicate balance of control actions from which neither player can deviate without losing the game. These properties are illustrated in Figure 9.2-1. Boundary control produces a velocity vector tangent to the boundary of the playable set, while unilateral control actions by either player drive the system to one side or the other.

The playable set for the **u** player is the region below the Σ surface and from this player's perspective, this boundary represents a controllability boundary for the set of initial states from which he or she can control the system to the target against the best efforts of the **v** player to prevent this from happening.

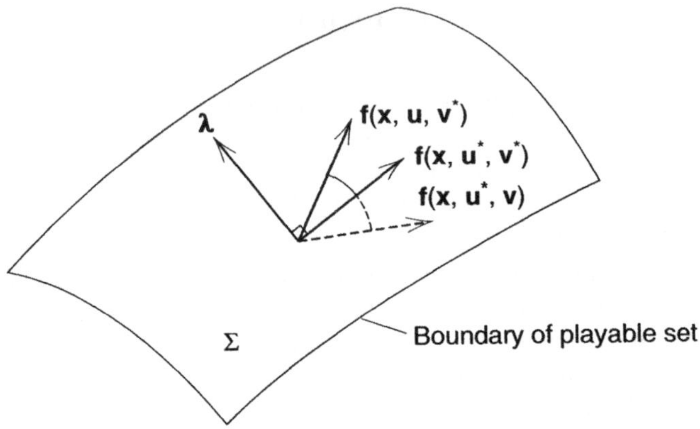

Figure 9.2-1 Possible control actions on the boundary of the playable set.

If $\boldsymbol{\lambda}$ is an outward directed normal to this controllability boundary, then no control $\mathbf{u} \in \mathcal{U}$ can drive the system into the playable set, that is, opposite of the direction of the outward normal vector $\boldsymbol{\lambda}$, as indicated in Figure 9.2-1. If this were not true then the surface Σ depicted in Figure 9.2-1 could not be a controllability boundary for the $\mathbf{u}$ player, since continuity of solutions $\mathbf{x}(t)$ to (9.2-1) with respect to initial conditions, for given controls $\mathbf{u}(\mathbf{x})$ and $\mathbf{v}^*(\mathbf{x})$, implies that the control $\mathbf{u}$ would drive points from outside the controllable set to inside, which is a contradiction. We conclude that, for all $\mathbf{u} \in \mathcal{U}$,

$$0 = \boldsymbol{\lambda}^T\mathbf{f}(\mathbf{x}, \mathbf{u}^*, \mathbf{v}^*) \leq \boldsymbol{\lambda}^T\mathbf{f}(\mathbf{x}, \mathbf{u}, \mathbf{v}^*). \tag{9.2-2}$$

Condition (9.2-2) is based on the assumption that $\mathbf{v}^*$ represents the most unfavorable control action by the $\mathbf{v}$ player. If $\mathbf{v} \in \mathcal{V}$ were any control action other than $\mathbf{v}^*$, a larger controllable set could be possible for the $\mathbf{u}$ player.

Likewise, as illustrated in Figure 9.2-1, if the $\mathbf{u}$ player is trying as hard as possible to drive the state inside of the playable set, then on the boundary of the playable set no control action by the $\mathbf{v}$ player can drive the system to the outside of the playable set in the direction of the outward normal vector $\boldsymbol{\lambda}$. Otherwise, by the same continuity argument as before, the control $\mathbf{v}$ would drive states from the inside of the playable set to the outside, and Σ would not be the controllability boundary for the $\mathbf{u}$ player against the most unfavorable control actions by the $\mathbf{v}$ player. It follows that, for all $\mathbf{v} \in \mathcal{V}$,

$$0 = \boldsymbol{\lambda}^T\mathbf{f}(\mathbf{x}, \mathbf{u}^*, \mathbf{v}^*) \geq \boldsymbol{\lambda}^T\mathbf{f}(\mathbf{x}, \mathbf{u}^*, \mathbf{v}). \tag{9.2-3}$$

The requirements as given by (9.2-1) and (9.2-2) taken together form what is known as a game-theoretic **saddle-point** condition [Isaacs, 1965]:

$$\boldsymbol{\lambda}^{\mathrm{T}}\mathbf{f}(\mathbf{x},\mathbf{u}^*,\mathbf{v}) \le \boldsymbol{\lambda}^{\mathrm{T}}\mathbf{f}(\mathbf{x},\mathbf{u}^*,\mathbf{v}^*) \le \boldsymbol{\lambda}^{\mathrm{T}}\mathbf{f}(\mathbf{x},\mathbf{u},\mathbf{v}^*). \tag{9.2-4}$$

This is a basic necessary condition that the controls $\mathbf{u}^*$ and $\mathbf{v}^*$ must satisfy while generating trajectories that lie in the boundary of the playable set. An equivalent statement of the game saddle-point condition is that

$$0 = \max_{\mathbf{v}\in\mathcal{V}} \min_{\mathbf{u}\in\mathcal{U}} \{\boldsymbol{\lambda}^{\mathrm{T}}\mathbf{f}(\mathbf{x},\mathbf{u},\mathbf{v})\} = \min_{\mathbf{u}\in\mathcal{U}} \max_{\mathbf{v}\in\mathcal{V}} \{\boldsymbol{\lambda}^{\mathrm{T}}\mathbf{f}(\mathbf{x},\mathbf{u},\mathbf{v})\}, \tag{9.2-5}$$

which displays the fact that for a game-theoretic saddle point, either sequence of optimizations, max min or min-max, must produce the same result.

If there exist controls $\mathbf{u}^*(\mathbf{x})$ and $\mathbf{v}^*(\mathbf{x})$ that will drive the system along a trajectory $\mathbf{x}(t)$ in the boundary Σ of the playable set, then it follows from Section 6.2 that a vector $\boldsymbol{\lambda}(t)$ satisfying the adjoint differential equations

$$\dot{\boldsymbol{\lambda}}^{\mathrm{T}} = -\boldsymbol{\lambda}^{\mathrm{T}}\left[\frac{\partial \mathbf{f}}{\partial \mathbf{x}} + \frac{\partial \mathbf{f}}{\partial \mathbf{u}}\frac{\partial \mathbf{u}}{\partial \mathbf{x}} + \frac{\partial \mathbf{f}}{\partial \mathbf{v}}\frac{\partial \mathbf{v}}{\partial \mathbf{x}}\right] \tag{9.2-6}$$

will be an outward normal to the boundary of the playable set, provided that $\boldsymbol{\lambda}(0)$ is an outward normal. Furthermore, if the boundary of the playable set and the boundary of the target set intersect, then the **target transversality condition** on $\boldsymbol{\lambda}$, developed in Section 6.2, must hold at a point where $\mathbf{x}(t)$ reaches the target.

The $\mathcal{H}$ Function

From the above arguments we can conclude that, if $\mathbf{u}^*(\mathbf{x})$ and $\mathbf{v}^*(\mathbf{x})$ are controls that generate a trajectory $\mathbf{x}(t)$ in the boundary of the playable set, then there must exist a nonzero continuous vector $\boldsymbol{\lambda} = [\lambda_1 \ \cdots \ \lambda_{n_x}]^{\mathrm{T}}$, which is a solution to the **adjoint equations**

$$\dot{\boldsymbol{\lambda}}^{\mathrm{T}} = -\frac{\partial \mathcal{H}}{\partial \mathbf{x}} - \frac{\partial \mathcal{H}}{\partial \mathbf{u}}\frac{\partial \mathbf{u}}{\partial \mathbf{x}} - \frac{\partial \mathcal{H}}{\partial \mathbf{v}}\frac{\partial \mathbf{v}}{\partial \mathbf{x}} \tag{9.2-7}$$

such that $\boldsymbol{\lambda} = \boldsymbol{\lambda}(t)$ is an outward normal to the boundary of the playable set, and for every point $\mathbf{x} = \mathbf{x}(t)$ along the boundary trajectory the controls $\mathbf{u}^* = \mathbf{u}^*(\mathbf{x})$ and $\mathbf{v}^* = \mathbf{v}^*(\mathbf{x})$ satisfy the **saddle-point** condition

$$0 = \mathcal{H}(\mathbf{x},\mathbf{u}^*,\mathbf{v}^*,\boldsymbol{\lambda}) = \min_{\mathbf{u}\in\mathcal{U}} \max_{\mathbf{v}\in\mathcal{V}} \mathcal{H}(\mathbf{x},\mathbf{u},\mathbf{v},\boldsymbol{\lambda}) = \max_{\mathbf{v}\in\mathcal{V}} \min_{\mathbf{u}\in\mathcal{U}} \mathcal{H}(\mathbf{x},\mathbf{u},\mathbf{v},\boldsymbol{\lambda}), \tag{9.2-8}$$

where the **Hamiltonian** function $\mathcal{H}(\cdot)$ is defined by

$$\mathcal{H}(\mathbf{x}, \mathbf{u}, \mathbf{v}, \boldsymbol{\lambda}) \triangleq \boldsymbol{\lambda}^{\mathrm{T}}\mathbf{f}(\mathbf{x}, \mathbf{u}, \mathbf{v}). \tag{9.2-9}$$

Parametric Min-Max Games

For a given value of $\mathbf{x}$ and $\boldsymbol{\lambda}$ at some point on the boundary of the playable set, the $\mathcal{H}$ function defined by (9.2-9) is a function of $\mathbf{u}$ and $\mathbf{v}$ only. According to (9.2-8), at every point $\mathbf{x}$ and corresponding $\boldsymbol{\lambda}$, we are to find a game-theoretic saddle point of this function with respect to $\mathbf{u}$ and $\mathbf{v}$ subject to the restrictions that $\mathbf{u} \in \mathcal{U}$ and $\mathbf{v} \in \mathcal{V}$. Necessary conditions for solving this problem are obtained using the min-max necessary conditions in Section 3.4.

We generally assume that a game saddle-point solution exists. Furthermore, to simplify the notation we refer to the game saddle-point condition simply as a min-max condition. From Weierstrass's theorem [Vincent and Grantham, 1981, p. 17] sufficient conditions for a game saddle-point solution to exist are that:

(i) The equations of motion are **separable** in $\mathbf{u}$ and $\mathbf{v}$, of the form

$$\dot{\mathbf{x}} = \mathbf{f}(\mathbf{x}, \mathbf{u}, \mathbf{v}) = \mathbf{F}(\mathbf{x}, \mathbf{u}) + \overline{\mathbf{F}}(\mathbf{x}, \mathbf{v}); \tag{9.2-10}$$

(ii) Neither of the constraint sets $\mathcal{U}$ and $\mathcal{V}$ depends on the other player's control; and

(iii) The control constraint sets are compact (closed and bounded).

These sufficient conditions are satisfied in most applications. However, we do not generally require that these conditions be satisfied. Instead, we simply assume that min-max solutions exist.

As in the parametric game case in Section 3.4, in order to apply parametric min-max game theory the sets $\mathcal{U}$ and $\mathcal{V}$ must be defined in terms of algebraic relations. In general $\mathcal{U} = \mathcal{U}(\mathbf{x})$ and $\mathcal{V} = \mathcal{V}(\mathbf{x})$ may be a function of $\mathbf{x}$. However, if $\mathcal{U}$ and $\mathcal{V}$ are just fixed constant sets, then considerable simplification of the necessary conditions occurs. In particular, as we show below, the second and third terms in (9.2-7) will vanish. Hence we assume that $\mathcal{U}$ and $\mathcal{V}$ are defined in terms of inequality constraints of the form

$$\mathcal{U} = \{\mathbf{u} \in R^{n_u} \,|\, \mathbf{h}(\mathbf{u}) \geq \mathbf{0}\}, \tag{9.2-11}$$

$$\mathcal{V} = \{\mathbf{v} \in R^{n_v} \,|\, \overline{\mathbf{h}}(\mathbf{v}) \geq \mathbf{0}\}, \tag{9.2-12}$$

where $\mathbf{h} = [h_1 \cdots h_{n_h}]^{\mathrm{T}}$ is a n_h-dimensional vector function of the control vector $\mathbf{u}$ and $\overline{\mathbf{h}} = [\overline{h}_1 \cdots \overline{h}_{n_{\overline{h}}}]^{\mathrm{T}}$ is a $n_{\overline{h}}$-dimensional vector function of the control vector $\mathbf{v}$. Every component of the $\mathbf{h}$ and $\overline{\mathbf{h}}$ functions is assumed to be continuous in its arguments and to have continuous partial derivatives as well. The functions are further restricted to that class that will produce *regular sets* for $\mathcal{U}$ and $\mathcal{V}$, as defined in Section 3.2. Under these assumptions we have:

Min-Max Necessary Conditions: *Let* $\mathbf{u}^*$ *be a regular point of* $\mathcal{U}$ and $\mathbf{v}^*$ *be a regular point of* $\mathcal{V}$. *If* $\mathcal{H}$ *takes on a min-max with respect to* $\mathbf{u}$ *at* $\mathbf{u}^*$ *and* $\mathbf{v}$ *at* $\mathbf{v}^*$, *then Lagrange multiplier vectors* $\boldsymbol{\gamma} = [\gamma_1 \ \cdots \ \gamma_{n_h}]^{\mathrm{T}}$ *and* $\overline{\boldsymbol{\gamma}} = [\overline{\gamma}_1 \ \cdots \ \overline{\gamma}_{n_{\overline{h}}}]^{\mathrm{T}}$ *exist such that*

$$\frac{\partial \mathcal{L}}{\partial \mathbf{u}} = \mathbf{0} \tag{9.2-13}$$

$$\frac{\partial \mathcal{L}}{\partial \mathbf{v}} = \mathbf{0} \tag{9.2-14}$$

$$\mathbf{h}(\mathbf{u}^*) \geq \mathbf{0} \tag{9.2-15}$$

$$\overline{\mathbf{h}}(\mathbf{v}^*) \geq \mathbf{0} \tag{9.2-16}$$

$$\boldsymbol{\gamma}^{\mathrm{T}}\mathbf{h}(\mathbf{u}^*) = 0 \tag{9.2-17}$$

$$\overline{\boldsymbol{\gamma}}^{\mathrm{T}}\overline{\mathbf{h}}(\mathbf{v}^*) = 0 \tag{9.2-18}$$

$$\boldsymbol{\gamma} \geq \mathbf{0} \tag{9.2-19}$$

$$\overline{\boldsymbol{\gamma}} \leq \mathbf{0} \tag{9.2-20}$$

where the **Lagrangian** *function is*

$$\mathcal{L}(\mathbf{x}, \mathbf{u}, \mathbf{v}, \boldsymbol{\lambda}, \boldsymbol{\gamma}, \overline{\boldsymbol{\gamma}}) \triangleq \mathcal{H}(\mathbf{x}, \mathbf{u}, \mathbf{v}, \boldsymbol{\lambda}) - \boldsymbol{\gamma}^{\mathrm{T}}\mathbf{h}(\mathbf{u}) - \overline{\boldsymbol{\gamma}}^{\mathrm{T}}\overline{\mathbf{h}}(\mathbf{v}) \tag{9.2-21}$$

and the partial derivatives are evaluated at $\mathbf{u} = \mathbf{u}^*$, $\mathbf{v} = \mathbf{v}^*$.

A Min-Max Principle for the Playable Set Boundary

To simplify the adjoint equations (9.2-7) we note from (9.2-13) and (9.2-14) that

$$\frac{\partial \mathcal{H}}{\partial \mathbf{u}} = \boldsymbol{\gamma}^{\mathrm{T}} \frac{\partial \mathbf{h}}{\partial \mathbf{u}}$$

$$\frac{\partial \mathcal{H}}{\partial \mathbf{v}} = \overline{\boldsymbol{\gamma}}^{\mathrm{T}} \frac{\partial \overline{\mathbf{h}}}{\partial \mathbf{v}}.$$

Thus the second and third terms on the right-hand side of (9.2-7) can be written as

$$\frac{\partial \mathcal{H}}{\partial \mathbf{u}} \frac{\partial \mathbf{u}}{\partial \mathbf{x}} = \boldsymbol{\gamma}^{\mathrm{T}} \frac{\partial \mathbf{h}}{\partial \mathbf{u}} \frac{\partial \mathbf{u}}{\partial \mathbf{x}} = \boldsymbol{\gamma}^{\mathrm{T}} \frac{d\mathbf{h}}{d\mathbf{x}}$$

$$\frac{\partial \mathcal{H}}{\partial \mathbf{v}} \frac{\partial \mathbf{v}}{\partial \mathbf{x}} = \overline{\boldsymbol{\gamma}}^{\mathrm{T}} \frac{\partial \overline{\mathbf{h}}}{\partial \mathbf{v}} \frac{\partial \mathbf{v}}{\partial \mathbf{x}} = \overline{\boldsymbol{\gamma}}^{\mathrm{T}} \frac{d\overline{\mathbf{h}}}{d\mathbf{x}}.$$

Since, from (9.2-11)–(9.2-12) and (9.2-17)–(9.2-20), $\boldsymbol{\gamma}^{\mathrm{T}}\mathbf{h}(\mathbf{u})$ has a stationary minimum at $\mathbf{u} = \mathbf{u}^*(\mathbf{x})$ and $\bar{\boldsymbol{\gamma}}^{\mathrm{T}}\bar{\mathbf{h}}(\mathbf{v})$ has a stationary maximum at $\mathbf{v} = \mathbf{v}^*(\mathbf{x})$, it follows that the right-hand sides of these last two equations are zero, and that the second and third terms on the right-hand side of (9.2-7) will be zero at $\mathbf{u}^* \in \mathcal{U}$ and $\mathbf{v}^* \in \mathcal{V}$. Thus we obtain the same simplification as in the optimal control case.

Playability Min-Max Principle: *Given the constraint sets (9.2-11) and (9.2-12), if the admissible controls* $\mathbf{u}^* = \mathbf{u}^*(\mathbf{x}) \in \mathcal{U}$ *and* $\mathbf{v}^* = \mathbf{v}^*(\mathbf{x}) \in \mathcal{V}$ *generate a trajectory* $\mathbf{x}(t)$ *in the boundary of the playable set for the dynamical system (9.2-1), then there exists a nonzero continuous vector* $\boldsymbol{\lambda} = \boldsymbol{\lambda}(t)$ *satisfying the adjoint equations*

$$\dot{\boldsymbol{\lambda}}^{\mathrm{T}} = -\frac{\partial \mathcal{H}}{\partial \mathbf{x}} \tag{9.2-22}$$

such that, at each point $\mathbf{x} = \mathbf{x}(t)$ *along the boundary,* $\boldsymbol{\lambda}(t)$ *is an outward normal to the boundary of the playable set and the controls* $\mathbf{u}^*$ *and* $\mathbf{v}^*$ *satisfy the min-max condition*

$$\min_{\mathbf{u} \in \mathcal{U}} \max_{\mathbf{v} \in \mathcal{V}} \mathcal{H}(\mathbf{x}, \mathbf{u}, \mathbf{v}, \boldsymbol{\lambda}) = \mathcal{H}(\mathbf{x}, \mathbf{u}^*, \mathbf{v}^*, \boldsymbol{\lambda}) = 0, \tag{9.2-23}$$

where

$$\mathcal{H}(\mathbf{x}, \mathbf{u}, \mathbf{v}, \boldsymbol{\lambda}) \triangleq \boldsymbol{\lambda}^{\mathrm{T}}\mathbf{f}(\mathbf{x}, \mathbf{u}, \mathbf{v}). \tag{9.2-24}$$

Furthermore, if $\mathbf{x}(t)$ *reaches a target set of the form*

$$\mathcal{X} = \{\mathbf{x} \in R^{n_x} \,|\, \mathbf{g}(\mathbf{x}) = \mathbf{0}\} \tag{9.2-25}$$

at a point $\mathbf{x}_f = \mathbf{x}(t_f)$, *where the gradient vectors* $\partial g_i(\mathbf{x}_f)/\partial \mathbf{x}$ *are linearly independent, then there exists a nonzero vector* $\boldsymbol{\rho} \in R^{n_g}$ *such that*

$$\boldsymbol{\lambda}_f^{\mathrm{T}} = \boldsymbol{\rho}^{\mathrm{T}} \frac{\partial \mathbf{g}(\mathbf{x}_f)}{\partial \mathbf{x}}, \tag{9.2-26}$$

where $\boldsymbol{\lambda}_f = \boldsymbol{\lambda}(t_f)$.

Example 9.2-1: Boundary Control for the Homicidal Chauffeur Playable Set In the homicidal chauffeur game both the pedestrian (through v) and the driver of the car (through u) affect the dynamics of the system given by equations (9.1-3). Our objective is to find that region of the (x_1, x_2) state space in which the pedestrian cannot avoid being hit by the car when the driver of the car is trying to run down the pedestrian. This region is the playable set for the driver of the car. We determine here the nature of the controls u and v for driving the system along the boundary of the playable set using the playability min-max principle. Application of these results is used in Exercise 9-4 to find the actual boundaries for two different car shapes.

From the definition (9.2-24) and the system equations (9.1-3) it follows that for this problem

$$\mathcal{H} = \lambda_1(S_v \sin v + ux_2) + \lambda_2(S_v \cos v - S_u - ux_1). \tag{9.2-27}$$

The adjoint equations defined by (9.2-22) are

$$\begin{aligned} \dot{\lambda}_1 &= u\lambda_2 \\ \dot{\lambda}_2 &= -u\lambda_1. \end{aligned} \tag{9.2-28}$$

Since the control v is unbounded a necessary condition for satisfying the min-max condition (9.2-23) is given by

$$\frac{\partial \mathcal{H}}{\partial v} = \lambda_1 S_v \cos v - \lambda_2 S_v \sin v = 0. \tag{9.2-29}$$

If we set $\mathcal{H} = 0$ as required by (9.2-18), we obtain

$$\lambda_1(S_v \sin v + ux_2) + \lambda_2(S_v \cos v - S_u - ux_1) = 0. \tag{9.2-30}$$

Equations (9.2-29) and (9.2-30) are two homogeneous equations in λ_1 and λ_2. For a nontrivial solution for λ_1 and λ_2 to exist, the determinant of the coefficients of these variables must be zero, which yields

$$S_v + ux_2 \sin v - (S_u + ux_1) \cos v = 0. \tag{9.2-31}$$

The control for v can be determined from (9.2-31) provided we know the current state, the parameters S_v and S_u, and the control u. The first two requirements are the same as in an optimal control problem, where the system parameters and state are assumed known. But requiring knowledge of the other player's control represents a new difficulty. This is expected, since, in general, optimal control by a given player will dependent on how the other player affects the system. Since we are only seeking controls that will drive the system along the boundary of the playable set, we can obtain the control for u (in general as a function of the current state, parameters, and control v) from the second part of the min-max principle, by minimizing $\mathcal{H}$ with respect to u.

Before doing this, however, we make note of one more relationship between u and v. Since (9.2-29) must be satisfied for all time ($\partial\mathcal{H}/\partial v \equiv 0$) we can take the time derivative of $\partial\mathcal{H}/\partial v$ and set the result equal to zero. After some manipulation we obtain

$$\lambda_1(u - \dot{v}) \sin v + \lambda_2(u - \dot{v}) \cos v = 0. \tag{9.2-32}$$

With (9.2-29) and (9.2-32) we again have two homogeneous equations in λ_1 and λ_2. Setting the determinant of the coefficients equal to zero yields

$$\dot{v} = u. \tag{9.2-33}$$

Thus on the boundary of the playable set the playability min-max principle is satisfied when the evading pedestrian turns his or her velocity vector at a rate equal to the rate that the pursuer turns the car. In nonrotating coordinates, this corresponds to the evader running in a straight line!

Since the pursuer's control u is bounded we examine the switching function

$$\sigma_u \triangleq \frac{\partial \mathcal{H}}{\partial u} = \lambda_1 x_2 - \lambda_2 x_1 \tag{9.2-34}$$

in order to obtain a necessary condition for u. We first examine the possibility of singular control for u. This requires $\sigma_u(t) \equiv 0$. Taking the first derivative of $\sigma_u(t)$ with respect to time, we obtain

$$\dot{\sigma}_u = (\lambda_1 S_v \cos v - \lambda_2 S_v \sin v) - S_u \lambda_1. \tag{9.2-35}$$

Since the term in parentheses on the right-hand side of (9.2-35) is $\partial \mathcal{H}/\partial v$, which is zero, it follows that singular control implies $\lambda_1(t) \equiv 0$. Since the time derivative of λ_1, given in (9.2-28), must therefore be zero, it follows that singular control corresponds to $u \equiv 0$. Note that it is not possible to have $\lambda_2(t) \equiv 0$ in (9.2-28) since this would imply $\boldsymbol{\lambda}(t) \equiv 0$, which is not allowed by the playability min-max principle. It follows from (9.2-34) that the singular trajectory is given by $x_1 \equiv 0$. Singular control corresponds to the case of the evader running along a straight line directly ahead or directly behind the car. We can summarize the control for the pursuer as follows:

$$u = \begin{cases} u_{\max} & \text{if } \sigma_u < 0 \\ 0 & \text{if } \sigma_u \equiv 0 \\ -u_{\max} & \text{if } \sigma_u > 0. \end{cases} \tag{9.2-36}$$

■

Actually finding the playable set for the above nonlinear example is rather detailed and is left as an exercise. However, for other problems, such as linear two-dimensional systems, finding the boundary of the playable set is often no more difficult than finding the boundary of the controllable set.

Example 9.2-2: Unstable Linear Second-Order System Consider the following linear system in which both the u and v players control the system in the same way [Vincent and Lin, 1995]:

$$\dot{x}_1 = x_2$$
$$\dot{x}_2 = -x_1 + 2x_2 + u + v,$$

where $|u| \leq 2$ and $|v| \leq 1$. Forming

$$\mathcal{H} = \lambda_1 x_2 + \lambda_2(-x_1 + 2x_2 + u + v),$$

it is clear from the playability min-max principle that λ_2 is a switching function for both the u player and the v player and that for $\min_u \max_v \mathcal{H}$, the v player's boundary control will be of opposite sign to the u player's control. Thus in order to find the boundary of the playable set, we can simply set $w = u + v$ with $|w| \leq 1$ and then determine the boundary using the same methods as in Chapter 6. The playable set for this example is illustrated in Figure 9.2-2. ■

9.3 QUALITATIVE DESIGN CONCEPTS

The playable set represents a useful qualitative performance criteria for comparing different control designs for $\mathbf{u}(\mathbf{x})$, such as saturating linear state feedback,

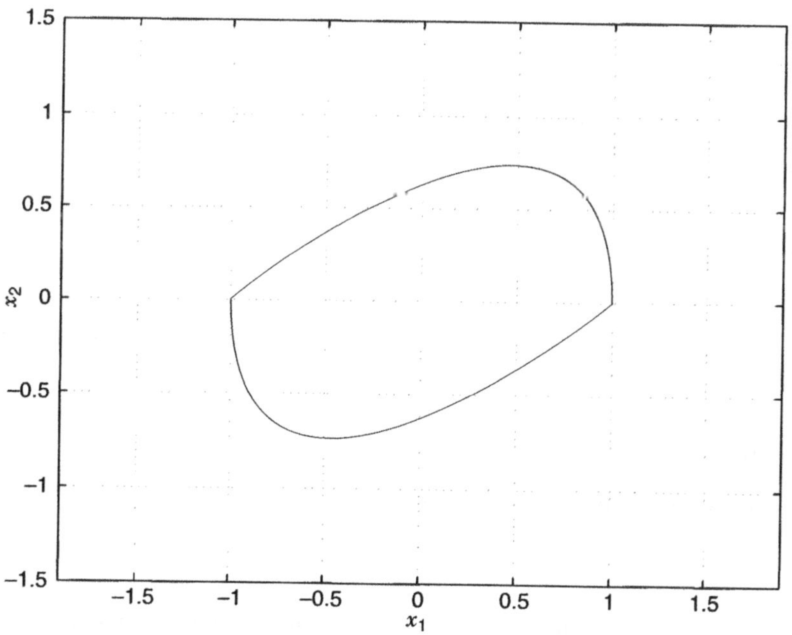

Figure 9.2-2 The playable set for Example 9.2-2.

LQR, or game-theoretic. For example, consider the problem of designing a controller for the nonlinear system (9.1-1), where $\mathbf{v}$ represents some uncertain input acting on the system. The objective is to achieve asymptotic stability for the equilibrium point $\mathbf{f}(\overline{\mathbf{x}}, \overline{\mathbf{u}}, \mathbf{0}) = \mathbf{0}$, where $\overline{\mathbf{x}}$ is an equilibrium point corresponding to a constant control input $\overline{\mathbf{u}}$ with no uncertain input. Ideally, in the presence of an uncertain input, the system is asymptotically stable to $\overline{\mathbf{x}}$ throughout the playable set, regardless of any admissible control action by $\mathbf{v}$. However, unless one is able to solve the complete quantitative differential game (see Section 9.4), this is not the case. More generally, the region of effectiveness for $\mathbf{u}(\mathbf{x})$ is a subset of the playable set. Since $\mathbf{v}$ represents an uncertain input, it may have the effect of destabilizing the system in the vicinity of the origin as well as reducing the domain of attraction.

v-Reachable and v-Attractive Sets

There are two sets we now define that are useful in comparing the performance of a given control design $\mathbf{u}(\mathbf{x})$ to one that is effective throughout the playable set.

A given point in state space is said to be $\mathbf{v}$-reachable from the equilibrium point $\overline{\mathbf{x}}$ if there exists an admissible control law for $\mathbf{v}$ such that, under a given admissible control law for $\mathbf{u}(\mathbf{x})$, the system can be driven from $\overline{\mathbf{x}}$ to the given point in finite time. The **v-reachable set** is the set of all $\mathbf{v}$-reachable points under the specified control law $\mathbf{u}(\mathbf{x})$. The reachability maximum principle from Chapter 6 can be used to find the $\mathbf{v}$-reachable set. The $\mathbf{v}$-reachable set is the region where the uncertain input could drive the state, despite $\mathbf{u}(\mathbf{x})$.

A given point in state space is said to be $\mathbf{v}$-attractive if, under a given admissible control law for $\mathbf{u}(\mathbf{x})$ and any admissible control law for $\mathbf{v}$, the system will be driven from the given point to an arbitrary small neighborhood of the $\mathbf{v}$-reachable set in finite time. The **v-attractive set** consists of all $\mathbf{v}$-attractive points under the specified control law for $\mathbf{u}$. In the $\mathbf{v}$-attractive set, the control $\mathbf{u}(\mathbf{x})$ transfers the state to (at least a neighborhood of) a point on the $\mathbf{v}$-reachable set, regardless of the uncertain input $\mathbf{v}$.

The $\mathbf{v}$-reachable set must be contained within the $\mathbf{v}$-attractive set for stability. Ideally, the $\mathbf{v}$-reachable set would be just the point $\overline{\mathbf{x}}$, and the $\mathbf{v}$-attractive set would be the playable set.

Linear Systems with Bounded Control

Consider the linear dynamical system of the form

$$\dot{\mathbf{x}} = \mathbf{A}\mathbf{x} + \mathbf{b}u + \mathbf{e}v \tag{9.3-1}$$

where $\mathbf{x}$ is an $n_x \times 1$ state vector, $\mathbf{A}$ is an $n_x \times n_x$ constant matrix, $\mathbf{b}$ and $\mathbf{e}$ are $n_x \times 1$ constant matrices, u is a scalar control input by the first player, and v is a scalar control input by the second player. The u player wants to transfer the state to some target set, usually the origin, while the v player wants to keep the state out of the target set. The control variables are assumed to be constrained by inequalities of the form

$$\begin{aligned} h_1 &= u_{\max} - u \geq 0 \\ h_2 &= u - u_{\min} \geq 0 \\ h_3 &= v_{\max} - v \geq 0 \\ h_4 &= v - v_{\min} \geq 0 \end{aligned} \qquad (9.3\text{-}2)$$

where $u_{\max} \geq u_{\min}$ and $v_{\max} \geq v_{\min}$ are constaints.

In order to employ the playability min-max principle, we first form the $\mathcal{H}$ function

$$\mathcal{H} = \boldsymbol{\lambda}^{\mathrm{T}}(\mathbf{A}\mathbf{x} + \mathbf{b}u + \mathbf{e}v), \qquad (9.3\text{-}3)$$

from which we obtain the adjoint equations

$$\dot{\boldsymbol{\lambda}}^{\mathrm{T}} = -\boldsymbol{\lambda}^{\mathrm{T}}\mathbf{A}.$$

Necessary conditions for $\mathcal{H}$ to take on a (zero) min-max value with respect to u and v are obtained by first forming the $\mathcal{L}$ function

$$\mathcal{L} = \mathcal{H} - \gamma_1 h_1 - \gamma_2 h_2 - \gamma_3 h_3 - \gamma_4 h_4,$$

where the $\mathcal{H}$ function is given by (9.3-3) and the h functions are given by (9.3-2). Setting the partial derivative of $\mathcal{L}$ with respect to u equal to zero and the partial derivative of $\mathcal{L}$ with respect to v equal to zero, we obtain

$$\frac{\partial \mathcal{H}}{\partial u} = \boldsymbol{\lambda}^{\mathrm{T}}\mathbf{b} + \gamma_1 - \gamma_2 = 0$$

$$\frac{\partial \mathcal{H}}{\partial v} = \boldsymbol{\lambda}^{\mathrm{T}}\mathbf{e} + \gamma_3 - \gamma_4 = 0.$$

Using the same procedures as in Section 7.4 we can check for the possibility of singular control for either u or v or both. Singular control corresponds to the interior control case ($\gamma_1 = 0$ and $\gamma_2 = 0$) or ($\gamma_3 = 0$ and $\gamma_4 = 0$) with $\sigma_u \triangleq \boldsymbol{\lambda}^{\mathrm{T}}\mathbf{b} \equiv 0$ or $\sigma_v \triangleq \boldsymbol{\lambda}^{\mathrm{T}}\mathbf{e} \equiv 0$. It follows that the control u will be bang-bang if $|\mathcal{P}_u| \neq 0$, where

$$\mathcal{P}_u = [\mathbf{b}, \mathbf{A}\mathbf{b}, \ldots, \mathbf{A}^{n_x - 1}\mathbf{b}]$$

and the control υ will be bang-bang if $|\mathcal{P}_\upsilon| \neq 0$, where

$$\mathcal{P}_\upsilon = [\mathbf{e}, \mathbf{Ae}, \ldots, \mathbf{A}^{n_x - 1}\mathbf{e}].$$

As in Section 8.1, the maximum number of switches along a min-max trajectory will depend on the eigenvalues of the **A** matrix.

Example 9.3-1: Inverted Pendulum The following system corresponds to a linearized inverted pendulum in which the nonlinear gravitational term has been replaced by an uncertain input υ [Soldatos et al., 1991; Vincent and Lin, 1995]:

$$\begin{aligned} \dot{x}_1 &= x_2 \\ \dot{x}_2 &= u + \upsilon, \end{aligned} \tag{9.3-4}$$

where x_1 is the angular displacement of the pendulum from the vertical and u is a control torque. We assume that $|u| \leq u_{\max} = 2$ and $|\upsilon| \leq \upsilon_{\max} = 1$. If we set $\upsilon = 0$ and seek a minimum-time control to the origin, then this problem becomes equivalent to the juggernaut problem of Chapter 7. Recalling the solution to that problem, the minimum-time controller is bang-bang with at most one switch. Furthermore, the system may be driven, in minimum time, to the origin from every point in state space. Thus, for $\upsilon = 0$, the controllable set is the entire state space. The playable set to the origin is also the entire state space. This follows from the facts that the min-max time game-theoretic control for υ is given by $\upsilon = -u$ (see Section 9.4) and that $u_{\max} > \upsilon_{\max}$. From the above definitions, it follows that there must exist a game-theoretic control $u(\mathbf{x})$ such that the υ-reachable set is just the origin and the υ-attractive set is the entire state space.

Let us now compute the υ-reachable set and the υ-attractive set using a controller based on saturating LQR design. If we choose **Q** equal to the two-dimensional identity matrix and $R = 1$, the LQR design method yields the following state variable feedback controller:

$$u = -\sigma,$$

where

$$\sigma = x_1 + 1.7321 x_2.$$

Under saturating state variable feedback design, the control law becomes

$$u = \begin{cases} -\sigma & \text{if } |\sigma| \leq u_{\max} \\ -u_{\max}\,\mathrm{sgn}(\sigma) & \text{otherwise.} \end{cases} \tag{9.3-5}$$

With the control law for u specified, the υ-reachable set is determined by simply finding the reachable set from the origin for the v player. From the reachability maximum principle, it follows that

$$\mathcal{H} = \lambda_1 x_2 + \lambda_2(u + v)$$

is maximized by the control $v = v_{\max}\,\mathrm{sgn}(\lambda_2)$, which for this problem implies

$$v = v_{\max}\,\mathrm{sgn}(x_2). \tag{9.3-6}$$

Integrating (9.3-4) forward in time starting near the origin, subject to the control u as given by (9.3-5) and the uncertain input v as given by (9.3-6), results in a trajectory that asymptotically approaches the boundary of the v-reachable set shown in Figure 9.3-1. By integrating backward in time starting from a point outside, but nearby, the v-reachable set, we obtain a trajectory that asymptotically approaches the boundary of the v-attractive set shown in Figure 9.3-2. ■

In the previous example we obtained a **v**-attractive set smaller than the playable set (which was the entire state space) and a nonzero **v**-reachable set.

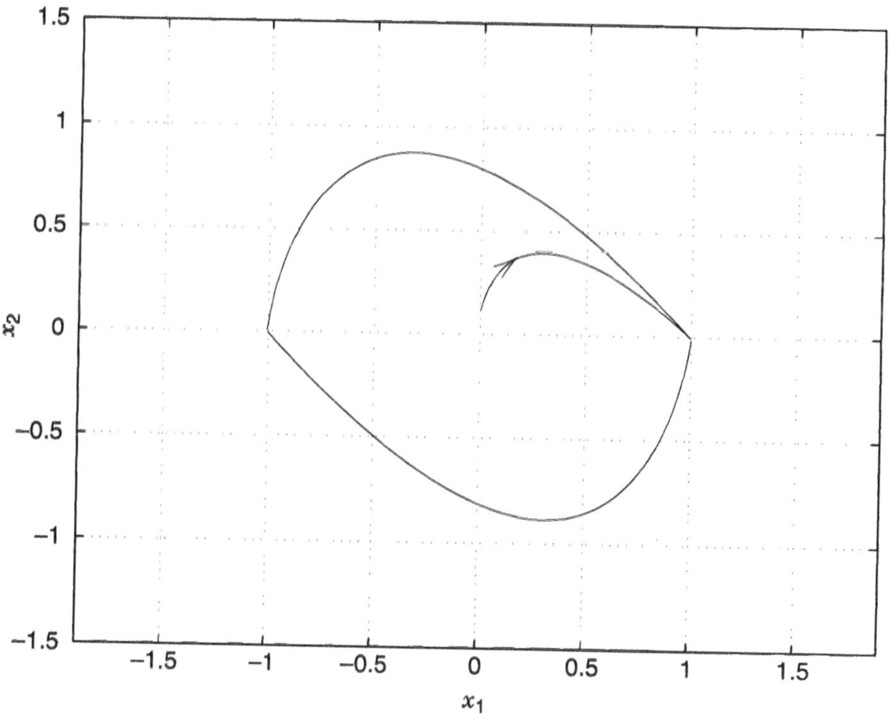

Figure 9.3-1 The v-reachable set for Example 9.3-1.

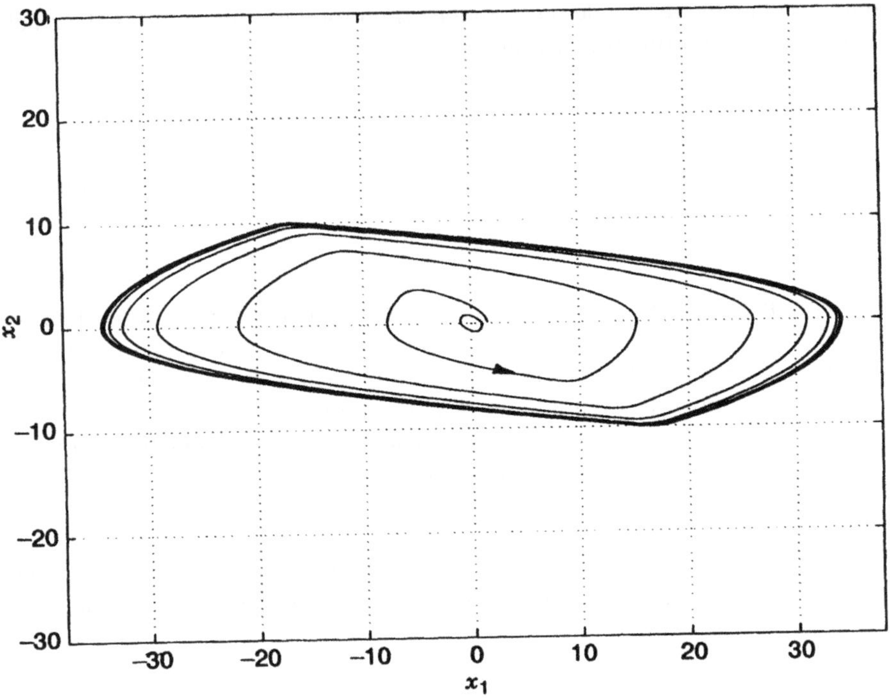

Figure 9.3-2 The v-attractive set for Example 9.3-1.

The size and shape of the **v**-reachable and **v**-attractive sets depend very much on the control law used for **u** (see Exercise 9-7). In fact, very poor control designs could result in situations where the **v**-reachable set is not contained in the **v**-attractive set. In such a situation it would always be possible for the uncertain input to destabilize the system. Note that we were able to obtain the **v**-attractive set by integrating the system of equations backward in time, just as we would do to find a domain of attraction. In fact, from the definition, we see that the **v**-attractive set is a **guaranteed domain of attraction** to the **v**-reachable set for the **u** player using the control **u**(**x**).

Example 9.3-2: Magnetic Suspension System An example of a linear system in a game setting is provided by a magnetic suspension system [Vincent, 1980] given by

$$\begin{aligned} \dot{x}_1 &= x_2 \\ \dot{x}_2 &= \beta_1 x_1 + \beta_2 u + v, \end{aligned} \tag{9.3-7}$$

where x_1 and x_2 are the position and velocity, respectively, of a platform magnetically suspended between two attraction magnets, with an unstable equilibrium. The control u, for stabilizing the system, corresponds to an additional magnetic force generated by control coils on the platform. The input v represents an additional uncertain force input to the system. Both parameters, β_1 and β_2, are positive constants. For this particular example we choose to set $\beta_1 = \beta_2 = 1$. The control input is assumed to be bounded, by

$$|u| \leq 1,$$

and we assume that the uncertain input is also bounded, by

$$|v| \leq v_{\max} < 1.$$

With $u = v = 0$, the system is unstable with hyperbolic trajectories $x_1^2 - x_2^2 =$ constant.

Saturating State Feedback It can be verified easily that, with $v = 0$, the system is completely controllable (rank $[\mathcal{P}_u] = 2$). This means that for this system there exist feedback gains k_1 and k_2 such that a linear feedback control law

$$u = -k_1 x_1 - k_2 x_2$$

will produce asymptotic stability of the origin starting from any point in state space, provided that $v = 0$ and u were unbounded. For $|u| \leq 1$, under **saturating state feedback** control

$$u(\mathbf{x}) = \begin{cases} -(k_1 x_1 + k_2 x_2) & \text{if } |k_1 x_1 + k_2 x_2| \leq 1 \\ -\text{sgn}(k_1 x_1 + k_1 x_2) & \text{if } |k_1 x_1 + k_2 x_2| > 1, \end{cases} \tag{9.3-8}$$

the domain of attraction to the origin, with $v = 0$, is no longer the entire state space. The domain of attraction can be made equal to the controllable set, $|x_1 + x_2| < 1$ (see Example 6.3-3), by choosing $k_1 = k_2$ sufficiently large, for example, $k_1 = k_2 = 3$. The boundaries of the domain of attraction in this case are straight lines of slope $-45°$ passing through $x_1 = \pm 1$ on the x_1-axis, as illustrated in Figure 9.3-3. Other choices (e.g., $k_1 \neq k_2$) will result in a domain of attraction smaller than the controllable set.

As in the previous example we determine the v-reachable and the v-attractive sets using the reachability maximum principle. It follows that

$$\mathcal{H} = \lambda_1 x_2 + \lambda_2(\beta_1 x_1 + \beta_2 u + v)$$

is maximized by the control

$$v = v_{\max}\, \text{sgn}(\lambda_2). \tag{9.3-9}$$

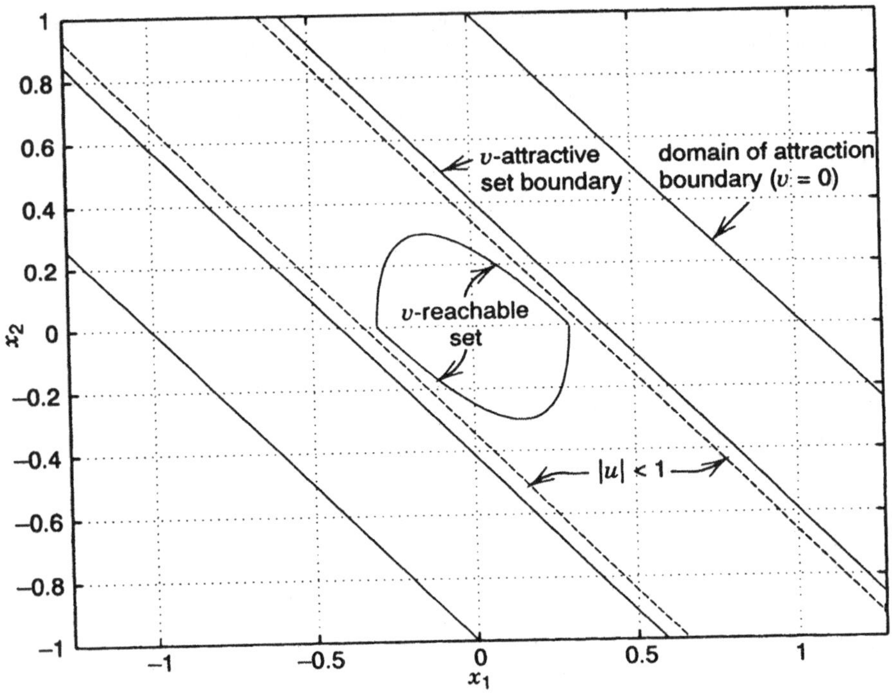

Figure 9.3-3 Regions of stability for Example 9.3-2 with $k_1 = k_2 = 3$. In this case the υ-attractive set and the playable set are identical.

Figure 9.3-3 illustrates the υ-reachable and the υ-attractive sets for the system (9.3-7) with $u(\mathbf{x})$ given by (9.3-8), with $k_1 = k_2 = 3$, and $\upsilon(\mathbf{x})$ given by (9.3-9), with $\upsilon_{\max} = 0.6$. In this "worst-case" situation it is possible for the uncertain force to produce the υ-reachable set boundary illustrated in the figure. Clearly the asymptotic stability that the origin had with $\upsilon = 0$ has been replaced with **Lagrange stability** (ultimate boundedness) for $\upsilon \neq 0$. With an uncertain input ($\upsilon \neq 0$) we can no longer speak of a domain of attraction to the origin. Rather, the domain of attraction to the origin with $\upsilon = 0$ has been replaced by the υ-attractive set, which is a domain of guaranteed attraction to the υ-reachable set. Since different control laws for $u(\mathbf{x})$ will produce different υ-reachable and the υ-attractive sets, one might seek a controller such that the υ-reachable set reduces to the origin.

Switching State Feedback In terms of the previous saturation approach for handling $|u| \leq 1$, consider the limiting case of an infinite-gain, switching-type control law based on $k_1 = k_2 \rightarrow \infty$. The resulting **switching feedback control** law is

$$u = \begin{cases} -\text{sgn}(x_1 + x_2) & \text{if } x_1 + x_2 \neq 0 \\ \in [-1, +1] & \text{if } x_1 + x_2 = 0. \end{cases} \tag{9.3-10}$$

A Lyapunov function can now be used to determine the region of stability for this control law. A useful Lyapunov function for this case is given by

$$V = \tfrac{1}{2}[(1 + 2\varepsilon)x_1^2 + 2x_1x_2 + x_2^2] = \tfrac{1}{2}[(x_1 + x_2)^2 + 2\varepsilon x_1^2], \tag{9.3-11}$$

where ε is a small positive number. This function is positive definite for all $\varepsilon > 0$ and its contours are elliptical. For small ε the ellipses are long and stretched out. In the limit as $\varepsilon \to 0$ the "contours" become pairs of straight lines $|x_1 + x_2| =$ constant, parallel to the boundaries encountered in the previous saturating state feedback case.

It follows, from (9.3-11) and the equations of motion (9.3-7) with $\beta_1 = \beta_2 = 1$, that

$$\dot{V} = (x_1 + x_2)^2 + 2\varepsilon x_1 x_2 + (x_1 + x_2)u + (x_1 + x_2)v.$$

Note that when $x_1 + x_2 = 0$, $\dot{V} < 0$. For $x_1 + x_2 \neq 0$, with u given by (9.3-10), we have

$$\dot{V} = (x_1 + x_2)^2 + 2\varepsilon x_1 x_2 - |x_1 + x_2| + (x_1 + x_2)v.$$

Thus $\dot{V} < 0$ requires

$$\frac{(x_1 + x_2)^2}{|x_1 + x_2|} + \frac{2\varepsilon x_1 x_2}{|x_1 + x_2|} + \frac{x_1 + x_2}{|x_1 + x_2|} v < 1.$$

It follows that the worst case for v, corresponding to $\max_{|v| \leq v_{\max}} \{\dot{V}\}$, is given by $v = v_{\max} \, \text{sgn}(x_1 + x_2)$. Thus to ensure stability we require

$$|x_1 + x_2| + \frac{2\varepsilon x_1 x_2}{|x_1 + x_2|} < 1 - v_{\max}.$$

It follows that for $\varepsilon \to 0$, the region of guaranteed asymptotic statability for the origin (the v-attractive set) is given simply by

$$|x_1 + x_2| < 1 - v_{\max}. \tag{9.3-12}$$

We conclude that under a switching control law given by (9.3-10) the system defined by (9.3-7) provides asymptotic stability for the origin, under the unknown but bounded external force v, in a region defined by (9.3-12). In other words, the v-reachable set is infinitesimally small and its domain of attraction is given by (9.3-12).

Min-Max Control As a final approach to designing a controller for this system, let us now consider a **min-max controller** for u based on game theory. That is, we seek the playable set for u. To do this we employ the playability min-max principle with the $\mathcal{H}$ function given by

$$\mathcal{H} = \lambda_1 x_2 + \lambda_2(x_1 + u + v).$$

By integrating the system equations backward from the two equilibrium points $x_1 = \pm(1 - v_{\max})$, $x_2 = 0$, with u and v determined from the playability min-max principle, we find the boundary of the playable set to be a region defined by $|x_1 + x_2| < 1 - v_{\max}$, which is the same as the v-attractive set obtained with the switching control law (9.3-10). Hence the previous control law, which corresponds to $\min_{|u|\leq 1} \max_{|v|\leq v_{\max}} \{\dot{V}\}$, is a perfectly good game-theoretic control law provided our objective is to simply guarantee that every point in the playable set be driven to the origin. Other game-theoretic control laws can be obtained for this problem by defining a **quantitative game**, involving some cost function, such as time required, and so on. For example, we may find trajectories within the playable set by finding min-max time trajectories, in which u seeks to minimize the time to the origin and v seeks to maximize the time to the origin (see Exercise 9-9). All such game-theoretic solutions will have the same playable set, regardless of the quantitative min-max performance measure. ■

9.4 QUANTITATIVE GAMES

The playability min-max principle is the game-theoretic analog to the controllability minimum principle. When the playability min-max principle is used to solve for the boundaries of the playable set one is solving a **game of kind** [Isaacs, 1965] or a **qualitative game** [Blaquière et al., 1969]. Within the playable set, the **u** player can guarantee to drive the state to a specified target, despite the actions of the **v** player. But there may be many ways to do so. In this section we learn not only how the **u** player can accomplish this guarantee, but also do so while optimizing a specified "cost" functional.

As might be expected, if a cost functional is defined that the **u** player wishes to minimize and the **v** player wishes to maximize, then a differential game min-max principle, analogous to Pontryagin's minimum principle for single-player optimal control problems, can be developed [Isaacs, 1965; Blaquière et al., 1969]. When we use the differential game min-max principle to solve for game-theoretic solutions, then we are said to be solving a **game of degree** [Isaacs, 1965] or a **quantitative game** [Blaquière et al., 1969].

The terms **game of kind** and **game of degree** were first coined by Isaacs in his pioneering book on differential games. However, we have adopted, for this book, the more commonly used terminology of **qualitative games** and **quantitative games**. In this regard, Rufus Isaacs, at one point during the US Navy workshop on differential games in 1973, referred to George Leitmann as "Hercules." This was not so much in reference to his stature but rather to the fact that, since Leitmann had used different names for these concepts, it was proper for him to use a different name for Professor Leitmann.

Zero-Sum Games

The type of **differential game** that we consider has the following integral "cost" functional:

$$J[\mathbf{u}(\cdot), \mathbf{v}(\cdot)] = \int_0^{t_f} f_0(\mathbf{x}, \mathbf{u}, \mathbf{v})dt \tag{9.4-1}$$

associated with transferring the state from a given initial point to a specified target set. It becomes a "zero-sum" game (what one player loses the other gains) if player 1 chooses $\mathbf{u}$ to minimize J and player 2 chooses $\mathbf{v}$ to maximize J. Specifically, we seek a control $\mathbf{u}(\mathbf{x}) \in \mathcal{U} \subseteq R^{n_v}$ and a control $\mathbf{v}(\mathbf{x}) \in \mathcal{V} \subseteq R^{n_v}$ that will min-max the cost integral for the $\mathbf{u}$ player, while transferring the state $\mathbf{x}(t) \in R^{n_x}$ of the dynamical system

$$\dot{\mathbf{x}} = \mathbf{f}(\mathbf{x}, \mathbf{u}, \mathbf{v}) \tag{9.4-2}$$

from a specified initial point

$$\mathbf{x}(0) = \mathbf{x}_i \tag{9.4-3}$$

to any final state $\mathbf{x}(t_f)$ in a **terminal set** $\mathcal{X} \subset R^{n_x}$,

$$\mathbf{x}(t_f) \in \mathcal{X} = \{\mathbf{x} \in R^{n_x} | \mathbf{g}(\mathbf{x}) = \mathbf{0}\}. \tag{9.4-4}$$

The final time t_f is generally unspecified in this problem formulation. It is determined as the first time that $\mathbf{x}(t)$ reaches $\mathcal{X}$. At each instant of time $0 \le t \le t_f$ the control vectors $\mathbf{u}$ and $\mathbf{v}$ are constrained to be elements of specified **control constraint sets** $\mathcal{U}$ and $\mathcal{V}$, respectively, which are possibly of the form $\mathcal{U}(\mathbf{x})$ and $\mathcal{V}(\mathbf{x})$ but usually of the form

$$\mathcal{U} = \{\mathbf{u} \in R^{n_u} | \mathbf{h}(\mathbf{u}) \ge \mathbf{0}\} \tag{9.4-5}$$

$$\mathcal{V} = \{\mathbf{v} \in R^{n_v} | \bar{\mathbf{h}}(\mathbf{v}) \ge \mathbf{0}\}. \tag{9.4-6}$$

In the above description $\mathbf{x} = [x_1 \cdots x_{n_x}]^T$ is an n_x-dimensional state vector, $\mathbf{u} = [u_1 \cdots u_{n_u}]^T$ is an n_u-dimensional control vector for player 1, and $\mathbf{v} = [v_1 \cdots v_{n_v}]^T$ is an n_v-dimensional control vector for player 2. The scalar-valued function $f_0(\mathbf{x}, \mathbf{u}, \mathbf{v})$ and each component of the vector functions $\mathbf{f}(\mathbf{x}, \mathbf{u}, \mathbf{v}) = [f_1 \cdots f_{n_x}]^T$, $\mathbf{g}(\mathbf{x}) = [g_1 \cdots g_{n_g}]^T$, $\mathbf{h}(\mathbf{u}) = [h_1 \cdots h_{n_h}]^T$, and $\overline{\mathbf{h}}(\mathbf{v}) = [\overline{h}_1 \cdots \overline{h}_{n_{\overline{h}}}]^T$ are assumed to be continuous and to have continuous partial derivatives with respect to their arguments. In addition, we assume that the gradient vectors $\partial g_i(\mathbf{x})/\partial \mathbf{x}$, $i = 1, \ldots, n_g$, are linearly independent, as are the gradient vectors $\partial h_i(\mathbf{u})/\partial \mathbf{u}$ and the gradient vectors $\partial \overline{h}_j(\mathbf{v})/\partial \mathbf{v}$ at any active inequalities $h_i(\mathbf{u}) = 0$, $\overline{h}_j(\mathbf{v}) = 0$. The control functions $\mathbf{u}(\mathbf{x})$ and $\mathbf{v}(\mathbf{x})$ may be discontinuous. However, we consider only trajectories $\mathbf{x}(t)$ for which $\mathbf{u}[\mathbf{x}(t)]$ and $\partial \mathbf{u}[\mathbf{x}(t)]/\partial \mathbf{x}$ are at worst piecewise continuous and $\mathbf{v}[\mathbf{x}(t)]$ and $\partial \mathbf{v}[\mathbf{x}(t)]/\partial \mathbf{x}$ are at worst piecewise continuous.

The min-max controls $\mathbf{u}(\mathbf{x})$ and $\mathbf{v}(\mathbf{x})$ that we seek are also called **strategies** for the two players since the values of the closed-loop controls $\mathbf{u}(\mathbf{x})$ and $\mathbf{v}(\mathbf{x})$ depend on $\mathbf{x}$, the state of the game. Open-loop controls $\mathbf{u}(t)$ and $\mathbf{v}(t)$ are not considered to be strategies since the control action of a given player would not account for changes in the predicted state, due to unpredicted control actions of the other player.

Min-Max Cost Surface

As with optimal control problems, it is convenient to convert the cost integral to an equivalent differential equation. We do so by introducing an additional quantity called the **accumulated cost**, designated by the scalar x_0 and given by the differential equation

$$\dot{x}_0 = f_0(\mathbf{x}, \mathbf{u}, \mathbf{v}), \tag{9.4-7}$$

where $f_0(\mathbf{x}, \mathbf{u}, \mathbf{v})$ does not depend on x_0. Since the initial value of the cost variable x_0 has no effect, we choose $x_0(0) = 0$.

Including the accumulated cost x_0 together with the state $\mathbf{x}$, we form an **augmented state** vector $\mathbf{X}^T = [x_0 \quad \mathbf{x}^T]$, governed by the differential equations

$$\dot{\mathbf{X}} = \mathbf{F}(\mathbf{X}, \mathbf{u}, \mathbf{v}), \tag{9.4-8}$$

where

$$\mathbf{F}(\mathbf{X}, \mathbf{u}, \mathbf{v}) \triangleq \begin{bmatrix} f_0(\mathbf{x}, \mathbf{u}, \mathbf{v}) \\ \mathbf{f}(\mathbf{x}, \mathbf{u}, \mathbf{v}) \end{bmatrix}. \tag{9.4-9}$$

Since we have introduced an additional controller in differential games, it is necessary to extend our definition of admissible controls. The controls $\mathbf{u}(\mathbf{x})$ and $\mathbf{v}(\mathbf{x})$ are **admissible** if they (a) are piecewise continuous and piecewise differentiable, (b) yield a solution $\mathbf{x}(t)$ to (9.4-2) from any initial state-space point of interest, and (c) satisfy $\mathbf{u} \in \mathcal{U}$ and $\mathbf{v} \in \mathcal{V}$ for any point along a trajectory generated by $\mathbf{u}(\mathbf{x})$ and $\mathbf{v}(\mathbf{x})$. We need to further expand this definition in order to make it applicable to the augmented state space and to satisfy the termination requirement (9.4-4). In particular, an admissible control $\mathbf{u}(\mathbf{x})$ is said to be **playable** at the point $\mathbf{x}(0) = \mathbf{x}_i$ if $\mathbf{x}(t_f) \in \mathcal{X}$ for every admissible control $\mathbf{v}(\mathbf{x})$. Thus a playable control $\mathbf{u}(\mathbf{x})$ generates a solution $\mathbf{x}(t)$ to (9.4-2) for any admissible control $\mathbf{v}(\mathbf{x})$ and guarantees that, from every point of the playable set to $\mathcal{X}$, the resulting trajectory reaches the target regardless of $\mathbf{v}(\mathbf{x})$. The combination of a playable control $\mathbf{u}(\mathbf{x})$ and an admissible control $\mathbf{v}(\mathbf{x})$ will be called a **playable pair**.

We seek a playable control $\mathbf{u}^*(\mathbf{x})$ that minimizes J and an admissible control $\mathbf{v}^*(\mathbf{x})$ that maximizes J, that is, that yields a min-max for $x_0(t_f)$. Such a control pair will be called **min-max controls** and a corresponding trajectory $\mathbf{x}(t)$ will be called a min-max trajectory. Since min-max controls must be sought among the class of playable controls for player 1, the trajectory $\mathbf{x}(t)$ for a differential game takes place entirely within the playable set to the target for the $\mathbf{u}$ player. Outside the set, as discussed in Section 9.2, player 1 cannot guarantee that the state reaches the target, since some control $\mathbf{v}(\mathbf{x})$ exists that can keep the state from reaching the target.

The total cost is a function of the initial state $\mathbf{x}(0)$ and the control functions $\mathbf{u}(\mathbf{x})$ and $\mathbf{v}(\mathbf{x})$, and it can be designated by the **cost functional** $V(\cdot)$ defined by

$$V[\mathbf{x}(0), \mathbf{u}(\mathbf{x}), \mathbf{v}(\mathbf{x})] \triangleq \int_0^{t_f} f_0[\mathbf{x}, \mathbf{u}(\mathbf{x}), \mathbf{v}(\mathbf{x})]dt. \tag{9.4-10}$$

Our problem is to find a playable pair of controls $\mathbf{u}(\mathbf{x})$ and $\mathbf{v}(\mathbf{x})$ such that V takes on a min-max. A playable control pair $\mathbf{u}^*(\mathbf{x})$ and $\mathbf{v}^*(\mathbf{x})$ are **min-max controls** for V at $\mathbf{x}_i$ if and only if

$$V[\mathbf{x}(0), \mathbf{u}^*(\mathbf{x}), \mathbf{v}(\mathbf{x})] \le V[\mathbf{x}(0), \mathbf{u}^*(\mathbf{x}), \mathbf{v}^*(\mathbf{x})] \le V[\mathbf{x}(0), \mathbf{u}(\mathbf{x}), \mathbf{v}^*(\mathbf{x})] \tag{9.4-11}$$

for all playable controls $\mathbf{u}(\mathbf{x})$ and $\mathbf{v}(\mathbf{x})$. Along a min-max trajectory generated by $\mathbf{u}^*(\mathbf{x})$ and $\mathbf{v}^*(\mathbf{x})$ we may consider any point $\mathbf{x} = \mathbf{x}(t)$ as an initial point. Then V may be thought of as a function of $\mathbf{x}$ only, using the definition

$$V^*(\mathbf{x}) \triangleq \int_t^{t_f} f_0[\mathbf{x}, \mathbf{u}^*(\mathbf{x}), \mathbf{v}^*(\mathbf{x})]dt. \tag{9.4-12}$$

In this form $V^*(\mathbf{x})$ is known as the **value** of the game. It is the **min-max cost-to-go** from any point on a min-max trajectory to the terminal set $\mathcal{X}$.

In the augmented state space a surface Σ, defined by

$$x_0 + V^*(\mathbf{x}) = c, \tag{9.4-13}$$

is a min-max cost surface. It is composed of points, along augmented trajectories $\mathbf{X}(t)$ generated by $\mathbf{u}^*(\mathbf{x})$ and $\mathbf{v}^*(\mathbf{x})$, that all have the same min-max cost-to-go.

9.5 ISAACS' MIN-MAX PRINCIPLE

Suppose that a min-max trajectory $\mathbf{x}(t)$, generated by min-max controls $\mathbf{u}^*(\mathbf{x})$ and $\mathbf{v}^*(\mathbf{x})$, has been determined from some initial point $\mathbf{x}_i$, to the terminal set $\mathcal{X}$. Let $\mathcal{S}$ be a state-space curve through $\mathbf{x}_i$ containing initial conditions that all have the same min-max cost-to-go to $\mathcal{X}$. Consider the augmented state-space trajectories from each point of $\mathcal{S}$. These trajectories define a surface Σ, illustrated in Figure 9.5-1. Consider now some point on the augmented trajectory starting from $\mathbf{X}_i^{\mathrm{T}} = [0 \quad \mathbf{x}_i^{\mathrm{T}}]$. If a nonoptimal (but playable) control $\mathbf{u}(\mathbf{x})$ were to be applied at some point, clearly the augmented trajectory would have to depart from the surface Σ at this point, in the direction of increased cost for player 1. Similarly, if a nonoptimal control $\mathbf{v}(\mathbf{x})$ were applied, the augmented trajectory would move below the surface Σ, corresponding to a reduced cost for player 1. In particular, the surface Σ is a **semipermeable surface**. Against $\mathbf{v}^*(\mathbf{x})$ no trajectory $\mathbf{X}(t)$ in augmented state-cost space generated by $\mathbf{u}(\mathbf{x})$ can penetrate Σ from above (higher cost) to below (lower cost). Against $\mathbf{u}^*(\mathbf{x})$ no trajectory $\mathbf{X}(t)$ in augmented state-cost space generated by $\mathbf{v}(\mathbf{x})$ can penetrate Σ from below Σ (lower cost) to above (higher cost).

This property of a min-max Σ surface can be stated more precisely. Let the vector $\mathbf{N} = [n_0 \quad n_1 \quad \cdots \quad n_{n_x}]^{\mathrm{T}}$ be an **upward normal** to Σ at any point along

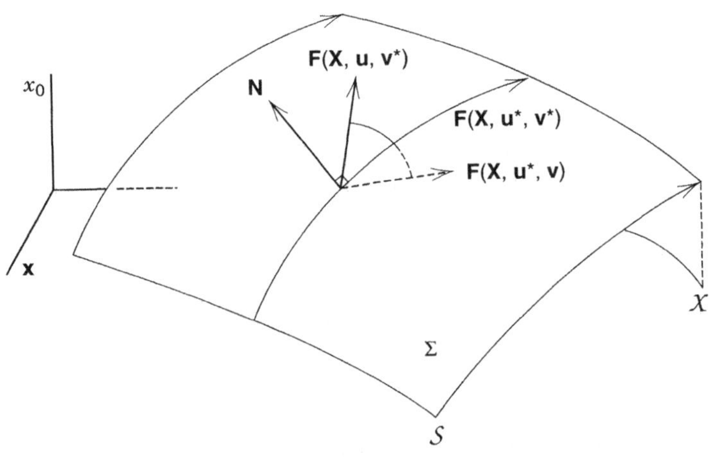

Figure 9.5-1 A min-max cost surface in augmented state space.

an optimal trajectory, directed toward increasing cost, as illustrated in Figure 9.5-1. It follows that $n_0 \geq 0$. As with optimal control problems, $n_0 = 0$ could correspond to a situation where a portion of the surface Σ is perpendicular to the state space. However, more commonly, the surface Σ has an edge and an optimal trajectory that runs along the edge of the surface Σ can do so with either $n_0 > 0$ (**N** normal to Σ) or $n_0 = 0$ (**N** normal to the edge and in the plane of Σ). This is a special case corresponding to the boundary of the playable set.

As a semipermeable surface, the surface Σ has the property that, for any $\mathbf{X} \in \Sigma$,

$$\mathbf{N}^{\mathrm{T}}\mathbf{F}(\mathbf{X}, \mathbf{u}^*, \mathbf{v}^*) = 0$$

and

$$\mathbf{N}^{\mathrm{T}}\mathbf{F}(\mathbf{X}, \mathbf{u}^*, \mathbf{v}) \leq \mathbf{N}^{\mathrm{T}}\mathbf{F}(\mathbf{X}, \mathbf{u}^*, \mathbf{v}^*) \leq \mathbf{N}^{\mathrm{T}}\mathbf{F}(\mathbf{X}, \mathbf{u}, \mathbf{v}^*) \tag{9.5-1}$$

for all $\mathbf{u} \in \mathcal{U}$ and $\mathbf{v} \in \mathcal{V}$, where $\mathbf{u}^* = \mathbf{u}^*(\mathbf{x})$ and $\mathbf{v}^* = \mathbf{v}^*(\mathbf{x})$ at $\mathbf{x}$. This property is a min-max game-theoretic saddle-point condition. At every point of an optimal trajectory the dot product between the upward normal vector **N** to the Σ surface and the augmented velocity vector **F** must take on a min-max with respect to the controls $\mathbf{u} \in \mathcal{U}$ and $\mathbf{v} \in \mathcal{V}$, with zero being the min-max value for the dot product. This property follows from the fact that trajectories starting on Σ and generated by $\mathbf{u}(\mathbf{x})$ against $\mathbf{v}^*(\mathbf{x})$ cannot go below Σ, since this would correspond to a total cost less than the min-max cost. Similarly, trajectories starting on Σ and generated by $\mathbf{v}(\mathbf{x})$ against $\mathbf{u}^*(\mathbf{x})$ cannot go above Σ, since this would correspond to a total cost greater than player 1's "worst-case" min-max cost. The fact that $\mathbf{N}^{\mathrm{T}}\mathbf{F}(\mathbf{X}, \mathbf{u}^*, \mathbf{v}^*)$ is zero follows from the requirement that min-max trajectories that start on Σ stay on it.

Consider an **augmented adjoint vector** $\mathbf{\Lambda}^{\mathrm{T}} = [\lambda_0 \quad \boldsymbol{\lambda}^{\mathrm{T}}] = [\lambda_0 \quad \lambda_1 \quad \cdots \quad \lambda_{n_x}]$, where the λ_0 component of $\mathbf{\Lambda}$ corresponds to the x_0 component of the augmented state vector **X** and the **adjoint vector** $\boldsymbol{\lambda}$ corresponds to the state vector **x** for the dynamic system (9.4-2). From the results in Chapter 6, if $\mathbf{\Lambda}(0)$ is an upward normal to Σ at $\mathbf{X}(0)$ and if $\mathbf{\Lambda}(t)$ is defined as the solution to the augmented adjoint differential equations

$$\dot{\mathbf{\Lambda}}^{\mathrm{T}} = -\mathbf{\Lambda}^{\mathrm{T}}\left[\frac{\partial \mathbf{F}}{\partial \mathbf{X}} + \frac{\partial \mathbf{F}}{\partial \mathbf{u}}\frac{\partial \mathbf{u}}{\partial \mathbf{X}} + \frac{\partial \mathbf{F}}{\partial \mathbf{v}}\frac{\partial \mathbf{v}}{\partial \mathbf{X}}\right], \tag{9.5-2}$$

then $\mathbf{\Lambda}(t)$ will be an upward normal to Σ at $\mathbf{X}(t)$. The initial condition for $\mathbf{\Lambda}$ must be chosen such that $\mathbf{\Lambda}(0)$ is in the direction of $\mathbf{N}(0)$. Thus $\lambda_0 \geq 0$.

Since $\mathbf{F} = \mathbf{F}[\mathbf{x}, \mathbf{u}(\mathbf{x}), \mathbf{v}(\mathbf{x})]$ is not a function of the x_0 component of **X**, equations (9.5-2) in partitioned form can be written equivalently as the **adjoint equations**

$$\dot{\lambda}_0 = 0 \Rightarrow \lambda_0 = \text{constant} \geq 0 \tag{9.5-3}$$

$$\dot{\boldsymbol{\Lambda}}^{\mathrm{T}} = -\boldsymbol{\Lambda}^{\mathrm{T}}\left[\frac{\partial \mathbf{F}}{\partial \mathbf{x}} + \frac{\partial \mathbf{F}}{\partial \mathbf{u}}\frac{\partial \mathbf{u}}{\partial \mathbf{x}} + \frac{\partial \mathbf{F}}{\partial \mathbf{v}}\frac{\partial \mathbf{v}}{\partial \mathbf{x}}\right]. \tag{9.5-4}$$

Furthermore, note that (9.5-2) is homogeneous (of degree one) in $\boldsymbol{\Lambda}$. In other words, if $\boldsymbol{\Lambda}(t)$ is a solution, then so is $k\boldsymbol{\Lambda}(t)$ for $k \geq 0$. Thus we can scale $\boldsymbol{\Lambda}(t)$ in any convenient fashion. There are two cases, depending on whether $\lambda_0 = 0$ or $\lambda_0 > 0$. The "**abnormal control**" case, where $\lambda_0 = 0$, corresponds to the playability min-max principle. For $\lambda_0 > 0$ we scale $\boldsymbol{\Lambda}(t)$ by taking $\lambda_0 = 1$.

As with optimal control problems, there also exists a relationship between the vector $\boldsymbol{\Lambda}$ and the value function defined by (9.4-12). If the value function were known, then the Σ surface could be defined by

$$V^*(\mathbf{x}) + x_0 = c, \tag{9.5-5}$$

where c is the min-max cost associated with driving the system from $\mathcal{S}$ to the terminal set $\mathcal{X}$. Since $\boldsymbol{\Lambda}$ is in the direction of the upward normal to the Σ surface, it follows that we could scale $\boldsymbol{\Lambda}$ so that

$$\boldsymbol{\Lambda} = k \,\text{grad}[x_0 + V^*(\mathbf{x})],$$

where $k > 0$ is any suitable constant. If $\lambda_0 > 0$ and we take $k = 1$, we obtain

$$[\lambda_0 \quad \lambda_1 \quad \cdots \quad \lambda_{n_x}] = \left[1 \quad \frac{\partial V^*}{\partial x_1} \quad \cdots \quad \frac{\partial V^*}{\partial x_{n_x}}\right].$$

From this result we conclude that

$$\lambda_0 = 1 \tag{9.5-6}$$

and

$$\boldsymbol{\lambda}^{\mathrm{T}} = \frac{\partial V^*(\mathbf{x})}{\partial \mathbf{x}}. \tag{9.5-7}$$

In other words, for the normal case where $\lambda_0 = 1$, the adjoint vector $\boldsymbol{\lambda} = \text{grad}[V^*(\mathbf{x})]$ is the gradient of the differential game's value function (provided that the gradient exists). This observation, along with some assumptions about the terminal set, allows us to obtain conditions for the final value of the adjoint vector.

Transversality Conditions

Assume that the terminal set is given by a set of equality conditions of the form

$$\mathcal{X} = \{\mathbf{x} \in R^{nx} \,|\, \mathbf{g}(\mathbf{x}) = \mathbf{0}\}, \tag{9.5-8}$$

where $\mathbf{g} = [g_1 \cdots g_{n_g}]^{\mathrm{T}}$ is a continuous n_g-dimensional vector function of the state vector $\mathbf{x}$ and has continuous partial derivatives with respect to $\mathbf{x}$. If we further assume that the gradient vectors $\partial g_i/\partial \mathbf{x}$, $i = 1, \ldots, n_g$, are linearly independent at every point of $\mathcal{X}$, then the tangent vectors to $\mathcal{X}$ at any point of $\mathcal{X}$ are exactly those vectors $\mathbf{e}$ that satisfy [Vincent and Grantham, 1981]

$$\frac{\partial \mathbf{g}(\mathbf{x})}{\partial \mathbf{x}} \mathbf{e} = \mathbf{0}. \tag{9.5-9}$$

Thus the **tangent set** to $\mathcal{X}$ is defined by

$$\mathcal{T} = \left\{ \mathbf{e} \in R^{nx} \,\middle|\, \frac{\partial \mathbf{g}(\mathbf{x})}{\partial \mathbf{x}} \mathbf{e} = \mathbf{0} \right\}. \tag{9.5-10}$$

Since the min-max cost-to-go is zero on the terminal set, the terminal set $\mathcal{X}$ is a surface of constant value for $V^*(\mathbf{x})$. Thus the gradient of the min-max cost-to-go on the terminal set must be normal to the tangent set $\mathcal{T}$ to $\mathcal{X}$, which in turn implies that the adjoint vector is also normal to the tangent set $\mathcal{T}$; that is,

$$\frac{\partial V^*}{\partial \mathbf{x}} \mathbf{e} = 0 \Rightarrow \boldsymbol{\lambda}^{\mathrm{T}} \mathbf{e} = 0 \tag{9.5-11}$$

for all $\mathbf{e} \in \mathcal{T}$. The implication of (9.5-9) and (9.5-11) is that the vector $\boldsymbol{\lambda}$ is a linear combination of the row vectors in the matrix $\partial \mathbf{g}/\partial \mathbf{x}$ [Vincent and Grantham, 1981]. Thus at the terminal time t_f, with $\mathbf{x}_f \triangleq \mathbf{x}(t_f)$ and $\boldsymbol{\lambda}_f \triangleq \boldsymbol{\lambda}(t_f)$, we have

$$\boldsymbol{\lambda}_f^{\mathrm{T}} = \boldsymbol{\rho}^{\mathrm{T}} \frac{\partial \mathbf{g}(\mathbf{x}_f)}{\partial \mathbf{x}} = \sum_{i=1}^{n_g} \rho_i \frac{\partial g_i(\mathbf{x}_f)}{\partial \mathbf{x}}, \tag{9.5-12}$$

where $\boldsymbol{\rho} = [\rho_1 \cdots \rho_{n_g}]^{\mathrm{T}}$ is an n_g-dimensional vector. This condition is the same **transversality condition** as in Chapter 7. The information supplied by the transversality condition provides just enough additional information to solve the quantitative differential game.

Information on $\boldsymbol{\lambda}$ is determined from (9.5-12) at the terminal time, whereas information on $\mathbf{x}$ is generally known at the initial time. Thus solving the adjoint equations together with the system equations results in solving a **two-point boundary value problem**.

The Differential Game *H* Function

As before, the dot product between the adjoint vector $\boldsymbol{\lambda}$ and the state-space velocity vector is designated by $\mathcal{H}(\mathbf{x},\mathbf{u},\mathbf{v},\boldsymbol{\lambda}) = \boldsymbol{\lambda}^{\mathrm{T}}\mathbf{f}(\mathbf{x},\mathbf{u},\mathbf{v})$. It will, however, be convenient to define the **min-max *H* function** as

$$\begin{aligned} H(\mathbf{x},\mathbf{u},\mathbf{v},\boldsymbol{\lambda},\lambda_0) &\triangleq \lambda_0 f_0(\mathbf{x},\mathbf{u},\mathbf{v}) + \boldsymbol{\lambda}^{\mathrm{T}}\mathbf{f}(\mathbf{x},\mathbf{u},\mathbf{v}) \\ &= \lambda_0 f_0(\mathbf{x},\mathbf{u},\mathbf{v}) + \mathcal{H}(\mathbf{x},\mathbf{u},\mathbf{v},\boldsymbol{\lambda}) \\ &= \boldsymbol{\Lambda}^{\mathrm{T}}\mathbf{F}(\mathbf{x},\mathbf{u},\mathbf{v}). \end{aligned} \tag{9.5-13}$$

With this definition, the observations made in the previous sections can now be restated in a somewhat simpler form. In particular, we will use (9.5-4) along with the fact that $\boldsymbol{\Lambda}$ will be a vector normal to Σ to restate (9.5-1). If $\mathbf{u}^*(\mathbf{x})$ and $\mathbf{v}^*(\mathbf{x})$ is a min-max control pair for the integral-cost min-max differential game problem defined in Section 9.4, then there must exist a nonzero continuous vector function $\boldsymbol{\Lambda}^{\mathrm{T}}(t) = [\lambda_0 \quad \boldsymbol{\lambda}^{\mathrm{T}}(t)]$, with constant $\lambda_0 \geq 0$, and a constant multiplier vector $\boldsymbol{\rho} = [\rho_1 \ \cdots \ \rho_{n_g}]^{\mathrm{T}}$, where from (9.5-4) $\boldsymbol{\lambda} = [\lambda_1 \ \cdots \ \lambda_{n_x}]^{\mathrm{T}}$ is a solution to

$$\dot{\boldsymbol{\lambda}}^{\mathrm{T}} = -\frac{\partial H}{\partial \mathbf{x}} - \frac{\partial H}{\partial \mathbf{u}}\frac{\partial \mathbf{u}}{\partial \mathbf{x}} - \frac{\partial H}{\partial \mathbf{v}}\frac{\partial \mathbf{v}}{\partial \mathbf{x}} \tag{9.5-14}$$

satisfying

$$\boldsymbol{\lambda}^{\mathrm{T}}(t_f) = \boldsymbol{\rho}^{\mathrm{T}}\frac{\partial \mathbf{g}[\mathbf{x}(t_f)]}{\partial \mathbf{x}}, \tag{9.5-15}$$

such that

$$\min_{\mathbf{u}\in\mathcal{U}}\max_{\mathbf{v}\in\mathcal{V}} H(\mathbf{x},\mathbf{u},\mathbf{v},\boldsymbol{\lambda},\lambda_0) = H(\mathbf{x},\mathbf{u}^*,\mathbf{v}^*,\boldsymbol{\lambda},\lambda_0) = 0 \tag{9.5-16}$$

for every point $\mathbf{x}$ of a min-max trajectory generated by $\mathbf{u}^* = \mathbf{u}^*(\mathbf{x})$ and $\mathbf{v}^* = \mathbf{v}^*(\mathbf{x})$.

Min-Max Conditions for the *H* Function

For a given value of $\boldsymbol{\lambda}$ at some point $\mathbf{x}$ along an optimal trajectory, the H function is a function of $\mathbf{u}$ and $\mathbf{v}$ only. This function is to be minimized with respect to $\mathbf{u} \in \mathcal{U}$ and maximized with respect to $\mathbf{v} \in \mathcal{V}$. Necessary conditions for solving this type of problem are available from **parametric min-max game theory**, previously developed in Chapter 3. The only requirement is that the sets $\mathcal{U}$ and $\mathcal{V}$ must be defined in terms of algebraic relations. In general $\mathcal{U}$ and $\mathcal{V}$ may be either fixed sets $\mathcal{U} \subseteq R^{n_u}$ and $\mathcal{V} \subseteq R^{n_v}$, or sets $\mathcal{U}(\mathbf{x})$ and $\mathcal{V}(\mathbf{x})$ that are functions of $\mathbf{x}$. If $\mathcal{U}$ and $\mathcal{V}$ are sets independent of $\mathbf{x}$, then as we show shortly the second and third terms on the right-hand side of (9.5-14) are identically zero. When this is not the case, use of the results obtained so far will be difficult to apply. This is because the feedback control laws, required to satisfy (9.5-14)–(9.5-16), could not be evaluated from these equations unless both control laws were known.

Since we wish to use (9.5-14)–(9.5-16) as a set of constructive necessary conditions for determining $\mathbf{u}^*(\mathbf{x})$ and $\mathbf{v}^*(\mathbf{x})$, we assume that $\mathcal{U}$ and $\mathcal{V}$ are defined in terms of **state-independent control inequality constraints** of the form

$$\begin{aligned} \mathcal{U} &= \{\mathbf{u} \in R^{n_u} \mid \mathbf{h}(\mathbf{u}) \geq \mathbf{0}\} \\ \mathcal{V} &= \{\mathbf{v} \in R^{n_v} \mid \bar{\mathbf{h}}(\mathbf{v}) \geq \mathbf{0}\}, \end{aligned} \tag{9.5-17}$$

where $\mathbf{h} = [h_1 \ \cdots \ h_{n_h}]^{\mathrm{T}}$ and $\bar{\mathbf{h}}_1 \ \cdots \ h_{n_{\bar{h}}}]^{\mathrm{T}}$ are assumed to be continuous and continuously differentiable in their respective arguments. The functions are further restricted to that class that will produce **regular** sets for $\mathcal{U}$ and $\mathcal{V}$. In other words, the tangent vectors to $\mathcal{U}$ and $\mathcal{V}$ are exactly those vectors $\mathbf{e}_u$ and $\mathbf{e}_v$ that satisfy the conditions

$$\begin{aligned} \frac{\partial h_i}{\partial \mathbf{u}} \mathbf{e}_u &\geq 0 \\ \frac{\partial \bar{h}_j}{\partial \mathbf{v}} \mathbf{e}_v &\geq 0 \end{aligned} \tag{9.5-18}$$

for all $\mathbf{h}$ components $h_i(\mathbf{u})$ such that $h_i(\mathbf{u}) = 0$ and for all $\bar{\mathbf{h}}$ components $\bar{h}_j(\mathbf{v})$ such that $\bar{h}_j(\mathbf{v}) = 0$. Having imposed the regularity conditions on $\mathcal{U}$ and $\mathcal{V}$, we now have the following necessary conditions for a min-max of $H(\mathbf{x}, \mathbf{u}, \mathbf{v}, \boldsymbol{\lambda}, \lambda_0)$ with respect to $\mathbf{u}$ and $\mathbf{v}$ (see Section 3.4):

Min-Max Necessary Conditions: *Let* $\mathbf{u}^*$ *be a regular point of* $\mathcal{U}$ *and* $\mathbf{v}^*$ *be a regular point of* $\mathcal{V}$. *If* H *takes on a min-max with respect to* $\mathbf{u}$ *at* $\mathbf{u}^*$ *and* $\mathbf{v}$ *at* $\mathbf{v}^*$, *then Lagrange multipliers* $\boldsymbol{\gamma} = [\gamma_1 \ \cdots \ \gamma_{n_h}]^{\mathrm{T}}$ *and* $\bar{\boldsymbol{\gamma}} = [\bar{\gamma}_1 \ \cdots \ \bar{\gamma}_{n_{\bar{h}}}]^{\mathrm{T}}$ *exist such that*

$$\frac{\partial L}{\partial \mathbf{u}} = \mathbf{0}^{\mathrm{T}} \tag{9.5-19}$$

$$\frac{\partial L}{\partial \mathbf{v}} = \mathbf{0}^{\mathrm{T}} \tag{9.5-20}$$

$$\mathbf{h}(\mathbf{u}^*) \geq \mathbf{0} \tag{9.5-21}$$

$$\bar{\mathbf{h}}(\mathbf{v}^*) \geq \mathbf{0} \tag{9.5-22}$$

$$\boldsymbol{\gamma}^{\mathrm{T}}\mathbf{h}(\mathbf{u}^*) = 0 \tag{9.5-23}$$

$$\bar{\boldsymbol{\gamma}}^{\mathrm{T}}\bar{\mathbf{h}}(\mathbf{v}^*) = 0 \tag{9.5-24}$$

$$\boldsymbol{\gamma} \geq \mathbf{0} \tag{9.5-25}$$

$$\bar{\boldsymbol{\gamma}} \leq \mathbf{0}, \tag{9.5-26}$$

where

$$L(\mathbf{x}, \mathbf{u}, \mathbf{v}, \boldsymbol{\lambda}, \lambda_0, \boldsymbol{\gamma}) \triangleq H(\mathbf{x}, \mathbf{u}, \mathbf{v}, \boldsymbol{\lambda}, \lambda_0) - \boldsymbol{\gamma}^{\mathrm{T}}\mathbf{h}(\mathbf{u}) - \bar{\boldsymbol{\gamma}}^{\mathrm{T}}\bar{\mathbf{h}}(\mathbf{v}^*) \tag{9.5-27}$$

and the partial derivatives are evaluated at $\mathbf{u} = \mathbf{u}^*$ *and* $\mathbf{v} = \mathbf{v}^*$.

The Min-Max Principle

For state-independent control constraint sets of the form (9.5-17), the necessary conditions developed in the previous section for a min-max of the H function also allow us to greatly simplify the adjoint equations (9.5-14).

It follows from (9.5-19) that

$$\frac{\partial H}{\partial \mathbf{u}} = \boldsymbol{\gamma}^{\mathrm{T}} \frac{\partial \mathbf{h}}{\partial \mathbf{u}}. \tag{9.5-28}$$

Thus the second term on the right-hand side of (9.5-14) may be written as

$$\frac{\partial H}{\partial \mathbf{u}} \frac{\partial \mathbf{u}}{\partial \mathbf{x}} = \boldsymbol{\gamma}^{\mathrm{T}} \frac{\partial \mathbf{h}}{\partial \mathbf{u}} \frac{\partial \mathbf{u}}{\partial \mathbf{x}} = \boldsymbol{\gamma}^{\mathrm{T}} \frac{d\mathbf{h}}{d\mathbf{x}}. \tag{9.5-29}$$

From (9.5-17), (9.5-23), and (9.5-25), $\boldsymbol{\gamma}^{\mathrm{T}}\mathbf{h}(\mathbf{u})$ has a stationary minimum at $\mathbf{u} = \mathbf{u}^*(\mathbf{x})$. Thus it follows that the left-hand side of (9.5-29) is zero.

Similarly, since $\bar{\boldsymbol{\gamma}}^{\mathrm{T}}\bar{\mathbf{h}}(\mathbf{v})$ has a stationary maximum at $\mathbf{v} = \mathbf{v}^*(\mathbf{x})$, we conclude that

$$\frac{\partial H}{\partial \mathbf{v}} \frac{\partial \mathbf{v}}{\partial \mathbf{x}} = \bar{\boldsymbol{\gamma}}^{\mathrm{T}} \frac{\partial \bar{\mathbf{h}}}{\partial \mathbf{v}} \frac{\partial \mathbf{v}}{\partial \mathbf{x}} = \bar{\boldsymbol{\gamma}}^{\mathrm{T}} \frac{d\bar{\mathbf{h}}}{d\mathbf{x}} = \mathbf{0}^{\mathrm{T}}. \tag{9.5-30}$$

The results obtained so far can now be summarized in terms of a differential game min-max principle [Isaacs, 1965]:

Min-Max Principle: *Given the constraint sets (9.5-17), if* $\mathbf{u}^*(\mathbf{x}) \in \mathcal{U}$ *and* $\mathbf{v}^*(\mathbf{x}) \in \mathcal{V}$ *are min-max controls for the differential game (9.4-1)–(9.4-4), then there must exist a continuous and piecewise differentiable vector function* $\boldsymbol{\lambda}(t) = [\lambda_1 \ \cdots \ \lambda_{n_x}]^{\mathrm{T}}$, *a constant* $\lambda_0 \geq 0$ [=0 *or* 1], *with* $(\lambda_0, \boldsymbol{\lambda}) \neq \mathbf{0}$, *and a constant multiplier vector* $\boldsymbol{\rho} = [\rho_1 \ \cdots \ \rho_{n_g}]^{\mathrm{T}}$ *such that* $\boldsymbol{\lambda}(t)$ *satisfies the adjoint equations*

$$\dot{\boldsymbol{\lambda}}^{\mathrm{T}} = -\frac{\partial H}{\partial \mathbf{x}}, \tag{9.5-31}$$

where, at the terminal set

$$\mathbf{g}[\mathbf{x}(t_f)] = \mathbf{0}, \tag{9.5-32}$$

$\boldsymbol{\lambda}(t_f)$ *satisfies the transversality conditions*

$$\boldsymbol{\lambda}^{\mathrm{T}}(t_f) = \boldsymbol{\rho}^{\mathrm{T}} \frac{\partial \mathbf{g}[\mathbf{x}(t_f)]}{\partial \mathbf{x}}, \tag{9.5-33}$$

and such that H takes on a global min-max value with respect to $\mathbf{u} \in \mathcal{U}$ *and* $\mathbf{v} \in \mathcal{V}$ *at every point* $\mathbf{x}$ *along the trajectory* $\mathbf{x}(t)$ *generated by* $\mathbf{u}^*(\mathbf{x})$ *and* $\mathbf{v}^*(\mathbf{x})$. *Furthermore, the min-max value of H at every such point is zero:*

$$0 = H(\mathbf{x}(t), \mathbf{u}^*[\mathbf{x}(t)], \mathbf{v}^*[\mathbf{x}(t)], \boldsymbol{\lambda}(t), \lambda_0) = \min_{\mathbf{u} \in \mathcal{U}} \max_{\mathbf{v} \in \mathcal{V}} H(\mathbf{x}(t), \mathbf{u}, \mathbf{v}, \boldsymbol{\lambda}(t), \lambda_0), \tag{9.5-34}$$

where

$$H(\mathbf{x}, \mathbf{u}, \mathbf{v}, \boldsymbol{\lambda}, \lambda_0) \triangleq \lambda_0 f_0(\mathbf{x}, \mathbf{u}, \mathbf{v}) + \boldsymbol{\lambda}^{\mathrm{T}} \mathbf{f}(\mathbf{x}, \mathbf{u}, \mathbf{v}) \tag{9.5-35}$$

and the partial derivatives in (9.5-31) are evaluated at $\mathbf{u} = \mathbf{u}^*$ *and* $\mathbf{v} = \mathbf{v}^*$.

Example 9.5-1: Homicidal Chauffeur Consider again the homicidal chauffeur problem introduced in Section 9.1. We now make it a qualitative game by introducing the time of collision as a cost for the pursuer:

$$J = \int_0^{t_f} dt.$$

In terms of the rotating coordinate system used in Figure 9.5-2, the motion of the pursuer with respect to the evader is given by

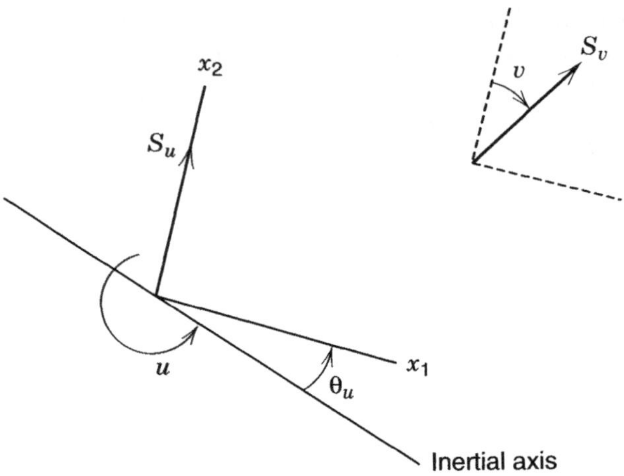

Figure 9.5-2 The homicidal chauffeur.

$$\begin{aligned} \dot{x}_1 &= S_v \sin v + u x_2 \\ \dot{x}_2 &= S_v \cos v - S_u - u x_1, \end{aligned} \qquad (9.5\text{-}36)$$

where $u = \dot{\theta}_u$ is the turning rate of the pursuer (and the coordinate system) in the counterclockwise direction, with

$$|u| \le u_{\max}, \qquad (9.5\text{-}37)$$

and v is the *relative* heading angle of the evader with respect to the pursuer, measured clockwise from the positive x_2-axis. If $v = 0$, then the pursuer and the evader are heading in the same direction. Also, $u = u_{\max}$ corresponds to a hard left turn by the pursuer, while $u = -u_{\max}$ corresponds to a hard right turn.

Except for a constant term $\lambda_0 = 1$, the H function

$$H = \lambda_0 + \lambda_1(S_v \sin v + u x_2) + \lambda_2(S_v \cos v - S_u - u x_1) = \lambda_0 + \mathcal{H}$$

is the same as the $\mathcal{H}$ function used for generating the playable set boundary. The adjoint equations are the same as before, given by

$$\begin{aligned} \dot{\lambda}_1 &= u\lambda_2 \\ \dot{\lambda}_2 &= -u\lambda_1, \end{aligned}$$

with the general terminal transversality condition that

$$\boldsymbol{\lambda}_f^T \mathbf{e} \le 0 \tag{9.5-38}$$

for all **e** tangent to the car (to allow for targets more generally shaped than via (9.5-32), as in Exercises 9-4 and 9-5). In addition, we also have the requirement that

$$H_f = 0.$$

Minimizing H with respect to u yields

$$u = \begin{cases} -u_{\max} & \text{if } \sigma > 0 \\ \text{singular} & \text{if } \sigma \equiv 0 \\ u_{\max} & \text{if } \sigma < 0, \end{cases}$$

where the switching function is

$$\sigma = \frac{\partial H}{\partial u} = \lambda_1 x_2 - \lambda_2 x_1.$$

The possibility of singular control remains to be examined.

Maximizing H with respect to v requires

$$0 = \frac{\partial H}{\partial v} = \lambda_1 S_v \cos v - \lambda_2 S_v \sin v$$

and

$$0 \ge \frac{\partial^2 H}{\partial v^2} = -\lambda_1 S_v \sin v - \lambda_2 S_v \cos v.$$

From these conditions (with $S_v > 0$) we get

$$\sin v^* = \frac{\lambda_1}{\sqrt{\lambda_1^2 + \lambda_2^2}}, \qquad \cos v^* = \frac{\lambda_2}{\sqrt{\lambda_1^2 + \lambda_2^2}}. \tag{9.5-39}$$

Then the H function becomes

$$0 = H = \lambda_0 + \sigma u + \sqrt{\lambda_1^2 + \lambda_2^2} - S_u \lambda_2. \tag{9.5-40}$$

Singular u control requires

$$0 = \sigma = \lambda_1 x_2 - \lambda_2 x_1 \tag{9.5-41}$$

and

$$0 = \dot{\sigma} = \lambda_1 (S_v \cos v^* - S_u) - \lambda_2 S_v \sin v^*,$$

which, from (9.5-39), reduces to

$$0 = \dot{\sigma} = -S_u \lambda_1. \tag{9.5-42}$$

Differentiating once more yields

$$0 = \ddot{\sigma} = -u S_u \lambda_2. \tag{9.5-43}$$

From (9.5-40)–(9.5-43), since $(\lambda_0, \lambda_1, \lambda_2) \neq \mathbf{0}$, we get the singular conditions

$$u \equiv 0, \qquad \lambda_1 \equiv 0, \qquad x_1 \equiv 0.$$

Having determined the nature of singular control, the optimal control for the pursuer is given by

$$u^* = \begin{cases} -u_{\max} & \text{if } \sigma > 0 \\ 0 & \text{if } \sigma \equiv 0 \\ u_{\max} & \text{if } \sigma < 0. \end{cases} \tag{9.5-44}$$

This control is used, along with v^* from (9.5-39), to integrate the state and adjoint equations backwards in time from points on the car, using the terminal transversality condition (9.5-38) along with $H_f = 0$.

In order to continue, the shape of the car must be specified. While we know the nature of the min-max control, the process of obtaining solutions for a particular car shape is quite detailed and difficult. The original form of the homicidal chauffeur problem [Isaacs, 1965] used a circular target. Results for this case are available [Breakwell and Merz, 1969]. Solutions for the case of a "flat car" (i.e., a line-segment target) are left as an exercise. ■

9.6 LINEAR SYSTEM-QUADRATIC COST

Consider a linear system of the form

$$\dot{\mathbf{x}} = \mathbf{A}\mathbf{x} + \mathbf{B}\mathbf{u} + \mathbf{E}\mathbf{v}, \tag{9.6-1}$$

where $\mathbf{x}$ is an n_x-dimensional state vector, $\mathbf{u}$ is an unbounded n_u-dimensional control input, $\mathbf{v}$ is an unbounded n_v-dimensional **uncertain input**, and $\mathbf{A}$, $\mathbf{B}$, and $\mathbf{E}$ are all constant matrices of appropriate dimension. We wish to design a **robust controller** $\mathbf{u}(\mathbf{x})$ whose objective is to stabilize the origin despite the uncertain input $\mathbf{v}$. Toward this end, consider modifying the cost function used for LQR design in Chapter 8 by introducing an additional term associated with the uncertain input:

$$x_0 = \frac{1}{2}\int_0^{t_f} (\mathbf{x}^{\mathrm{T}}\mathbf{Q}\mathbf{x} + \mathbf{u}^{\mathrm{T}}\mathbf{R}\mathbf{u} - \gamma^2\mathbf{v}^{\mathrm{T}}\mathbf{v})dt, \tag{9.6-2}$$

with $t_f \to \infty$. Here the symmetric matrices $\mathbf{Q} \geq 0$ and $\mathbf{R} > 0$ have the same requirements and play the same role in contributing to the cost function as in the LQR cost function, only now we have introduced another player who wants to maximize the cost function by driving $\mathbf{x}(t)$ away from zero. Note that each player is penalized in terms of the amount of control used (otherwise we could not avoid introducing control bounds). We examine this integral from the point of view that the $\mathbf{u}$-player wants to minimize the integral and the $\mathbf{v}$-player wants to maximize the integral. As we see shortly, the parameter $\gamma^2 > 0$ plays an important role in the solution. It cannot be made too small, for otherwise it may allow the (unbounded) uncertain input $\mathbf{v}$ to dominate the resulting feedback control input $\mathbf{u}(\mathbf{x})$ and make the system unstable.

Min-Max Control

Using the min-max principle, we first form the H function

$$H = \frac{\lambda_0}{2}(\mathbf{x}^{\mathrm{T}}\mathbf{Q}\mathbf{x} + \mathbf{u}^{\mathrm{T}}\mathbf{R}\mathbf{u} - \gamma^2\mathbf{v}^{\mathrm{T}}\mathbf{v}) + \boldsymbol{\lambda}^{\mathrm{T}}(\mathbf{A}\mathbf{x} + \mathbf{B}\mathbf{u} + \mathbf{E}\mathbf{v}).$$

Necessary conditions for minimizing H with respect to $\mathbf{u}$ and maximizing H with respect to $\mathbf{v}$ are given by

$$\frac{\partial H}{\partial \mathbf{u}} = \lambda_0\mathbf{u}^{\mathrm{T}}\mathbf{R} + \boldsymbol{\lambda}^{\mathrm{T}}\mathbf{B} = \mathbf{0}^{\mathrm{T}}$$

$$\frac{\partial H}{\partial \mathbf{v}} = -\lambda_0\gamma^2\mathbf{v}^{\mathrm{T}} + \boldsymbol{\lambda}^{\mathrm{T}}\mathbf{E} = \mathbf{0}^{\mathrm{T}}.$$

Setting $\lambda_0 = 1$, we obtain

$$\mathbf{u} = -\mathbf{R}^{-1}\mathbf{B}^{\mathrm{T}}\boldsymbol{\lambda}$$

$$\mathbf{v} = \frac{1}{\gamma^2}\mathbf{E}^{\mathrm{T}}\boldsymbol{\lambda}.$$

Substituting these results into the H function, and collecting similar terms, we obtain

$$0 = H = \frac{1}{2}\mathbf{x}^T\mathbf{Q}\mathbf{x} - \frac{1}{2}\boldsymbol{\lambda}^T\left(\mathbf{B}\mathbf{R}^{-1}\mathbf{B}^T - \frac{1}{\gamma^2}\mathbf{E}\mathbf{E}^T\right)\boldsymbol{\lambda} + \boldsymbol{\lambda}^T\mathbf{A}\mathbf{x}.$$

At any state position $\mathbf{x}$ we can think of the requirement $H = 0$ as an equation in $\boldsymbol{\lambda}$. The symmetry of $H = 0$ suggests assuming a solution of the form

$$\boldsymbol{\lambda} = \mathbf{S}\mathbf{x},$$

where $\mathbf{S}$ is a constant symmetric $n_x \times n_x$ matrix. Substituting this expression for $\boldsymbol{\lambda}$ into $H = 0$, we obtain

$$\mathbf{x}^T\left[\mathbf{Q} + 2\mathbf{S}\mathbf{A} - \mathbf{A}\left(\mathbf{B}\mathbf{R}^{-1}\mathbf{B}^T - \frac{1}{\gamma^2}\mathbf{E}\mathbf{E}^T\right)\mathbf{S}\right]\mathbf{x} = 0.$$

Since this relation must be true for all $\mathbf{x}$ along a trajectory, the fact that the quadratic form must be zero implies that the symmetric part of the matrix in brackets must be zero. Thus we have the **algebraic matrix Riccati equation**

$$\mathbf{Q} + \mathbf{S}\mathbf{A} + \mathbf{A}^T\mathbf{S} - \mathbf{S}\left(\mathbf{B}\mathbf{R}^{-1}\mathbf{B}^T - \frac{1}{\gamma^2}\mathbf{E}\mathbf{E}^T\right)\mathbf{S} = \mathbf{0}. \qquad (9.6\text{-}3)$$

As with the LQR design, we must first solve the (quadratic) Riccati equation (9.6-3) for a positive definite matrix $\mathbf{S}$ (see Section 7.6). The MATLAB command "**are**" is useful in this regard. However, if γ^2 is chosen less than some critical value $\bar{\gamma}^2$, either there will not exist a positive definite solution $\mathbf{S}$ to the Riccati equation, or the control law

$$\mathbf{u} = -\mathbf{K}\mathbf{x}, \qquad (9.6\text{-}4)$$

where

$$\mathbf{K} = \mathbf{R}^{-1}\mathbf{B}^T\mathbf{S}, \qquad (9.6\text{-}5)$$

will make the system unstable. What value for γ^2 should we use? Suppose we start with γ^2 very large. In this case, the min-max solution for $\mathbf{v}$ will be approximately $\mathbf{v}(\mathbf{x}) = \mathbf{0}$, so that the solution for $\mathbf{u}(\mathbf{x})$ will be nearly identical to the LQR design. In order to bring in the effect of the uncertain input, we must decrease the value for γ^2. Using a value of γ^2 only slightly larger than $\bar{\gamma}^2$ will allow for the largest uncertain input and yet still maintain system stability.

It is possible to obtain an expression that allows us to determine $\bar{\gamma}^2$. First substitute the control law (9.6-4) into the equation of motion (9.6-1) to obtain

$$\dot{\mathbf{x}} = \hat{\mathbf{A}}\mathbf{x} + \mathbf{E}\mathbf{v},$$

where $\hat{\mathbf{A}} = \mathbf{A} - \mathbf{BK}$ is asymptotically stable (and thus has an inverse). Designate by $\bar{\mathbf{x}}$ the equilibrium solution obtained under a constant control input $\mathbf{v}(t) \equiv \bar{\mathbf{v}}$,

$$\bar{\mathbf{x}} = -\hat{\mathbf{A}}^{-1}\mathbf{E}\bar{\mathbf{v}},$$

with the corresponding constant value for the control

$$\bar{\mathbf{u}} = \mathbf{K}\hat{\mathbf{A}}^{-1}\mathbf{E}\bar{\mathbf{v}}.$$

Then use these results to evaluate the resulting constant rate of change of cost, from (9.6-2), to yield

$$\dot{x}_0 = -\bar{\mathbf{v}}^{\mathrm{T}}\mathbf{M}\bar{\mathbf{v}},$$

where

$$\mathbf{M} \triangleq \gamma^2\mathbf{I} - \boldsymbol{\Gamma} \tag{9.6-6}$$

and

$$\boldsymbol{\Gamma} \triangleq \mathbf{E}^{\mathrm{T}}(\hat{\mathbf{A}}^{-1})^{\mathrm{T}}(\mathbf{Q} + \mathbf{K}^{\mathrm{T}}\mathbf{R}\mathbf{K})\hat{\mathbf{A}}^{-1}\mathbf{E} \tag{9.6-7}$$

are $n_v \times n_v$ and symmetric, and $\mathbf{I}$ is the $n_v \times n_v$ identity matrix. It follows that, for a min-max saddle-point solution to exist for the differential game, the $\mathbf{M}$ matrix must be at least positive semidefinite. Otherwise, some constant uncertain input $\mathbf{v}$ could always guarantee an infinite cost to the $\mathbf{u}$ player by making $\dot{x}_0 > 0$ for all $t \to \infty$.

For $\mathbf{M} \geq 0$ all of its eigenvalues β must be nonnegative, specifically its smallest eigenvalue β_{min}. Since, from (9.6-6), the eigenvalues β of $\mathbf{M}$ and the eigenvalues μ of $\boldsymbol{\Gamma}$ are related by

$$\mu = \gamma^2 - \beta,$$

the $\mathbf{M} \geq 0$ condition $\beta_{\text{min}} \geq 0$ is equivalent to

$$\mu_{\text{max}}(\boldsymbol{\Gamma}) \leq \gamma^2, \tag{9.6-8}$$

where $\mu_{\max}(\boldsymbol{\Gamma})$ is the **spectral radius** (largest eigenvalue) of $\boldsymbol{\Gamma}$. The critical value of γ^2 is obtained when $\mu_{\max}(\boldsymbol{\Gamma}) = \gamma^2$, which is a nonlinear equation in γ^2 since $\mathbf{K} = \mathbf{K}(\gamma^2)$, and thus $\boldsymbol{\Gamma}$, depend on γ^2.

For simplicity, consider the case of a scalar ($n_v = 1$) uncertain input, so that in (9.6-7) $\mathbf{E}$ is $n_x \times 1$ and $\boldsymbol{\Gamma}$ and $\mathbf{M}$ are scalars. Then from (9.6-6) and (9.6-7), the existence of a min-max solution, requiring $\mathbf{M} \geq 0$, yields the scalar relation

$$\gamma^2 \geq \mathbf{E}^{\mathrm{T}}(\hat{\mathbf{A}}^{-1})^{\mathrm{T}}(\mathbf{Q} + \mathbf{K}^{\mathrm{T}}\mathbf{R}\mathbf{K})\hat{\mathbf{A}}^{-1}\mathbf{E}. \tag{9.6-9}$$

This will be satisfied for large γ^2, and it follows that the critical value $\gamma^2 = \bar{\gamma}^2$ is obtained when

$$\gamma^2 = \mathbf{E}^{\mathrm{T}}(\hat{\mathbf{A}}^{-1})^{\mathrm{T}}(\mathbf{Q} + \mathbf{K}^{\mathrm{T}}\mathbf{R}\mathbf{K})\hat{\mathbf{A}}^{-1}\mathbf{E}. \tag{9.6-10}$$

Unfortunately, we cannot use (9.6-10) to evaluate γ^2 directly since a value for γ^2 must be specified before we can evaluate $\mathbf{K} = \mathbf{K}(\gamma^2)$. However, (9.6-10) can be used in an iterative fashion to determine $\bar{\gamma}^2$ by doing the following: select a value for γ^2, solve the Riccati equation (9.6-3) for a positive definite matrix $\mathbf{S}$, determine $\mathbf{K}$ from (9.6-5), and then evaluate the right-hand side of (9.6-10), yielding a value $\hat{\gamma}^2$. If the result of this evaluation agrees with the original selection ($\hat{\gamma}^2 = \gamma^2$), then $\hat{\gamma}^2 = \bar{\gamma}^2$. If not, γ^2 is either decreased in value (if $\gamma^2 > \hat{\gamma}^2$) or increased in value (if $\gamma^2 < \hat{\gamma}^2$ or if there does not exist a positive definite solution to the Riccati equation), and the process is repeated.

Example 9.6-1: Inverted Pendulum Consider again the inverted pendulum system introduced in Example 9.3-1. In this case the system matrices are given by

$$\mathbf{A} = \begin{bmatrix} 0 & 1 \\ 0 & 0 \end{bmatrix}, \qquad \mathbf{B} = \begin{bmatrix} 0 \\ 1 \end{bmatrix}, \qquad \mathbf{E} = \begin{bmatrix} 0 \\ 1 \end{bmatrix}.$$

If we choose

$$\mathbf{Q} = \begin{bmatrix} 1 & 0 \\ 0 & 1 \end{bmatrix}, \qquad R = 1,$$

it follows, by inspection, that there will be no real-valued solution to the Riccati equation when $\gamma^2 \leq 1$. Table 9.6-1 compares the controllers we obtain for some values of $\gamma^2 > 1$. Note that $\gamma^2 \to \infty$ corresponds to LQR design. As we decrease the value of γ^2, the control design, in terms of the eigenvalues for $\hat{\mathbf{A}}$, changes from an underdamped second-order system to a greatly overdamped second-order system.

Table 9.6-1 Controller Designs for Example 9.6-1

	Assumed γ^2	Gains K	Eigenvalues for $\hat{A}$	γ^2 from (9.6–10)
LQR	NA	1.0000, 1.7321	$-0.8660 \pm i0.5000$	NA
Min-max	100	1.0050, 1.7437	$-0.8718 \pm i0.4949$	1.9900
Min-max	5	1.1180, 2.0112	$-1.0056 \pm i0.3267$	1.8000
Min-max	2	1.4142, 2.7671	-0.6764, -2.0907	1.5000
Min-max	1.1	3.3166, 9.1633	-0.3775, -8.7858	1.0909
Min-max	1.05	4.5826, 14.6106	-0.3207, -14.2899	1.0476
Min-max	1.01	10.0499, 46.1636	-0.2187, -45.9448	1.0099

We see that as $\gamma^2 \to 1$, large gains are obtained, which means that for any practical application the control will saturate almost immediately, rendering the calculated eigenvalues meaningless. Nevertheless, we can choose a min-max controller close to the "optimal" design, for comparison with the LQR design. We make this comparison based on the v-reachable set, assuming saturating LQR and min-max controls with $\gamma^2 = 1.1$. As in Example 9.3-1 assume that $|u| \leq 2$ and $|v| \leq 1$. These sets are illustrated in Figure 9.6-1. The outer curve is the boundary of the v-reachable set under saturating LQR control and the inner curve is the boundary of the v-reachable set under saturating min-max control. In both cases $v(\mathbf{x})$ is determined using

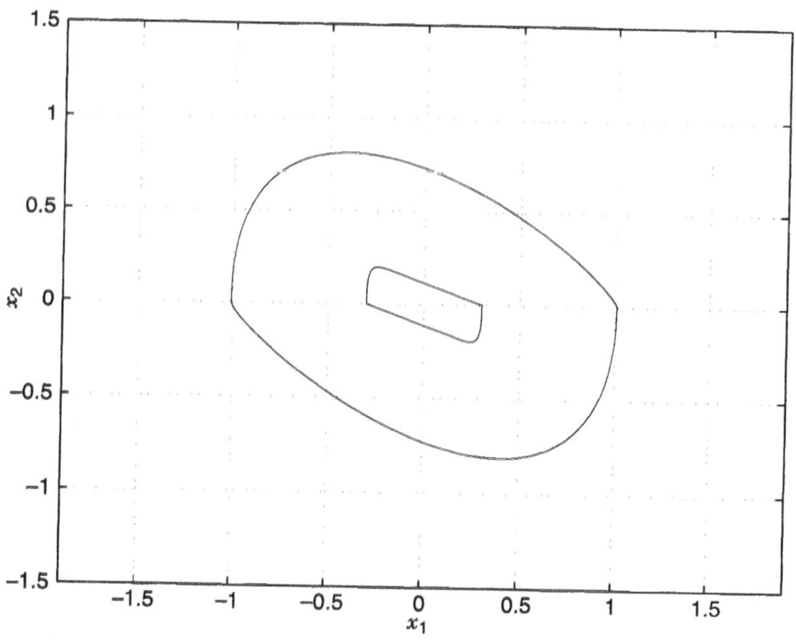

Figure 9.6-1 The v-reachable sets obtained using LQR and min-max design.

$$v = v_{max}\,\text{sgn}(x_2),$$

with $v_{max} = 1$. See Example 9.3-1 for details of how the boundaries are obtained using system trajectories. The initial portions of these trajectories are not shown. Clearly, the min-max controller decreases the size of the v-reachable set, illustrating the greater robustness of the system with respect to the uncertain input. ■

Example 9.6-2: Modified Inverted Pendulum Consider the previous example with the system defined in the same way, except the uncertain input enters each equation. That is,

$$\mathbf{E} = \begin{bmatrix} 1 \\ 1 \end{bmatrix}.$$

Now it is not possible to tell, by inspection, what the critical value of γ^2 should be. Table 9.6-2, which includes LQR design for comparison, illustrates the process of estimating $\bar{\gamma}^2$. Again, because of immediate control saturation, rather than using the estimated value of $\bar{\gamma}^2 = 3.42$, a more practical design would correspond to, say, $\gamma^2 = 4$.

Connections with H_∞ Control

There is a connection between min-max control for a linear system with quadratic cost and H_∞ control theory [Francis, 1987]. However, the connection is not exact for the problem as formulated above. Both theories deal with the problem of determining robust controllers for a linear system with unbounded control and unbounded uncertain inputs. The reason the connection is not exact is because the problem formulated above is not equivalent to a problem in standard H_∞ theory. However, with additional notation and a change in the way the cost function is formulated, game-theoretic control and H_∞ control can be shown to produce equivalent results [Başar and Bernhard, 1991].

Table 9.6-2 Iterative Estimation of γ^2 for Example 9.6-2

	Assumed γ^2	Gains K	Eigenvalues for $\hat{A}$	γ^2 from (9.6–10)
LQR	NA	1.0000, 1.7321	$-0.8660 \pm i0.5000$	NA
Min-max	100	1.0386, 1.7766	$-0.8883 \pm i0.4995$	9.1474
Min-max	5	3.2679, 4.3887	−0.9505, −3.4382	4.7191
Min-max	4	6.9974, 8.7943	−0.8847, −7.9096	3.9592
Min-max	3.5	37.3231, 44.6949	−0.8513, −43.8436	3.4989
Min-max	3.42	199.9754, 237.2952	−0.8457, −236.4494	3.4200

H_∞ control was originally developed by G. Zames as a frequency domain optimization method for designing robust control systems. In H_∞ control theory, stable and proper (see Section 1.5) closed-loop transfer functions $G(s)$, relating an uncertain input v to a controlled output z, are compared according to their ∞-norm, defined for single-input single-output systems by

$$\|G\|_\infty \triangleq \sup_\omega |G(i\omega)|, \tag{9.6-11}$$

where $i = \sqrt{-1}$. That is, the ∞-norm is the peak value obtained on a Bode magnitude plot of the transfer function versus frequency. More generally, for multi-input multi-output $n_z \times n_v$ transfer matrices $\mathbf{G}(s)$, relating $\mathbf{v}$ to $\mathbf{z}$, we have

$$\|\mathbf{G}\|_\infty \triangleq \sum_\omega [\sigma_{\max}\{\mathbf{G}(i\omega)\}], \tag{9.6-12}$$

where $\sigma_{\max}\{\mathbf{G}(i\omega)\}$ is the largest of the **singular values** of $\mathbf{G}(i\omega)$, which are defined by

$$\sigma_k = \sqrt{\mu_k\{\mathbf{G}^{\mathrm{T}}(-i\omega)\mathbf{G}(i\omega)\}} \tag{9.6-13}$$

with $\mu_k\{\cdot\}$ denoting the kth eigenvalue.

Since we have not introduced frequency domain methods in this book (except briefly, with harmonic balance in Section 1.5), a direct comparison between min-max and H_∞ control cannot be made here without considerable additional background material. We simply note that if we introduce an n_z-dimensional **controlled output** vector of the form

$$\mathbf{z} = \mathbf{C}\mathbf{x} + \mathbf{D}\mathbf{u},$$

where $\mathbf{C}$ and $\mathbf{D}$ are constant matrices of appropriate dimension, and a cost function of the form

$$x_0 = \tfrac{1}{2}\int_0^{t_f} (\mathbf{z}^{\mathrm{T}}\mathbf{z} - \gamma^2\mathbf{v}^{\mathrm{T}}\mathbf{v})dt,$$

then equivalent results can be obtained. In other words, seeking a min-max control for x_0, using the same approach as we used above, will result in the same Riccati equation for determining $\mathbf{u}(\mathbf{x})$ as one obtains from H_∞ theory. However, more general results are available using the H_∞ methods, since that theory does not require that every state variable be measured directly and without noise. As a consequence, state estimation plays a major role in H_∞ control theory.

We refer the interested reader to [Francis, 1987; Mustafa and Glover, 1990; Basar and Bernhard, 1991; Shahain and Hassul, 1993] and to a related approach known as H_2 optimal control [Saberi et al., 1995].

9.7 MISSILE GUIDANCE

In this section we investigate missile guidance pursuit-evasion problems in a more general setting than Example 7.3-4 with the objective being to develop and investigate various guidance laws. Reconsider the missile guidance problem of Example 7.3-4, except that here we will relax the tail-chase or head-on assumptions and also allow the evader to be maneuverable. The situation is depicted in Figure 9.7-1. Motion is confined to a plane and both the evader's velocity S_v and the pursuer's velocity S_u are assumed constant. The nonrotating coordinate system is attached to the pursuer. The angles that the velocity vectors make with the x_1-axis are indirectly controlled by each player. The evader's control v is the rate of change of x_4 and the pursuer's control u is the rate of change of x_3. The line of sight angle γ, defined relative to the fixed x_2-direction, is also shown in the figure. The motion of the evader with respect to the pursuer is given by

$$\begin{aligned}\dot{x}_1 &= S_v \cos x_4 - S_u \cos x_3 \\ \dot{x}_2 &= S_v \sin x_4 - S_u \sin x_3 \\ \dot{x}_3 &= u \\ \dot{x}_4 &= v,\end{aligned} \tag{9.7-1}$$

where

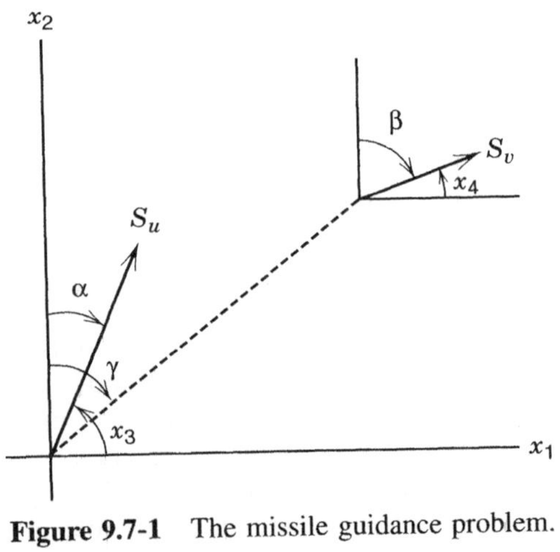

Figure 9.7-1 The missile guidance problem.

$$|u| \leq u_{\max}$$
$$|v| \leq v_{\max}.$$

The line-of-sight angle is given by

$$\tan \gamma = \frac{x_1}{x_2}$$

and $\dot{\gamma}$ can be determined by differentiating this result, which yields

$$\dot{\gamma} \sec^2 \gamma = \frac{x_2 \dot{x}_1 - x_1 \dot{x}_2}{x_2^2}.$$

Noting from Figure 9.7-1 that $\sec \gamma = r/x_2$ where $r = \sqrt{x_1^2 + x_2^2}$, we obtain

$$r^2 \dot{\gamma} = x_2 \dot{x}_1 - x_1 \dot{x}_2 \tag{9.7-2}$$

or, alternately from the equations of motion,

$$\dot{\gamma} = \frac{1}{r^2} [x_2(S_v \cos x_4 - S_u \cos x_3) - x_1(S_v \sin x_4 - S_u \sin x_3)]. \tag{9.7-3}$$

Before examining this problem from a game-theoretic point of view, let us first consider two other approaches to guidance.

Direct Pursuit

Under direct pursuit, it is assumed that $u_{\max} = v_{\max} \rightarrow \infty$, so that each of the heading angles can be instantaneously set. In this case we use the angles $\alpha = \pi/2 - x_3$ and $\beta = \pi/2 - x_4$ as the control variables so that the equations of motion become

$$\begin{aligned} \dot{x}_1 &= S_v \sin \beta - S_u \sin \alpha \\ \dot{x}_2 &= S_v \cos \beta - S_u \cos \alpha. \end{aligned} \tag{9.7-4}$$

Under direct pursuit the pursuer points the velocity vector directly at the evader in a "steepest descent" fashion (see Section 5.3); that is, α is chosen so that

$$\begin{aligned} \sin \alpha &= \frac{x_1}{r} \\ \cos \alpha &= \frac{x_2}{r}. \end{aligned} \tag{9.7-5}$$

Since the distance r between the pursuer and the evader is given simply by

$$r^2 = x_1^2 + x_2^2 \tag{9.7-6}$$

we can calculate the *closing rate* $\dot{r}$. In particular from (9.7-4) and (9.7-6) we obtain

$$r\dot{r} = S_v(x_1 \sin\beta + x_2 \cos\beta) - S_u(x_1 \sin\alpha + x_2 \cos\alpha). \tag{9.7-7}$$

Using (9.7-5) to substitute for x_1 and x_2 in (9.7-7) and dividing by r yields

$$\dot{r} = S_v(\sin\alpha \; \sin\beta + \cos\alpha \; \cos\beta) - S_u$$

or, alternately,

$$\dot{r} = S_v \cos(\alpha - \beta) - S_u. \tag{9.7-8}$$

Collision Course

It follows from (9.7-8) that, if $S_u > S_v$, then capture is always possible, since direct pursuit would yield $\dot{r} < 0$. However, capture may also occur even if the pursuer is slower than the evader. Indeed, the inefficiency of the direct pursuit algorithm is demonstrated by considering a case where $S_u > S_v$ is clearly not required for capture. For example, if the pursuer and the evader are moving at constant speed and direction (i.e., constant velocity vectors) and are on a collision course, as depicted in Figure 9.7-2, then capture by the pursuer is obtained with no control action by the pursuer!

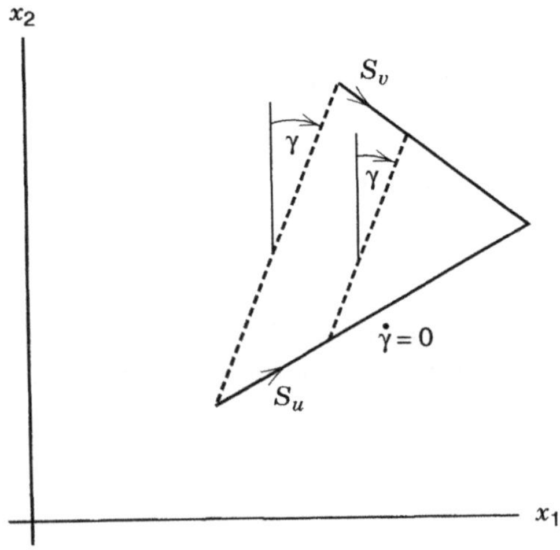

Figure 9.7-2 Pursuer and evader on a collision course.

Since the velocities are constant for the collision course case, the time of capture can be expressed either in terms of x_1 motion or x_2 motion:

$$\text{Time of capture} = \frac{x_1}{\dot{x}_1} = \frac{x_2}{\dot{x}_2}. \tag{9.7-9}$$

It follows from (9.7-2) and (9.7-9) that $\dot{\gamma} = 0$. This result is a necessary but not sufficient condition for collision for two objects moving at constant velocity in a plane. As one might imagine it is a useful "rule of thumb." For example, a ship's captain might use $\dot{\gamma}$ to assess the danger of collision with other ships in crowded shipping lanes.

Starting from the collision course situation depicted in Figure 9.7-2, if the pursuer is slower than the evader and uses direct pursuit rather than maintaining the velocity vector constant, capture need not take place. Our final observation with respect to direct pursuit is that it cannot be used as a guidance law in any realistic physical system because of the assumption of infinite turning rates. For this reason, it is not examined further. Nevertheless, direct pursuit is useful in providing insight into the role $\dot{\gamma}$ plays in the guidance laws that follow.

Proportional Navigation

The observation that the line of sight angle does not change on a collision course leads one to consider a guidance law that will not change the heading angle of the pursuer whenever $\dot{\gamma} = 0$. Proportional navigation is such a guidance law. We have previously demonstrated, in Example 7.3-4, that proportional navigation is an optimal guidance law for a tail-chase or head-on intercept of a nonmaneuvering target when the accumulated cost was the square of the control action. The system dynamics were the same as (9.7-1) except that the "evader" was not maneuvering ($v \equiv 0$) and a small angles assumption (α and β small) was made. The fact that proportional navigation is "optimal" for this situation does not necessarily mean that it is an ideal guidance law for this or other situations (maneuvering evader, not tail-chase, more realistic dynamics, etc.). However, due to the extensive use of proportional navigation, it is of interest to examine this guidance law in the context of the general guidance equations (9.7-1).

For the situation depicted in Figure 9.7-1, a **proportional navigation** guidance law for the pursuer would be of the form

$$u = k\dot{\gamma} \tag{9.7-10}$$

where k is a "navigation" constant. In practice, k may be adjusted during the maneuver. Nevertheless, we consider k to be constant in this analysis.

Since the control is bounded we implement **saturating proportional navigation** in the form

$$u = \begin{cases} u_{\max} & \text{if } k\dot{\gamma} \;\; > u_{\max} \\ k\dot{\gamma} & \text{if } |k\dot{\gamma}| < u_{\max} \\ -u_{\max} & \text{if } k\dot{\gamma} \;\; < -u_{\max}. \end{cases} \tag{9.7-11}$$

In order to implement this control law, we need only measure the single quantity $\dot{\gamma}$. The simplicity of this guidance law, as well as the ability to measure $\dot{\gamma}$ in real systems, helps account for its popularity. When proportional navigation is used in simulations, $\dot{\gamma}$ is calculated using (9.7-3).

Game-Theoretic Guidance

In order to develop a guidance law based on game theory for the general system (9.7-1), we use time of intercept as the performance index. That is, the pursuer (u) wants to minimize the time of capture and the evader (v) wants to maximize it. Using the methods of Section 9.5, the H function is given by

$$H = 1 + \lambda_1(S_v \cos x_4 - S_u \cos x_3) + \lambda_2(S_v \sin x_4 - S_u \sin x_3) + \lambda_3 u + \lambda_4 v \tag{9.7-12}$$

with the adjoint equations

$$\begin{aligned} \dot{\lambda}_1 &= 0 \\ \dot{\lambda}_2 &= 0 \\ \dot{\lambda}_3 &= -\lambda_1 S_u \sin x_3 + \lambda_2 S_u \cos x_3 \\ \dot{\lambda}_4 &= \lambda_1 S_v \sin x_4 - \lambda_2 S_v \cos x_4. \end{aligned}$$

Applying the min-max principle to (9.7-12) results in the following control algorithms:

$$u = \begin{cases} -u_{\max} & \text{if } \lambda_3 > 0 \\ \text{singular} & \text{if } \lambda_3 \equiv 0 \\ u_{\max} & \text{if } \lambda_3 < 0 \end{cases}$$

$$v = \begin{cases} -v_{\max} & \text{if } \lambda_4 < 0 \\ \text{singular} & \text{if } \lambda_4 \equiv 0 \\ v_{\max} & \text{if } \lambda_4 > 0, \end{cases}$$

where the possibility of singular control will now be examined. We observe from the adjoint equations that

$$\lambda_3 \equiv 0 \Rightarrow \dot{\lambda}_3 \equiv 0 \Rightarrow \tan x_3 = \frac{\lambda_2}{\lambda_1} \Rightarrow x_3 = \text{ constant}$$

$$\lambda_4 \equiv 0 \Rightarrow \dot{\lambda}_4 \equiv 0 \Rightarrow \tan x_4 = \frac{\lambda_2}{\lambda_1} \Rightarrow x_4 = \text{ constant}$$

and from the state equations

$$x_3 = \text{ constant } \Rightarrow \dot{x}_3 = 0 \Rightarrow u = 0$$
$$x_4 = \text{ constant } \Rightarrow \dot{x}_4 = 0 \Rightarrow v = 0.$$

It follows from 9.7-1 that

$$\frac{\dot{x}_1}{\dot{x}_2} = \frac{S_v \cos x_4 - S_u \cos x_3}{S_v \sin x_4 - S_u \sin x_3}$$

or alternately

$$\dot{x}_2(S_v \cos x_4 - S_u \cos x_3) - \dot{x}_1(S_v \sin x_4 - S_u \sin x_3) = 0.$$

Under singular control by both players, x_3 and x_4 are constant, so that this equation may be integrated to yield

$$x_2(S_v \cos x_4 - S_u \cos x_3) - x_1(S_v \sin x_4 - S_u \sin x_3) = \text{constant}. \qquad (9.7\text{-}13)$$

Comparing (9.7-13) with (9.7-3) we see that this result is equivalent to

$$r^2\dot{\gamma} = \text{constant}.$$

It also follows from (9.7-13) that the constant is zero for any trajectory that passes through the origin. Consider now any solution that ends in an interception, that is, trajectories defined by the equations of motion (9.7-1) that pass through the origin. It follows from the transversality conditions that, since $x_3(t_f)$ and $x_4(t_f)$ are unspecified, $\lambda_3(t_f)$ and $\lambda_4(t_f)$ must both be equal to zero. Since both λ_3 and λ_4 are switching functions, this strongly suggests that singular control will be used at termination (but, of course, this need not be the case for all situations).

We now know the nature of the control segments (maximum control, minimum control, and singular control), but not the order in which they will be used. However, by noting the role of $\dot{\gamma}$ in defining conditions for a terminating singular arc, the problem is simple enough to make a good guess as to the nature of the game-theoretic solution without actually having to solve the game problem.

The following is an algorithm that focuses on the control actions of the pursuer. It is based on the results obtained above but, strictly speaking, it does not represent a "solution" provided by satisfying all of the theoretical necessary conditions. If it does indeed satisfy game theory, then as with any closed-loop game solution for one player, it must be valid regardless of the control actions of the other player. If the other player does not use a game-theoretic solution, then capture will simply occur sooner. Simply stated, the algorithm is designed to provide control actions for the pursuer, using maximum and minimum control, until such time as the pursuer lines up with the evader, at which time the pursuer switches to singular control. Specifically, a simplified version of the algorithm is given by

$$u = \begin{cases} u_{\max} & \text{if } \dot{\gamma} \quad > \varepsilon \\ \dot{\gamma} & \text{if } |\dot{\gamma}| \ < \varepsilon \\ -u_{\max} & \text{if } \dot{\gamma} \quad < -\varepsilon, \end{cases} \tag{9.7-14}$$

where ε is a small positive number, chosen to avoid chattering, but small enough so that approximate singular control is obtained. Additional refinements would include changing the sign of the control when the evader is inside the turning radius of the pursuer and adding an arbitrary turn for the case when $\dot{\gamma} = 0$ and $\dot{r} > 0$.

This algorithm has a lot in common with saturating proportional navigation in that $\dot{\gamma}$ plays a main role. The main difference is in the bang-bang nature of the control. However, along with the above mentioned refinements, this contoller will produce capture for a larger region of the state space than is possible with saturating proportional navigation.

EXERCISES

9-1. Derive equations (9.1-2) for the nonmaneuvering homicidal chauffeur game.

9-2. Starting with $\min_{\mathbf{v} \in \mathcal{V}} G(\mathbf{u}, \mathbf{v}) \le G(\mathbf{u}, \mathbf{v})$ prove that

$$\max_{\mathbf{u} \in \mathcal{U}} \{\min_{\mathbf{v} \in \mathcal{V}} G(\mathbf{u}, \mathbf{v})\} \le \min_{\mathbf{v} \in \mathcal{V}} \{\max_{\mathbf{v} \in \mathcal{U}} G(\mathbf{u}, \mathbf{v})\}.$$

9-3. Verify the results given by (9.2-32) and (9.2-33). Verify the claim that in nonrotating coordinates (9.2-33) corresponds to the evader running in a straight line. Explain why (9.2-33) is plausible.

9-4. Challenging Problem! For $S_u = 2$, $S_v = 1$, and $u_{\max} = 1$, determine the boundary of the playable set for the homicidal chauffeur problem introduced in Example 9.2-1 using a "flat car" target defined by

$$-1 < x_{1f} < 1$$
$$x_{2f} = 0.$$

Note that the points $|x_{1f}| = 1$ are not included in the definition of capture. This allows the evader to get by the car with a "grazing touch" at the ends of the car without terminating the game. It is also implicitly required that capture involve a nonzero $\dot{x}_2$ at the car's interior. That is, a trajectory that is just tangent to the car does not constitute capture or collision.

9-5. Challenging Problem! For $S_u = 2$, $S_v = 1$, and $u_{\max} = 1$, determine the boundary of playable set for the homicidal chauffeur problem introduced in Example 9.2-1 using a circular car target defined by

$$x_{1f}^2 + x_{2f}^2 \leq 1.$$

9-6. Verify the results obtained in Figure 9.2-2.

9-7. Given the system

$$\dot{x}_1 = x_2$$
$$\dot{x}_2 = -x_1 + 2x_2 + u + v$$

with $|u| \leq 2$, $|v| \leq 1$, design a saturating LQR controller for $u(\mathbf{x})$ and then determine the v-reachable and v-attractive sets for this control design. Try a different design and compare results. Compare the v-attractive sets with the playable set obtained in Problem 9-6.

9-8. Find the boundary of the v-reachable set and the boundary of the v-attractive set for the magnetic suspension example 9.3-2 if $k_1 = 2$, $k_2 = 4$, and $v_{\max} = 0.3$.

9-9. Solve the magnetic suspension problem defined in Example 9.3-2 as a quantitative game:

$$\min_u \max_v \int_0^{t_f} dt$$

subject to

$$\dot{x}_1 = x_2$$
$$\dot{x}_2 = x_1 + u + v$$

with

$$|u| \le 1$$
$$|v| \le 1,$$

where $x_{1f} \triangleq x_1(t_f) = 0$ and $x_{2f} \triangleq x_2(t_f) = 0$. Plot the trajectories numerically by integrating backwards from the origin starting with various values of the adjoint vector $\boldsymbol{\lambda}(t_f)$.

9-10. Verify the results obtained in Figure 9.6-1.

9.11. Challenging Problem! Compare the LQR design with the min-max design ($\gamma^2 = 4$) in terms of v-reachable and v-attractive sets for Example 9.6-2.

9-12. Numerically solve (9.7-1) for the following situations, assuming that $x_1(0) = 5$ and $x_2(0) = 10$:

(a) $S_u = 2$, $S_v = 1$, $x_4 = 90°$, u uses direct pursuit.

(b) $S_u = 1$, $S_v = 2$, $x_4 = 180°$, $x_3 = 0°$.

(c) $S_u = 1$, $S_v = 2$, $x_4 = 180°$, u uses direct pursuit.

9-13. From (9.1-3) it follows that the equations of motion for a stationary evader are given by

$$\dot{x}_1 = ux_2$$
$$\dot{x}_2 = -S_u - ux_1.$$

Find the region of state space within which a pursuer using saturated proportional navigation (9.7-11) can capture the stationary target. Let $S_u = 5$, $u_{\max} = 2$, and $k = 3$. In what region of state space will the pursuer's control be saturated?

9-14. Determine a feedback control law for the pursuer of Exercise 9-13 so that he or she can capture the stationary evader from the entire state space.

9-15. Challenging Problem! let $S_u = 5$, $S_v = 2$, and $u_{\max} = 2$ for the homicidal chauffeur dynamical system given by (9.1-3). Determine a feedback control law for the pursuer that will guarantee capture of the evader throughout the x_1, x_2 state space with $\mathbf{x}_f = 0$.

REFERENCES

Anderson, M. J. and W. J. Grantham, "Lyapunov Optimal Feedback Control of an Inverted Pendulum," *J. of Dynamic Systems, Measurement, and Control*, Vol. 111, No. 4, 1989, pp. 554–558.

Andronov, A. A. and L. S. Pontryagin, "Systèmes Grossiers," *Dokl. Akad. Nauk. SSSR*, Vol. 14, 1937, pp. 247–251.

Ardema, M., "Singular Perturbations and the Sounding Rocket Problem," *Proc. of the Joint American Controls Conf.*, Denver, CO, 1979, pp. 901–907.

Arrow, K. J. and L. Hurwicz, "Gradient Methods for Constrained Maxima," in *Studies in Resource Allocation Processes*, Edited by K. J. Arrow and L. Hurwicz, Cambridge University Press, Cambridge, 1977, pp. 146–153.

Baker, E. L., *Modeling and Optimization of Shaped Charge Liner Collapse and Jet Formation*, Ph.D. dissertation, Washington State University, 1992.

Barbishin, E. A., *Introduction to the Theory of Stability*, Wolters-Noordhoff, Groningen, The Netherlands, 1970.

Barbishin, E. A. and N. N. Krasovskiĭ, "Concerning the Stability of Motion as a Whole," *Dokl. Akad. Nauk. SSSR*, Vol. 36, No. 3, 1953.

Barmish, B. R. and G. Leitmann, "On Ultimateboundedness Control of Uncertain Systems in the Absence of Matching Conditions," *IEEE Trans. on Automatic Control*, Vol. AC-27, 1982, pp. 153–157.

Barmish, B. R., M. Corless, and G. Leitmann, "A New Class of Stabilizing Controllers for Uncertain Dynamical Systems," *SIAM J. on Control and Optimization*, Vol. 21, 1983, pp. 246–255.

Başar, T. and P. Bernhard, *H_∞-Optimal Control and Related Minimax Design Problems*, Birkhäuser, Boston, 1991.

Bellman, R. E., *Dynamic Programming*, Princeton University Press, Princeton, NJ, 1957.

Bellman, R. E. and S. E. Dreyfus, *Applied Dynamic Programming*, Princeton University Press, Princeton, NJ, 1964.

Blaquière, A. and G. Leitmann, "On the Geometry of Optimal Processes," in *Topics in Optimization*, Edited by G. Leitmann, Academic Press, New York, 1967, pp. 265–371.

Blaquière, A., F. Gérard, and G. Leitmann, *Quantitative and Qualitative Games*, Academic Press, New York, 1969.

Bliss, G. A., *Lectures on the Calculus of Variations*, University of Chicago Press, Chicago, 1946.

Breakwell, J. V. and A. W. Merz, "Toward a Complete Solution of the Homicidal Chauffeur Game," *Proc. of the Int'l Conf. on the Theory and Application of Differential Games*, University of Massachusetts, Amherst, MA, 1969.

Brogan, W. L., *Modern Control Theory*, Prentice-Hall, Englewood Cliffs, NJ, 1982.

Bryson, A. E., Jr. and Y. C. Ho, *Applied Optimal Control*, Wiley, New York, 1975.

Bushaw, D. W., "Optimal Discontinuous Forcing Terms," *Contributions to the Theory of Nonlinear Oscillations*, Vol. IV, Princeton University Press, Princeton, NJ, 1958, pp. 29–52.

Chamberland, M. and A. S. Lewis, "Contours of Liapunov Functions," *J. of Optimization Theory and Applications*, Vol. 80, No. 1, January 1994, pp. 149–160.

Clark, C. W., *Mathematical Bioeconomics: The Optimal Management of Renewable Resources*, Wiley, New York, 1976.

Cliff, E. M. and T. L. Vincent, "An Optimal Policy for a Fish Harvest," *J. of Optimization Theory and Applications*, Vol. 12, No. 5, 1973, pp. 485–496.

Coddington, E. A. and N. Levinson, *Theory of Ordinary Differential Equations*, McGraw-Hill, New York, 1955.

Corless, M. and G. Leitmann, "Continuous State Feedback Guaranteeing Uniform Ultimate Boundedness for Uncertain Dynamical Systems," *IEEE Trans. on Automatic Control*, Vol. AC-26, No. 5, 1981, pp. 1139–1144.

Cottrell, R. G., T. L. Vincent, and S. H. Sadati, "Minimizing Interceptor Size Using Neural Networks for Terminal Guidance Law Synthesis," *J. of Guidance, Control, and Dynamics*, Vol. 19, No. 3, May-June 1996, pp. 557–562.

Crutchfield, J. P., J. D. Farmer, N. H. Packard, and R. S. Shaw, "Chaos," *Scientific American*, Vol. 255, No. 6, 1986, pp. 46–57.

Davis, H. T., *Introduction to Nonlinear Differential and Integral Equations*, Dover Publications, New York, 1962.

Davison, E. J., "On Pole Assignment in Multivariable Linear Systems," *IEEE Trans. on Automatic Control*, Vol. AC-13, No. 6, December 1968, pp. 747–748.

Devaney, R. L., *An Introduction to Chaotic Dynamical Systems*, Benjamin/Cummings Publishing Co., Menlo Park, CA, 1986.

Doeblin, E. O., *Measurement Systems*, McGraw-Hill, New York, 1966.

Dorling, C. M. and A. S. I. Zinober, "Two Approaches to Hyperplane Design in Multivariable Structure Systems," *Int'l. J. of Control*, Vol. 44, No. 1, 1986, pp. 65–82.

Ehrhard, P. and U. Müller, "Dynamical Behavior of Natural Convection in a Single-Phase Loop," *J. Fluid Mechanics*, Vol. 217, 1990, pp. 487–518.

Evans, W. R., "Graphical Analysis of Control Systems," *Trans. AIEE*, Vol. 67, 1948, pp. 547–551.

Faddeev, D. K. and V. N. Faddeeva, *Computational Methods of Linear Algebra*, W. H. Freeman and Co., San Francisco, CA, 1963.

Farmer, J. D., E. Ott, and J. A. Yorke, "The Dimension of Chaotic Attractors," *Physica 7D*, 1983, pp. 153–180.

Fetter, A. L. and J. D. Walecka, *Theoretical Mechanics of Particles and Continua*, McGraw-Hill, New York, 1980.

Filippov, A. F., "Differential Equations with Discontinuous Right-Hand Side," *Amer. Math. Soc. Transl., Series 2*, Vol. 42, 1964, pp. 199–231.

Flügge-Lotz, I., *Discontinuous and Optimal Control*, McGraw-Hill, New York, 1968.

Francis, B. A., *A Course in H_∞ Control*, Lecture Notes in Control and Information Sciences, Vol. 88, Springer-Verlag, New York, 1987.

Fritchman, B. M., *A Variational Approach to Generating Optimal Control Switching Surfaces*, Ph.D. dissertation, Washington State University, Pullman, WA, 1985.

Gayek, J. E. and T. L. Vincent, "On the Asymptotic Stability of Boundary Trajectories," *Int'l. J. of Control*, Vol. 41, No. 4, 1985, pp. 1077–1086.

Gayek, J. E. and T. L. Vincent, "On the Intersection of Controllable and Reachable Sets," *J. of Optimization Theory and Applications*, Vol. 50, No. 2, August 1986, pp. 267–278.

Gayek, J. E. and T. L. Vincent, "An Existence Theorem for Boundary Trajectories," *J. of Mathematical Analysis and Applications*, Vol. 132, No. 1, May 1988, pp. 290–299.

Gear, C. W., "The Numerical Integration of Ordinary Differential Equations," *Mathematics of Computation*, Vol. 21, No. 98, April 1967, pp. 146–156.

Gelfand, I. M. and S. V. Fomin, *Calculus of Variations*, Prentice-Hall, Englewood Cliffs, NJ, 1963.

Gill, P. E., W. Murray, M. A. Saunders, and M. H. Wright, "Model Building and Practical Aspects of Nonlinear Programming," in *Computational Mathematical Programming*, Edited by K. Schittkowski, NATO ASI Series, Vol. F15, Springer-Verlag, Heidelberg, 1985, pp. 209–247.

Goh, B. S., "Necessary Conditions for Singular Extremals Involving Multiple Control Variables," *SIAM J. Control*, Vol. 4, 1966, pp. 716–731.

Goh, B. S., "Optimal Control of a Fish Resource," *Malayan Scientist*, Vol. 5, 1969, pp. 65–70.

Goh, B. S., *Management and Analysis of Biological Systems*, Elsevier Scientific, Amsterdam, 1980.

Goh, C. J., N. J. Edwards, and A. Y. Zomaya, "Feedback Control of Minimum-Time Optimal Control Problems Using Neural Networks," *Optimal Control Applications and Methods*, Vol. 14, No. 1, 1993, pp. 1–16.

Grace, A., *Optimization Toolbox for Use with MATLAB*, The MathWorks, Inc., Matick, MA, 1992.

Grantham, W. J., *A Controllability Minimum Principle*, Ph.D. dissertation, Aerospace and Mechanical Engineering, University of Arizona, Tucson, AZ, 1973.

Grantham, W. J., "Estimating Controllability Boundaries for Uncertain Systems," in

Renewable Resource Management, Edited by T. L. Vincent and J. M. Skowronski, Springer-Verlag, New York, 1981*a*, pp. 151–161.

Grantham, W. J., "Estimating Reachable Sets," *J. of Dynamic Systems, Measurement, and Control*, Vol. 103, 1981*b*, pp. 420–423.

Grantham, W. J., "Controllability Conditions and Effective Controls for Nonlinear Systems," *Proc. IEEE Conf. on Decision and Control*, San Diego, CA, December 1981*c*.

Grantham, W. J., "Some Necessary Conditions for Steepest Descent Controllability," *Proc. American Control Conf.*, Alexandria, VA, June 1982.

Grantham, W. J., "ODESSYS: Ordinary Differential Equations Simulation System for Nonlinear Optimization, Optimal Controls, and Differential Games," *Proc. Third IEEE Computer Aided Control Systems Design Conf.*, Arlington, VA, August 1986.

Grantham, W. J. and A. O. Chingcuanco, "Lyapunov Steepest Descent Control of Constrained Linear Systems," *IEEE Trans. on Automatic Control*, Vol. AC-29, No. 8, August 1984, pp. 740–743.

Grantham, W. J. and B. S. Lee, "A Chaotic Limit Cycle Paradox," *Dynamics and Control*, Vol. 3, No. 2, April 1993, pp. 159–173.

Grantham, W. J. and T. L. Vincent, "A Controllability Minimum Principle," *J. of Optimization Theory and Applications*, Vol. 17, Nos. 1/2, October 1975, pp. 93–114.

Grantham, W. J. and T. L. Vincent, *Modern Control Systems Analysis and Design*, Wiley, New York, 1993.

Greenwood, D. T., *Classical Dynamics*, Prentice-Hall, Englewood Cliffs, NJ, 1977.

Guckenheimer, J. and P. Holmes, *Nonlinear Oscillations, Dynamical Systems, and Bifurcations of Vector Fields*, Springer-Verlag, New York, 1983.

Gutman, S. and G. Leitmann, "Stabilizing Feedback Control for Dynamical Systems with Bounded Uncertainty," *Proc. IEEE Conf. on Decision and Control*, 1976, pp. 94–99.

Hahn, W., *Stability of Motion*, Springer-Verlag, New York, 1967.

Hájek, O., "Discontinuous Differential Equations, I," *J. of Differential Equations*, Vol. 32, 1979, pp. 149–170.

Haken, H. "At Least One Lyapunov Exponent Vanishes if the Trajectory of an Attractor Does Not Contain a Fixed Point," *Physics Letters 94A*, 1983, p. 71.

Hartman, P., *Ordinary Differential Equations*, Wiley, New York, 1964.

Hestenes, M. R., *The Calculus of Variations and Optimal Control Theory*, Wiley, New York, 1966.

Heymann, M., "Comments on Pole Assignment in Multi-Input Controllable Linear Systems," *IEEE Trans. on Automatic Control*, Vol. AC-13, No. 6, December 1968, pp. 748–749.

Hildebrand, F. B., *Advanced Calculus with Applications*, Prentice-Hall, Englewood Cliffs, NJ, 1962.

Hindmarsh, A. C., "ODEPACK, A Systematized Collection of ODE Solvers," in *Scientific Computing*, Edited by R. S. Stepleman et al., North-Holland, Amsterdam, 1983, pp. 55–64.

Hirsch, M. W. and S. Smale, *Differential Equations, Dynamical Systems, and Linear Algebra*, Academic Press, New York, 1974.

Hostetter, G. H., M. S. Santina, and P. D'Carpio-Montalvo, *Analytical, Numerical, and Computational Methods for Science and Engineering*, Prentice-Hall, Englewood Cliffs, NJ, 1991.

Isaacs, R., *Differential Games*, Wiley, New York, 1965.

Itkin, U., *Control Systems of Variable Structure*, Wiley, New York, 1976.

Johnson, C. D. and J. E. Gibson, "Singular Solutions in Problems of Optimal Control," *IEEE Trans. on Automatic Control*, Vol. AC-8, No. 1, January 1963, pp. 4–15.

Kailath, T., *Linear Systems*, Prentice-Hall, Englewood Cliffs, NJ, 1980.

Kalman, R. E. and J. E. Bertram, "Control System Analysis and Design Via the 'Second Method' of Lyapunov: I Continuous-Time Systems," *Trans. ASME*, Vol. 82, Series D, No. 2, June 1960, pp. 371–393.

Kalman, R. E., Y. C. Ho, and K. S. Narendra, "Controllability of Linear Dynamical Systems," *Contributions to Differential Equations*, Vol. 1, No. 2, 1963, pp. 189–213.

Kaplan, J. L. and J. A. Yorke, "Chaotic Behavior of Multidimensional Difference Equations," in *Functional Differential Equations and Approximation of Fixed Points*, Edited by H. O. Peitgen and H. O. Walther, Lecture Notes in Mathematics, Vol. 730, Springer-Verlag, New York, 1979, pp. 228–237.

Kelly, H. J., "Method of Gradients," in *Optimization Techniques*, Edited by G. Leitmann, Academic Press, New York, 1962, pp. 205–254.

Kelley, H. J., R. E. Kopp, and H. G. Moyer, "Singular Extremals," in *Topics in Optimization*, Edited by G. Leitmann, Academic Press, New York, 1967, pp. 63–101.

Kellog, O. D., *Foundations of Potential Theory*, Dover, New York, 1953.

Khalil, H. K., *Nonlinear Systems*, Macmillian, New York, 1992.

Kleinman, D. L., "On an Iterative Technique for Riccati Equation Computations," *IEEE Trans. on Automatic Control*, Vol. AC-13, No. 1, 1968, pp. 114–115.

Krasovskiĭ, N. N., "On Stability with Large Initial Perturbations," *Amer. Math. Soc. Transl., Series 2*, Vol. 18, 1961, pp. 275–287.

Krasovskiĭ, N. N., *Stability of Motion*, Stanford University Press, Stanford, 1963.

Kreyszig, E., *Advanced Engineering Mathematics*, 5th Ed., Wiley, New York, 1983.

Kuo, B. C., *Digital Control Systems*, Holt, Rinehart, and Winston, New York, 1980.

Kuo, B. C., *Automatic Control Systems*, 4th ed., Prentice-Hall, Englewood Cliffs, NJ, 1982.

LaSalle, J. and S. Lefschetz, *Stability by Liapunov's Direct Method, with Applications*, Academic Press, New York, 1961.

Lee, B. S., *Chaos in an Optimal Control System*, Ph.D. dissertation, Washington State University, Pullman, WA, August 1991.

Lee, B. S., W. J. Grantham, and M. E. Fisher, "Differential Equation for Continuous Normalization," *KSME J.*, Vol. 8, No. 1, 1994, pp. 63–68.

Lee, E. B. and L. Marcus, *Foundations of Optimal Control Theory*, Wiley, New York, 1967.

Leitmann, G., "Optimum Thrust Programming for High-Altitude Rockets," *Aeronautical Engineering Review*, June 1957, pp. 63–66.

Leitmann, G., *An Introduction to Optimal Control*, McGraw-Hill, New York, 1966.

Leitmann, G., *The Calculus of Variations and Optimal Control*, Plenum Press, New York, 1981*a*.

Leitmann, G., "On the Efficacy of Nonlinear Control in Uncertain Linear Systems," *J. of Dynamic Systems, Measurement, and Control*, Vol. 102, June 1981*b*, pp. 95–102.

Leitmann, G. and W. E. Schmitendorf, "Profit Maximization Through Advertising: A Nonzero Sum Differential Game Approach," *IEEE Trans. on Automatic Control*, Vol. AC-23, No. 4, 1978, p. 645.

Letov, A. M., *Stability in Nonlinear Control Systems*, Princeton University Press, Princeton, NJ, 1961.

Lichtenberg, A. J. and M. A. Lieberman, *Regular and Stochastic Motion*, Springer-Verlag, New York, 1983.

Lorenz, E. N., "Deterministic Nonperiodic Flow," *J. of the Atmospheric Sciences*, Vol. 20, 1963, pp. 130–141.

Luenberger, D. G., "An Introduction to Observers," *IEEE Trans. on Automatic Control*, Vol. AC-16, No. 6, December 1971, pp. 596–602.

Luenberger, D. G., *Introduction to Dynamic Systems*, Wiley, New York, 1979.

Mandelbrot, B. B., *The Fractal Geometry of Nature*, W. H. Freeman and Co., New York, 1983.

Meditch, J. S., *Stochastic Optimal Linear Estimation and Control*, McGraw-Hill, New York, 1969.

Mees, A. I., *Dynamics of Feedback Systems*, Wiley, New York, 1981.

Miloh, T., *Determination of Critical Maneuvers for Collision Avoidance Using the Theory of Differential Games*, Institut für Schiffbau der Universität Hamburg, Bericht No. 219, 1974.

Minorsky, N., *Nonlinear Oscillations*, Van Nostrand, Princeton, NJ, 1962.

Mustafa, D. and K. Glover, *Minimum Entropy H_∞ Control*, Lecture Notes in Control and Information Sciences, Vol. 146, Springer-Verlag, New York, 1990.

Naber, G. L., *Topological Methods in Euclidean Spaces*, Cambridge University Press, London, 1980.

Nayfeh, A. H. and D. T. Mook, *Nonlinear Oscillations*, Wiley, New York, 1979.

Newton, T. A., "Two-Dimensional Homogeneous Quadratic Differential Systems," *SIAM Review*, Vol. 20, No. 1, January 1978, pp. 120–138.

Ogata, K., *Modern Control Engineering*, 2nd ed., Prentice-Hall, Englewood Cliffs, NJ, 1990.

Parker, T. S. and L. O. Chua, *Practical Numerical Algorithms for Chaotic Systems*, Springer-Verlag, New York, 1989.

Pars, L. A., *A Treatise on Analytical Dynamics*, Wiley, New York, 1968.

Pesch, H. J. and R. Bulirsch, "The Maximum Principle, Bellman's Equation, and Carathéodory's Work," *J. of Optimization Theory and Applications*, Vol. 80, No. 2, 1994, pp. 199–225.

Petzold, L. R., "Automatic Selection of Methods for Solving Stiff and Nonstiff Systems of Ordinary Differential Equations," *SIAM J. Sci. Stat. Comput.*, Vol. 4, 1983, pp. 136–148.

Pontryagin, L. S., V. G. Boltyanskii, R. V. Gamkrelidze, and E. F. Mishchenko, *The Mathematical Theory of Optimal Processes*, Wiley, New York, 1962.

Press, W. H., B. P. Flannery, S. A. Teukolsky, and W. T. Vetterling, *Numerical Recipes*, Cambridge University Press, Cambridge, 1986.

Reklaitis, G. V., A. Ravindran, and K. M. Ragsdell, *Engineering Optimization Methods and Applications*, Wiley, New York, 1983.

Rodden, J. J., "Numerical Applications of Lyapunov Stability Theory," *Proc. 1964 Joint Automatic Control Conf.*, Stanford, CA, June 1964, pp. 261–268.

Rosen, J. B., and Suzuki, S., "Construction of Nonlinear Programming Test Problems" *Commun. ACM*, Vol. 8, No. 113, 1965, p. 113.

Rosenberg, R. M., *Analytical Dynamics of Discrete Systems*, Plenum Press, New York, 1977.

Rössler, O. E., "An Equation for Continuous Chaos," *Physics Letters*, Vol. 57A, 1976, p. 397.

Rothschild, D. and A. Jameson, "Comparison of Four Numerical Algorithms for Solving the Liapunov Matrix Equation," *Int'l. J. of Control*, Vol. 11, No. 2, 1970, pp. 181–198.

Saberi, A., P. Sannuti, and B. M. Chen, *H_2 Optimal Control*, Prentice-Hall, New York, 1995.

Sabersky, R. H. and A. J. Acosta, *Fluid Flow*, Macmillan, New York, 1964.

Schmitt, T. J., *Double Pendulum Simulation*, Masters report, Aerospace and Mechanical Engineering, University of Arizona, Tucson, 1992.

Schroeder, P. B., "Plotting the Mandelbrot Set," *Byte*, Vol. 11, No. 13, 1986, pp. 207–210.

Schultz, D. G., "The Generation of Liapunov Functions," in *Advances in Control Systems*, Vol. 2, Edited by C. T. Leondes, Academic Press, New York, 1965, pp. 1–64.

Schuster, H. G., *Deterministic Chaos: An Introduction*, Second Revised Edition, VCH Publishers, New York, NY, 1988.

Seywald, H. and E. M. Cliff, "Goddard Problem in Presence of a Dynamic Pressure Limit," *J. of Guidance, Control, and Dynamics*, Vol. 16, No. 4, July–August 1993, pp. 776–781.

Shahain, B. and M. Hassul, *Control System Design using MATLAB*, Prentice-Hall, Englewood Cliffs, NJ, 1993.

Šiljak, D. D., *Nonlinear Systems*, Wiley, New York, 1969.

Skelton, R. E., *Dynamic Systems Control*, Wiley, New York, 1988.

Skowronski, J. M. and T. L. Vincent, "Playability with and Without Capture," *J. of Optimization Theory and Applications*, Vol. 36, 1982, pp. 111–128.

Smart, D. R., *Fixed Point Theorems*, Cambridge University Press, London, 1974.

Snow, D. R., "Determining Reachable Regions and Optimal Controls," in *Advances in Control Systems*, Vol. 5, Edited by C. T. Leondes, Academic Press, New York, 1967.

Soldatos, A. G., M. Corless, and G. Leitmann, "Stabilizing Uncertain Systems with Bounded Control," in *Mechanics and Control*, Edited by J. M. Skowronski, H. Flashner, R. S. Guttalu, Lecture Notes in Control and Information Sciences, Vol. 151, Springer-Verlag, New York, 1991, pp. 415–428.

Sørensen, P. R., "Fractals," *Byte*, Vol. 9, No. 10, 1984, pp. 157–172.

Sparrow, C., "The Lorenz Equations: Bifurcations, Chaos, and Strange Attractors," *Appl. Math. Sci.*, Vol. 41, 1982, pp. 1–269.

Swan, G. W. and T. L. Vincent, "Optimal Control Analysis in the Chemotherapy of IgG Multiple Myeloma," *Bulletin of Mathematical Biology*, Vol. 39, 1977, pp. 317–337.

Takahashi, Y., M. J. Rabins, and D. M. Auslander, *Control and Dynamic Systems*, Addison-Wesley, Reading, MA, 1970.

Teo, K. L., C. J. Goh, and K. H. Wong, *A Unified Computational Approach to Optimal Control Problems*, Longman, New York, 1991.

Tsiotras, P. and H. J. Kelley, "Goddard Problem with Constrained Time of Flight," *J. of Guidance, Control, and Dynamics*, Vol. 15, No. 2, March-April 1992, pp. 289–296.

Vincent, T. L., "Avoidance of Guided Projectiles," in *The Theory and Application of Differential Games*, Edited by J. D. Grote, D. Reidel, Dordrecht, 1975, pp. 267–279.

Vincent, T. L., "Control Design for Magnetic Suspension," *Optimal Control Applications and Methods*, Vol. 1, Jan/Mar 1980. pp. 41–53.

Vincent, T. L. and W. J. Grantham, *Optimality in Parametric Systems*, Wiley, New York, 1981.

Vincent, T. L. and C. S. Lee, "Parametric Optimization with Dynamic Systems," *Applied Mathematics and Computation*, Vol. 12, 1983, pp. 169–185.

Vincent, T. L. and Y. C. Lin, "Some Performance Measures Useful in the Design of Controllers for Two Dimensional Dynamical Systems Subject to Uncertain Inputs," *Dynamics and Control*, Vol. 5, 1995, pp. 69–98.

Vincent, T. L. and Z. Y. Wu, "Estimating Projections of the Controllable Set," *J. of Guidance, Control, and Dynamics*, Vol. 13, No. 3, 1990, pp. 572–575.

Vincent, T. L. and Yu, J. Z., "Control of a Chaotic System," *Dynamics and Control*, Vol. 1, 1991, pp. 35–52.

Vincent, T. L., E. M. Cliff, W. J. Grantham, and W. Y. Peng, "Some Aspects of Collision Avoidance," *AIAA J.*, Vol. 12, 1974, pp. 3–4.

Vincent, T. L., Y. C. Lin, and S. P. Joshi, "Controlling a Flexible Plate to Mimic a Rigid One," in *Control and Dynamic Systems, Vol. 35: Advances in Control Mechanics, Part 2 of 2*, Edited by C. T. Leondes, Academic Press, New York, 1990, pp. 87–135.

Vincent, T. L., B. S. Goh, and K. L. Teo, "Trajectory-Following Algorithms for Min-Max Optimization Problems," *J. of Optimization Theory and Applications*, Vol. 75, No. 3, December 1992, pp. 501–519.

Wolf, A., J. B. Swift, H. L. Swinney, and J. A. Vastano, "Determining Lyapunov Exponents from a Time Series, *Physica 16D*, 1985, pp. 285–317.

Wonham, W. M., "On Pole Assignment in Multi-Input Controllable Linear Systems," *IEEE Trans. on Automatic Control*, Vol. AC-12, No. 6, December 1967, pp. 660–665.

Wonham, W. M., *Linear Multivariable Control*, 3rd ed., Springer-Verlag, New York, 1985.

Zubov, V. I., *Methods of A. M. Lyapunov and Their Application*, U.S. Atomic Energy Commission, Report AEC-tr-4439, October 1961.

INDEX

Printed and bound by CPI Group (UK) Ltd, Croydon, CR0 4YY
16/09/2026
14963456-0007